CISM COURSES AND LECTURES

Series Editors:

The Rectors of CISM
Sandor Kaliszky - Budapest
Mahir Sayir - Zurich
Wilhelm Schneider - Wien

The Secretary General of CISM
Giovanni Bianchi - Milan

Executive Editor
Carlo Tasso - Udine

The series presents lecture notes, monographs, edited works and proceedings in the field of Mechanics, Engineering, Computer Science and Applied Mathematics.
Purpose of the series is to make known in the international scientific and technical community results obtained in some of the activities organized by CISM, the International Centre for Mechanical Sciences.

INTERNATIONAL CENTRE FOR MECHANICAL SCIENCES

COURSES AND LECTURES - No. 411

ROLLING CONTACT PHENOMENA

EDITED BY

BO JACOBSON
LUND UNIVERSITY

JOOST J. KALKER
DELFT UNIVERSITY OF TECHNOLOGY

This volume contains 259 illustrations

SPIN 10789509

In order to make this volume available as economically and as
rapidly as possible the authors' typescripts have been
reproduced in their original forms. This method unfortunately
has its typographical limitations but it is hoped that they in no
way distract the reader.

Springer-Verlag Wien New York

PREFACE

The contact mechanical study of rolling with creep started in 1926 when Carter published the exact solution in the 2D case. In 1958 this was followed by Johnson's studies on rolling with creep and spin in the 3D case. This study was approximate, which was the reason for Kalker's preoccupation with the problem. He succeeded in solving the problem numerically. Whereas Kalker's studies were virtually confined to the half-space geometry, Wriggers turned his attention to the FEM, in which much more general types of geometry could be treated, at the same time as non-linear deformations also could be studied.

Whereas the above studies were based on elasticity, Johnson and his pupils turned their attention towards plastic deformation contact problems, where permanent deformation of the material remained when the loading was removed. Such plastic deformations, in the form of contact surface corrugations, were later studied by Knothe.

The phenomenon of corrugation is fascinating and the formation of quasi-periodic irregularities on contacting surfaces has important applications both for rail bound traffic and for normal roads. The corrugations influence the ground contact force variations and thereby the vibrations and noise generation. For railway and road traffic the vibrations generated in the rolling contacts are one of the major reasons for the high level of disturbing noise transmitted to the surroundings. Stanworth studied some of these problems, especially the noise transmission from railways.

Whereas the above studies were on the basis of massive single material bodies, Pacejka studied inflated bodies like tyres. He studied the phenomenological behaviour of the automotive tyre so it became possible to make accurate theoretical models of the tyre properties, and thus predict their mechanical behaviour on a car or a truck.

All the above studies had been on the basis of dry contacts between the deformed bodies, but in rolling contacts where the friction and wear should be low the surfaces must be separated by a thin lubricant film. The lubricant then transmits the contact forces from one body to the other by absorbing very high pressures. Jacobson studied this type of elastohydrodynamic lubrication and made the first calculation of the oil film for a point contact.

During the week Monday October 4 through Friday October 8, 1999, Kalker, Johnson, Wriggers, Knothe, Stanworth and Jacobson lectured in a course "Rolling Contact Phenomena" at CISM, the International Centre for Mechanical Sciences in Udine, Italy. The course was organised by Professors Kalker and Jacobson. The lecture notes from the different speakers were updated and written in a form, similar for all lecturers. Jacobson and Kalker have put the notes together, and have chosen to put them in the same order in this book, as they had in the original announcement for the course. The course itself was lectured in a different order to make it easier for the students to follow, and for the teachers to give the many lectures.

Bo Jacobson
Joost J. Kalker

CONTENTS

Page

Preface

ROLLING CONTACT PHENOMENA
LINEAR ELASTICITY

J.J. Kalker

Delft University of Technology, Delft, The Netherlands

Abstract. In this paper, we treat the rolling contact phenomena of linear elasticity, with special emphasis on the elastic half-space.

Section 1 treats the basics; rolling is defined, the distance between the deformable bodies is calculated, the slip velocity between the bodies is defined and calculated; a very brief recapitulation of the theory of elasticity follows, and the boundary conditions are formulated.

Section 2 treats the half-space approximation. The formulae of Boussinesq-Cerruti are given, and the concept of quasiidentity is introduced. Then follows a brief description of the linear theory of rolling contact for Hertzian contacts, with numerical results, and of the theory of Vermeulen-Johnson for steady-state rolling. Finally, some examples are given.

Section 3 is devoted to the simplified theory of rolling contact.

In Section 4, the variational, or weak theory of contact is considered. First, we set up the virtual work inequality, and it is shown that it is implied by the boundary conditions of contact. Then the complementary virtual work inequality is postulated, and it is shown that it implies the boundary conditions of contact. Elasticity is introduced into both inequalities, and the potential energy and the complementary energy follow. Finally, surface mechanical principles are derived.

In Section 5, we return to the exact half-space theory. The problem is discretized, and solved by means of the CONTACT algorithm. Finally, results are shown in Section 6.

1 Basics

In this Section, rolling is defined. Also, the distance between the bodies is calculated, the slip velocity between the bodies is defined and computed; a very brief recapitulation of the theory of elasticity follows, and the boundary conditions are formulated.

1.1 Definition of rolling

Consider two bodies of revolution. They are distinguished from each other by attaching to them the numbers 1 and 2. They are, for the time being considered as rigid. They are pressed together so that they touch in a point; line contact will be treated presently. They are rotated, so that the contact point moves over the bodies. Then there are two possibilities: either the velocity $\mathbf{v}^1$ of the contact point over body 1 equals the velocity $\mathbf{v}^2$ of the contact point over body 2, or this is not so. In the former case (equal velocities) one speaks of *rolling*, in the latter case one speaks of *sliding* or *rolling with sliding*.

We consider the case that the bodies contact each other along a line. Again, the bodies are rotated, and the line moves over the bodies. Rolling occurs when the velocities of the contacting line over the bodies are equal at each point of the line, otherwise we speak of sliding or rolling with sliding.

We now consider that the bodies are deformable. First we have to define what we mean by that. We assume that each body is made up of particles that are glued together to form a continuum. Stresses and strains may be present in such a body. We count the displacement from an unstressed state in a manner which will be explained presently. The bodies are pressed together so that a contact patch forms between them, and they are rotated. When the velocity $\mathbf{v}^1$ of the contact patch over body 1 almost equals the velocity $\mathbf{v}^2$ of the contact patch over body 2 we speak of *rolling*, otherwise of *sliding* or *rolling with sliding*:

$$|\mathbf{v}^1 - \mathbf{v}^2| \ll |\mathbf{v}^1 + \mathbf{v}^2| \tag{1}$$

(rolling, otherwise rolling with sliding).

As we said, we count the displacement from the unstressed state, that is, the state in which there are no stresses acting in the bodies. A local Cartesian coordinate system is attached to each body. Each particle corresponds to a point $\mathbf{y}$ of the local coordinate system. The body is deformed; the displacement of the point $\mathbf{y}$ is denoted by $\mathbf{w}$, function of $\mathbf{y}$ and the time t. In the deformed state the particle lies in

$$\mathbf{y} + \mathbf{w}, \quad \mathbf{w} = \mathbf{w}(\mathbf{y}, t) \tag{2}$$

We assume $\mathbf{w}$ small as well as its gradients:

$$|\mathbf{w}| \ll |\mathbf{y}|; \quad |\partial \mathbf{w}/\partial \mathbf{y}| \ll 1 \tag{3}$$

We assume that inertia terms may be neglected and that the deformation is elastic: then we arrive at a linearly elastostatic theory.

We have two bodies; then we call the position $\mathbf{y}^a$ in the a-th coordinate system, and the displacement $\mathbf{w}^a$.

We want to compare the quantities of the bodies. To that end, we must refer them to a single coordinate system. So we introduce a third, global coordinate system in which the particles in the undeformed state are given by $\mathbf{x}^a$, and in the deformed state by $\mathbf{x}^a + \mathbf{u}^a$, that is, in the global coordinate system we also distinguish between the particles of body 1 and body 2.

The global coordinate system is connected to the two local systems by rotation matrices $A(t)$ and the distance between the origins $\mathbf{R}(t)$ which are functions of the time. The rotation matrices are orthogonal matrices. Their columns are denoted by $\mathbf{n}^a, \mathbf{t}^a, \mathbf{b}^a$. Let $(1, 0, 0)^T$ be given in the local system; then $\mathbf{n}^a$ is that same vector in the global system. Similarly, $\mathbf{t}^a$ is the global representation of the vector $(0, 1, 0)^T$ in the local system, and $\mathbf{b}^a$ is the global representation of $(0, 0, 1)^T$. It is easy to see that $\mathbf{n}^a, \mathbf{t}^a, \mathbf{b}^a$ indeed form an orthonormal system.

The following connection exists between the $\mathbf{y}^a + \mathbf{w}^a$ and the $\mathbf{x}^a + \mathbf{u}^a$:

$$\mathbf{x}^a + \mathbf{u}^a = A^a(\mathbf{y}^a + \mathbf{R}^a) + A^a\mathbf{w}^a \tag{4}$$

We identify

$$\mathbf{x}^a = (A^a\mathbf{y}^a + \mathbf{R}^a) \quad \Rightarrow \quad \mathbf{y}^a = A^{aT}(\mathbf{x}^a - \mathbf{R}^a) \tag{5}$$

$$\mathbf{u}^a = A^a\mathbf{w}^a(\mathbf{y}^a, t) \quad \Rightarrow \quad \mathbf{w}^a = A^a\mathbf{u}^a \tag{6}$$

In particular, (5) gives the global coordinate system in its dependence on the local coordinate system, and vice versa, and (6) defines the global displacement $\mathbf{u}^a$ in terms of the local displacement $\mathbf{w}^a$, and vice versa.

It is clear from (5) that a variable belonging to body a can be written as a function of $\mathbf{x}^a$ and t or of $\mathbf{y}^a$ and t.

Let $\mathbf{m}^a$ be the outer normal on body a at $\mathbf{y}^a$; since $\mathbf{w}^a$ is small with small gradients, $\mathbf{m}^a$ is also the normal on the deformed body a at $\mathbf{y}^a + \mathbf{w}^a$. To see this, we assume that the surface of the body is given by $F(\mathbf{y}) = 0$. Let $\mathbf{z} = \mathbf{y} + \mathbf{w}$ be the position of $\mathbf{y}$ in the deformed state; that is, $\mathbf{y} = \mathbf{z} - \mathbf{w}$. The deformed surface is then given by $F(\mathbf{z} - \mathbf{w}) = 0$. The normal $\mathbf{m}$ on the undeformed body at $\mathbf{y}$ is given by

$$\mathbf{m} = \partial F/\partial\mathbf{y} \tag{7}$$

The normal on the deformed body is

$$\mathbf{m}' = \partial F/\partial\mathbf{z} = (\partial F/\partial\mathbf{y})(\partial\mathbf{y}/\partial\mathbf{z}).$$

But

$$\partial\mathbf{y}/\partial\mathbf{z} = I - \partial\mathbf{w}/\partial\mathbf{z} \approx I,$$

where I is the 3×3 unit matrix. From this proposition it follows that $\mathbf{m} \approx \mathbf{m}'$. In the global coordinates, $\mathbf{m}^a$ becomes $\mathbf{n}^a$, with

$$\mathbf{n}^a = A^a\mathbf{m}^a \tag{8}$$

The points $\mathbf{y}^1 + \mathbf{w}^1$ and $\mathbf{y}^2 + \mathbf{w}^2$ are in contact with one another when

$$\mathbf{x}^1 + \mathbf{u}^1 = \mathbf{x}^2 + \mathbf{u}^2, \text{ and } \mathbf{n}^1 = -\mathbf{n}^2 \tag{9}$$

Example: A rail surface consists of the union of a number of circular cylinders with parallel axes that touch one another. That is, the cylinders intersect and at the intersection the first derivatives are continuous.

The equation of one circular cylinder is, in a Cartesian coordinate system $(O;x,y,z)$

$$(x - a)^2 + (z - b)^2 - R^2 = 0, \quad \mathbf{n} = q(2(x - a), 0, 2(z - b))$$

with the scalar q chosen so that $\mathbf{n}$ is a unit vector; then $\mathbf{n}$ is the unit normal on the rail.

1.2   The Distance

We assume that the bodies are not in contact at the point $\mathbf{x}$, but that they are $O(\mathbf{u})$ apart, and that the surface is smooth, all in the neighborhood of the point $\mathbf{x}$, while the relation $\mathbf{n}^1 = \mathbf{n}^2$ is approximately valid. In a small neighborhood of $\mathbf{x}$ the bodies may be visualized as two parallel slabs, see Fig. 1.

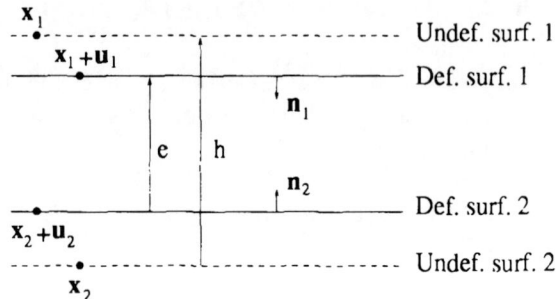

Figure 1: The Distance

Let the distance between the slabs, (2) to (1), be h in the unstressed state, and e in the deformed state. Then we have

$$h = \mathbf{n}^{2T}(\mathbf{x}^1 - \mathbf{x}^2), \quad e = \mathbf{n}^{2T}\{(\mathbf{x}^1 + \mathbf{u}^1) - (\mathbf{x}^2 + \mathbf{u}^2)\} \tag{10}$$

We simplify the expression for e. As we will see later on, the displacement occurs in a linearized contact problem such as we have here (small displacements, small displacement gradients) only in the form $\mathbf{u}^1 - \mathbf{u}^2$. We call this the **displacement difference**, and we denote it by $\mathbf{u}$. Introduction of the displacement difference into the expression for e, we find

$$e = h + \mathbf{n}^{2T}\mathbf{u} = h - \mathbf{n}^{1T}\mathbf{u} \tag{11}$$

e: deformed, h: undeformed distance, (2) to (1).
We analyze the deformed distance e.

- $e > 0$: there is a gap between the bodies at $\mathbf{x}^1$, $\mathbf{x}^2$; $\mathbf{x} = (\mathbf{x}^1 + \mathbf{x}^2)/2$

- $e = 0$: the bodies are in contact at $\mathbf{x}$;

- $e < 0$: the bodies penetrate at $\mathbf{x}$. Impossible.

In sum, only $e(\mathbf{x}) \geq 0$ is possible.

In addition to the distance there is the play of forces. Contact tractions $\mathbf{p}^1$, $\mathbf{p}^2$ (dimension: N/m^2) act on bodies (1) and (2). According to Newton's Third Law,

$$\mathbf{p}^1 = -\mathbf{p}^2 \doteq \mathbf{p} \tag{12}$$

Then the normally directed traction on (1) at $\mathbf{x}$, positive if compressive, is given by

$$p_N = \mathbf{n}^{2T}(\mathbf{x})\mathbf{p} \qquad \text{Normal component} \tag{13}$$

and the tangential traction is given by

$$\mathbf{p}_T = \mathbf{p} - \mathbf{n}^2 p_N(\mathbf{x}) \quad \text{Tangential component} \tag{14}$$

When the bodies do not attract each other, the normal component of the traction vanishes outside contact and is positive (=compressive) inside, while the deformed distance is positive outside contact and vanishes inside. We can summarize this as follows:

$$e(\mathbf{x}) \ge 0: \quad \text{either a gap or contact;} \tag{15}$$
$$p_N(\mathbf{x}) \ge 0: \quad \text{either no, or a compressive normal traction;} \tag{16}$$

$$e p_N = 0 \tag{17}$$

In words, in contact, p_N may be positive, and $e = 0$;
outside contact, the normal traction vanishes, and $e > 0$.

We formalize this as follows. We choose a potential contact zone (also called the potential contact area, pot.con.) which is such that

• The potential contact zone encompasses the real contact zone completely;

• In the potential contact zone $p_N \ge 0$, $e \ge 0$, $p_N e = 0$.

Note that the pot.con. may be chosen freely, as long as the above is satisfied.

Example: Consider a so-called Winkler bedding, i.e. a rigid flat plate upon which are mounted springs. The springs are tangentially unconnected; they are equally long and have the same positive spring constant per unit area $k\ m^3/N$ in the normal direction:

$\mathbf{n}^{2T}\mathbf{u}^1 = k p_N$, $\mathbf{n}^{2T}\mathbf{u}^2 = -k p_N$, with $k > 0$
$\Rightarrow e = h + \mathbf{n}^{2T}\mathbf{u} = h + \mathbf{n}^{2T}(\mathbf{u}^1 - \mathbf{u}^2) = h + 2k p_N$

Suppose that $h > 0$. Since $p_N \ge 0$ we must have that $e \ge h > 0$, and hence $p_N = 0$.
Then $e = h > 0$, and $\mathbf{n}^{2T}\mathbf{u}^a = 0$, $a = 1, 2$.
Suppose that $h < 0$. Since $e \ge 0$ we must have that $2k p_N = e - h > 0 \Rightarrow e = 0$ and $2k p_N = -h > 0$. Hence $\mathbf{n}^{2T}\mathbf{u}^1 = -\mathbf{n}^{2T}\mathbf{u}^2$. The solution is shown in Fig. 2.

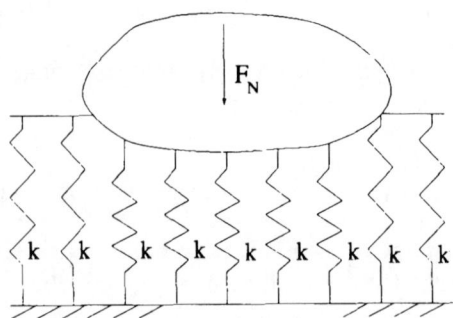

Figure 2: Compression of a Winkler bedding (normal contact)

It is seen that the inequalities play an crucial role in determining the elastic field. This is typical of a contact problem.

1.3   The Slip

We determine the relative velocity of a particle of body 1 with respect to body 2.

Suppose that the particle $\mathbf{y}^1$ is in contact with the particle $\mathbf{y}^2$ at the time t, that is

$$\mathbf{x}^1(t) + \mathbf{u}^1(\mathbf{x}^1, t) = \mathbf{x}^2(t) + \mathbf{u}^2(\mathbf{x}^2, t)$$

If we denote time differentiation by a high dot, the velocity of particle 1 with respect to the global coordinate system is

$$\mathbf{v}^1 = \dot{\mathbf{x}}^1(t) + \dot{\mathbf{u}}^1(\mathbf{x}^1, t) = \dot{\mathbf{x}}^1(t) + (\partial \mathbf{u}^1/\partial \mathbf{x}^1)\dot{\mathbf{x}}^1 + \partial \mathbf{u}^1/\partial t \tag{18}$$

$\mathbf{v}^2$ is similarly defined.

The slip $\mathbf{s}$ is the relative velocity of two particles in contact, that is,

$$\mathbf{s} = \mathbf{v}^1 - \mathbf{v}^2 = \{\dot{\mathbf{x}}^1(t) - \dot{\mathbf{x}}^2(t)\} + \{\dot{\mathbf{u}}^1(\mathbf{x}^1, t) - \dot{\mathbf{u}}^2(\mathbf{x}^2, t)\} \tag{19}$$

Now,

$$\mathbf{x}^1 = \mathbf{x}^2 + \mathbf{u}^2 - \mathbf{u}^2 \quad \Rightarrow \quad \mathbf{x}^1 \approx \mathbf{x}^2 \approx \mathbf{x} \doteq (\mathbf{x}^1 + \mathbf{x}^2)/2 \tag{20}$$

so that, owing to the smallness of $\mathbf{u}^a$, and if $\dot{\mathbf{x}}^1 \approx \dot{\mathbf{x}}^2$

$$\mathbf{s} = \dot{\mathbf{x}}^1 - \dot{\mathbf{x}}^2 + (\partial \mathbf{u}^1/\partial \mathbf{x} - \partial \mathbf{u}^2/\partial \mathbf{x})\dot{\mathbf{x}} + \partial \mathbf{u}^1/\partial t - \partial \mathbf{u}^2/\partial t \tag{21}$$

If $\dot{\mathbf{x}}^1$ is not approximately equal to $\dot{\mathbf{x}}_2$, then $(\dot{\mathbf{x}}^1 - \dot{\mathbf{x}}^2)$ is large with respect to $\{(\partial \mathbf{u}^a/\partial \mathbf{x}^a)\dot{\mathbf{x}}^a - \partial \mathbf{u}^a/\partial t\}$, and $\mathbf{s}$ is also given by (21), albeit that the second and third terms are negligible with respect to the first.

We call

$$\mathbf{c} = \dot{\mathbf{x}}^1 - \dot{\mathbf{x}}^2 \qquad \text{creep} \qquad (22)$$

$$\mathbf{v} = -\dot{\mathbf{x}} \qquad \text{rolling velocity} \qquad (23)$$

$$\mathbf{u} = \mathbf{u}^1 - \mathbf{u}^2 \quad \text{displacement difference} \qquad (24)$$

$$\mathbf{s} = \mathbf{c} - (\partial\mathbf{u}/\partial\mathbf{x})\mathbf{v} + \partial\mathbf{u}/\partial t \qquad \text{slip} \qquad (25)$$

The minus sign in the definition of the rolling velocity calls for comment, see Fig. 3:

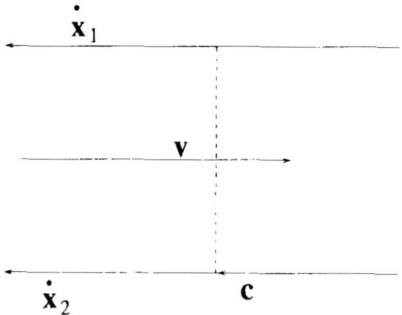

Figure 3: The definitions of creep and rolling velocity

Place the origin of the global coordinate system temporarily in $\mathbf{x}$. Then $\dot{\mathbf{x}}$ is the velocity of the particles near $\mathbf{x}$, and it is seen that the material flows backwards through the coordinate system, counter to the rolling direction. Hence the minus sign in the definition of the rolling velocity.

If a coordinate system may be found in which all quantities $\mathbf{c}$, $\mathbf{u}$, $\mathbf{v}$ are independent of the time t, one speaks of steady state rolling, otherwise of non-steady state rolling:

$$\mathbf{s} = \mathbf{c} - (\partial\mathbf{u}/\partial\mathbf{x})\mathbf{v} \qquad \text{steady state rolling} \qquad (26)$$

$$\mathbf{s} = \mathbf{c} - (\partial\mathbf{u}/\partial\mathbf{x})\mathbf{v} + \partial\mathbf{u}/\partial t \qquad \text{non-steady state (or transient) rolling} \qquad (27)$$

We finish this section on the slip by analyzing the creep for bodies of revolution that are rotated about their axes which are almost in the same plane. A number of interesting technological problems fall into this category.

1. Problems in which the contact area is almost flat. Examples:

    (a) A ball rolling over a plane;

    (b) An offset printing press: contact short in the rolling direction;

    (c) An automotive wheel rolling over a road.

2. Problems with contact short in the rolling direction, curved in the lateral. Examples:

(a) A railway wheel rolling over a rail;

(b) A ball rolling in a deep groove, as in ball bearings.

3. Problems in which the contact area is curved in the rolling direction, and conforming in the lateral direction. Example: a pin rolling in a hole.

1.4   Leading Edge, Trailing Edge

The contact region C has an edge. This edge consists of three parts: the leading edge, the trailing edge, and, possibly, a neutral edge. When the rolling velocity points outside the contact area at a point of the edge, this points belongs to the LEADING EDGE: particles move into the contact area with rolling velocity.

When the rolling velocity points into the contact region C at a point of the edge, the point belongs to the TRAILING EDGE: and particles leave the contact area at macroscopic rolling velocity.

When the rolling velocity vanishes, or is parallel to the edge, we speak of a NEUTRAL EDGE: particles move only into or outside the contact area through the elastic deformation, not through the rigid body motion.

1.5   Example

We give an example of a simplified wheel-rail system. see Fig. 4.

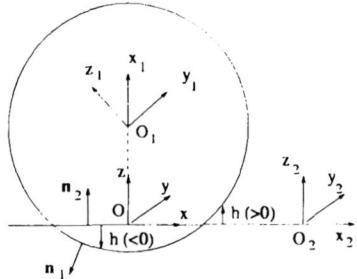

Figure 4: A simplified wheel-rail system

The origin O^1 of the wheel lies on the axis of the cylinder (body 1). The origin O^2 of the rail (body 2) lies on its surface. The origin O of the global coordinate system lies on the surface of the rail, perpendicularly below O^1. It is contact fixed. The y-axis of the wheel coincides with the axis of the cylinder; it is the rotation axis of the wheel. The y-axis of the global system coincides with the projection of the y-axis of the rail. The x-axis is in the same sense as the rolling direction. The z-axis of the global system points vertically upwards. The sense of the y-axis is so that the global system is right-handed. The wheel rotates about its y-axis, clockwise, so that the wheel rolls over the rail to the right.

We have

$$\mathbf{x}^a = A^a \mathbf{y}^a + \mathbf{R}^a \longleftrightarrow \mathbf{y}^a = A^{aT}(\mathbf{x}^a - \mathbf{R}^a).$$

$\mathbf{x}^a$ is the global coordinate of the particle $\mathbf{y}^a$. $\mathbf{R}^a$ is the position of O^a in the global coordinates. The particle-fixed velocity at $\mathbf{x}$ is

$$\dot{\mathbf{x}}^a = \dot{A}^a \mathbf{y}^a + \dot{\mathbf{R}}^a = \dot{A}^a A^{aT}(\mathbf{x}^a - \mathbf{R}^a) + \dot{\mathbf{R}}^a.$$

This means: substitute $\mathbf{x}^a, \mathbf{R}^a, \dot{\mathbf{R}}^a$ in the right-hand side, and you find $\dot{\mathbf{x}}^a$ without calculating $\mathbf{y}^a$.

Indeed we have:

BODY 1

$$\mathbf{R}^1 = (0, 0, r)^T \quad \Longrightarrow \quad \dot{\mathbf{R}}^1 = 0$$

$$A^1 = \begin{pmatrix} + \cos wt & 0 & + \sin wt \\ 0 & 1 & 0 \\ - \sin wt & 0 & + \cos wt \end{pmatrix};$$

$$\dot{A}^1 = w \begin{pmatrix} - \sin wt & 0 & + \cos wt \\ 0 & 0 & 0 \\ - \cos wt & 0 & - \sin wt \end{pmatrix};$$

$$\dot{A}^1 A^{1T} = w \begin{pmatrix} 0 & 0 & 1 \\ 0 & 0 & 0 \\ -1 & 0 & 0 \end{pmatrix};$$

$$\dot{\mathbf{x}}^1 = \dot{A}^1 A^{1T}(\mathbf{x}^1 - \mathbf{R}^1) = \begin{pmatrix} wz - wr \\ 0 \\ -wx \end{pmatrix}.$$

BODY 2

$$\mathbf{x}^2 = \mathbf{y}^2 + \begin{pmatrix} -vt \\ 0 \\ 0 \end{pmatrix} \Longrightarrow \dot{\mathbf{x}}^2 = \begin{pmatrix} -v \\ 0 \\ 0 \end{pmatrix}$$

When $\mathbf{x} = 0$:

$$\mathbf{c} = \dot{\mathbf{x}}^1 - \dot{\mathbf{x}}^2 = \begin{pmatrix} v - wr \\ 0 \\ 0 \end{pmatrix}; \quad \mathbf{v} = -(\dot{\mathbf{x}}^1 + \dot{\mathbf{x}}^2)/2 = \begin{pmatrix} (v + wr)/2 \\ 0 \\ 0 \end{pmatrix}$$

$\mathbf{c}$: creep at the origin, $\mathbf{v}$: rolling velocity at the origin.

1.6   Recapitulation of the linear theory of elasticity

In the linear theory of elasticity it is more convenient to work with index notation than with matrix-vector notation. In index notation, the displacement $\mathbf{u}^a$ ($a = 1, 2$: body number) is denoted by its components u_i^a, $i = 1, 2, 3$, in a Cartesian coordinate system. The body number a is omitted when no confusion may arise. Partial differentiation with respect to the coordinate x_j is denoted by $,j$:

$$u_{i,j} \doteq \partial u_i / \partial x_j, \quad \text{the displacement gradient} \tag{28}$$

summation over 1,2,3 of repeated subscripts is understood. This summation convention is undone by placing one or more subscripts between brackets.

As usual, we work in a small displacement, small displacement gradient theory. The linearized stresses are defined as

$$e_{ij} = (u_{i,j} + u_{j,i})/2 = e_{ji}, \tag{29}$$

and the stress is given by σ_{ij}. The equations of equilibrium are

$$\sigma_{ij} = \sigma_{ji} \quad \text{and} \quad \sigma_{ij,j} = 0 \tag{30}$$

where we disregard inertial forces such as gravity in the rightmost equation. The surface load, also called the surface traction, is $p_i = \sigma_{ij} n_j$, where n_j is the outer normal on the surface.

According to Hooke's Law there is a linear relationship between the stresses σ_{ij} and the strains e_{hk}. In the general case, this relation reads

$$\sigma_{ij} = E_{ijhk} e_{hk}, \quad E_{ijhk}: \text{elastic constants} \tag{31}$$

and in the case of an isotropic material

$$e_{ij} = \frac{1+\nu}{E} \sigma_{ij} - \frac{\nu}{E} \delta_{ij} \sigma_{kk}, \quad \sigma_{ij} = \frac{E e_{ij}}{1+\nu} + \frac{E \delta_{ij} e_{kk}}{(1+\nu)(1-2\nu)} \tag{32}$$

with

E: Young's Modulus, ν: Poisson's Ratio, (elastic constants)

δ_{ij}: Kronecker delta, $=0$ when $i \neq j$, $=1$ when $i = j$.

1.7   Friction

When two bodies slide over each other, it will often be observed that this motion is opposed by a force. This is the phenomenon of friction; the force is called the force of friction. Usually a finite compensating force is needed to set a body sliding, while in many experiments the friction force remains constant during sliding. So it is assumed that the shearing force is bounded by a force bound g, which depends on the normal force F_z, the magnitude of the sliding velocity V and other parameters, thus

$$g = g(F_z, V, ...), \quad ...: \text{ other parameters; } F_z: \text{ normal comp. of total contact force} \quad (33)$$

When the sliding velocity (the slip, in fact) vanishes, the tangential force may fall below the traction bound g in absolute value; when sliding occurs, the tangential force is at the traction bound, and it opposes the slip:

$$|F_\tau| < g(F_z, V, ...) \quad (34)$$

F_τ: tangential component of total contact force, $\tau = 1, 2$,

$$|F_\tau| = \sqrt{F_1^2 + F_2^2}$$

$$\text{if } V \neq 0: \quad F_\tau = -g v_\tau / V, \quad \text{Greek index: tangential comp.} \quad (35)$$

v_τ: tangential component of sliding velocity; $V = |v_\tau|$.

Coulomb [2] proposed that g is proportional to F_z with a constant of proportionality called the coefficient of friction:

$$g(F_z, V, ...) = f F_z \quad \text{Coulomb, 1785} \quad (36)$$

In order to interpret (36) Archard proposed in [1] that the friction was primarily caused by the adhesion of the bodies to each other. The adhesion takes place at the tips of the roughnesses of the bodies, which are called the asperities. At these tips the bodies are in contact and they form junctions, the real contact surface C_r, as opposed to the apparent contact area C which consists of the junctions and the region in between them. Archard showed that the area of the real contact surface is proportional to F_z, the normal compressive force.

At the real contact surface the bodies are welded together by interatomic forces. Owing to the sliding motion the welded asperities shear; eventually the welds break, and the asperities form new welds with different partners. The shearing of the asperities is accompanied by micro-plastic deformation, and also by the detachment of material particles from the bodies, thus leading to wear. From this it is seen that friction and wear are closely connected phenomena.

Despite this bolstering of Coulomb's Law, it is generally agreed by tribologists that (36) must be modified. The simplest modification is the introduction of two coefficients of friction, one for non-sliding (f_{stat}) and one for sliding (f_{kin}), with

$$f_{stat} > f_{kin} \quad (37)$$

This did not suffice for many researchers, and they proposed more complicated relationships for the coefficient of friction.

So far we considered the total contact force, and the global velocity in sliding. In contact mechanics, however, we need a local form of the friction law. A very simple theory suggests itself: instead of global quantities use local quantities, that is instead of **F** use **p**, the local traction, and instead of the global sliding velocity V use the slip **s**:

$$|p_\tau| \leq g(p_z, |s_\tau|, ...)$$

$$\text{if} \quad |s_\tau| \neq 0 \qquad \Longrightarrow \qquad p_\tau = -g s_\tau / |s_\tau| \tag{38}$$

p_τ: tangential traction component

s_τ: slip component, $\tau = 1, 2$

This law was stated and experimentally confirmed by Rabinowicz [7]. It was used earlier in theoretical work see [4], [6], [3], and [5]. The traction bound is taken as

$$g(p_z, |s_\tau|, ...) = f(|s_\tau|, ...)p_z \tag{39}$$

and f is usually taken constant.

1.8   Boundary conditions

We recall Hooke's Law:
$$\sigma_{ij} = E_{ijhk} e_{hk} \tag{40}$$
It is valid for all types of bodies. Sometimes it is possible to bring it in surface mechanical form:
$$u_i(\mathbf{x}) = \int_{\partial V} A_{ik}(\mathbf{x}, \mathbf{y}) p_k(\mathbf{y}) dS(\mathbf{y}) \tag{41}$$
where $A(\mathbf{x}, \mathbf{y})$ is the displacement at $\mathbf{x}$ due to a point load at $\mathbf{y}$; it is called the Influence Function. The influence function depends strongly on the form of the body. In 3D elasticity it has been calculated for a few forms, in particular the half-space.

The advantage of (41) over (40) resides in the fact that for a 3D body (41) is taken over its 2D boundary only, while (40) extends over its entire 3D iterior.

The surface of body a is divided into three parts.

- In A_p the surface load is prescribed as $\bar{\mathbf{p}}$.

- In A_u the surface displacement is prescribed as $\bar{\mathbf{u}}$

- A_c is the potential contact zone. It was already defined below eq. (17).

We give here a slightly different definition of the potential contact zone, which is roughly equivalent to the one given before. It reads as follows:

The potential contact zone can be freely chosen under the following conditions:

1. It must completely encompass the contact region;

2. In it, $\mathbf{x}^1 - \mathbf{x}^2 = O(\mathbf{u})$;

3. In it, $\mathbf{n}^1 \approx -\mathbf{n}^2$.

Then the following relations hold in the potential contact:

$$e(\mathbf{x}) = h(\mathbf{x}) + \mathbf{n}^{2T}\mathbf{u}(\mathbf{x}), \quad \mathbf{u}(\mathbf{x}) = \mathbf{u}^1(\mathbf{x}) - \mathbf{u}^2(\mathbf{x}), \quad e \geq 0 \tag{42}$$
$$p_N(\mathbf{x}) = \mathbf{n}^{2T}\mathbf{p}^1, \quad p_N \geq 0, \quad p_N e = 0 \tag{43}$$

where the fields in $\mathbf{x}^1$ and $\mathbf{x}^2$ are suitably extended to

$$\mathbf{x} = (\mathbf{x}_1 + \mathbf{x}_2)/2. \tag{44}$$

The region where $e = 0$ is called the contact zone (contact region, contact area, contact patch). It is denoted by C. In it, $p_N \geq 0$.
The region where $e > 0$ is called the exterior zone (.. region, .. area). It is denoted by E. In it, $p_N = 0$. (42),(43) are the mathematical description of $C \cup E = A_c$. As we saw in the Example of the Winkler bedding, (42),(43) determine the solution of the frictionless problem completely and uniquely. The inequalities of (42),(43) play a crucial role in the determination of the contact area and the elastic field.
The equation $p_N e = 0$ is very important. In conjunction with the inequalities it means that when $e > 0 \implies p_N = 0$ and vice versa.

We turn to the boudary conditions of friction.
We had defined the tangential component of the slip in the contact zone as $\mathbf{s} = \dot{\mathbf{x}}^1 - \dot{\mathbf{x}}^2 + \dot{\mathbf{u}}^1 - \dot{\mathbf{u}}^2$.
The contact area is divided into two parts, $viz.$ the stick area (area of adhesion, ..zone, ..region) H, where the tangential component of the slip vanishes, and the area (zone, region) of slip S, where this is not so. We have for the total boundary conditions in the potential contact area A_c:

$$\begin{array}{llll}
\text{in } H: & |\mathbf{s}_T| = 0, & |\mathbf{p}_T| \leq g & (45) \\
\text{in } S: & |\mathbf{s}_T| \neq 0, & \mathbf{p}_T = -g\mathbf{s}_T/|\mathbf{s}_T| & (46) \\
& S \cup H = C, & S \cap H = \varnothing & (47) \\
\text{in } C: & e = 0, & p_N \geq 0 & (48) \\
\text{in } E: & \mathbf{p} = 0, & e > 0 & (49) \\
& C \cup E = A_c, & C \cap E = \varnothing & (50)
\end{array}$$

References of Section 1

[1] J.F. ARCHARD (1957), *Elastic deformation and the laws of friction.* Proceedings of the Royal Society of London, Ser. **A 243**, p. 190-205.

[2] C.A. COULOMB (1785), *Théorie des machines simples*. Mémoire de Mathématique et de Physique de l'Académie Royale, p. 161-342.

[3] F.W. CARTER (1926), *On the Action of a Locomotive Driving Wheel*. Proceedings of the Royal Society of London **A 112** p. 151-157.

[4] C. CATTANEO (1938), *Sul contatto di due corpi elastici: distribuzione locale degli sforzi*. Accademia Nazionale Lincei Rendiconti Ser 6 **XXVII** p. 342-348, 434-436, 474-478.

[5] H. FROMM (1927), *Berechnung des Schlupfes beim Rollen deformierbaren Scheiben*. Zeitschrift für angewandte Mathematik und Mechanik **7** p. 27-58.

[6] R.D. MINDLIN (1949), *Compliance of elastic bodies in contact*. Journal of Applied Mechanics **16**, p. 353-383.

[7] E. RABINOWICZ, *Friction and Wear of Materials*. John Wiley and Sons, New York.

2 The half-space approximation

One of the ways to attack a contact problem is by means of the FEM. Especially when we deal with 3D problems this is very time consuming. The BEM of (1.41) is a much more promising option, if it is at all feasible.

This is so when we deal with a homogeneous, isotropic elastic 3D half-space. This is of great technological importance, as many problems may be approximately solved by using a half-space.

A half-space consists of all points on one side of a plane, the bounding plane. For instance, in a Cartesian coordinate system $(O; x, y, z)$ a half-space may be defined by $\{(x, y, z) : z \geq 0\}$. The contact field of two elastic bodies may be calculated by means of half-spaces when the maximum diameter of the contact area is small with respect to a typical dimension of the bodies, such as their diameter or the minimum radius of curvature near the contact. Then, the elastic field in the contact part may be calculated by replacing the body locally by a half-space. The boundary conditions are then of the real body, but the elasticity equations are solved for the half-space. The situation is shown in Fig. 5.
Examples of the cases where the half-space approximation is valid:

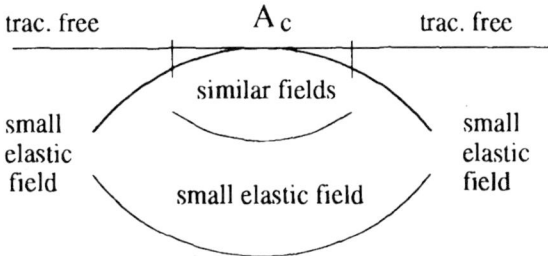

Figure 5: The half-space approximation

A ball pressed onto a thick slab, tread contact in wheels and rails, see Fig. 6.

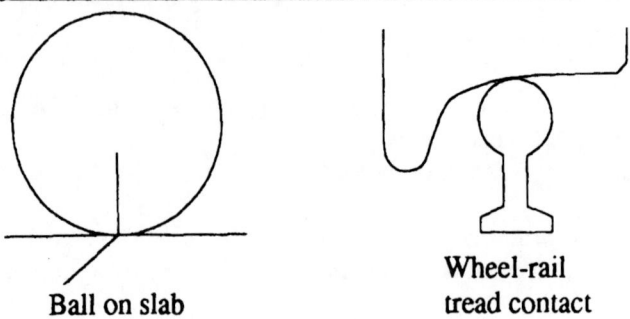

Ball on slab

Wheel-rail
tread contact

Figure 6: Examples of the half-space approximation

Properties of the half-space approximation are:

1. Many geometries are elastically alike.
   This is most important as it renders software written for half-spaces applicable to many geometries. The relative ease of the half-space approximation leads one to use it even where one is apt to make serious errors.

2. The influence numbers $A_{ik}(\mathbf{x}, \mathbf{y})$ can be calculated exactly by means of the formulae of Boussinesq [1] and Cerruti [2].
   Derivations of these formulae may be found in [6] and in [3].

We give these formulae. We denote the global coordinate system by $(O; x, y, z) = (O; x_i), i = 1, 2, 3.$ O lies in the surface of the half-space. The axes of x and y lie in the surface of the half-space. The axis if $x = x_1$ points in the rolling direction, the axis of $z = x_3$ points vertically upwards into body 1, and $y = x_2$ completes the right-handed coordinate system. Moreover, we denote the components of $\mathbf{u}^a$ by $u_i^a = (u^a, v^a, w^a)$ and the components of the surface traction by $\mathbf{p}^a$ by $p_i^a = (p_x^a, p_y^a, p_z^a)$.

Clearly, by Newton's Third Law,

$$p_i^1 = -p_i^2 \doteq p_i \tag{51}$$

(41) reads

$$u_i^a(\mathbf{x}) = \int_{-\infty}^{\infty} \int_{-\infty}^{\infty} A_{ij}^a \, p_j^a(x', y') \, dx' dy' \tag{52}$$

In body $a, a = 1, 2$ we have

$$A_{1,3}^a = \frac{1}{4\pi G^a} \{ \frac{(x - x')|z|}{r^3} - \frac{(1 - 2\nu^a)(x - x')}{r(|z| + r)} \} \tag{53}$$

$$A_{2,3}^a = \frac{1}{4\pi G^a} \{ \frac{(y - y')|z|}{r^3} - \frac{(1 - 2\nu^a)(y - y')}{r(|z| + r)} \} \tag{54}$$

$$A_{3,3}^a = \frac{-(-1)^a}{4\pi G^a} \{ \frac{z^2}{r^3} + \frac{2(1 - \nu^a)}{r} \} \tag{55}$$

$$r = \sqrt{(x - x')^2 + (y - y')^2 + z^2} \tag{56}$$

The factor $-(-1)^a$ in (55) calls for comment. When $a = 1$, it equals unity. Then the z-axis points into the body, and the component w^1 has the same direction as the concentrated load $p_3^1 = p_3$. When $a = 2$, we must have the same formulae but in a coordinate system in which the z-axis points vertically downwards, while the axes of x and y remain the same. Then one must invert the sign of z in all formulae (53)-(55), and one must reverse the sign of w as well.

So we see that the displacements in bodies 1 and 2 due to a normal loading obey the law

$$\begin{aligned}
u^1(x, y, z) &= u^2(x, y, -z) \\
v^1(x, y, z) &= v^2(x, y, -z) \\
w^1(x, y, z) &= -w^2(x, y, -z) \\
\text{if } p_x &= p_y = 0
\end{aligned}$$

The displacement differences, which are prescribed in the normal and tangential problems, are

$$\begin{aligned}
u(x, y) &= u^1(x, y, 0) - u^2(x, y, 0) = \\
&= \frac{1}{4\pi}\{\frac{1 - 2\nu^1}{G^1} - \frac{1 - 2\nu^2}{G^2}\} \int_{-\infty}^{\infty}\int_{-\infty}^{\infty} p_z(x', y')\frac{x' - x}{R^2}dx'dy' \\
v(x, y) &= v^1(x, y, 0) - v^2(x, y, 0) = \\
&= \frac{1}{4\pi}\{\frac{1 - 2\nu^1}{G^1} - \frac{1 - 2\nu^2}{G^2}\} \int_{-\infty}^{\infty}\int_{-\infty}^{\infty} p_z(x', y')\frac{y' - y}{R^2}dx'dy' \\
w(x, y) &= w^1(x, y, 0) - w^2(x, y, 0) = \\
&= \frac{1}{2\pi}\{\frac{1 - \nu^1}{G^1} + \frac{1 - \nu^2}{G^2}\} \int_{-\infty}^{\infty}\int_{-\infty}^{\infty} p_z(x', y')/Rdx'dy' \quad (57)
\end{aligned}$$

$$p_x = p_y = 0, \quad R = \sqrt{(x' - x)^2 + (y' - y)^2}$$

G^a :   modulus of rigidity, ν^a :   Poisson's Ratio:

Elastic   constants of body a

We combine ν^1, ν^2 and G^1, G^2 in the following manner:

$$\begin{aligned}
1/G &= \{(1/G^1) + (1/G^2)\}/2 \quad &(58) \\
\nu/G &= \{(\nu^1/G^1) + (\nu^2/G^2)\}/2 \quad &(59) \\
K &= (G/4)\{(1 - 2\nu^1)/G^1 - (1 - 2\nu^2)/G^2\} \quad &(60)
\end{aligned}$$

It is easy to see that G lies between G^1 and G^2. In the case of elastic symmetry (= both bodies having the same elastic constants),

$$G = G^1 = G^2, \quad \nu = \nu^1 = \nu^2, \quad K = 0 \quad (61)$$

The constant K vanishes when there is elastic symmetry, and also when both bodies are incompressible: Steel on Steel, Rubber on Steel. Its maximum is 0.5

but in practice it is mostly small, e.g. 0.03 for steel on brass, and 0.09 for steel on aluminium. In terms of the constants of (58)-(60), the displacement difference becomes:

$$u(x,y) \ = \ \frac{K}{\pi G}\int_{-\infty}^{+\infty} p_z(x',y')\frac{x'-x}{R}dx'dy' \tag{62}$$

$$v(x,y) \ = \ \frac{K}{\pi G}\int_{-\infty}^{+\infty} p_z(x',y')\frac{y'-y}{R}dx'dy' \tag{63}$$

$$w(x,y) \ = \ \frac{1-\nu}{\pi G}\int_{-\infty}^{+\infty} p_z(x',y')/Rdx'dy' \tag{64}$$

$$p_x \ = \ p_y = 0 \tag{65}$$

$$R \ = \ \sqrt{(x'-x)^2 + (y'-y)^2} \tag{66}$$

The procedure for the tangential traction/displacement is very nearly the same. We have for the displacement due to the load p_x, second index of A_{ij}^a equal to 1:

$$A_{11}^a \ = \ (-(-1)^a)\frac{1}{4\pi G^a}\{(1/r) +$$

$$+ \ \frac{1-2\nu^a}{|z|+r} + \frac{(x'-x)^2}{r^3} - \frac{(1-2\nu^a)(x'-x)^2}{r[|z|+r]^2}\} \tag{67}$$

$$A_{21}^a \ = \ (-(-1)^a)\frac{1}{4\pi G^a}\{\frac{(x'-x)(y'-y)}{r^3} +$$

$$- \ \frac{(1-2\nu^a)(x'-x)(y'-y)}{r[|z|+r]^2}\} \tag{68}$$

$$A_{31}^a \ = \ -\frac{1}{4\pi G^a}\{\frac{(x'-x)|z|}{r^3} + \frac{(1-2\nu^a)(x'-x)}{r[|z|+r]}\} \tag{69}$$

The displacement due to a load p_y in the y direction, second index of A_{ij}^a equal to 2 is found by the interchange of x and y, u and v, p_x and p_y. The factor $-(-1)^a$ calls for comment: we must take into account that the shearing tractions on the bodies have different signs, and that therefore u,v,w have different signs in the two bodies, but that w is taken in a coordinate system that has a z-axis with the other sign, so that the factor $-(-1)^a$ is neutralized for the vertical displacement w.

The total surface displacement differences $u(x,y)$, $v(x,y)$, $w(x,y)$ become (note that $z = 0$!)

$$u(x,y) \ = \ \frac{1}{\pi G}\int_{-\infty}^{\infty}\int_{-\infty}^{\infty}\{p_x(x',y')[\frac{1-\nu}{R} + \frac{(x'-x)^2}{R^3}] +$$

$$+ \ p_y(x',y')\frac{(x'-x)(y'-y)}{R^3} + Kp_z(x',y')\frac{x'-x}{R^2}\}dx'dy' \tag{70}$$

$$v(x,y) \ = \ \frac{1}{\pi G}\int_{-\infty}^{\infty}\int_{-\infty}^{\infty}\{p_x(x',y')\frac{(x'-x)(y'-y)\nu}{R^3} +$$

$$+ \quad p_y(x',y')[\frac{1-\nu}{R} + \frac{(y'-y)^2}{R^3}] + Kp_z(x',y')\frac{y'-y}{R^2}\}dx'dy'$$

$$w(x,y) = \frac{1}{\pi G} \int_{-\infty}^{\infty}\int_{-\infty}^{\infty}\{ -Kp_z(x',y')\frac{x'-x}{R^2} +$$

$$- \quad Kp_y(x',y')\frac{y'-y}{R^2} + p_z(x',y')\frac{1-\nu}{R}\}dx'dy' \tag{71}$$

3. Quasiidentity.

The half-spaces in contact are called quasiidentical, when $K = 0$:

$$\text{Quasiidentity} \quad \Longleftrightarrow K = 0 \tag{72}$$

When $K = 0$, the displacement differences u and v are not influenced by the normal traction p_z, and the displacement diference w is not influenced by the tangential tractions p_x and p_y. Yet there is a coupling through the traction bound, viz.

$$|(p_x, p_y)| \le fp_z, \quad f: \text{coefficient of friction.} \tag{73}$$

Thus we must act as follows:

$$\text{Determine} \quad p_z \quad \text{when } p_x = p_y = 0 \text{ (Normal problem)}$$
$$\text{Determine} \quad p_x, \quad p_y \quad \text{when } g = fp_z \text{ (Tangential problem)} \tag{74}$$

This is called the Johnson process. It is approximately valid when K is small. When $K \ne 0$, we must act differently. We use the so-called Panagiotopoulos process:

(a) Set $I = 0$, and $p_x^I = p_y^I = 0$

(b) Determine the compressive traction p_z^I with (p_x^I, p_y^I) as the tangential traction, by means of a normal (= compressive traction) contact algorithm.

(c) Determine the next tangential traction (p_x^{I+1}, p_y^{I+1}) with p_z^I as the compressive traction, by means of a tangential contact algorithm.

(d) If (p_x^{I+1}, p_y^{I+1}) is close enough to (p_x^I, p_y^I), we stop, else we set $I = I + 1$, and we restart at b.

The Panagiotopoulos process was used by Oden and Pires [7] to prove the existence of the elastic field for elastostatic contact. I use it myself for 3D half-space contact, and for 2D contact of elastic and viscoelastic layered media. It can, apparently, always be made to converge, but sometimes one must perturb the discretization of the contact problem.

Remark: The Johnson process is a one-step Panagiotopoulos process.

4. Slip in the Half-Space.

We had seen in Section 1 that the slip was given by

$$\mathbf{s} = \mathbf{c} - (\partial \mathbf{u}/\partial \mathbf{x})\mathbf{v} + \partial \mathbf{u}/\partial t$$

with   s   the slip

            c   the creep

            u   the displacement difference

            v   the rolling velocity

            x   the position

            t   the time.

In half-space rolling, we assume that the rolling takes place along the positive x-axis with constant velocity V, and that the motion takes place in the contact plane, so that

$$\mathbf{c} = V(v_x - \varphi y, v_y + \varphi x)^T \tag{75}$$
$$\mathbf{v} = (V, 0)^T \tag{76}$$

and

$$s_x = V(v_x - \varphi y - \partial u/\partial x + \partial u/\partial (Vt)) \tag{77}$$
$$s_y = V(v_y + \varphi x - \partial v/\partial x + \partial v/\partial (Vt)) \tag{78}$$

v_x is called the longitudinal creepage, v_y is called the lateral creepage, and φ is called the spin. Another word for creepage is creep ratio.

We identify

$$\int_0^t V t' dt' = q, \quad \text{distance traversed} \tag{79}$$

and divide (77) and (78) by V, where we call

$$S = s/V, \quad \text{the relative slip} \tag{80}$$

When we replace the time variable t by the distance traversed q, and the slip s by the relative slip $\mathbf{S}$, then we see that the relative slip is independent of the rolling velocity V. Since the relative slip has the same direction as the real slip, and the formula for the tangential traction in sliding depends only on that direction when the coefficient of friction is independent of the rolling velocity (as we will assume), the entire problem becomes independent of the rolling velocity, or, more precisely, depends on the rolling velocity only through the coefficient of friction.

2.1 The Hertz Problem

Consider two paraboloids with parallel axes, numbered 1 and 2, see Fig. 7.

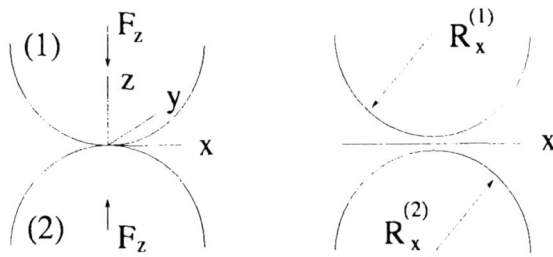

Figure 7: The Hertz problem

They are brought together so that their tips touch. A Cartesian coordinate system is introduced in which the plane of x and y is the common tangent plane, the origin O lies in the tangent point, and the z-axis points vertically upwards into paraboloid 1. We assume for simplicity that the planes of principal curvature coincide with one another and with the planes of x and y, as is often the case in rolling. The radii of curvature in the plane of x are $R_x^{(1)}$ and $R_x^{(2)}$ and those in the plane of y are $R_y^{(1)}$, $R_y^{(2)}$. We count them positive when the centre of curvature lies inside the corresponding paraboloid. The bodies are compressed over a distance q. The problem is:
FIND THE CONTACT AREA AND THE PRESSURE DISTRIBUTION WHEN FRICTION IS ABSENT OR IN THE CASE OF QUASIIDENTITY.

We choose the unstressed state so, that the displacements and the stresses vanish at infinity. Then the equation of the surface of body a is

$$z^{(a)} = -(-1)^a \{ \frac{x^2}{2R_x^{(a)}} + \frac{y^2}{2R_y^{(a)}} \} - q^{(a)} \tag{81}$$

In the contact region, we have that

$$h = z^{(1)} - z^{(2)} = Ax^2 + By^2 - q \tag{82}$$

with

$$A = \frac{1}{2R_x^{(1)}} + \frac{1}{2R_x^{(2)}} \tag{83}$$

$$B = \frac{1}{2R_y^{(1)}} + \frac{1}{2R_y^{(2)}} \tag{84}$$

$$q = q^{(1)} - q^{(2)} \tag{85}$$

$$\frac{1}{R} = \frac{A+B}{2} \tag{86}$$

Hertz [9], [10] assumed that the contact area was elliptic in the half-space approximation, with

$$C = \{(x,y,0) | (x/a)^2 + (y/b)^2 \le 1\} \tag{87}$$

Then it may be shown that the normal pressure is semi-ellipsoidal

$$p_z(x,y) = G f_{00} \sqrt{1 - (x/a)^2 - (y/b)^2} \tag{88}$$

where G and ν are the combined modulus of rigidity and the Poisson's ratio, see (58), (59).

The total normal force is found by integration of (88)

$$N = \int \int_C p_z dx dy = (2/3)\pi ab G f_{00}, \quad f_{00} = \frac{3N}{2\pi ab G} \tag{89}$$

The minor and the major semi-axes of the contact ellipse are denoted by S and L, and the axial ratio by $m' = S/L$:

$$m' = S/L, \quad S = \min(a,b), \quad L = \max(a,b) \tag{90}$$

The excentricity is signed. We have

$$|m| = \sqrt{1 - m'^2}, \quad m > 0 \quad \text{if} \quad a < b, \quad m < 0 \quad \text{if} \quad a > b \tag{91}$$

Then q, A and B are given by

$$q = \frac{3N(1-\nu)S\mathbf{K}}{2\pi ab G} \tag{92}$$

$$A(|m|) = B(-|m|) = \frac{3N(1-\nu)(\mathbf{D} - m^2\mathbf{C})}{2\pi ab S G} \tag{93}$$

$$B(|m|) = A(-|m|) = \frac{3N(1-\nu)(1 - m^2)\mathbf{D}}{2\pi ab S G} \tag{94}$$

where $\mathbf{K}$, $\mathbf{B}$ and $\mathbf{D}$ are complete elliptic integrals. We will encounter more complete elliptic integrals. They can all be expressed as a linear combination of two of them, for which we take $\mathbf{C}$ and $\mathbf{D}$:

$$\mathbf{B} = \int_0^{\pi/2} \frac{\cos^2 t}{\sqrt{1 - m^2 \sin^2 t}} dt, \mathbf{B} = \mathbf{D} - m^2\mathbf{C} \tag{95}$$

$$\mathbf{C} = \int_0^{\pi/2} \frac{\sin^2 t \cos^2 t}{\sqrt{1 - m^2 \sin^2 t}^3} dt \tag{96}$$

$$\mathbf{D} = \int_0^{\pi/2} \frac{\sin^2 t}{\sqrt{1 - m^2 \sin^2 t}} dt \tag{97}$$

$$\mathbf{E} = \int_0^{\pi/2} \sqrt{1 - m^2 \sin^2 t}\, dt, \quad \mathbf{E} = (2 - m^2)\mathbf{D} - m^2\mathbf{C} \tag{98}$$

$$\mathbf{K} = \int_0^{\pi/2} \frac{1}{\sqrt{1 - m^2 \sin^2 t}} dt, \quad \mathbf{K} = 2\mathbf{D} - m^2\mathbf{C} \tag{99}$$

We give these functions in their dependence on m' in Table 1, taken from [4].

m'	B	C	D	E	K	m^2
↓ 0	1	-2+ln 4/m'	-1+ln 4/m'	1	+ln 4/m'	1.00
0.1	0.9889	1.7351	2.7067	1.0160	3.6956	0.99
0.2	0.9686	1.1239	2.0475	1.0505	3.0161	0.96
0.3	0.9451	0.8107	1.6827	1.0965	2.6278	0.91
0.4	0.9205	0.6171	1.4388	1.1507	2.3593	0.84
0.5	0.8959	0.4863	1.2606	1.2111	2.1565	0.75
0.6	0.8719	0.3929	1.1234	1.2763	1.9953	0.64
0.7	0.8488	0.3235	1.0138	1.3456	1.8626	0.51
0.8	0.8267	0.2706	0.9241	1.4181	1.7508	0.36
0.9	0.8055	0.2292	0.8491	1.4933	1.6546	0.19
1.0	$.7864=\frac{\pi}{4}$	$.1964=\frac{\pi}{16}$	$.7854=\frac{\pi}{4}$	$1.571=\frac{\pi}{2}$	$1.571=\frac{\pi}{2}$	0.00

Table 1: Complete elliptic integrals (Jahnke-Emde, 1943)[4]

It is seen from Table 1 that $\mathbf{D} > \mathbf{C}$, and it follows from (93) and (94) that $A(|m| = B(-|m|) \geq A(-|m|) = B(|m|)$, so that we have

$$A \geq B \implies m \geq 0, \quad a \leq b \tag{100}$$
$$A \leq B \implies m \leq 0, \quad a \geq b \tag{101}$$

$A, B :$ see (86)

In order to find the excentricity of the contact ellipse, we set with Hertz [9]

$$\cos t = \frac{|A - B|}{A + B}, \quad \text{with } R = \frac{2}{A + B} \tag{102}$$

and it follows from (93),(94), the fact that $\mathbf{D} > \mathbf{C}$ and the expression of $\mathbf{E}$ in $\mathbf{D}$ and $\mathbf{C}$ that

$$\cos t = m^2(\mathbf{D} - \mathbf{C})/\mathbf{E} \tag{103}$$

We give the axial ratio m' as a function of t in Table 2.

t	90°	80°	70°	60°	50°	40°	30°	20°	10°	0°
$m' = S/L$	1.00	.79	.62	.47	.36	.26	.18	.10	.05	0

Table 2: The axial ratio of the contact ellipse as a function of t.

Table 2 is taken from [6] (page 197). We see from (86) to (103) that the shape of the contact ellipse depends only on the radii of curvature of the bodies, and not on the applied load nor the elastic properties. The size of the contact area does, however:

$$A + B = \frac{2}{R} = \frac{3N(1 - \nu)\mathbf{E}}{2G\pi abS} \tag{104}$$

or

$$3N(1 - \nu)R\mathbf{E} = 4\pi abSG \tag{105}$$

A frequently used quantity is f_{00}, see (88). Gf_{00} is the maximum value of the ellipsoidal pressure distribution $p_z(x, y)$. It is

$$f_{00} = \frac{3N}{2\pi abG} = \frac{2S}{(1-\nu)RE} \tag{106}$$

Finally we determine the deepest penetration of the bodies, see (82):

$$q = (1-\nu)KSf_{00} = \frac{2S^2K}{RE} \tag{107}$$

2.1.1 The linear theory of rolling contact for Hertzian contacts

We recall the formulae for the relative slip s/V in steady state rolling:

$$S_x = s_x/V = v_x - \varphi y - \partial u(x, y)/\partial x \tag{108}$$
$$S_y = s_y/V = v_y + \varphi x - \partial v(x, y)/\partial x \tag{109}$$

with

$$
\begin{array}{rcl}
V & : & \text{the rolling velocity} \\
v_x, v_y, \varphi & : & \text{longitudinal, lateral, spin creepage} \\
(u, v) & : & \text{tangential displacement difference, body (1)-(2)} \\
x\text{-axis} & : & \text{rolling direction} \\
z\text{-axis} & : & \text{vertical direction pointing into body (1)} \\
y\text{-axis} & : & \text{completes right-handed system} \\
O & : & \text{origin in centre of contact patch (ellipse)}
\end{array}
$$

When the creepages are very small, it is easy for (u, v) to compensate them, without violating the traction bound. That is, almost the entire contact patch will be covered by the stick region. Instead of saying that the creepages are small, one can also say that the coefficient of friction goes up to infinity. Then the traction bound will not be violated as long as (p_x, p_y) is finite. Now for finite (u, v) the traction is finite except maybe on the edge of the elliptic contact area, as it may be shown that the traction has inverse square root behaviour near the edge:

$$(p_z(x, y), p_y(x, y)) = O(\sqrt{1 - (x/a)^2 - (y/b)^2}^{-1}) \tag{110}$$

so that the slip is confined to part of the edge of the contact area.

Hence inside the contact patch we have

$$\partial u(x, y)/\partial x = v_x - \varphi y, \quad \partial v(x, y)/\partial x = v_y + \varphi x \tag{111}$$

Integrating with respect to x we find

$$
\begin{array}{rcll}
u(x, y) & = & v_x x - \varphi xy + k(y) & \text{in } C \tag{112} \\
v(x, y) & = & v_y x + \varphi x^2/2 + l(y) & \text{in } C \tag{113} \\
p_z(x, y) & = & p_y(x, y) = 0 & \text{in } E \tag{114}
\end{array}
$$

where k and l are arbitrary functions of y.

The question arises, how to determine k and l. We observe that (112) to (114) fully determine the contact problem together with the proper behaviour at infinity, where the elastic field has died out according to the half-space hypothesis.
To that end, we observe that at the leading edge of the contact ellipse the traction must vanish. For at the leading edge unloaded material flows into the contact patch. During transit of the contact patch, traction builds up, which is suddenly released at the trailing edge. So at the leading edge the traction must vanish, but it need not do so at the trailing edge. Hence k and l must be determined so that the traction vanishes at the leading edge, which is given by

$$\{(x,y,z)|x \geq 0, z = 0, |y| = b\sqrt{1 - (x/a)^2}\} \quad \text{leading edge} \qquad (115)$$

We explain why this theory is called the Linear Theory. When we consider (112)-(114), and multiply v_x, v_y, φ, k, l with the same constant D, then it is clear that $D(p_x, p_y)$ satisfy the no-slip contact conditions with $D(p_x, p_y) = (0,0)$ on the leading edge. So the no-slip tangential traction corresponding to $D(v_x, v_y, \varphi)$ is $D(p_x, p_y)$ – a linear relationship.

At this place, we will not enter into the details of the calculation, but merely mention the result. We tabulate the total tangential force (F_x, F_y) and the twisting moment M_z in terms of the creepages, as follows:

$$F_x = \int\int_C p_x dx dy, F_y = \int\int_C p_y dx dy, \qquad (116)$$

$$M_z = \int\int_C (xp_y - yp_x) dx dy \qquad (117)$$

$$F_x = -GabC_{11}v_x, \quad F_y = -GabC_{22}v_y - G(ab)^{1.5}C_{23}\varphi, \qquad (118)$$

$$M_z = -G(ab)^{1.5}C_{32}v_y - G(ab)^2 C_{33}\varphi \qquad (119)$$

with

$$F_x, F_y, M_z \quad : \quad \text{total force and moment on body 1} \qquad (120)$$

$$v_x, v_y, \varphi \quad : \quad \text{creepage (relative rigid slip) of body 1 wrt.body 2} \qquad (121)$$

This form of the total forces and the twisting moment follows from symmetries, while it is an empirical fact that

$$C_{11} > 0, C_{22} > 0, C_{23} = -C_{32} > 0, C_{33} > 0 \qquad (122)$$

These creepage and spin coefficients are tabulated in Table 3. The creepage and spin coefficients can be calculated in a purely numerical way. They can also be calculated more accurately in a semi-analytical way; these more accurate results are shown here.

a/b	C_{11}		C_{22}		$C_{23} = -C_{32}$		C_{33}	
	$\nu = 0$	0.5	$\nu = 0$	0.5	$\nu = 0$	0.5	$\nu = 0$	0.5
0.1	2.51	4.85	2.51	2.53	.334	.731	6.42	11.7
0.2	2.59	4.81	2.59	2.66	.483	.809	3.46	5.66
0.3	2.68	4.80	2.68	2.81	.607	.889	2.49	3.72
0.4	2.78	4.82	2.78	2.98	.720	.977	2.02	2.77
0.5	2.88	4.83	2.88	3.14	.827	1.07	1.74	2.22
0.6	2.98	4.91	2.98	3.31	.930	1.18	1.56	1.86
0.7	3.09	4.97	3.09	3.48	1.03	1.29	1.43	1.60
0.8	3.19	5.05	3.19	3.65	1.13	1.40	1.34	1.42
0.9	3.29	5.12	3.29	3.82	1.23	1.51	1.27	1.27
1.0	3.40	5.20	3.40	3.98	1.33	1.63	1.21	1.16
1/.9	3.51	5.30	3.51	4.16	1.44	1.77	1.16	1.06
1/.8	3.65	5.42	3.65	4.39	1.58	1.94	1.10	.954
1/.7	3.82	5.58	3.82	4.67	1.76	2.18	1.05	.852
1/.6	4.06	5.80	4.06	5.04	2.01	2.50	1.01	.751
1/.5	4.37	6.11	4.37	5.56	2.35	2.96	.958	.650
1/.4	4.84	6.57	4.84	6.31	2.88	3.70	.912	.549
1/.3	5.57	7.34	5.57	7.51	3.79	5.01	.868	.446
1/.2	6.96	8.82	6.96	9.79	5.72	7.89	.828	.341
1/.1	10.7	12.9	10.7	16.0	12.2	18.0	.795	.228

Table 3: Table of the creepage and spin coefficients

2.1.2 The theory of Vermeulen and Johnson for steady state rolling

In 1964, Vermeulen and Johnson [8] gave a theory for steady-state rolling for pure creepage ($\varphi = 0$). They assumed that the stick area was also bounded by an ellipse, with the same axial ratio as the contact ellipse and the same orientation, see Fig. 8, but touching the contact ellipse at its foremost point.

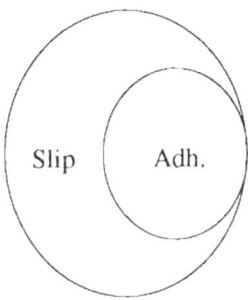

Figure 8: Areas and slip and adhesion according to Vermeulen and Johnson

Based upon that, they derived the creepage-force law shown in Fig. 9, broken line. In the left figure, the dots and crosses are the experimental values of Vermeulen and Johnson. Kalker improved the theoretical curve a little by making use of the creepage coefficients, and the result is shown in the left and right figure 9 by a drawn line. In the left figure, the dots and crosses are the experimental results of Vermeulen and Johnson, and in the right figure they are a comparison with the results of the numerical programs Fastsim (see Section 3) and Contact (see Section 4).

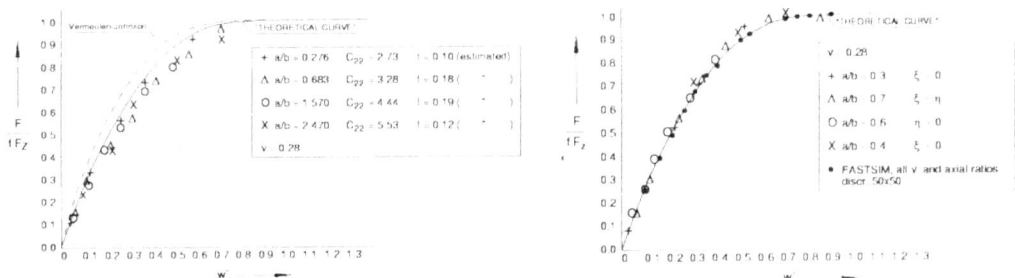

Figure 9: Results of Vermeulen and Johnson. Left, comparison with the experiment; right, comparison with Fastsim and Contact, Kalker (1990)

It is seen that the drawn line is a near perfect fit to the experiment, Contact and Fastsim.
It is of interest to have the equation of the drawn line.
Let

$$\xi' \doteq -\frac{abGC_{11}v_x}{3fF_z} \tag{123}$$

$$\eta' \doteq -\frac{abGC_{22}v_y}{3fF_z} \quad \text{with} \tag{124}$$

C_{ii} :   Creepage coefficient of the linear theory, see Table 3   (125)

a, b :   Semiaxes of contact ellipse, a in rolling direction   (126)

$$G \quad : \quad \text{Modulus of rigidity} \tag{127}$$
$$f \quad : \quad \text{Coefficient of friction} \tag{128}$$
$$F_z \quad : \quad \text{Total compressive normal force} \tag{129}$$

We set

$$w' = |(\xi', \eta')| \tag{130}$$

and it equals

$$
\begin{aligned}
w' &= \quad 1 - [1 - (F/fF_z)]^{1/3} \quad \text{if } F \le fF_z, \quad F = |(F_x, F_y)| \\
&\ge \quad 1 \qquad\qquad\qquad\qquad \text{if } F > fF_z
\end{aligned}
\tag{131}
$$

We have for the "theoretical line":

$$(F_x, F_y) \quad = \quad (F/w')(\xi', \eta') \tag{132}$$
$$F \quad = \quad fF_z[1 - (1 - w')^3] \quad \text{if } w' \le 1 \tag{133}$$
$$= \quad fF_z \qquad\qquad\qquad \text{if } w' \ge 1 \tag{134}$$

The only difference between the definition of Kalker's line and the line of Vermeulen-Johnson is that Vermeulen-Johnson derive their line analytically on the basis of their division of the contact area in regions of stick and slip, and arrive then at a different set of creepage coefficients, whereas Kalker maintains the form of the line of Vermeulen and Johnson, but gives it the correct slope at the origin $w' = 0$.

Indeed we have for the linear theory:

$$F \quad = \quad \frac{\partial F}{\partial w'}\Big|_{w'=0} w' = 3fF_z w' \Longrightarrow$$
$$(F_x, F_y) \quad = \quad (F/w')(\xi', \eta') = 3fF_z(\xi', \eta') = -abG(C_{11}v_x, C_{22}v_y)$$

as it should be.

2.2   Examples

2.2.1   Klingel rolling

Klingel rolling is a phenomenon that takes place in railway wheel sets when the wheels are conical. Then the wheel set may be modelled by a rigid double cone, the rails are modelled by two parallel straight rigid knives $2b$ apart, upon which the wheel set is moving. The theory can be extended to more realistic cases.

It is observed that the wheel set is not moving in a straight line, but has rather a sinusoidal motion. The situation is shown in Fig. 10. IT IS REQUIRED TO GIVE AN EXPLANATION OF THIS PHENOMENON, WHICH IS CALLED THE KLINGEL MOTION.

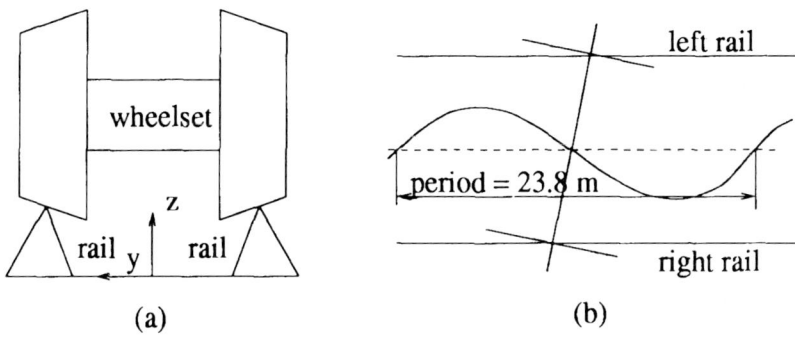

Figure 10: a: The wheel set as a double cone on knife-like rails   b: The sinusoidal motion of the centre of the wheel set

A coordinate system is introduced in which the origin lies in the middle between of the rails, the x-axis points in the rolling direction along the rails, the z-axis points vertically upwards, and the y-axis completes the right-handed system, $i.e.$ it points in the lateral, left direction if one stands looking in the direction of x. Let the centre of the wheel set have coordinates (x, y, R), where we neglect the height difference of the centre of the wheel set due to the shift of the mid position over a distance y with respect to the rails. The apex angle of the double cone is 2γ, a small angle, the radius of the double cone in the mid-position $y = 0$ is R, so that the radius of the wheels are

$$\text{radius of left wheel on rail} \;=\; R + \gamma y$$
$$\text{radius of right wheel on rail} \;=\; R - \gamma y$$

$2b$ is the width of the track.

The angular velocity of the wheel set about the y-axis is $\omega > 0$. The angle that the axis of the wheel set makes with the y-axis is α, a small angle. We have, when there is no slip between the wheel set and the rails:

$$\dot{y}/V = \alpha; \quad \dot{\alpha} = -\omega\gamma y/b \quad \Rightarrow \quad \ddot{y} = -\omega\gamma y V/b = -\omega^2\gamma y R/b$$

with

$$V = \omega R = \text{rolling velocity}$$

Consequently,

$$y = y_0 \cos(\omega t \sqrt{\gamma R/b})$$

where y_0 is the amplitude (integration constant), the phase is arbitrary, and the angular frequency is $\omega\sqrt{\gamma R/b}$, hence the frequency is $(\omega\sqrt{\gamma R/b})/(2\pi)$ Hz. The period on the rails is $2\pi R/\sqrt{\gamma R/b}$, and therefore independent of the angular velocity of the wheel set. A pilot numerical value of the period on the rails is found by setting

$$\gamma = 1/40, \quad R = 0.5 \; m, \quad b = 0.715 \; m \quad \Longrightarrow \quad \text{period} = 23.8 \; m$$
$$\text{Frequency at 100 km/h} = 1.17 \; Hz$$

2.2.2 A ball between two surfaces

Consider two steel surfaces that are mounted on an axle so that they make an angle of 2γ with one another. A steel ball with radius R is placed between them. The flat surfaces are rotated in such a way that the angle 2γ is conserved; a coordinate system is introduced in which the z-axis points vertically upwards, the y-axis points horizontally outwards, the x-axis completes the right-handed system, and the origin O lies in the centre of the ball. The motion of the planes is so that the origin of the coordinate system is stationary.

IT IS REQUIRED TO FIND THE TOTAL FORCE IN Y-DIRECTION, WHEN THE RADIAL POSITION OF THE ORIGIN OF THE COORDINATE SYSTEM IS GIVEN. The situation is sketched in Fig. 11.

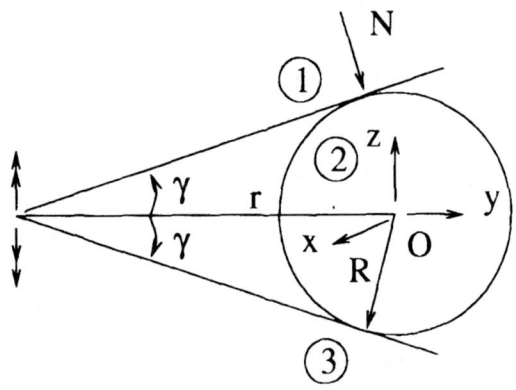

Figure 11: A ball between two surfaces

This problem was investigated by Johnson [5] in 1959.

The angular velocity of the upper plane is ω and that of the lower plane is $-\omega$. The ball does not rotate about the vertical axis z, by the symmetry of the two contacts. The spin creepage at the upper contact, (2)-(1), is $-\omega/V = -1/r$, where r is the distance between the axis of rotation of the two surfaces and the centre of the ball, see Fig. 11 and $\gamma = R/r$ is taken small. V is the velocity of rolling at the upper contact. The spin creepage, (2)-(3), at the lower contact is the same. The radius of the circular contact patch is a, the total normal force is $N = 16a^3G/(3R(1 - \nu))$, where R is the radius of the ball. So the total force in y-direction is, according to the linear theory,

$$
\begin{aligned}
F &= 2N\gamma + 2Ga^3C_{23}/r = N\gamma[2 - 2C_{23}a^3G/(NR)] = \\
  &= N\gamma[2 - 2*3(1 - \nu)/16C_{23}] = 1.6N\gamma
\end{aligned}
$$

when we take care to stop the lateral creepage.

2.2.3 Creepage and spin for a railway wheel set

A railway wheel set consists of two wheels mounted on an axle. The wheel surfaces consist of a flange, a tread, and a throat in between; the tread has (almost) the form of a cone, as well as the flange; the throat is more or less circular, see Fig. 12. The rail consists of foot, a web and a rail head, see also Fig. 12.

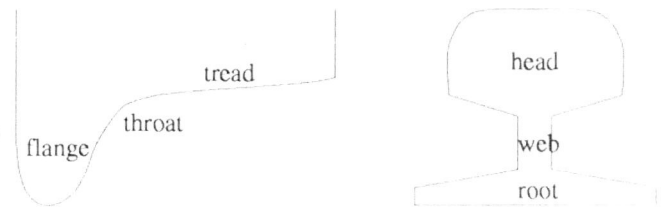

Figure 12: A railway wheel and rail surface

Fig. 13 schematically shows the wheel set standing on the rails. A coordinate system is introduced in which the origin lies in the centre of the wheel set, the z-axis points vertically upwards, the x-axis points along the rails in the rolling direction from left to right, and the y-axis completes the right-handed coordinate system, pointing to the left if one is facing the rolling direction.

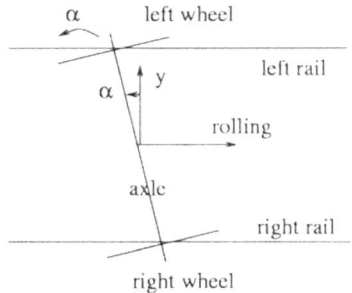

Figure 13: The wheel set standing on the rails

We start with the lateral creepage. Let α be the small angle between the y-axis and the centre line of the wheel set, shown positive in Fig. 13. The lateral creepage consists of two components, one due to α and one due to $\dot{y}$, the lateral velocity of the centre of the wheel set. We have that $\dot{y}/V$ is compensated in part by α:

$$v_y = \dot{y}/V - \alpha \qquad (135)$$

Note that in Klingel rolling, the lateral creepage vanishes. Next we consider the longitudinal creepage. It consists of three parts: the creepage due to the difference of the rolling radii of the wheels, and the rotation of the wheel set about a vertical axis through its centre, and the braking or accelerating motion of the wheel set as a whole. The longitudinal creepage depends on whether we consider the left or the right wheel. The first component reads $\omega(R_{left} - R_{right})/V$, where R_{left} and R_{right} are the radii of left and right wheel, which depend on the lateral coordinate y, and V is the rolling velocity of the centre of the wheel set. It compensates the longitudinal creepage, and it changes sign with the wheel.

The second component reads $-\dot{\alpha}b/V$; it changes sign with the wheel.

The third component reads $(V - \omega R_{mean})/V$; it does not change sign with the wheel. In total, we have for the longitudinal creepage in the left wheel

$$v_{x,left} = \frac{-\omega(R_{left} - R_{right}) - \dot{\alpha}b + (V - \omega R_{mean})}{V} \tag{136}$$

and in the right wheel

$$v_{x,right} = \frac{\omega(R_{left} - R_{right}) + \dot{\alpha}b + (V - \omega R_{mean})}{V} \tag{137}$$

Finally we consider the spin. It consists of two components, the kinematic spin and the geometric spin.

The kinematic spin is due to the rotation of the wheel set about the vertical axis. It equals $\dot{\alpha}/V$, and it is the same for both wheels.

The geometric spin is due to the conicity of the wheels, see Fig. 14. The conicity of the wheels is γ_{left}, γ_{right}. The normal at the contact point of the right wheel makes an angle γ_{right} with the z-axis, hence the angular velocity about this normal is $\omega\gamma_{right}$, and the contribution to the spin is $\omega\gamma_{right}/V = \gamma_{right}/R$. Similarly, the angular velocity about the normal at the contact point of the left wheel is $-\omega\gamma_{left}$, and the contribution to the spin on the left wheel is $-\gamma_{left}/R$. The total spin on the wheels is

$$\varphi_{left} = \dot{\alpha}/V - \gamma_{left}/R \tag{138}$$
$$\varphi_{right} = \dot{\alpha}/V + \gamma_{right}/R \tag{139}$$

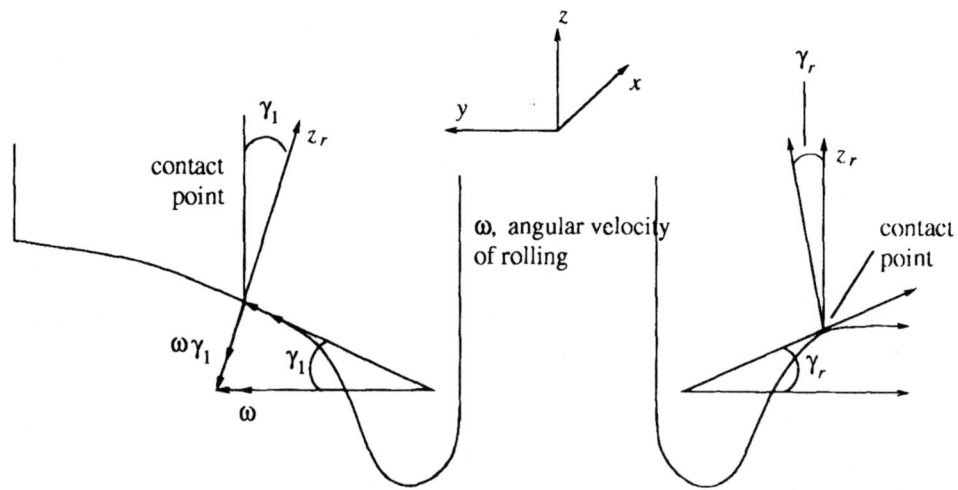

Figure 14: The geometric spin

References of Section 2

[1] J. BOUSSINESQ (1885), *Application des potentiels à l'équilibre et du mouvement des solides élastiques*. Paris, Gauthier-Villars.

[2] V. CERRUTI (1882), *Accademia dei Lincei, Roma. Mem. fis. mat.*

[3] G.M.L. GLADWELL (1980), *Contact problems in the classical theory of elasticity*. Sijthoff and Noordhoff, Alphen a/d Rijn, the Netherlands.

[4] E. JAHNKE, F. EMDE (1943), *Tables of functions*. Dover publications, New York.

[5] K.L. JOHNSON (1959), *The influence of elastic deformation upon the motion of a ball rolling between two surfaces*. Proceedings of the Institution of Mechanical Engineers **173**, p. 795-810.

[6] A.E.H. LOVE (1926), *A treatise on the theory of elasticity*. 4th Ed. Cambridge University Press.

[7] J.T. ODEN, E.B. PIRES (1983), *Nonlocal and nonlinear friction laws and variational principles for contact problems in elasticity*. Journal of Applied Mechanics **50**, p. 67-76.

[8] P.J. VERMEULEN, K.L. JOHNSON (1964), *Contact of nonspherical bodies transmitting tangential forces*. Journal of Applied Mechanics **31**, p. 338-340.

[9] H. HERTZ (1882A), *Über die Berührung fester elastischer Körper*. In: H. Hertz, Gesammelte Werke, Band 1. Leipzig: J.A. Barth, (1895), p. 155-173.

[10] H. HERTZ (1882B), *Über die Berührung fester elastischer Körper und über die Härte*. In: H. Hertz, Gesammelte Werke, Band 1. Leipzig: J.A. Barth, (1895), p. 174-196.

3   The simplified theory of rolling contact

3.1   Discretization of the slip

We recall the definition of the slip, see (25):

$$\mathbf{s} = \mathbf{c} - (\partial \mathbf{u}/\partial \mathbf{x})\mathbf{v} + \partial \mathbf{u}/\partial t \tag{140}$$

with

$$\mathbf{s} = \text{the slip: the velocity of (1) over (2)}$$
$$\mathbf{c} = \text{the creep: the rigid velocity of (1) over (2)} = \dot{\mathbf{x}}^1 - \dot{\mathbf{x}}^2$$
$$\mathbf{u} = \text{the surface displacement difference} = \mathbf{u}^1 - \mathbf{u}^2$$
$$\mathbf{x} = \text{the position}$$
$$\mathbf{v} = \text{the rolling velocity} = -(\dot{\mathbf{x}}^1 + \dot{\mathbf{x}}^2)$$
$$t = \text{the time}$$
$$\dot{\mathbf{x}}^a = \text{velocity of particle } \mathbf{y}^a \text{ with respect to contact patch}$$

We consider $\mathbf{u}(\mathbf{x} + k\mathbf{v}, t - k), k > 0$ which we expand about $k = 0$, taking along only the first two terms:

$$\mathbf{u}(\mathbf{x} + k\mathbf{v}, t - k) = \mathbf{u}(\mathbf{x}, t) + k\{(\partial \mathbf{u}/\partial \mathbf{x})\mathbf{v} - \partial \mathbf{u}/\partial t\} + O(k^2)$$

If we neglect the $O(k^2)$, then we obtain

$$\mathbf{s} = \mathbf{c} + (\mathbf{u} - \mathbf{u}')/k \tag{141}$$

where

$$\mathbf{u} = \mathbf{u}(\mathbf{x}, t)$$
$$\mathbf{u}' = \mathbf{u}(\mathbf{x} + k\mathbf{v}, t - k)$$

In principle, $\mathbf{u}'$ is known in a non-steady state, where $\mathbf{u}$ evolves in time under a non-constant creep $\mathbf{c}(t)$.
In a steady state, $\mathbf{u}$ and $\mathbf{u}'$ are independent of explicit time, and

$$\mathbf{u} = \mathbf{u}(\mathbf{x}), \mathbf{u}' = \mathbf{u}(\mathbf{x} + k\mathbf{v}), \tag{142}$$

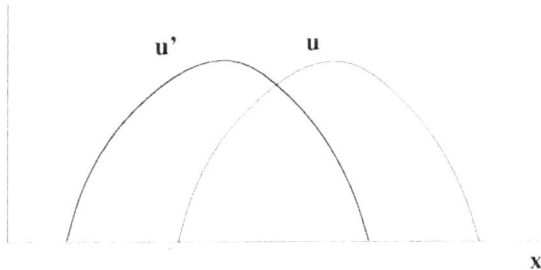

Figure 15: **u** and **u'** in a steady state

see Fig. 15. (141) is the promised discretization of the slip. It holds for both steady and non-steady rolling. In the steady state, **u'** is known and **u** is unknown, in the steady state both are unknown, but the solution is independent of explicit time.

3.2 Simplified theory

In the simplified theory of rolling contact we approximate the relation between the tangential surface displacement (u^a, v^a) of body a and the tangential surface traction on body a, viz. (p_x^a, p_y^a) by

$$(u^a, v^a) = L^a (p_x^a, p_y^a) \tag{143}$$

where L^a is a parameter called the **FLEXIBILITY**. The flexibility is comparable to $1/E^a$, where E^a is the modulus of elasticity of body a.

To imagine this load displacement law, we think of a bed of springs, see Fig. 16.

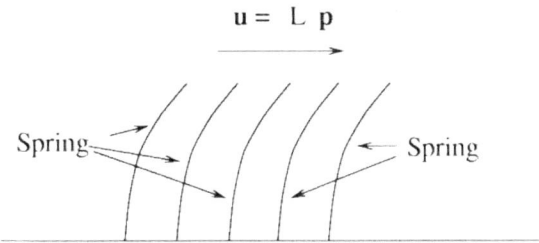

Figure 16: The body as a bed of springs

This is the basic hypothesis of the Simplified Theory.

Important remark. We note that the normal displacement is not approximated by a relationship like (143). This is because the lack of accuracy in the normal simplified relationship. Instead, we use the Hertz theory, whereby the contact becomes elliptical and the normal traction distribution semi-ellipsoidal.
Later, we will go further and approximate the semi-ellipsoidal normal traction by a

paraboloidal one, on the same axes, and with the same total normal force.

We consider the relation between the surface traction (p_x^1, p_y^1) and (p_x^2, p_y^2), and we prove that they are opposite:

$$(p_x^1, p_y^1) = -(p_x^2, p_y^2) \doteq (p_x, p_y) \qquad (144)$$

Indeed, outside contact $(p_x^1, p_y^1) = -(p_x^2, p_y^2) = (0,0)$, while inside contact, by Newton's Third Law, $(p_x^1, p_y^1) = -(p_x^2, p_y^2)$. This proves proposition (144).

As we saw before, a key role is played by the displacement difference. It is

$$(u, v) \doteq (u^1 - u^2, v^1 - v^2) = (L^1 + L^2)(p_x, p_y) \doteq L(p_x, p_y) \qquad (145)$$

We consider the flexibility L. It is comparable to $1/E$, where E is Young's modulus. Like E, it depends on the material of the bodies (steel, e.g.) but unlike E it depends on the form of the bodies, and also on the loading:

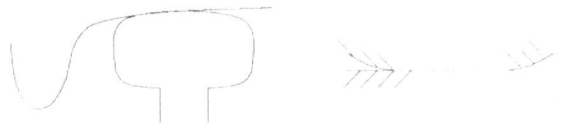

Figure 17: Flat bodies of various forms

$$
\begin{aligned}
L &= L(\text{depth/contact length}, (p_x, p_y), \text{ material}) \\
E &= E(\text{material}), \text{ by contrast}
\end{aligned}
$$

Flat bodies of various forms are shown in Fig. 17. We consider the wheel-rail system. The contact area is about 1 cm long; the wheel is about 10 cm wide, the rail head is about 5 cm thick; depth/contact length $> 5 \approx \infty$ in elasticity! The material is always steel, so $L = L[(p_x, p_y)]$. Analysis of (p_x, p_y): p_x and p_y depend strongly on the form and size of the contact patch, that is on the semi-axes of the contact ellipse. Also, we will consider three special loadings, (1),(2),(3), with flexibilities $L_1(a, b), L_2(a, b)$, $L_3(a, b)$. From these we will later make one single $L(a, b, (1), (2), (3))$.

Summarized so far. We consider two bodies in contact. Only surface quantities are of interest. We have a normal problem $(w^a; p_z^a)$ and a tangential problem $(u^a, v^a; p_x^a, p_y^a)$. The normal problem is solved by Hertz: the contact patch is elliptical, with semi-axes a, b; the pressure distribution is semi-ellipsoidal.

The tangential problem approximately satisfies the hypothesis of the simplified theory:

$$(u^a, v^a) = L^a(p_x^a, p_y^a)$$

but

$$\begin{aligned}
(p_x^1, p_y^1) &= -(p_x^2, p_y^2) = (p_x, p_y) \implies \\
(u^1, v^1) &= L^1(p_x, p_y) \\
(u^2, v^2) &= -L^2(p_x, p_y)
\end{aligned}$$

and the displacement difference is

$$\begin{aligned}
(u, v) &\doteq (u^1 - u^2, v^1 - v^2) = (L^1 + L^2)(p_x, p_y) \doteq L(p_x, p_y) \\
L &= L(a, b, (1), (2), (3))
\end{aligned}$$

with (1),(2),(3) three special loadings.

3.2.1 Coulomb's Law

We recall Coulomb's Law:

$$\begin{aligned}
g &= fp_z; \ |(p_x, p_y)| < g \implies \\
(S_x, S_y) &= \text{local velocity of (1) over (2)} = (0,0); & (146) \\
|(S_x, S_y)| &> 0 \implies \\
(p_x, p_y) &= -g(S_x, S_y)/|(S_x, S_y)| & (147)
\end{aligned}$$

with

$$\begin{aligned}
(S_x, S_y) &= (v_x - \varphi y, v_y + \varphi x) - \partial(u,v)/\partial x + \partial(u,v)/\partial t \\
&= (v_x - \varphi y, v_y + \varphi x) + [(u,v) - (u',v')]/k & (148) \\
u &= u((x,y),t); u' = u((x+k,y), t-k); (\mathbf{v}^T = (1,0,0)). & (149)
\end{aligned}$$

3.2.2 The linear theory

In the linear theory, the stick area H covers the entire contact area C; as then $|(p_x, p_y)| < g$, this corresponds to small tractions. In steady state rolling we have

$$\begin{aligned}
(0,0) &= (S_x, S_y) = (v_x - \varphi y, v_y + \varphi x) - \partial(u,v)/\partial x \implies \\
u &= v_x x - \varphi xy + k(y) & (150) \\
v &= v_y x + \varphi x^2/2 + l(y) & (151)
\end{aligned}$$

where k and l are arbitrary functions of y which have the character of integration constants: their derivative with respect to x vanishes.
With the simplified theoretic hypothesis $(u, v) = L(p_x, p_y)$ we obtain

$$\begin{aligned}
(p_x, p_y) &= (v_x x - \varphi xy + k(y), v_y x + \varphi x^2/2 + l(y))/L, \ \mathbf{x} \in C \\
&= (0,0), \ \mathbf{x} \notin C & (152)
\end{aligned}$$

The question arises how to determine k and l. It is answered as follows.
A particle lies in front of the leading edge of the contact patch, and is unloaded. It moves until it reaches the leading edge, still unloaded. It enters the contact patch, and traction starts building up from zero. Traction keeps building up until the traction bound is reached. Then the traction remains on the traction bound, and slip sets in. At the trailing edge, slip prevails and the particle leaves the contact patch, again unloaded. In linear theory, the traction bound is never reached, but at the trailing edge the traction is suddenly released till zero.

We denote the leading edge by $x_L = a(y) > 0$, and the trailing edge is then $x_T = -a(y) < 0$. At the leading edge we have

$$
\begin{aligned}
p_x(a(y), y) &= v_x a(y) - \varphi a(y) y + k(y) = 0 \Longrightarrow \\
k(y) &= -v_x a(y) + \varphi a(y) y \Longrightarrow \\
p_x(x, y) &= v_x[x - a(y)] - \varphi[x - a(y)]y \ \ \text{inside } C \\
&= 0 \ \ \text{outside } C
\end{aligned}
\tag{153}
$$

Similarly

$$
\begin{aligned}
p_y(x, y) &= v_y[x - a(y)] + \varphi[x^2 - a(y)^2]/2 \ \ \text{inside } C \\
&= 0 \ \ \text{outside } C
\end{aligned}
\tag{154}
$$

where

$$
\begin{aligned}
C &= \{(x, y, z)| z = 0, (x/a)^2 + (y/b)^2 \le 1\} \tag{155} \\
a(y) &= a\sqrt{1 - (y/b)^2} \tag{156}
\end{aligned}
$$

We can calculate the total force due to these loadings:

$$
F_x = \int_{-b}^{b} \int_{-a(y)}^{a(y)} p_x(x, y) dx dy = -8a^2 b v_x / (3L) \tag{157}
$$

$$
\begin{aligned}
F_y &= \int_{-b}^{b} \int_{-a(y)}^{a(y)} p_y(x, y) dx dy = \\
&= -8a^2 b v_y / (3L) - \pi a^3 b \varphi / (4L) \tag{158}
\end{aligned}
$$

This is the force calculated by the simplified theory. On the other hand, F_x and F_y have been calculated by the true theory of elasticity. Such a theory will be termed the Exact or Complete theory. We had found

$$F_x = -abGC_{11}v_x$$
$$F_y = -abGC_{22}v_y - (ab)^{1.5}GC_{23}\varphi \qquad (159)$$

where the C_{ij} are the creepage and spin coefficients tabulated in Table 3 of Section 2 which depend only on (a/b) and Poisson's ratio ν. G is the modulus of rigidity which can be expressed in Young's modulus E and ν:

$$G = \frac{E}{2(1+\nu)} \qquad (160)$$

3.2.3   The flexibility parameter

We had seen that the linear theory provides a link between the simplified and the exact theories. We can use that link to calculate three values of the flexibility parameter L. Indeed,

$$
\begin{aligned}
\text{Simplified theory:} \quad F_x &= -8a^2bv_x/(3L) \\
F_y &= -8a^2bv_y/(3L) - \pi a^3b\varphi/(4L) \\
\text{Exact theory:} \quad F_x &= -abGC_{11}v_x \\
F_y &= -abGC_{22}v_y - (ab)^{1.5}GC_{23}\varphi
\end{aligned}
$$

Equating the coefficients of v_x, v_y, φ in simplified and exact theory gives three values of L:

$$(v_x) \quad : \quad L_1 = \frac{8a}{3GC_{11}} \qquad (161)$$

$$(v_y) \quad : \quad L_2 = \frac{8a}{3GC_{22}} \qquad (162)$$

$$(\varphi) \quad : \quad L_3 = \frac{\pi a^2}{4G\sqrt{ab}C_{23}} \qquad (163)$$

We add a small table valid for $\nu = 0.25$; we see that these values differ considerably.

a/b	0.1	0.3	1.0	1/.3	1/.1
$GL_1/a = 8/(3C_{11})$	0.806	0.775	0.647	0.421	0.228
$GL_2/a = 8/(3C_{22})$	1.06	0.970	0.784	0.417	0.208
$GL_3/a = \pi a/[4(ab)^{0.5}C_{23}]$	0.525	0.602	0.534	0.352	0.170

Table 4: $L(a, b, (1), (2), (3))$

The dependence on a/b is very marked; but between the L_i for constant a/b there are also large differences, especially between L_2 and L_3 which both refer to F_y. We form a single value of L as a weighted mean of the L_i:

$$L = \frac{L_1|v_x| + L_2|v_y| + L_3|\varphi|\sqrt{ab}}{\sqrt{v_x^2 + v_y^2 + ab\varphi^2}} \qquad (164)$$

Clearly, when

$$v_x = v_y = 0 \quad \Rightarrow \quad L = L_3$$
$$v_y = \varphi = 0 \quad \Rightarrow \quad L = L_1$$
$$\varphi = v_x = 0 \quad \Rightarrow \quad L = L_2$$

as it should be.

3.2.4 The traction bound

According to the Hertz theory, the normal traction has the following form:

$$p_z(x,y) = Z_0\sqrt{1 - (x/a)^2 - (y/b)^2}$$

with Z_0 constant; the pressure is semi-ellipsoidal.

   The simplified theoretic analogue of the Hertz theory exists, but it has grave defects; yet its normal pressure is extremely interesting, because, with it, the simplified theory becomes a consistent whole. To find this adapted pressure, we take a and b from the Hertz theory, and the form of the normal pressure from simplified theory:

$$p_z'(x,y) = Z_0'\{1 - (x/a)^2 - (y/b)^2\} \quad \text{(no root!)} \qquad (165)$$

Z_0 is known from the Hertz theory; Z_0' must be adapted. We do this so that the total compressive forces F_z and F_z' are equal:

$$\begin{aligned}
F_z &= \int\int_C Z_0\sqrt{1 - (x/a)^2 - (y/b)^2}\,dxdy = \\
&= 2\pi ab Z_0/3 \Longrightarrow \\
Z_0 &= \frac{3F_z}{2\pi ab} \\
F_z' &= \int\int_C Z_0'\{1 - (x/a)^2 - (y/b)^2\}\,dxdy = \\
&= \pi ab Z_0'/2 \longrightarrow \\
Z_0' &= \frac{2F_z}{\pi ab}
\end{aligned}$$

So

$$p_z(x,y) = \frac{3F_z}{2\pi ab}\sqrt{1 - (x/a)^2 - (y/b)^2} \qquad (166)$$

$$p_z'(x,y) = \frac{2F_z}{\pi ab}\{1 - (x/a)^2 - (y/b)^2\} \qquad (167)$$

Both can be used in the traction bound $g = fp_z$. We choose p_z' for reasons of consistency – once simplified theory, always simplified theory.

3.2.5 An analytical solution

As we saw, the tangential traction due to pure longitudinal creepage ($v_y = \varphi = 0$) has the following form in no-slip theory:

$$p_x(x,y) = (x - a(y))v_x/L \tag{168}$$
$$a(y) = a\sqrt{1 - (y/b)^2}$$

The traction bound reads

$$g = \frac{2fF_z}{\pi ab}\{[1 - (y/b)^2] - (x/a)^2\} =$$
$$= \frac{2fF_z}{\pi a^3 b}\{a(y)^2 - x^2\} \tag{169}$$

So, the picture is as shown in Fig. 18:

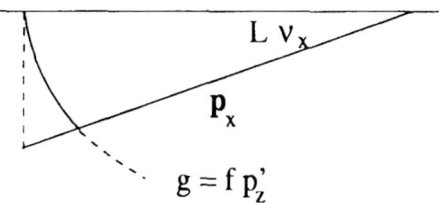

Figure 18: Traction distribution for pure longitudinal creepage in simplified theory

Near the trailing edge $-a(y)$ Coulomb's Law is broken by the linear theory. So the linear theory is never really valid, but only approximately so when $v_x \to 0$. Also shown in the figure is the nonlinear simplified theory. The exact solution is shown in Fig. 19. The solution for pure and combined lateral and longitudinal creepage is analogous.

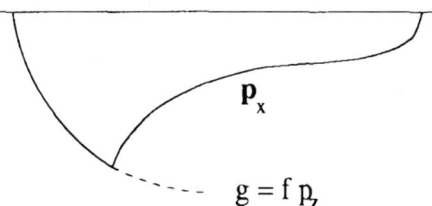

Figure 19: Exact theoretic traction distribution for pure longitudinal creepage

Summarizing, the simplified theoretic solution for the case $\varphi = 0$ is analytically known. For the case of nonvanishing φ we need a numerical theory, F A S T S I M.

3.2.6   Fastsim

FASTSIM is an algorithm to determine the traction in the general case (notably non-vanishing φ). We recall Eq. (141), where we insert the simplified theoretic hypothesis:

$$\begin{aligned} \mathbf{s} &= \mathbf{c} + (\mathbf{u} - \mathbf{u}')/k \\ &= \mathbf{c} + L(\mathbf{p} - \mathbf{p}')/k, \end{aligned}$$

where

$$\begin{aligned} \mathbf{c} &= \text{creep } \dot{\mathbf{x}}_1 - \dot{\mathbf{x}}_2 \\ \mathbf{p} &= \text{present traction} = (p_x[(x,y),t], p_y[(x,y),t])^T \\ \mathbf{p}' &= \text{previous traction} = \mathbf{p}(\mathbf{x} + \mathbf{v}k), t - k) \end{aligned}$$

We follow a particle on its path. At the leading edge $a(y,t)$, $\mathbf{p} = 0$, that is, known. We work by induction. Suppose $\mathbf{p}'$ is known. What is $\mathbf{p}$?

$$\mathbf{p} = \mathbf{p}' + (k/L)[\mathbf{s} - \mathbf{c}]$$

On the right-hand side, only the slip $\mathbf{s}$ is unknown. The rigid slip $\mathbf{c}$ is, of course, known. Tentatively we set $\mathbf{s} = 0$.

$$\mathbf{p}_H \doteq \mathbf{p}(\mathbf{x})_{s=0} = \mathbf{p}' - (k/L)\mathbf{c}$$

We determine $|\mathbf{p}_H|$, and compare with (known)g.

- $\mathbf{p}_H \leq g$: set $\mathbf{p} = \mathbf{p}_H$;
  then $|\mathbf{p}| \leq g$, and $\mathbf{s} = 0$: COULOMB.

- $\mathbf{p}_H > g$: set $\mathbf{p} = (g/|\mathbf{p}_H|)\mathbf{p}_H$. Then $|\mathbf{p}| = g$, and

$$\begin{aligned} \mathbf{s} &= \mathbf{c} + (L/k)(\mathbf{p} - \mathbf{p}') = \\ &= (L/k)(-\mathbf{p}_H + \mathbf{p}) \\ &= -(L/k)\mathbf{p}_H(1 - [g/|\mathbf{p}_H|]) \end{aligned}$$

so that $\mathbf{s}$ is exactly opposite $\mathbf{p}$:   COULOMB.

SUMMARY OF THE ALGORITHM for steady-state rolling:
$\mathbf{p}(\mathbf{x}, t) = \mathbf{p}(\mathbf{x})$.

$$\begin{aligned} \text{given} \quad & : \\ \mathbf{x} &= (x, y) \\ \mathbf{x} + k\mathbf{v} &= (x + q, y) \\ m &= \text{number of } x\text{-intervals} \\ n &= \text{number of } y\text{-intervals} \\ V &= |\mathbf{v}| = \text{rolling velocity}, > 0 \\ L &= \text{flexibility}, > 0 \end{aligned}$$

$$
\begin{aligned}
g &= g(x,y) \quad \text{traction bound at } (x,y) \in C \\
c(x,y) &= \text{rigid slip at } (x,y) \in C \\
a(y) &= \text{leading edge} \\
-a(y) &= \text{trailing edge} \\
a,b &: \text{semi-axes of contact ellipse} \\
\mathbf{p}_H, \mathbf{p} &: \text{tangential traction, to be calculated} \\
\mathbf{F} &: \text{total tangential force, to be calculated.}
\end{aligned}
$$

THE ALGORITHM:

1. set $r = 2b/n$; $y = b - r/2$; $\mathbf{F} = 0$ (Initiation of programme and y-loop)

2. $q = 2a(y)/m$; $x = a(y) - q$; $p = p(x + q, y) = 0$ (Initiation of x-loop).

3. $\mathbf{p}' = \mathbf{p}$ ($\mathbf{p}'$ is $\mathbf{p}$ just found).

4.

$$
\mathbf{p} = \mathbf{p}' - [q/VL]\mathbf{c}(x + q/2, y)
$$

(Form $\mathbf{p}_H$; $(x + q/2)$ is point between x and $x + q$)

5. If $|\mathbf{p}| > g$ then $\mathbf{p} = (g/|\mathbf{p}|)\mathbf{p}$
   (Form $\mathbf{p}$ if the traction bound is exceeded).
   Otherwise, $|\mathbf{p}| \le g$, we are in the stick area, and
   $\mathbf{p}(x,y) = \mathbf{p} = \mathbf{p}_H$.

6. $\mathbf{p}(x,y) = \mathbf{p}$; $\mathbf{F} = \mathbf{F} + qr\mathbf{p}$
   (Fill in $\mathbf{p}(x,y)$; update $\mathbf{F}$).

7. $x = x - q$; if $x > -a(y)$ go to 3 (Test of x-loop)

8. $y = y - r$; if $y > -b$ go to 2 (Test of y-loop)

9. READY

3.2.7   Results

We show some results.

First: The regions of stick (A) and slip (S) for various creepages and spins (quali-tative). These regions were obtained with a parabolic traction bound (Fig. 20). They are quite close to the areas of slip and stick obtained with the exact theory. Using an ellipsoidal traction bound gives bad results.

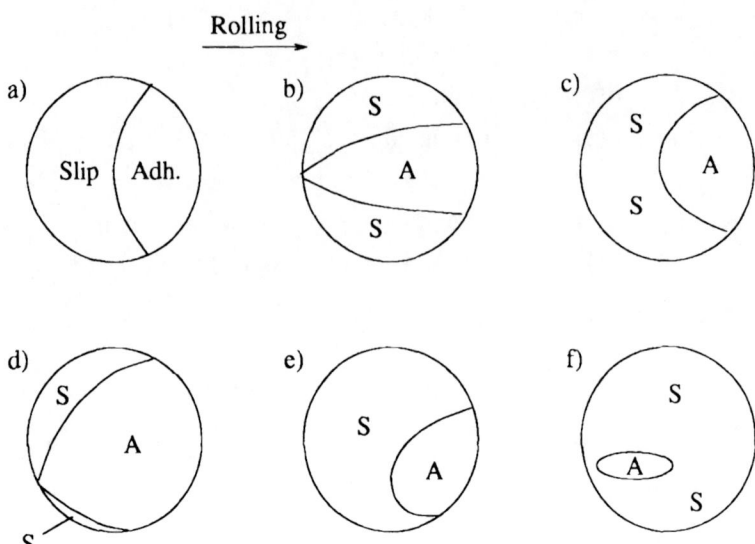

Figure 20: Areas of stick (A) and slip (S). a): Pure creepage ($\varphi = 0$); b): Pure spin, small; ($v_x = v_y = 0$); c): Lateral creepage with spin ($v_x = 0$); d): Longitudinal creepage with spin ($v_y = 0$); e): General case; f): Large pure spin. (Source: Kalker [1])

Next, we show a specific case of a calculation by Fastsim. We see areas of slip (shaded) and stick, an element division, and the traction (arrows). It is a case of pure spin (Fig. 21).

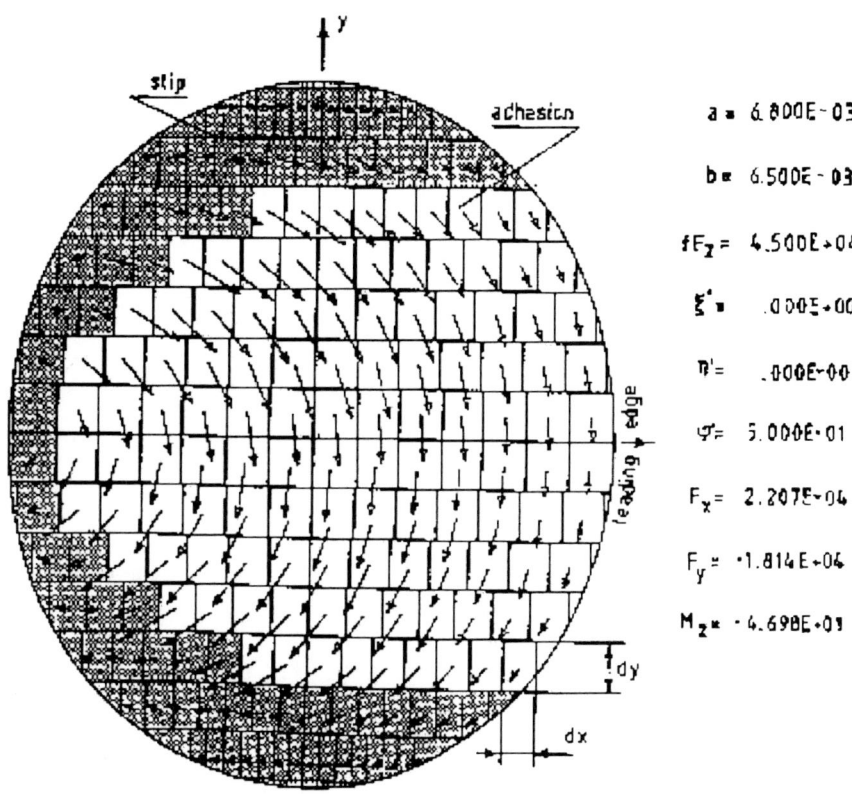

Figure 21: Result of a calculation by Fastsim. a=6.8 mm. b=6.5 mm. fF_z = 45000 N. $v_x = v_y = 0$; $\varphi = 0.00054mm^{-1}$. $F_x = 0$, $F_y = -1.812e4N$. (Source: Kalker and Piotrowski [2])

Next, we show the total force due to longitudinal creepage and lateral creepage without spin in comparison with the exact theory (Fig. 22), and the same for pure spin (Fig. 23). The curves have been scaled so that in each figure the slope in the origin is the same. Indeed we have:

$$\xi' = -\frac{abGC_{11}v_x}{3fF_z}$$

$$\eta' = -\frac{abGC_{22}v_y}{3fF_z}$$

$$w' = \sqrt{\xi'^2 + \eta'^2}$$

$$\psi = -\frac{(ab)^{1.5}GC_{23}\varphi}{fF_z}$$

The correspondence is striking, especially in the pure creepage case (Fig. 22).

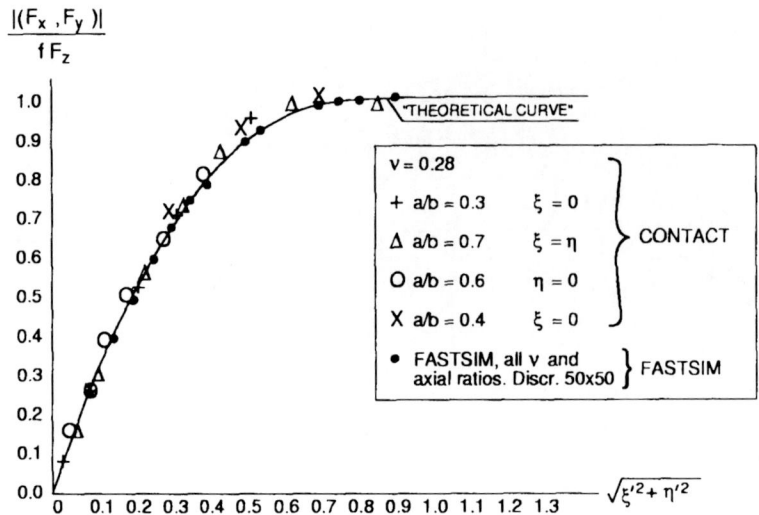

Figure 22: The total force due to pure creepage. Comparison of Fastsim and exact theory. (Source: Kalker [1]).

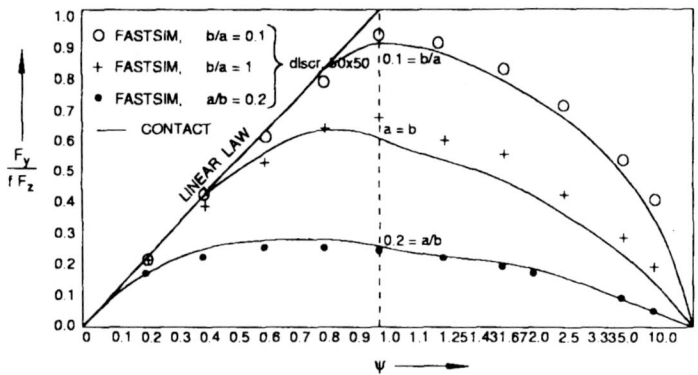

Figure 23: The total force due to pure spin. Comparison of Fastsim (dots) and exact theory (Contact) (Source: Kalker [1]).

Finally we show the total force in some cases where only one of the creepages vanishes, Fig. 24. It is seen that $v_y = -\varphi$, $v_x = 0$ is pretty bad, but we think that it is one of the worst cases. The others are quite good.

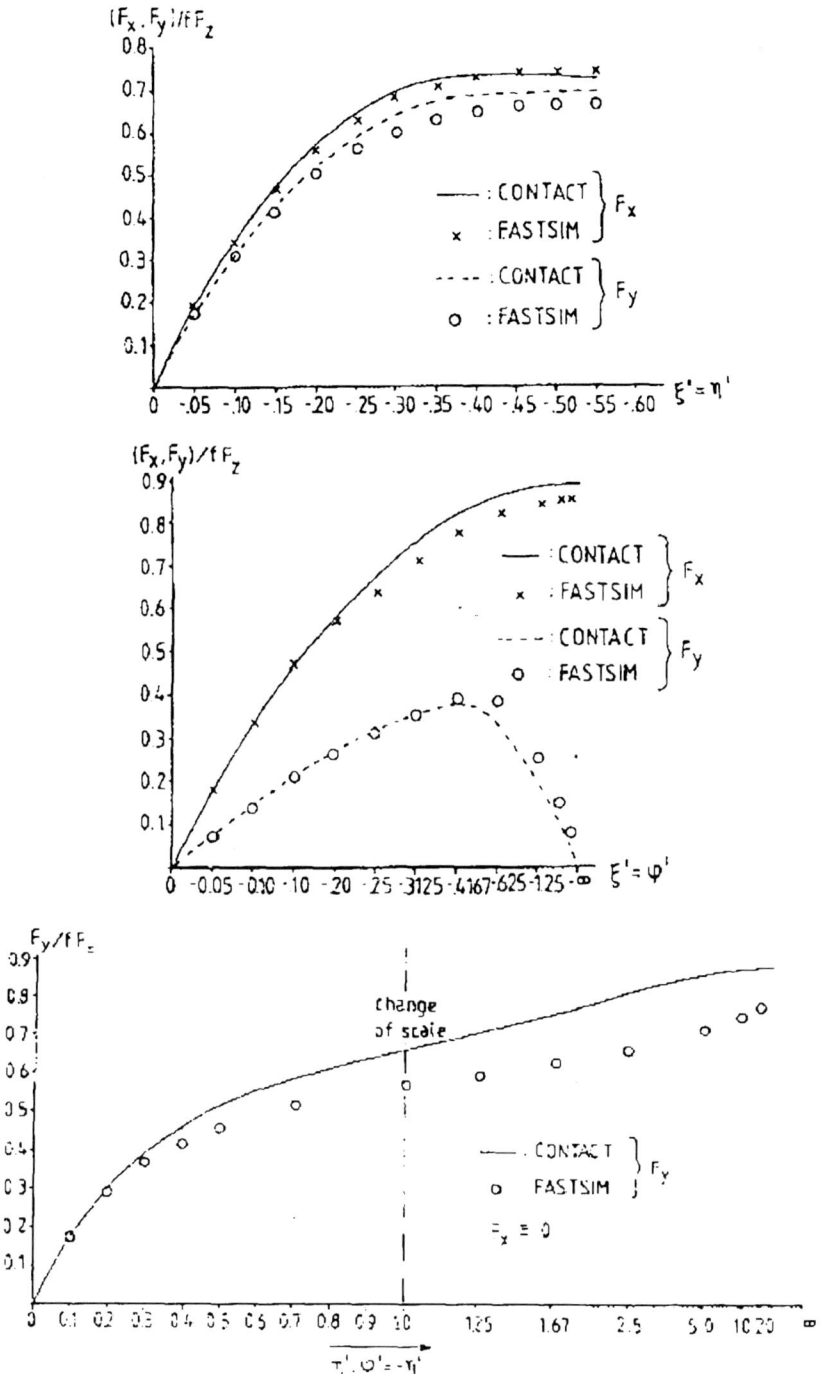

Figure 24: Some curves when only one of the creepages vanishes. The total force for $f = 1$, $F_z = 1$, $a = b = 1$, $G = 1$, $\nu = 0.25$, and (1) $\xi' = \eta'$, $\psi = 0$; (2) $\xi' = \psi$, $\eta' = 0$; (3) $\eta' = -\psi$, $\xi' = 0$. (Source: Kalker [1]).

3.2.8   Summary of Simplified Theory

We split the problem into two subproblems, *viz.* the normal problem and the tangential problem. The normal problem is solved by Hertz, the tangential problem by simplified theory.

In Simplified Theory, we set $\mathbf{u} = L\mathbf{p}$, pictured by a bed of springs (brush model, mattress model). Coulomb's Law is adopted.

Linear theory, *i.e.* the theory in which the slip vanishes, was studied as an example of simplified theory. Its development closely parallels the exact theoretic model. The comparison of simplified and exact linear theories yields the value of L, the flexibility parameter, in its dependence on the modulus of rigidity G, the semi-axes of the contact ellipse a and b, and the creepages v_x, v_y and φ:

$$L = L(G, a, b : v_x, v_y, \varphi).$$

In fact, the linear theory is only approximate, it holds for small creepages.

When the spin vanishes, one can develop an analytical solution of simplified theory, indicated in the text. When the spin does not vanish, one needs a numerical method: FASTSIM.

Results show the division of the contact area into regions of stick and slip; the behaviour of the total force when only one creepage component is non-zero, and some cases when only one component of the creepage vanishes. We see that Simplified Theory may contain an error of roughly 10-15 percent of the maximum force fF_z, accurate enough for many needs.

IT NEEDS TO BE EMPHASIZED THAT SIMPLIFIED THEORY CAN ONLY BE USED FOR QUASIIDENTICAL BODIES.

References of Section 3

[1] J.J. KALKER (1990), *Three-dimensional elastic bodies in rolling contact*. Kluwer, Dordrecht.

[2] J.J. KALKER, J. PIOTROWSKI (1989), *Some new results in rolling contact*. Vehicle System Dynamics **18**, p. 223-242.

4 The variational theory of contact

In this Section, we treat the variational theory of contact. It was introduced by Fichera [2] and Duvaut and Lions [1]. We follow the description of Kalker's book [4]. The variational theory of contact is an alternative to the conventional boundary condition description of contact of Section 1. In fact, the conventional description follows from it, and vice versa. We base ourselves on the Principle of Virtual Work and its dual, the Principle of Complementary Virtual Work from which we derive variational inequalities. The point of departure for both principles are the equations of equilibrium:

$$\sigma_{ij,j} + f_i - \rho\ddot{u}_i = 0, i,j = 1,2,3, \text{ summation convention} \qquad (170)$$

where

$$
\begin{array}{rcl}
u_i & : & \text{the displacement} \\
\sigma_{ij} & : & \text{the linearized stress} \\
f_i & : & \text{the body force per unit volume} \\
(\dot{\ }) & = & d/dt, \quad t\text{: time; particle fixed time differentiation} \\
,i & = & \partial/\partial x_i
\end{array}
$$

in two bodies that occupy the volumes V^1 and V^2.

4.1 The Variation of a Function

Consider a function f of one or more variables. f is subject to certain constraints. If it satisfies these constraints, it is called feasible. Consider another function δf; this is a function called the variation of f. It may be arbitrary (bilateral variation), or arbitrary nonnegative or nonpositive (nonnegative or nonpositive unilateral variation). At any rate, when f is feasible, $f + \epsilon.\delta f$ must be feasible also for ϵ small enough, positive.

Example. Let f be subject to the constraint $f \geq 0$ and let actually $f > 0$. Then for any δf, $f + \epsilon.\delta f > 0$ for ϵ small enough. So the variation δf is bilateral. Now let actually $f = 0$. Then for ϵ small enough, positive δf must be nonnegative: i.e. δf is nonnegative, unilateral.

Example. Let f be fully prescribed. Then $\delta f = 0$.

4.2 Virtual Work

We multiply the equation (170) with minus the variation of u_i: $-\delta u_i$ and integrate over V^1, V^2:

$$0 = -\sum_{a=1,2} \int_{V^a} (\sigma_{ij,j} + f_i - \rho\ddot{u}_i)\delta u_i dV \qquad (171)$$

The integrand on the right-hand side is the virtual work done on a particle of volume dV, so that (171) is the virtual work equation. This is equivalent to

$$0 = \sum_{a=1,2} [- \int_{V^a} (\sigma_{ij,j} + f_i - \rho \ddot{u}_i)\delta u_i dV + \int_{\partial V^a} p_i \delta u_i dS]$$

$$- \sum_{a=1,2} \int_{\partial V^a} p_i \delta u_i dS \qquad (172)$$

with

dV : element of volume

dS : element of surface

$p_i = \sigma_{ij}n_j$, surface load on ∂V, body number omitted

n_j : outer normal on V at ∂V.

In the third term of (172) we introduce the boundary conditions of Section 1, subsection 1.7,

$$u_i = \bar{u}_i, \text{ prescribed displacement in region } A_u \subset \partial V$$
$$\Rightarrow \delta u_i = 0 \text{ on } A_u \qquad (173)$$
$$p_i = \bar{p}_i, \text{ prescribed load in region } A_p \subset \partial V \qquad (174)$$

In the potential contact area $A_c^1 \approx A_c^2 \approx A_c$ we have:

$$p_i^1 = -p_i^2 \doteq p_i \text{ in } A_c, \text{ (Newton's Third Law)}$$
$$\Rightarrow p_i^1 \delta u_i^1 + p_i^2 \delta u_i^2 = p_i \delta u_i \qquad (175)$$

with $u_i \doteq (u_i^2 - u_i^2)$, displacement difference.

This gives for (172):

$$0 = \sum_{a=1,2} [- \int_{V^a} (\sigma_{ij,j} + f_i - \rho \ddot{u}_i)\delta u_i dV + \int_{\partial V^a} p_i \delta u_i dS] +$$

$$- \sum_{a=1,2} [\int_{A_p^a} \bar{p}_i \delta u_i dS] - \int_{A_c} p_i \delta u_i dS \qquad (176)$$

In the potential contact A_c we introduce a right-handed orthogonal curvilinear coordinate system x, y: they are represented by Greek indices which run through the values x, y. We introduce a coordinate z along the inner normal to body 1 at (x, y). dS is the element of area at (x, y). Then we may write

$$p_i \delta u_i = p_z \delta u_z + p_\tau \delta u_\tau \qquad (177)$$

with

p_z : normal pressure, positive if compressive

p_τ : tangential traction

We consider the deformed distance and the slip.

The deformed distance $e = h + u_z$; h is prescribed, so $p_z \delta u_z = p_z \delta e$. Now, as we saw in Section 1, if $e > 0$ then $p_z = 0$ (outside contact). If $e = 0$ (inside contact), then $p_z \geq 0$ (compression). e cannot be negative, so, if $e = 0$ then $\delta e \geq 0$, since varied quantities must be feasible. Thus if the contact conditions are satisfied, then

$$p_z \delta e \geq 0 \quad \text{in } A_c \ \text{ sub } e \geq 0 \tag{178}$$

where 'sub' means 'subject to the auxiliary condition(s)'.
The contact formation can be summarized as follows:

$$p_z \geq 0, \quad p_z e = 0, \quad e \geq 0 \tag{179}$$

The contact area does not occur explicitly in the formulations (178) and (179).

The slip (*i.e.* the velocity of body 1 over body 2) is given by

$$s_\tau = w_\tau + \dot{u}_\tau \tag{180}$$

with

$$u_\tau = u_\tau^1 - u_\tau^2 \tag{181}$$
$$w_\tau = \dot{x}_\tau^1 - \dot{x}_\tau^2 \tag{182}$$

u_τ is called the displacement difference, and w_τ is the rigid slip, which is defined as the local velocity of body 1 over body 2 when both are regarded as rigid.

We integrate (180) from time t' to time t, where $t' < t$. We call

$$\int_{t'}^{t} s_\tau(x_\alpha; q) dq \ = \ E_\tau \approx (t - t') s_\tau \quad \text{(local) shift} \tag{183}$$

$$\int_{t'}^{t} w_\tau(x_\alpha; q) dq \ = \ W_\tau \approx (t - t') w_\tau \quad \text{(local) rigid shift} \tag{184}$$

and we denote the displacement difference

$$u_\tau = u_\tau(x_\alpha; t) \tag{185}$$
$$u_\tau' = u_\tau(x_\alpha; t') \tag{186}$$

Note that u_τ' is the displacement difference at time t', not a derivative, and that the coordinate system is particle fixed.
The integral of (180) is

$$E_\tau = W_\tau + u_\tau - u_\tau' \tag{187}$$

We consider an evolution: we proceed stepwise, from t' to t. That means: u_τ', the rigid slip w_τ and the rigid shift E_τ are all known: they control the evolution from t' to t. So we have

$$\delta E_\tau = \delta u_\tau \ \Rightarrow \ p_\tau \delta u_\tau = p_\tau \delta E_\tau \tag{188}$$

Assume there is slip ($|s_\tau| \neq 0$ so that $|E_\tau| \neq 0$), and

$$p_\tau = -gE_\tau/|E_\tau|, \quad |E_\tau| = \sqrt{E_1^2 + E_2^2} \tag{189}$$

where g is the traction bound, and we have adapted Coulomb's law to shifts. So:

$$|E_\tau| \neq 0 \ \Rightarrow \ p_\tau \delta E_\tau = -gE_\tau \delta E_\tau/|E_\alpha| = -g\delta|E_\tau| \tag{190}$$

Assume that there is no slip:

$$|E_\tau| = 0 \ \Rightarrow \ |p_\tau| \leq g \tag{191}$$

By Schwartz's inequality and (191) we have

$$p_\tau \delta E_\tau \geq -|p_\alpha||\delta E_\tau| \geq -g|\delta E_\tau| \tag{192}$$

Since $E_\tau = 0$, $\tau = 1, 2$, we have

$$|\delta E_\tau| = |E_\tau + \delta E_\tau| - |E_\tau| = \delta|E_\tau| \ \text{ if } (|E_\tau| = 0) \tag{193}$$

so that by (188), (192, 193, 190)

$$
\begin{aligned}
p_\tau \delta u_\tau \ &= \ p_\tau \delta E_\tau \geq -g\delta|E_\tau| \\
&\Leftrightarrow \ p_\tau \delta u_\tau = -g|\delta E_\tau| + \eta \\
\eta \ &\geq \ 0 \ \text{ for slip and no-slip}
\end{aligned}
\tag{194}
$$

We note that (178) holds both inside and outside the contact, while (194) holds both in the slip area ($|E_\tau| \neq 0$) and in the adhesion zone ($|E_\tau| = 0$). So (178) and (194) will lead to a uniform formulation of the contact conditions on A_c in which neither the unknown contact area nor the unknown areas of slip and adhesion are mentioned explicitly, but note that A_c is known *a priori*.

We conclude from (176),(178),(194) that a necessary condition for contact is

$$
\begin{aligned}
0 \ = \ &\sum_{a=1,2} [-\int_{V^a} (\sigma_{ij,j} + f_i - \rho\ddot{u}_i)dV + \int_{\partial V^a} p_i \delta u_i dS] + \\
&- \sum_{a=1,2} \int_{A_p^a} \bar{p}_i \delta u_i dS + \int_{A_c} g\delta|E_\tau| dS - \text{ nonnegative} \\
&\forall \ \delta u_i \ \text{sub } u_i = \bar{u}_i \text{ in } A_u^a; \ e \geq 0 \text{ in } A_c
\end{aligned}
\tag{195}
$$

Hence, by Gauss's Theorem

$$
\begin{aligned}
0 \ \leq \ \delta U \doteq \ &\sum_{a=1,2} [-\int_{V^a} (\sigma_{ij,j} - \rho\ddot{u}_i + f_i)\delta u_i)dV + \\
&+ \int_{A_p^a} (p_i - \bar{p}_i)\delta u_i dS] + \int_{A_c} [p_i \delta u_i + g\delta|E_\tau|]dS =
\end{aligned}
\tag{196}
$$

$$= \sum_{a=1,2} [\int_{V^a} (\sigma_{ij}\delta u_{i,j} + \rho\ddot{u}_i - f_i\delta u_i dS) +$$

$$- \int_{A_p^a} \bar{p}_i\delta u_i dS] + \int_{A_c} g\delta|E_\tau| dS \tag{197}$$

$$\forall \quad \delta u_i \quad \text{sub } u_i = \bar{u}_i \text{ in } A_u^a; \; e = h + u_z \geq 0 \text{ in } A_c \tag{198}$$

with h, e: distance between opposing points in the undeformed and deformed state.

It may be shown that the conditions (196) or (197) sub (198) are not only necessary, but also sufficient for the contact conditions. We will not prove that here; we will give a similar proof when we consider the complementary virtual work inequality.

4.3 Complementary Virtual Work

We start from the equilibrium equations (170), which we take as auxiliary conditions which must always be satisfied. We consider the quasistatic case that the density $\rho = 0$, that is, accelerations are not taken into account. Also, the body force f_i is prescribed. Together with (170) $\rho = 0$ this implies that

$$\delta(\sigma_{ij,j} + f_i - \rho\ddot{u}_i) = \delta\sigma_{ij,j} = 0 \tag{199}$$

We multiply (199) by the displacement u_i, and integrate:

$$0 = \sum_{a=1,2} \int_{V^a} u_i\delta\sigma_{ij,j} dV \tag{200}$$

and we find in much the same manner as before, after some calculation

$$0 \geq \delta C \doteq \sum_{a=1,2} [\int_{V^a} u_i\delta\sigma_{ij,j} dV +$$

$$- \int_{\partial V^a} u_i\delta p_i dS + \int_{A_u^a} \bar{u}_i\delta p_i dS] +$$

$$- \int_{A_c} [h\delta p_z + |E_\tau|\delta g + (W_\tau - u_\tau')\delta p_\tau] dS \tag{201}$$

$$= \sum_{a=1,2} [\int_{V_a} -e_{ij}\delta\sigma_{ij} dV + \int_{A_u^a} \bar{u}_i\delta p_i dS]$$

$$- \int_{A_c} [h\delta p_z + |E_\tau|\delta g + (W_\tau - u_\tau')\delta p_\tau] dS \tag{202}$$

$$\forall \quad \delta p_i, \delta\sigma_{ij} \quad \text{sub } \sigma_{ij,j} + f_i = 0 \text{ in } V_a$$

$$p_i \quad = \bar{p}_i \text{ in } A_p^a; \; p_z \geq 0, |p_\tau| \leq g \text{ in } A_c. \tag{203}$$

In (202) the term $-e_{ij}\delta\sigma_{ij}$ appears instead of $-u_{i,j}\delta\sigma_{ij}$. These expressions are equal because $\sigma_{ij} = \sigma_{ji}$ and $e_{ij} = (u_{i,j} + u_{j,i})/2$.

The conditions (201)-(203) are implied by the contact conditions. Conversely, the contact problem is implied by those equations. We prove this.

To that end we restrict ourselves to the conditions in A_c. The other conditions are classical.
We start from (201), which is equivalent to (202). (203) is also valid. In (201) we note that $\delta\sigma_{ij,j} = 0$ in V^a; if we set $\delta p_i = 0$ on ∂V^a outside A_c, then

$$0 \leq \int_{A_c} [u_z\delta p_z + u_\tau\delta p_\tau + h\delta p_z + |E_\tau|\delta g + (W_\tau - u_\tau')\delta p_\tau]dS$$

1. Set $\delta p_\tau = \delta g = 0$; this is the normal contact problem. We obtain, by the independence of the δp_z

$$0 \leq (u_z + h)\delta p_z = e\delta p_z \quad \text{sub } p_z \geq 0.$$

If $p_z > 0 \Rightarrow \delta p_z$ is bilateral, and $e = 0$ (contact).
If $p_z = 0 \Rightarrow \delta p_z \geq 0$, and $e \geq 0$ (no contact).
Here we define the contact ares as the region where $p_z > 0$ ("Force" definition). It then appears that the deformed distance e is nonnegative outside contact, and $= 0$ inside.

2. Now we set $\delta p_z = 0$. We are left with

$$
\begin{aligned}
0 &\leq (u_\tau + W_\tau - u_\tau')\delta p_\tau + |E_\tau|\delta g = \\
&= E_\tau\delta p_\tau + |E_\tau|\delta g
\end{aligned}
$$

sub $g - |p_\tau| \geq 0$.

- If $|p_\tau| < g$ ("force" definition of the area of adhesion), then δp_τ and δg are independent and bilateral, so that

$$E_\tau = |E_\tau| = 0$$

- If $|p_\tau| = g$ ("force" definition of the area of slip), then $\delta g - \delta|p_\tau| \geq 0$. We decompose E_τ and δp_τ into components E_τ^p, δp_τ^p parallel to the vector (p_τ), and into components E_τ^o, p_τ^o orthogonal to that vector.

We set $\delta p_\tau^p = \delta g = 0$. Then $E_\tau^o\delta p_\tau^o \geq 0$.
Now δp_τ^o is bilateral, since in first order it does not contribute to

$$\epsilon\delta|p_\tau| = \sqrt{|p_\tau + \epsilon\delta p_\tau^p|^2 + |\epsilon\delta p_\tau^o|^2} - |p_\tau|$$

Thus $E_\tau^o = 0$, that is, the slip is parallel to the tangential traction:

$$E_\tau = \pm|E_\tau|p_\tau/|p_\tau|.$$

Suppose $E_\tau = +|E_\tau|p_\tau/|p_\tau|$. Then

$$
\begin{aligned}
0 &\leq |E_\tau|(p_\tau \delta p_\tau/|p_\tau|) + |E_\tau|\delta g = \\
&= |E_\tau|(\delta|p_\tau| + \delta g)
\end{aligned}
$$

Now take $\delta g = 0$. As $|p_\tau| = g$, $\delta|p_\tau| \leq 0$, and

$$0 \leq |E_\tau|\delta|p_\tau| \leq 0 \Rightarrow |E_\tau| = 0$$

Evidently this does not correspond to an area of slip, and anyway this situation (and much more) is also described by

$$E_\tau = -|E_\tau|p_\tau/|p_\tau|, \quad \leftrightarrow \quad p_\tau = -g E_\tau/|E_\tau|$$

Then we have

$$0 \geq |E_\tau|(\delta g - \delta|p_\tau|) = |E_\tau|\delta(g - |p_\tau|)$$

Now, $\delta(g - |p_\tau|) \geq 0$, unilateral, hence $|E_\tau| \geq 0$, which corresponds to slip opposite the traction p_τ when it is at the traction bound.

We have established the normal contact formation conditions, *viz.* $p_z \geq 0$, hence:

1. If $p_z > 0$ then $e = 0$ (contact)

2. If $p_z = 0$ then $e \geq 0$ (no contact)

and we have established Coulomb's Law, *viz.* $|p_\tau| \leq g$, hence:

1. If $|p_\tau| < g$ then $E_\tau = 0$

2. If $|p_\tau| = g$ then $E_\tau = -|E_\tau|p_\tau/|p_\tau|$, or, equivalently, $p_\tau = -g E_\tau/|E_\tau|$.

These constitute the contact conditions.

4.4   Application to elasticity

We assume elasticity:

$$
\begin{aligned}
\frac{1}{2}(u_{i,j} + u_{j,i}) &= e_{ij} = e_{ji}, \quad \text{linearized strain;} \\
\sigma_{ij} &= E_{ijhk}e_{hk}, \quad \text{stress-strain relations;} \quad \sigma_{ij} = \sigma_{ji}, \quad \text{stress;} \\
E_{ijhk}(\mathbf{y}) &= E_{jihk} = E_{hkij}, \quad \text{elastic constants;} \\
\frac{1}{2}E_{ijhk}e_{ij}e_{hk} &= \quad \text{elastic energy/unit volume} > 0 \\
&\quad\quad \text{unless } e_{ij}e_{ij} = 0.
\end{aligned}
\tag{204}
$$

The elastic constants are prescribed, but they may be position dependent. We can invert the stress-strain relations:

$$
\begin{aligned}
e_{hk} &= S_{hkij}(\mathbf{y})\sigma_{ij} \\
S_{hkij} &= S_{ijhk} = S_{khij} \\
S_{ijhk}\sigma_{ij}\sigma_{hk} &> 0 \quad \text{if } \sigma_{ij}\sigma_{ij} \neq 0
\end{aligned}
\tag{205}
$$

WE SET

$$\rho = 0: \quad \text{elastostatics} \tag{206}$$

$$\delta g = 0, \quad \text{that is, } g \text{ is given } a \text{ } priori \tag{207}$$

in order to be able to define a potential energy and a complementary energy of the system. We have, by (197)

$$0 \leq \delta U = \sum_{a=1,2} [\int_{V^a} (E_{ijhk} u_{h,k} \delta u_{i,j} - f_i \delta u_i) dV - \int_{A_p^a} \bar{p}_i \delta u_i] dS +$$

$$+ \int_{A_c} g \delta |E_\tau| dS =$$

$$= \delta [\sum_{a=1,2} (\int_{V^a} (\frac{1}{2} E_{ijhk} u_{i,j} u_{h,k} - f_i u_i) dV$$

$$- \int_{A_p^a} \bar{p}_i u_i dS) + \int_{A_c} g |E_\tau| dS]$$

sub (206), (207), and (198).

This is equivalent to

$$\delta U \geq 0, \quad \text{sub}(206),(207),(198), \text{ with}$$

$$U = \sum_{a=1,2} [-\int_{V^a} (\frac{1}{2} E_{ijhk} u_{h,k} u_{i,j} - f_i u_i) dV +$$

$$- \int_{A_p^a} \bar{p}_i u_i dS] + \int_{A_c} g |E_\tau| dS \tag{208}$$

$$\tag{209}$$

U is called the potential energy of the system. In the same way,

$$\delta C \leq 0 \text{ sub } (203), (206), (207), \text{ with}$$

$$C = \sum_{a=1,2} [-\int_{V^a} \frac{1}{2} S_{ijhk} \sigma_{ij} \sigma_{hk} dV + \int_{A_u^a} \bar{u}_i p_i dS] +$$

$$- \int_{A_c} [h p_z + (W_\tau - u_\tau') p_\tau] dS \tag{210}$$

C is called the complementary energy.

It may be shown (see [4] (Section 4.2.1)) that these conditions characterize

1. The global minimality of U at the solution;

2. The global maximality of C at the solution;

3. The equality of U and C at the solution;

4. The uniqueness of the solution,

all under the rather restrictive conditions (206) and (207).

4.5 The case that $\delta g \neq 0$

According to (207) the theory of the previous subsections does not seem to apply when δg is not constrained to be zero, that is, when g is not prescribed beforehand. We saw in Section 2 that the normal pressure is independent of the tangential tractions for symmetry of all 3D bodies, and for quasiidentical half-spaces. As the normal problem is not influenced by g, we can determine the normal problem regardless of g in those cases; thereafter $g = fp_z$, g fixed, f: coefficient of friction, and we have $\delta g = 0$. So in these cases the theory is exactly verified.

A process to deal with $\delta g \neq 0$ is the Panagiotopoulos process [7], see Section 2. This is an iterative method, and it results in the exact solution when it converges, which is not always. It is designed in such a way that the theories of minimal potential energy and maximal complementary energy can be used.

The Panagiotopoulos process works as follows:

1. Set $m = 0$. Assume that $p_\tau^{(0)} = 0$.

2. Determine $p_z^{(m)}$ with $p_\tau^{(m)}$ as tangential traction.
   This is a normal contact problem with fixed tangential traction, so that g does not play a role.

3. Determine $p_\tau^{(m+1)}$ with $p_z^{(m)}$ as normal traction, and $g = fp_z^{(m)}$ as traction bound.
   Now g is given, and hence $\delta g = 0$, and we can apply our theories.

4. When $p_\tau^{(m+1)}$ is close enough to $p_\tau^{(m)}$, stop; else reiterate.

4.6 Existence-uniqueness theory

The principle of virtual work has been used to establish existence and uniqueness theorems of the contact mechanical field for several types of bulk material:

- In 1964 Fichera [2] established the existence and uniqueness of the normal contact problem of frictionless contact $g = 0$.

- In 1972 Duvaut and Lions [1] established existence-uniqueness for the tangential field for linear viscoelastic and dynamic fields when g is a function of position only, independent of time and other quantities.

- In 1983 Oden and Pires [6] proved the existence of the linear elastic field due to normal contact and friction.

- In the foregoing analysis we have considered contact problems in which a single step is taken from a "previous" instant t' to the present time t.
  When we have a contact evolution, it is not *a priori* clear whether the solution exists and is continuous as a function of time. Under certain conditions an affirmative answer was given by Klarbring *e.a.* [5].

- Another problem is the uniqueness-existence of a steady state in a continuous evolution. Kalker [3] proved this for quasiidentical, 2D no–slip half-space rolling contact under the condition that the normal compressive force and the creepage were constant from a certain instant of time onwards.

4.7   Surface mechanical principles

We express the principles in surface mechanical form, *i.e.* a form in which the volume integral is absent. To that end we take test functions in the principles of minimum potential and maximum complementary energies which satisfy all elasticity equations as well as the homogeneous boundary conditions:

$$u_i = \bar{u}_i = 0 \text{ in } A_u^a, \text{ and } p_i = \bar{p}_i = 0 \text{ in } A_p^a. \tag{211}$$

We assume no body force, and elastostatics, so that the equilibrium conditions are

$$\sigma_{ij,j} = 0 \tag{212}$$

Finally,

$$\sigma_{ij} = E_{ijhk}(\mathbf{y})u_{i,j} \tag{213}$$

Then we obtain for the two principles, after some calculation

$$min_{u,p}U = \int_{A_c} [\frac{1}{2}p_z u_z + (\frac{1}{2}p_\tau u_\tau + g|E_\tau|)]dS \tag{214}$$

sub (211)-(213), and $e = h + u_z \geq 0$,

$$max_{u,p}C = -\int_{A_c} [(h + \frac{1}{2}u_z)p_z + (W_\tau + \frac{1}{2}u_\tau - u'_\tau)p_\tau]dS \tag{215}$$

sub (211)-(213), and $p_z \geq 0$, $|p_\tau| \leq g$,
which lack volume integrals.
Note that they are valid only when $\rho = 0$, and $\delta g = 0$. When $\delta g \neq 0$, one can use the Panagiotopoulos process. The principle (215) has been used extensively in numerical work (CONTACT, since 1982).

4.8   Complementary or potential energy in numerical work?

A disadvantage of the method of potential energy is that the integral over A_c at one stage or another contains the variation $\delta|s_\tau|$, while the derivative of $|s_\tau|$ is discontinuous when $|s_\tau| = 0$.

The method of maximum complementary energy does not have this disadvantage, but it can only be used in statics, and the equations of equilibrium have to be satisfied inside the bodies. This is no problem when one can use the surface mechanical method, as its test functions are required to do just that. So in that case the method of maximum complementary energy is to be preferred.

4.9   Conclusion

The variational, or weak formulation of the contact problem has been presented. It consists of two variational inequalities, one derived from the principle of virtual work and one from the principle of complementary virtual work. These principles are very fruitful guidances for deriving algorithms for the contact problem.

References of Section 4

[1] G. DUVAUT, J.-L. LIONS (1972), *Les inéquations en mécanique et en physique.* Paris, Dunod.

[2] G. FICHERA (1964), *Problemi elastostatici con vincoli unilaterale: il probleme di Signorini con ambigue condizioni al contorno.* Memorie Accademie Nazionale dei Lincei Ser. 8, 7, p. 91-140.

[3] J.J. KALKER (1970), *Transient phenomena in two elastic cylinders rolling over each other with dry friction.* Journal of Applied Mechanics 37, p. 677-688.

[4] J.J. KALKER (1990), *Three-dimensional, elastic bodies in rolling contact.* Kluwer Academic Publishers Dordrecht, Boston, London.

[5] A. KLARBRING, A. MIKELIČ, M. SHILLÓR (1990), *A global existence result for the quasistatic frictional contact problem with normal compliance.* Proc. IV. Meeting on Unilateral Problems in Structural Analysis, Ed. F. Maceri *et al.*, Birkhäuser Verlag, Zürich.

[6] J.T. ODEN, E.B. PIRES (1983), *Nonlocal and nonlinear friction laws and variational principles for contact problems in elasticity.* Journal of Applied Mechanics 50 p. 67-76.

[7] P.D. PANAGIOTOPOULOS (1975), *A nonlinear programming approach to the unilateral contact and friction -boundary value problem in the theory of elasticity.* Ingenieur Archiv 44, p. 421-432.

5   Numerical analysis: Exact theory

5.1   Discretization

We choose the potential contact region as a rectangle with sides parallel to the $x-$ and y-axis in the surface of the half-space. The x-axis is in the direction of rolling. The z-axis points vertically upwards into body 1, and the y-axis completes the right-handed Cartesian coordinate system. The vertex of the potential contact with the lowest x and y values has the coordinates $(x_0, y_0, 0)$. The potential contact is subdivided into N equal and equally oriented subrectangles (elements) The elements are numbered from 1 to N; they have height $2\Delta y$ and width $2\Delta x$. Their centroids are denoted by

$$\mathbf{x}_I = (x_I, y_I, 0) \tag{216}$$

where a capital latin index runs from 1 to N.

The origin of the potential contact does not lie necessarily in its centre. The situation is shown in Figure 25.

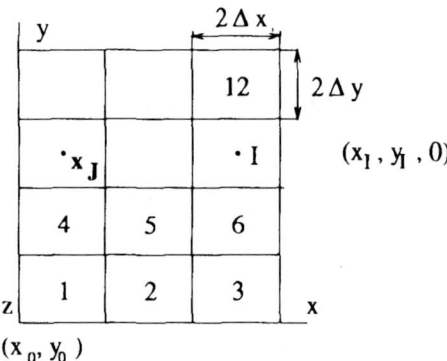

Figure 25: The potential contact area (pot.con.)

We take the traction $\mathbf{p} = (p_j)$ constant in each element, so that the traction is piecewise constant. This traction would seem to be representative for the force/unit area at the centroid of the element in question.
At the time t it is denoted by (p_{Jj}), and at the time t' it is denoted by (p'_{Jj}).
We are interested in the surface displacement difference $u(x, y, 0)$ in the surface point $\mathbf{x} = (x, y, 0)$. To that end we integrate the representation of Boussinesq-Cerruti over each element. We use the following notation: $(B_{iJj})(x, y)$ is a 3-dimensional array. i corresponds to the component of the displacement difference $u_i(x, y)$, and Jj to the component of the traction p_{Jj}. $(x, y, 0)$ is the position where the displacement difference is taken. Then

$$(B_{iJj}(x, y)) = \frac{1}{\pi G} \int_{x_J - \Delta x}^{x_J + \Delta x} dx' \int_{y_J - \Delta y}^{y_J + \Delta y} dy' \cdot$$

$$\begin{pmatrix} \frac{1-\nu}{R} + \frac{(x'-x)^2}{R^3} & \frac{\nu(x'-x)(y'-y)}{R^3} & \frac{K(x'-x)}{R^2} \\ \frac{\nu(x'-x)(y'-y)}{R^3} & \frac{(1-\nu)}{R} + \frac{(y'-y)^2}{R^3} & \frac{K(y'-y)}{R^2} \\ -\frac{K(x'-x)}{R^2} & -\frac{K(y'-y)}{R^2} & \frac{1-\nu}{R} \end{pmatrix} \qquad (217)$$

with $R = \sqrt{(x'-x)^2 + (y'-y)^2}$.

There are six different integrands in this formula, $viz.$

$$J_1 = \frac{1}{R}$$

$$J_2 = \frac{(x'-x)^2}{R^3}$$

$$J_3 = \frac{(x'-x)(y'-y)}{R^3}$$

$$J_4 = \frac{(y'-y)^2}{R^3}$$

$$J_5 = \frac{x'-x}{R^2}$$

$$J_6 = \frac{y'-y}{R^2}$$

Note that $J_1 = J_2 + J_4$, that the integration over J_2 can be derived from the integration over J_4, and that the integration over J_5 can be derived from the integration over J_6, all by symmetry. So we need only integrate over J_3, J_4 and J_6. We write the result using the notation

$$[[f(x,y)]] = [[f(x,y)]]_{x-x_J-\Delta x,\, y-y_J-\Delta y}^{x-x_J+\Delta x,\, y-y_J+\Delta y} \qquad (218)$$

The result is:

$$(\pi G)B_{1J1}(x,y) = [[y \ln(x+r) + (1-\nu)x\ln(y+r)]] \qquad (219)$$

$$(\pi G)B_{1J2}(x,y) = (\pi G)B_{2J1} = [[-\nu r]] \qquad (220)$$

$$(\pi G)B_{1J3}(x,y) = -(\pi G)B_{3J1} = [[K(y \ln r + x \arctan(y/x))]] \qquad (221)$$

$$(\pi G)B_{2J2}(x,y) = [[x \ln(y+r) + (1-\nu)y\ln(x+r)]] \qquad (222)$$

$$(\pi G)B_{2J3}(x,y) = -(\pi G)B_{3J2} = [[K(x \ln r + y \arctan(x/y))]] \qquad (223)$$

$$(\pi G)B_{3J3}(x,y) = [[(1-\nu)(x\ln(y+r) + y\ln(x+r))]], \qquad (224)$$

with $r = \sqrt{x^2 + y^2}$, and

$$\frac{1}{G} = \frac{1}{2}(\frac{1}{G^1} + \frac{1}{G^2}) \quad 1,2:\ \text{body numbers} \qquad (225)$$

$$\frac{\nu}{G} = \frac{1}{2}(\frac{\nu^1}{G^1} + \frac{\nu^2}{G^2}) \qquad (226)$$

$$K = \frac{G}{4}(\frac{1-2\nu^1}{G^1} - \frac{1-2\nu^2}{G^2}) \qquad (227)$$

We sample the displacement difference in $(\mathbf{x}_I, t)$: $(\mathbf{u})$ and in $(\mathbf{x}_I + \mathbf{v}(t - t'), t')$: $(\mathbf{u}')$, present and previous displacement differences:

$$\mathbf{u}_I = (u_{Ii}) = (u, v, w) \quad \text{at} \quad (\mathbf{x}_I, t) \qquad (228)$$

$$\mathbf{u}'_I = (u'_{Ii}) = (u, v, w) \quad \text{at} \quad (\mathbf{x}_I + \mathbf{v}(t - t'), t') \qquad (229)$$

To that end we define

$$A_{IiJj} = B_{iJj}(\mathbf{x}_I) \tag{230}$$
$$C_{IiJj} = B_{iJj}(\mathbf{x}_I + \mathbf{v}(t - t')) \tag{231}$$
$$\mathbf{p} = (p_{Jj}) = \mathbf{p}(\mathbf{x}_J, t) \tag{232}$$
$$\mathbf{p}' = (p'_{Jj}) = \mathbf{p}(\mathbf{x}_J, t') \tag{233}$$

NOTE that in non-steady rolling $p'_{Jj} \neq p_{Jj}$, while p'_{Jj} is known; on the other hand, in steady state rolling, $p'_{Jj} = p_{Jj}$, so that p'_{Jj} is unknown.

We have:

$$u_{Ii} = \sum_{J=1}^{N}\sum_{j=1}^{3} A_{IiJj}p_{Jj} \tag{234}$$

$$u'_{Ii} = \sum_{J=1}^{N}\sum_{j=1}^{3} C_{IiJj}p'_{Jj} \tag{235}$$

NOTE that in non-steady rolling u'_{Ii} is known, while in steady-state rolling it is unknown, but it equals u_{Ii} shifted over a distance $\mathbf{v}(t - t')$.

5.2 The NORM algorithm

We will now treat the algorithm for the normal contact. We assume that the tangential traction is prescribed, as well as the undeformed distance h: $p_{J\tau}$, $\tau = 1,2$ is the tangential traction, function of the position $\mathbf{x}_J$; $p_{Jz} = p_z(\mathbf{x}_J)$ is the normal pressure, both on body 1. The normal pressure is positive in the contact region, and vanishes outside it.

The displacement difference, defined as $\mathbf{u}_I = \mathbf{u}^1(\mathbf{x}_I) - \mathbf{u}^2(\mathbf{x}_I)$ is, in components, $\mathbf{u}_I = (u_I, v_I, w_I)$. The undeformed distance h_I is the scalar distance between body 1 and body 2 at $\mathbf{x}_I$ in the undeformed state, measured from body 2 to body 1. The deformed distance e_I is similarly defined, but in the deformed state. We have:

$$e_I = h_I + w_I \tag{236}$$

The deformed distance is positive outside the contact area, and vanishes inside it. We take the displacement fixed at infinity, at zero. So the boundary conditions become:

$$p_{Iz} \geq 0, \quad w_I \geq 0, \quad p_{Iz}w_I = 0 \tag{237}$$
$$\mathbf{u}^a \to 0 \quad \text{if } |\mathbf{x}| \to \infty, \quad a = 1,2: \text{ body number} \tag{238}$$

The condition at infinity is automatically satisfied if we use the Boussinesq-Cerruti formulae of the previous subsection. The condition $p_z = 0$ outside the potential contact is satisfied if we simply set the normal traction 0 where there are no elements, i.e. on $z = 0$, outside the pot.con. Inside the pot.con. we act as follows.

1. We set

$$h_I^* = h_I + \sum_{J,\tau} A_{I3J\tau} p_{J\tau} \qquad \forall \mathbf{x}_I \in Q$$

   Explanation. The given tangential tractions are taken into account by modifying the undeformed distance (h^* instead of h).
   Q is the entire pot.con.

2. Set $p_{Jz} = 0 \quad \forall J$: Clear all normal tractions.

3. Initially, the exterior $E = Q$, and the contact area $C = \emptyset$. During the process, E and C are modified until they correspond to the solution.

4.

$$e_I = h_I^* + \sum_{J} A_{I3J3} p_{Jz} \qquad \forall \mathbf{x}_I \in Q \tag{239}$$

   Equations: $e_I = 0 \ \forall \mathbf{x}_I \in C$, $p_{Jz} = 0 \ \forall \mathbf{x}_J \in E$.
   These are N linear equations for the N p_{Jz} which can be solved.

5. Are all pressures $p_{Jz} \geq 0$ in C? If "no", go to Point 6. If "yes", go to Point 7.

   Explanation. It is checked whether the pressures are all positive in C.
   If "yes", go to Point 7.
   If "no", place all elements with negative pressure in the Exterior E, where the pressure will be annihilated according to the equations of Point 4.

   Note. If we place only one such element in the exterior E, the algorithm can be proved mathematically by means of variational calculus. We have not succeeded in doing so when we place all such elements in the exterior E. Yet we never have had a failure!

6. Restore. If $p_{Jz} < 0$, element J is placed in Exterior E, and p_{Jz} is set equal to zero. Go to Point 4.

7. Are all deformed distances $e_I > 0$ in E? If "yes", WE ARE READY.
   If "no", place all elements with deformed distance < 0 in C; go to Point 4.

   Explanation. It is checked whether all deformed distances are positive in E.
   If "yes", all pressures are positive in C and vanish in E, while all deformed distances are positive in E and vanish in C. All contact conditions are met; we are READY.
   If "no", we place all elements with negative deformed distance in C and go to Point 4, where care is taken that the deformed distances in the new C are annihilated. Here also we should reform only one element in order to get a proof of the algorithm.

5.2.1   Normal force prescribed

There is an important variant of the NORM algorithm as we described it here. To state it, we first introduce the notion of the approach of the bodies.

Let h_I be the undeformed distance between the bodies; then $h_I - d$ is the distance, but with the bodies moved a distance d towards each other. d is called the approach of the bodies. In the algorithm above we set $d = 0$, or better, we give it a fixed value.

If we want to prescribe the total normal force, we regard d as a variable of the problem, but instead we demand that that the total normal force

$$F_z = 4\Delta x \Delta y \sum_{J\in Q} p_{Jz} = F_z' \tag{240}$$

is prescribed as $F_z' > 0$.

We have to adapt NORM to cope with this variant.

1. $h_I^* = h_I + \sum_{J,\tau} (A_{I3J\tau} p_{J\tau} - d$
   Initially, set $d = 0$. Then find the minimum of h_I^* over the entire pot.con Q: minimizer I'. Now, take d so that $h_{I'}^* = 0$.

2. Set $p_{I'z} = \frac{F_z'}{4\Delta x \Delta y} > 0$. Set the other $p_{Iz} = 0$, for all other $I \in Q$.

3. Set C = Element I'. Set E = all other elements of Q.
   E and C are modified until they correspond to the solution.

4. $e_I = h_I^* + \sum_J A_{I3J3} p_{Jz}$. Note that h_I^* contains d in a linear fashion. The equations are: $e_I = 0\ \forall I \in C$; (p_{Jz}, d variable;) $p_{Jz} = 0$ in E; $4\Delta x \Delta y \sum_J p_{Jz} = F_z'$
   ($N + 1$ linear equations with $N + 1$ unknowns, $viz.$ $N\ p_{Iz}$, and d.

From here on we follow the ordinary NORM (points 5. to 7.).
This algorithm will break down when $F_z' < 0$, as then at least one p_{Iz} must be negative.

5.3   The TANG algorithm

We will now treat the algorithm for the tangential contact. We assume that the normal traction is prescribed, as well as the creepages: $p_{Jz} = p_z(\mathbf{x}_J)$, $c_I = c(\mathbf{x}_I)$. In non-steady state rolling, the previous displacement difference $(u,v)_I' = (u,v)(\mathbf{x}_I + \mathbf{v}(t - t'), t')$ is known, in steady state rolling it is equal to $(u,v)_I' = (u,v)(\mathbf{x}_I + \mathbf{v}(t - t'))$, as the displacement difference $(u)_I$ is independent of explicit time in the contact fixed coordinate system. Therefore, $(u,v)_I'$ is "as unknown" as $(u,v)_I$ itself.

The displacement differences $\mathbf{u}$, $\mathbf{u}'$ depend on the surface traction field $\mathbf{p}_J = \mathbf{p}(\mathbf{x}_J, t)$ and the previous traction field $\mathbf{p}_J' = \mathbf{p}(\mathbf{x}_J, t')$ in the following manner:

$$u_I = \sum_J \sum_j A_{I1Jj} p_{Jj} \tag{241}$$

$$v_I = \sum_J \sum_j A_{I2Jj} p_{Jj} \quad \text{at } (\mathbf{x}_I, t), \tag{242}$$

$$u'_I = \sum_J \sum_j C_{I1Jj} p'_{Jj} \tag{243}$$

$$v'_I = \sum_J \sum_j C_{I2Jj} p'_J \quad \text{at } (\mathbf{x}_I + \mathbf{v}(t - t'), t') \tag{244}$$

The slip is defined as follows:

$$\mathbf{s} = \mathbf{c} - \frac{\partial \mathbf{u}}{\partial \mathbf{x}} \mathbf{v} + \frac{\partial \mathbf{u}}{\partial t}, \quad \text{or, discretized:} \tag{245}$$

$$s_I = c_I + \frac{(\mathbf{u}_I - \mathbf{u}'_I)\mathbf{v}}{t - t'} \tag{246}$$

The traction bound is given as $g_I = fp_{Iz}$, with f the coefficient of friction. We have:

$$|(p_{I1}, p_{I2})| \leq g_I; \tag{247}$$

$$\text{if the inequality holds: } s_I = 0 \tag{248}$$

$$\text{if the equality holds: } s_I = -|s_I|(p_{I1}, p_{I2})/g_I \tag{249}$$

and the half-space is fixed at infinity.

The algorithm runs as follows:

1. Set $(p_{I1}, p_{I2}) = (0,0)$: Clear the tangential tractions.

2. Initially, the slip area $S = \emptyset$, and the adhesion area H (= stick area) = the complete contact patch C. During the process, S and H are modified, but always in such a way that $H \bigcup S = C$, until they correspond to the solution.

3. Solve the following equations:

$$s_I = 0 \text{ in Adhesion area } H; \text{ these are linear equations} \tag{250}$$

$$|(p_{J1}, p_{J2})| = g_J \text{ in slip area } S. \text{ (Non-linear equations)} \tag{251}$$

Remark. These equations are nonlinear. They may be solved, e.g., by means of the Newton-Raphson method.

4. If $|p_{I1}, p_{I2}| > g_I$, for any element in H, place element I in Slip area S.
   If this has happened at least once, go back to Point 3.
   If it has not happened, go to Point 5.

5. If the slip s_I is in the same sense as the tangential traction (p_{I1}, p_{I2}), rather than opposite as it should be in the slip area, then place element I in the adhesion area H.
   If this has happened at least once, go back to Point 3, else we are READY.

5.3.1   Total force components prescribed

There is an important variant of the TANG algorithm as we described it here. It is that one or two of the total tangential force components are prescribed. In order to do that, either v_x and/or v_y must be left free, and the total tangential force F_x and/or F_y must be prescribed. This goes in much the same way as it was done in NORM, and we will not give any details.

Convergence becomes bad when $|(F_x, F_y)|$ approaches fF_z; the solution does not exist when this quantity exceeds fF_z. The convergence in notably threatened when two components of the total tangential force are prescribed.

6   Results

In the present Section we show some results of the numerical work.

In Subsection 1, we show numerical results of steady-state rolling in two dimensions, both for symmetrical and asymmetrical bodies. In Subsection 2, we show results on non-steady state rolling in 2D. In Subsection 3 we show results of 3D steady-state rolling, and in Subsection 4 results of non-steady state 3D rolling.

6.1   2D Steady-state rolling

6.1.1   Symmetrical bodies

Consider two infinitely long cylinders with parallel axes which are made of the same homogeneous, isotropic elastic material: modulus of rigity G, Poisson's ratio ν. The cylinders are pressed together so that a contact strip forms between them; the strip is bounded by two parallel lines. A coordinate system is introduced of the form $(O; x, y, z)$ in which the origin O lies on the centre line of the strip, the axis of x points in the direction perpendicular to the axis of the strip, in the plane of contact, and in the rolling direction, and the axis of z is also perpendicular to the centre line of the strip, but points upwards into the upper cylinder (body 1), while the axis of y lies along the centre line of the strip, in such a way that the coordinate system $(O; x, y, z)$ is right-handed. An alternative notation for (x, y, z) is $\mathbf{x} = (x_1, x_2, x_3)$.

The rolling velocity and the creepage were introduced before. When the lateral creepage $v_y = 0$, and the spin also vanishes, then the motion and the elastic deformation are two-dimensional and in plane strain. Then the state of the traction distribution was determined analytically by Carter [6], and by Fromm [7]. Carter uses the half-space approximation, while Fromm does not. The normal traction distribution is Hertzian/semi-elliptical, and the tangential traction distribution is given in Fig. 26. Shown are the numerical values obtained by CONTACT. The coincidence is perfect.

A particle enters the contact area from the right. It is unloaded at the time. It is seen that the stick area borders on the leading edge of the contact strip, and that the traction increases till it reaches the traction bound; there, slip sets in and the traction remains on the traction bound till the loaded particle leaves the contact area, and loses its loading.

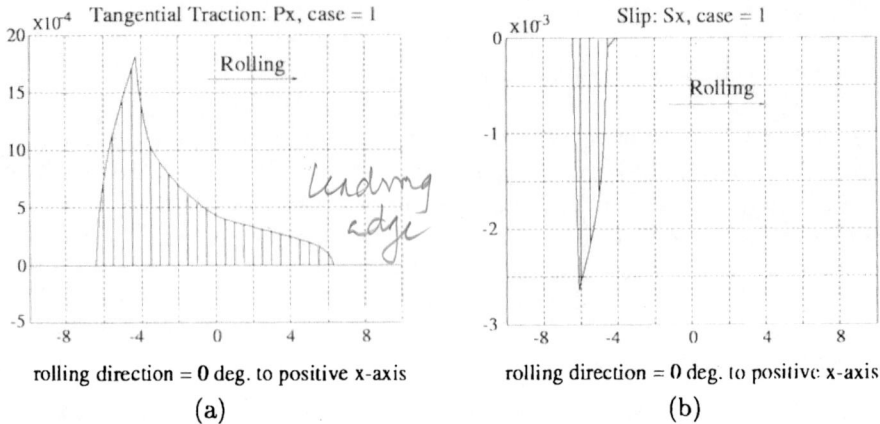

Figure 26: Symmetric rolling. Left: the traction distribution according to Carter (1926) and to CONTACT. Right: the slip distribution. Leading edge on the right, rolling from left to right.

6.1.2 2D asymmetric rolling

Now, we have the same geometry, but the elastic constants of the cylinders differ. Again, we set the lateral creepage and the spin equal to zero, and the motion and the elastic deformation are two-dimensional in plane strain. The state of traction was determined numerically by Bentall and Johnson [2] for so-called free rolling, that is, rolling with no net tangential force (dots), see Fig. 27. Also shown are the numerical values obtained by CONTACT (line). The coincidence is good, considering that both methods are numerical. The slip is shown in the figure on the right (CONTACT).

It is worth while to analyze the solution more deeply. Rolling takes place from left to right. The lower body is perfectly rigid, the upper body has a modulus of rigidity of unity. The Poisson ratio of the lower body (1) is irrelevant, as the body does not deform at all; the Poisson ratio of the upper body (2) is 0.286, a quite normal value. The total tangential force F_x vanishes.
Slip in the direction of rolling takes place in the interval $[0.410, 0.625]$ and in $[-0.625, -0.575]$. Slip counter to the direction of rolling takes place in $[-0.505, -0.300]$. In the remainder of the contact area sticking takes place. The traction is shown in the left figure, the slip in the right figure. Note the difference in the first derivative of the slip at the leading and the trailing edge of a slip area!

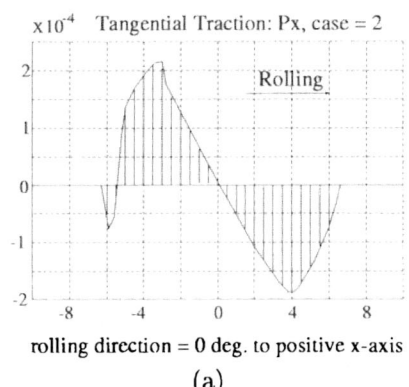

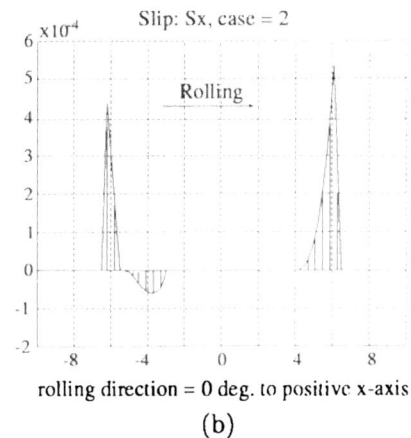

(a)   (b)

Figure 27: Free rolling of asymmetric bodies, from left to right. Parameters: $G_1 = 1, G_2 = \infty, \nu_1 = 0.286, \nu_2 = $ irrelevant, $f = 0.1, F_x = 0$. Left figure: the traction. Right figure: the slip. Rolling is from left to right.

We also calculated some cases of tractive rolling. In Fig. 28 we compare two cases which are the same, except that the material constants of bodies 1 and 2 have been interchanged, while the tangential force has changed sign. The upper are the constants of Fig. 27. The right figures are the slip corresponding to the left ones.

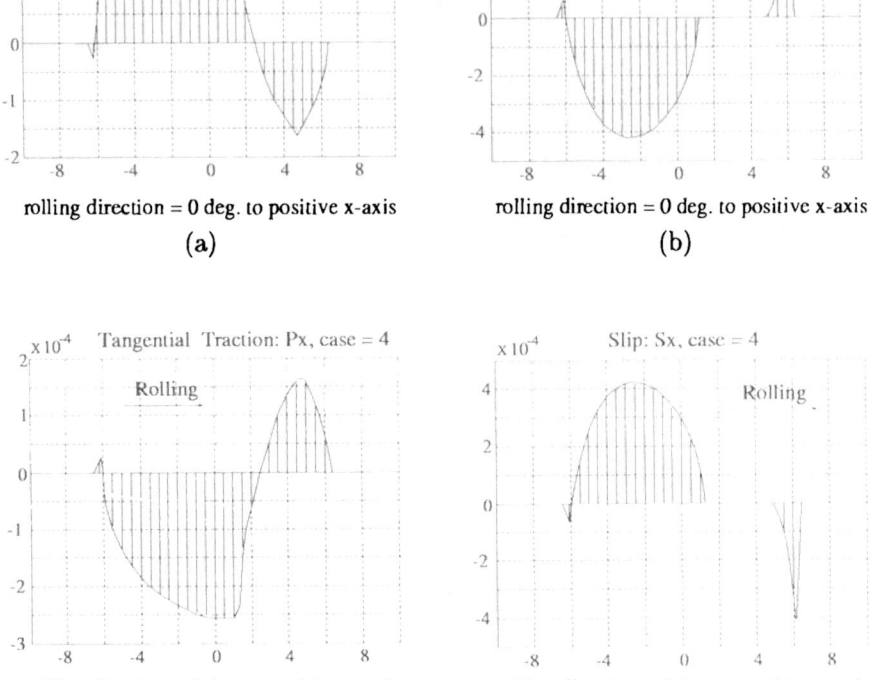

Figure 28: Tractive, asymmetric rolling. Comparison of interchanged material constants and inverse tangential force. p_x (and also the slip) merely change sign. Rolling is from left to right.

Another contrast is found in Fig. 29: The material constants are the same as in Fig. 27, but the total tangential force is $-0.5fF_z$. Also shown is the slip corresponding to the values of the left figure. Fig. 28 and Fig. 29 look totally different, although the absolute value of the total tangential force is exactly the same!

6.2   2D Transient rolling

The results from this section are taken from Kalker [11]. The notation of that paper differs slightly fom ours: in Kalker [11] we use X for p_x, and the slip is defined as the local velocity of body 2 over body 1, and is just the opposite from what we always have. We will show only traction distributions, as we are more interested in qualitative than in quantitative behaviour.

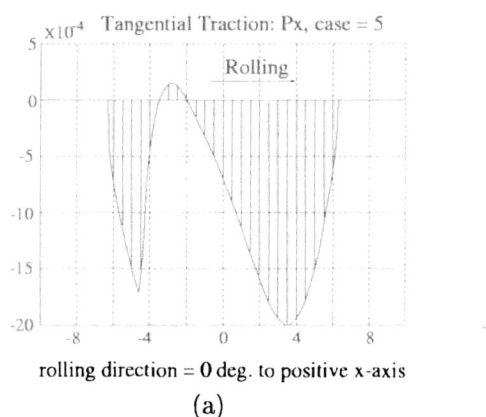
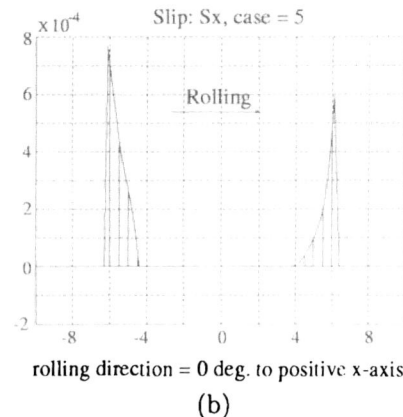

(a)   (b)

Figure 29: Traction and slip for the material constants of Fig. 27, but with total tangential force $-0.5fF_z$. Rolling is from left to right.

6.2.1   From Cattaneo to Carter

Two similar cylinders are pressed together. Then, opposing couples are applied to their axes, so that the cylinders start rolling over each other at time $t = 0$. The evolution is shown in Fig. 30. At the time $t = 0 + \epsilon$, $\epsilon \downarrow 0$ the traction distribution looks as as shown in Fig. 30(a). It is the so-called Cattaneo distribution, of shifting rather than rolling.

$F_x/fF_z = 0.75$; the half-circles represent the Coulomb traction bound. The dots are a comparison with a previous theory; it is seen that the coincidence is quite good. The contact width $2a = 2$. The rolling velocity $V = 1$.

Three phases can be distinguished in the evolution.

1. The first phase is the initiation, in this case the Cattaneo shift, Fig. 30(a).

2. In the second phase, the old traction distribution is shoved out of the contact area, and replaced with a traction distribution already akin to the steady state traction distribution. In the present case of Fig. 30, this phase takes place from $t = 0$ to $t = 1$ (Fig. 30(b) to Fig. 30(e)).
   Immediately after rolling starts, the traction at the point A (Fig. 30(a) and (b)) leaves the traction bound temporarily; it drops and rises again till it reaches the traction bound again at $t = 0.4$ (Fig. 30(b)). At that time, the stick area breaks into two. The left hand stick area vanishes fast; it has gone at $t = 0.58$.
   The vertical tangent at B moves inward with rolling velocity, together with the particle carrying it. It vanishes at $t = 1$, and this marks the end of phase 2.

3. In phase 3, the traction distribution adapts itself to the steady state, which is virtually reached when $t = 2$, i.e $>$ after one contact length has been traversed.

It is remarkable how fast the transience from a rather arbitrary initial traction distribution to that of steady state rolling takes place, once the total force is held constant.

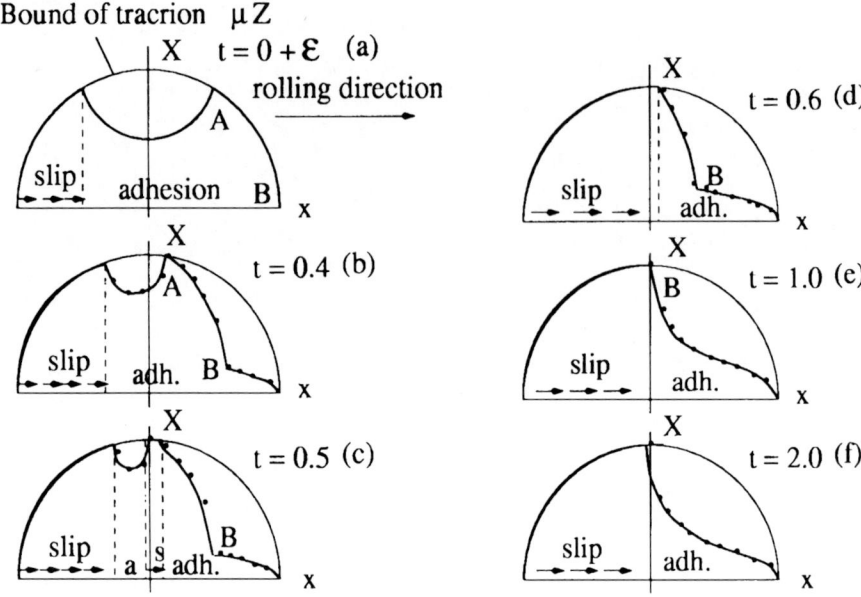

Figure 30: From Cattaneo to Carter (shift to steady state rolling). (Source: Kalker [11])

6.2.2   Rolling of similar cylinders under varying normal force

Two similar cylinders are rolled over each other under the action of a constant tangential force F_x. The normal force is so that the half contact width has the following behaviour:

$$a(t) = 1 + \frac{0.4}{\pi} \sin(2\pi t)$$

while

$V(t) = 1$, $K = 0$ (similarity), $f = 1$, $F_x = 0.255\pi$, $F_z = \frac{1}{2}\pi a^2$. In a quasi-steady state the initial traction distribution is immaterial.

The tangential tractions are shown in Fig. 31, in which are represented
$t = 1, 1.2, 1.4, 1.6, 1.8$, while $t = 2.0$ (full period after $t = 1$) is represented by dots in the figure corresponding to $t = 1$. The semi-circles represent the Coulomb traction. Slip takes place wherever the Coulomb traction is attained, opposite the traction.
From the coincidence of the traction at $t = 2$ with that at $t = 1$ it is seen that at $t = 1$ the quasi-steady state is already reached, from which it is seen that the transience proceeds extremely fast.

6.2.3   Harmonic rolling velocity and tangential force

Two similar cylinders are rolled over each other with a harmonic rolling velocity and a harmonic tangential force. The normal pressure and the friction coefficient remain constant, while a quasi-steady state is investigated in which the initial traction distribution

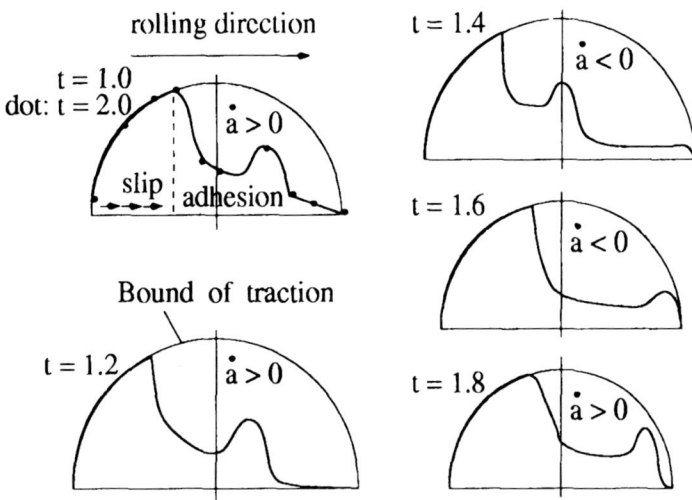

Figure 31: Rolling of similar cylinders under varying normal force. (Source: Kalker [11])

is immaterial.

The governing parameters of the program are:
$V(t) = \pi \cos(\pi t)$, $F_x = -0.355\pi \sin \pi t$,
$a = 1$, $K = 0$, $f = 1$, $F_z = \frac{1}{2}\pi$.

Half a period, from $t = 1$ to $t = 2$, is shown in Fig. 32. The leading edge at each instant is marked by L, the trailing edge by T. The traction distribution at $t = 2$ is the traction distribution at $t = 1$, multiplied geometrically by (-1) with the origin as centre of multiplication. The circles, as usual, represent the Coulomb traction; slip is present wherever the Coulomb traction is attained, and it is opposite the traction.

It is seen that at $t = 1.5$, when the rolling velocity changes sign, a vertical tangent is introduced at the edge $x = 1$, which moves inward with rolling velocity. At the same time the slip vanishes near the leading edge $x = 1$ and slip starts at the trailing edge $x = -1$.

Transience was completed in half a period from $t = 0$ to $t = 1$.

6.2.4 Frictional compression followed by transient rolling of dissimilar cylinders

Two dissimilar cylinders ($G_1 = 1$, $G_2 = \infty$, $\nu_1 = 0.286$, ν_2 immaterial, coefficient of friction $f = 0.15$) are brought into contact and compressed. The resulting traction distribution has been described by Spence [16]. The resulting tangential traction is shown in Fig. 33(a). Subsequently they are rolled; the semi-contact width $a = 1$. Here also, the three phases are clearly visible. The first phase, initiation, is given by Fig. 33(a); the second, that of the transience to almost the distribution of steady-state rolling is given

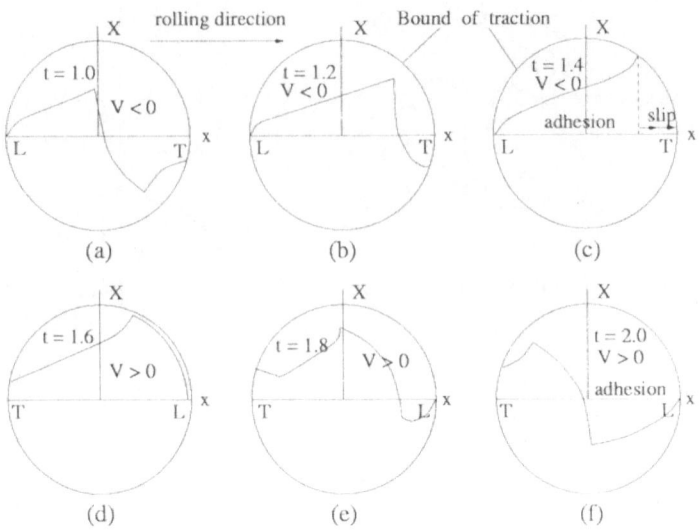

Figure 32: Harmonic rolling velocity and tangential force. (Source: Kalker [11])

by Figs. 33(b)-(d), while at $t = 1.8$ the steady state has virtually set in Fig. 33(e).

6.3   3D Steady state rolling

6.3.1   Quasiidentical steady state rolling: surface stresses

We recall the areas of slip and adhesion. They were described in the Section on simplified theory, *viz.* Section 3, and shown in Fig. 20. Here we will start with showing the contact stresses in quasiidentical steady state rolling with a circular contact area (Fig. 34). In Fig. 34, A signifies an adhesion area (stick area), and S a slip area. The arrows are in the direction of the traction, the circles are the boundary of the contact areas, the lines inside the contact areas are the slip-stick boundaries. In Figs. 34(a') and 34(d') are shown the absolute value of the tangential traction on the line $x - x$ in Figs. 34(a) and 34(d). The dot in the contact areas indicates the so-called spin pole: $(v_y/\varphi, -v_x/\varphi)$; the creep forms a rotating field around this point. This is the reason why it is lacking in Fig. 34(d) (pure longitudinal creepage). The rolling direction is throughout from left to right. Fig. 34 is Fig. 5.19 of Kalker [12], Chapter 5.

In Fig. 34(a) and 34(a') we show the case of pure spin, $v_x = v_y = 0$. The stick area has its characteristic pointed form. The traction forms nearly a rotating field, but it is clear that a net lateral force results. On the line $x - x$, which is the path of a particle, the traction distribution $|(p_x, p_y)|$ looks very much like the 2D Carter distribution (Fig. 26 of this Section), with as vertical tangent at the stick-slip boundary, see Fig. 34(a').

Fig. 34(b) shows the case of combined longitudinal creepage and spin, $v_y = 0$. Here it is clear that there will be a net longitudinal and lateral component of the tangential force. Fig. 34(c) shows the case of combined lateral creepage and spin ($v_x = 0$); the traction becomes more directed along the y-axis, as compared to Fig. 34(a).

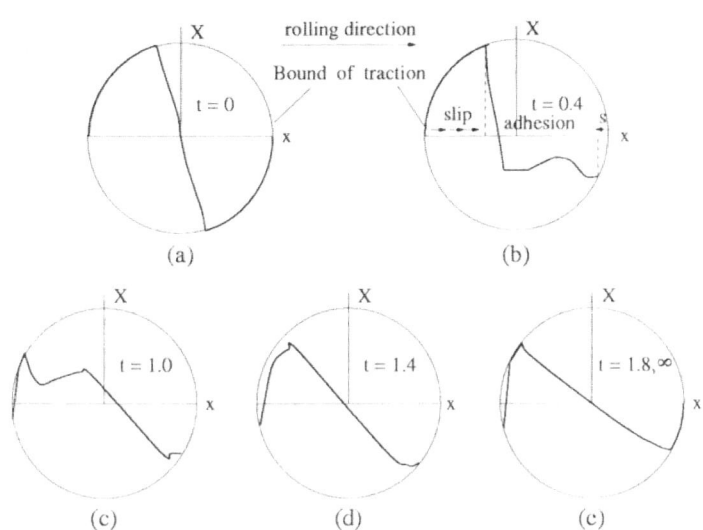

Figure 33: Frictional compression followed by rolling of dissimilar cylinders. (Source: Kalker [11])

Fig. 34(d) shows the case of pure longitudinal creepage, $v_y = \varphi = 0$; all tractions are almost parallel to the x-axis.

Fig. 34(d') shows $|p_x|$ along the line $x - x$; again the impression is one of Carter's traction distribution. Note that with almost the same stick-slip bound at $x - x$, the present graph is "thinner".

Fig. 34(e), finally, shows the case of large spin, with almost no stick area; the field is fully rotating, and the lateral force is lower than in Fig. 34(a), which also represents pure spin.

6.3.2   Quasiidentical steady state rolling: subsurface stresses

The subsurface stresses are important in strength and endurance calculations. They were first calculated by Ahmadi [1]. They can be computed by a special module of CONTACT. In Kalker [12], an algorithm is presented to calculate the displacements and the displacement gradients on and inside the elastic half-space $z \geq 0$ due to a uniform load of arbitrary direction acting on a rectangle on the surface of a half-space. The half-space is homogeneous and isotropic, with modulus of rigidity G and Poisson's ratio ν.

The most important quantities that determine the strength of the material are the first and second invariants of the stress, $viz.$

$$\sigma_{ii} \quad : \quad \text{first invariant} \tag{252}$$

$$s_{ij}s_{ij} \quad : \quad \text{second invariant, with}$$

$$s_{ij} = \sigma_{ij} - \sigma_{hh}\delta_{ij}/3 \quad : \quad \text{stress deviator} \tag{253}$$

The second invariant yield an ideal stress σ_I,

$$\sigma_I = \sqrt{s_{ij}s_{ij}} \tag{254}$$

The von Mises yield criterion of plasticity is a bound on the ideal stress:

$$\sigma_I \leq k, \quad \text{with } k \text{ the yield stress} \tag{255}$$

Fig. 35 shows σ_I and $-\sigma_{ii}$ on the z-axis for a uniform load on a square, centered at the origin, and loaded

1. by a uniform traction in the z-direction, of unit intensity;

2. by a purely shearing traction of unit intensity, in the direction of a side of the square.

Poisson's ratio $\nu = 0.28$, $G = 1$.

In case 1, the purely normal load, shown by the full lines in Fig. 35, the ideal stress σ_I has a maximum of 0.55 at about 0.4 side-length under the surface. This behaviour is well-known. $-\sigma_{ii}$ has a boundary maximum of 2.6, and drops rapidly to where its value meets the falling-off part of the curve of σ_I at $z \doteq 1.40$.

In case 2 $\sigma_{ii} = 0$. σ_I (the broken line in Fig. 35) behaves like $-\sigma_{ii}$ in case 1, starting at $\sigma_I = 1.42$ on $z = 0$, but dropping much more rapidly so that it has virtually vanished at $z = 1.67$ side length.

Fig. 35 can be used to assess the quality of the half-space approximation. The stress behaviour is dominated by the normal pressure from a depth of about 1.40 times the contact area diameter. Looking at the drawn lines of Fig. 35 we see that at a depth of about three contact area diameters the the loads the ideal stress has dropped to 10 % of its maximum value, and $-\sigma_{ii}$ to about 2.5 % of its maximum value. So it seems safe to assume that the stresses have almost died out at that level. This supports the statement that the half-space approximation is justified when the diameter of contact is less than 1/3 of the diameter of the contacting bodies.

At a depth of 5 times the diameter of contact, the numbers are 1 % and 0.25 %, respectively.

6.3.3   The total force transmited in rolling

A very important quantity of rolling contact in technology is the total force transmitted. It finds application in vehicle dynamics, both in rail vehicles and in motor cars. As to motor cars, more will be heard of it in the lectures of Prof. Pacejka. The car tyre is, however, so complicated that a rigorous theory of it is still beyond our ken. In rail vehicles, we have the advantage that we can regard the contact as quasiidentical, and fortunately the theory for that is well developed.

Speed is of the essence in the total force theories. In vehicle dynamics, an something like a million steps must be taken in order to advance the real time by one second. And in other applications speed is a pleasant advantage because one does not have to wait a long time before getting accurate results.

We have the following theories.

- **The linear theory** is valid for longitudinal, lateral and spin creepage which are infinitesimally small. The theory is described in Section 2. The quantities determining the total force components are called the creepage and spin coefficients; they are tabulated in Section 2, Table 3; they hold only for quasiidentical, Hertzian bodies.

- **The theory of Vermeulen and Johnson** is valid for unrestricted longitudinal and lateral creepage and zero spin. It is an empirical formula with a theoretical background. It is described in Section 2. It holds only for quasiidentical, Hertzian bodies.

- **FASTSIM** is an algorithm based on the simplified theory of rolling contact. It is very fast but approximate. It can be used for unrestricted creepage and spin, but is confined to quasiidentity, and to Hertzian bodies. It is described in Section 3.

- **CONTACT** is an algorithm based on the true theory of elasticity. Owing to the fact that a discretization is used it is approximate. It is about 2000 times as slow as FASTSIM. It is the only algorithm which can be used for non quasiidentical and non-Hertzian bodies. It is described in the present Section 5.

- **Table Book.** A book of tables has been constructed in which one can interpolate linearly to obtain the total force and the twisting moment from the longitudinal, lateral and spin creepages and the ratio of the axes of the Hertzian contact ellipse. The construction has taken place with CONTACT, and comprises 115,000 entries (4.5 MByte). Poisson's ratio has been fixed to 0.28, the value for steel, for the Table Book is intended for railway use. The programme using the Table Book is about 8 times as fast as FASTSIM, but it needs this enormous amount of storage space, while FASTSIM needs only 41 kB and can be used for unrestricted Poisson's ratio. But in many cases FASTSIM's error exceeds that of the Table Book. The Table Book is only for quasiidentical, Hertzian bodies with Poisson's ratio equal to 0.28.

6.4   3D Transient rolling

6.4.1   Quasiidentical bodies

The present section is based on Kalker (1990) sec. 5.2.2.5. We consider Kalker's Fig. 22, Section 5, (here Fig. 36): Quasiidentical transient rolling. The calculations have been made by CONTACT. Considerable smoothing and editing of the figures has taken place; the figure is an interpretation of the true result. Two identical spheres are compressed and rolled over each other.

Radius spheres: $R = 337.5$, $G_1 = G_2 = G = 1$, $\nu_1 = \nu_2 = \nu = 0.28$. $F_z = \text{Constant} =$

$0.4705 = (7/9)^3$; $a = b = 3.5$. $f = 0.4013$.

Radius stick area "Cattaneo": $0.7a = 2.45$. $F_x = $ Constant throughout $= -f \times F_z \times$ $0.657 = -0.1240$, $v_y = \varphi = 0$.

After the Cattaneo shift the spheres roll with constant force $(F_x, 0, F_z)$ without spin, with velocity $V = 1$. Step is $Vt = 0.5$.

Elements: squares with sides 1.

Potential contact: 9×9, center in origin.

In the figures p_x is shown for various values of y (upper 4 rows). Full line: p_x, dotted line: traction bound fp_z.

In the lowest row, we show the areas of slip (S) and adhesion (A). The columns correspond to $Vt = 0, 1, 3, 3, 5, 7 = \infty$. The traction distributions are very much like those in the two-dimensional case (Fig. 30).

6.4.2 Non-quasiidentical bodies

We start from the Spence compression, *viz.* a sphere with radius 243, modulus of rigidity $G = 2$, and Poisson's ratio $\nu = 0$ pressed onto a flat, rigid slab. The friction coefficient is $f = 0.4013$. The final radius of contact is 3.5 units. Then rolling starts in the x-direction, with creepage and spin kept zero. The surface is discretized into squares with side 1; the potential contact is a square with side 7. The distance traversed $V(t - t')$ is discretized into steps of 0.2 units. The results are shown in Fig. 37. This figure is similar to Fig. 36. In the basic figures the absolute value of the tangential traction $|p_\tau|$ is shown drawn, together with the traction bound fp_z (broken line), as a function of the rolling coordinate x, with the lateral coordinate y as a parameter. The tangential traction is mirror symmetric about the x-axis. The arrows under the x-axis represent the direction of the traction. The lowest row represents the contact area which is taken circular, and its division into regions of Slip (S) and Adhesion (A). The columns depict the situation when the distance traversed is $Vt = 0, 1, 2, 3, 4, 5, 7, 10, 13$. At the final position, which represents almost 2 contact widths traversed, the steady state has been virtually attained.

Fig. 37 was made after considerable smoothing and editing.

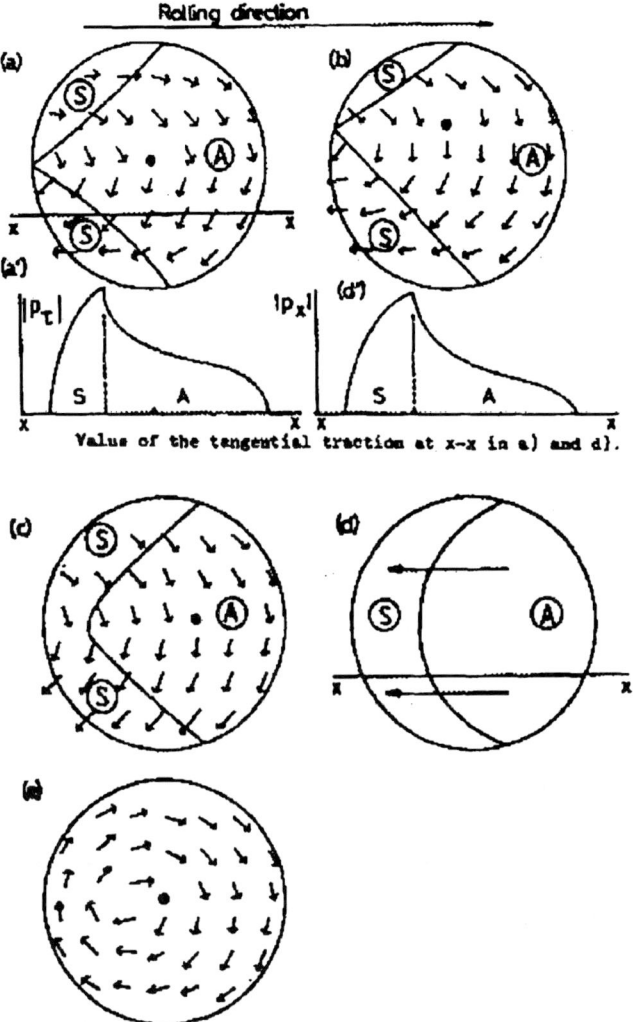

Figure 34: Contact stresses in quasiidentical steady state rolling with circular contact area. (Source: Kalker [12])

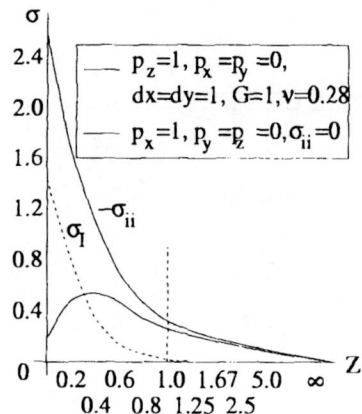

Figure 35: Subsurface stresses in the half-space $z \geq 0$. (Source: Kalker [12] Fig. 20, Section 5)

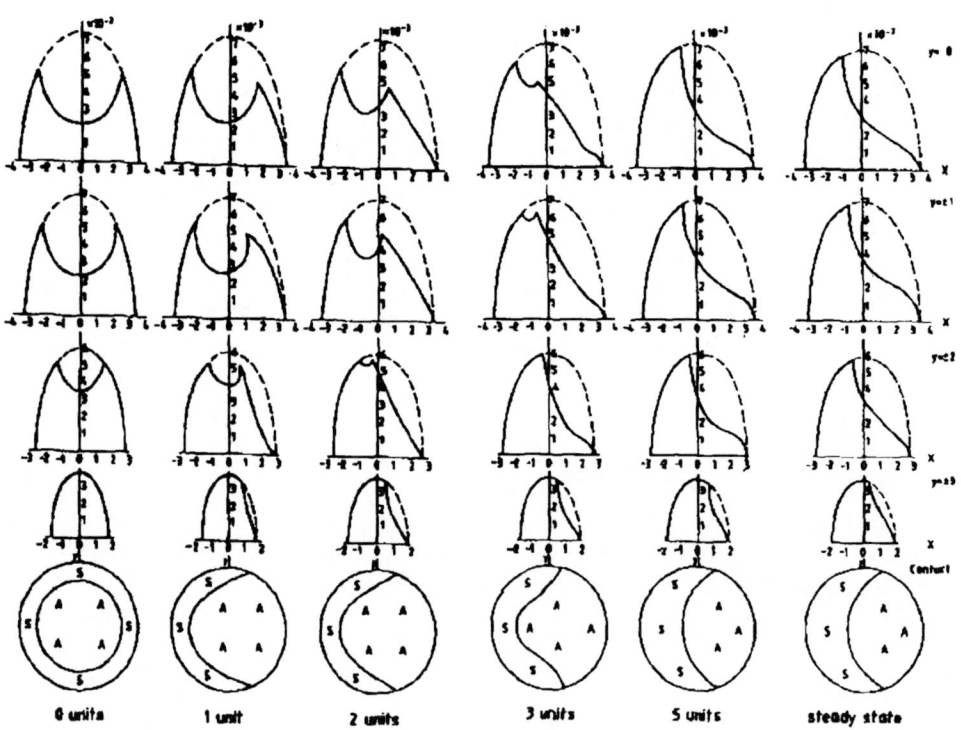

Figure 36: 3D quasiidentical tansient rolling

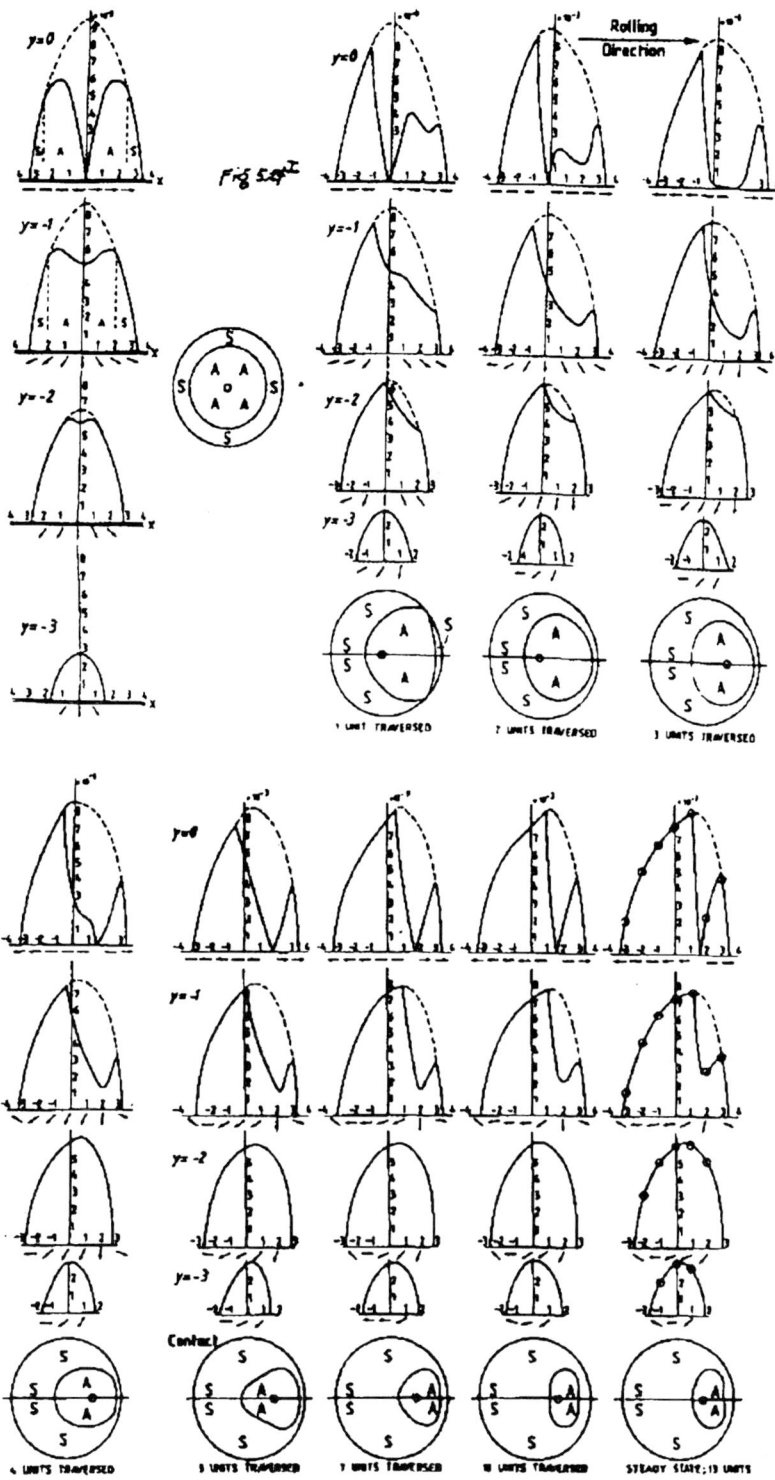

Figure 37: 3D non-quasiidentical transient rolling

7   Further and alternate research

1. The work of Bufler [5].
   In this work, Bufler solves the non-quasi-identical problem of two 2-D bodies rolling
   over each other with infinite friction (or infinitesimal creepage). An analytical,
   closed form solution is found.

2. The works of Bentall and Johnson ([2] and [3]).
   The work from 1967 considers two non-quasi-identical 2-D cylinders rolling over
   each other with finite creepage. This work is, in content, a generalisation of Bufler's
   work (see 1.). The method is a boundary element method in which the traction
   elements are continuous and piecewise linear. The work from 1968 concerns a strip
   between two rolling cylinders. Here also, the traction elements are continuous and
   piecewise linear. The solutions are numerical with a partly heuristic algorithm.

3. The work of Braat, Kalker and Saes.
   This work was published by Kalker[13] and Braat and Kalker [4]. It concerns rolling
   bodies consisting of different flat layers which are glued together. The geometry is
   2-D. Examples are: Rubber on steel, steel on aluminium, various kinds of plastic
   glued onto each other, etc. The rolling bodies are cylinders which are approximated
   by layered half-spaces (2-D). Only steady-state rolling is considered, but the layers
   may be elastic or viscoeleastic.
   The solution is found by the boundary element method and the CONTACT algorithm. The Panagiotopoulos method is used throughout, as the rolling bodies are
   not symmetric. The element equations are found with the aid of an Airy function
   $H(x, z)$ which must satisfy a bipotential equation

$$H_{,xxxx} + 2H_{,xxzz} + H_{,zzzz} = 0$$

   where x is the coordinate in rolling direction and the axis of z points vertically
   into the upper cylinder. The displacements and the stresses are derivatives of H
   with respect to x and z. A Fourier transform is performed in the x-direction, and
   the bipotential equation reduces to an ordinary differential eauation in z which can
   be solved. Inverse transformation then yields the element equations, which can be
   inserted into the CONTACT algorithm.

4. The work of Gross-Thebing [8].
   This work concerns the linear theory of rolling contact when the bodies are 3-
   D and the creepage is a harmonic function of the time and friction is infinite.
   Consequently, the tangential contact force is likewise harmonic, but the creepage
   and the force are not in phase. The theory will be covered extensively in a lecture
   by Knothe during this course.

5. Kalker's thesis work [10].
   This concerns 3-D rolling bodies, quasi-identical, and with an elliptic Hertzian contact area. As distinct from the works considered up to now, it is not a boundary

element method with rectangular elements which carry a constant or continuous, piecewise linear load, but inside the contact ellipse the displacements are polynomials; it can then be shown that the tractions which vanish outside the contact ellipse are polynomials of the same degree as the displacements, but multiplied with a certain function. This is applied to find the connection between the creepage and spin on the one hand and the total force and the twisting moment on the other hand (creepage-force law) for infinite friction, or in other terms infinitesimal creepage and spin. It is also applied to find an approximate creepage-force law for finite creepage and spin, but the process involved does not always converge.

6. Nielsen's thesis work [15].
   This concerns 2-D rolling bodies, quasi-identical, with a contact area in the form of one or more strips. It is a generalisation of Kalker's work, see 5. The importance of the work lies in the applications: apart from Carter's work and an elementary generalisation of it Nielsen treats corrugation (among other things the growth or diminishing of a corrugation field by abrasive wear), velocity dependent coefficient of friction, asperity fields (in which we have multiple contacts), and non-steady 2-D contact.

7. The strip theory (Haines/Ollerton [9]).
   The work of Carter [6] and Nielsen [?] may also be applied as a strip theory, in which 2-D solutions are placed beside one another to form a 3-D approximate solution. The idea to do this is due to Haines and Ollerton, who did experiments to verify their theory. The work of Nielsen and Haines/ Ollerton is confined to longitudinal creepage. Kalker [10] extended the work of Haines and Ollerton to lateral creepage and spin. Notably the spin solution is approximate. Note that the simplified theory, treated in Section 3, is also a strip theory.

8. Kalker and Piotrowski [14] investigated the case that the friction coefficient has two values, *viz.* $f = f_{kin}$ when there is slip, and $f = f_{stat}$ when there is no slip. Fastsim was used. The effect on the total tangential force appeared to be small ($f = f_{kin}$).

References of Section 6

[1] N. AHMADI (1982), *Non-Hertzian normal and tangential loading of elastic bodies in contact.* Ph.D Thesis, Appendix C2. Northwestern University, Evanston IL, USA.

[2] R.H. BENTALL, K.L. JOHNSON (1967), *Slip in the rolling contact of two dissimilar elastic cylinders.* International Journal of Mechanical Sciences 9, p. 389-404.

[3] R.H. BENTALL, K.L. JOHNSON (1968), *An elastic strip in plane rolling contact.* International Journal of Mechanical Sciences 10, p. 637.

[4] G.F.M. BRAAT, J.J. KALKER (1993), *Theoretical and experimental analysis of the rolling contact between two cylinders coated with multilayered viscoelastic rub-*

ber, in: A.H. Aliabadi and C.A. Brebbia (Eds.) Contact Mechanics, Computational Techniques; Contact Mechanics Publications, pp. 119-126.

[5] H. BUFLER (1959), *Zur Theorie der rollenden Reibung.* Ingenieur Archiv **27**, p. 137.

[6] F.C. CARTER (1926), *On the action of a locomotive driving wheel.* Proceedings of the Royal Society of London, **A112**, p. 151-157.

[7] H. FROMM (1927), *Berechnung des Schlupfes beim Rollen deformierbarer Scheiben.* Zeitschrift für angewandte Mathematik und Mechanik **7**, p. 27-58.

[8] A. GROSS-THEBING (1992), *Lineare Modellierung des instationären Rollkontaktes von Rad und Schiene.* Fortschritt Berichte VDI, Reihe 12, Nr. 199, VDI-Verlag, Düsseldorf.

[9] D.J. HAINES, E. OLLERTON (1963), *Contact stresses in flat elliptical contact surfaces which support radial and shearing forces during rolling.* Proceedings of the Institute of Mechanical Engineering Science, **179**, Part 3.

[10] J.J. KALKER (1967), *On the rolling contact of two elastic bodies in the presence of dry friction.* Thesis Delft.

[11] J.J. KALKER (1971), *A minimum principle of the law of dry friction with application to elastic cylinders in rolling contact.* Journal of Applied Mechanics, **38**, p. 875-880, 881-887.

[12] J.J. KALKER (1990), *Three-dimensional elastic bodies in rolling contact.* Kluwer Academic Publishers (Dordrecht) 314+XXVI pp.

[13] J.J. KALKER (1991), *Viscoelastic multilayered cylinders rolling with dry friction.* Journal of Applied Mechanics, **58** p. 666-679.

[14] J.J. KALKER, J. PIOTROWSKI (1989), *Some new results in rolling contact.* Vihicle System Dynamics **18**, p. 223-242.

[15] J.B. NIELSEN (1998), *New developments in the theory of wheel/rail contact mechanics.* Thesis Lyngby IMM-PHD-1998-51.

[16] D.A. SPENCE (1975), *The Hertz problem with finite friction.* Journal of Elasticity, **5**, p. 297-319.

FINITE ELEMENT METHODS FOR ROLLING CONTACT

P. Wriggers
University of Hannover, Hannover, Germany

Abstract

This contribution is concerned with finite-element-formulation of rolling contact problems. For this, first the theoretical background of continuum mechanics and contact kinematics is given for steady and non-steady rolling processes. This includes remarks on the implementation of time-dependent materials.

Next the basic finite element formulation for large deformation processes is presented for fixed and moving reference frames. The development of the discretization of contact contributions follows. Here standard approaches and new C^1-continuous contact elements are discussed for the case of frictional and frictionless contact. Examples show the performance of the different formulations and discretizations.

1   Introduction

Boundary value problems involving contact are of great importance in industrial applications in mechanical and civil engineering. In numerous technical applications one body rolls on another one. This is true for car tires, railwail wheels or rolls in several mechanical engineering constructions such as printing machines. In these cases often contact can be treated in a special way. This contribution will give discuss as well theoretical background as numerical methods for the solution of rolling contact problems. Due to the fact that often nonlinear effects have to be included we will formulate the theory in terms of finite deformations. For most industrial applications numerical methods have to be applied since the contacting bodies have complex geometries or undergo large deformations. Here we focus on the finite element method knowing that there are of course also different other possibilities, e. g. as described in the good overview in [Kal90]. Here extensive studies can be found of steady–state and transient rolling contact problems under the assumption of infinitesimal deformations of linearly elastic bodies. The classes of rolling contact problems with large deformations were analyzed by [Bat80], [OL86] or [PZ84], just to name a few.

The following introductory remarks are related to the steps which have to be followed when treating rolling contact problems within the finite element method.

1. Kinematics for rolling contact. Since many technical contact problems involve large deformations of the bodies being in contact we will formulate all contact relations for finite deformations. Then, in general, two steps have to be followed to set up the contact geometry: the search for contact and development of the local kinematical relations. Within the global contact search one has to find where the contact takes place and then locally the contact constraint equations have to be formulated for the situation of rolling contact.

There are basically two ways to describe the kinematics of bodies which are rolling. These are the classical Lagrangian description of the motion of solids and the reference to a configuration which is rotating with the body. The second possibility is also named arbitrary Lagrangian–Eulerian description (ALE). Both kinematical descriptions have advantages and disadvantages. Thus it depends on the problem at hand which one has to be selected.

The ALE formulation is very useful when the axis of rotation does not change and when the motion is stationary. Then it makes sense to introduce a rotating reference frame. Technical examples are rolling of car tires with constant velocity or wheels of a train, see Fig. 1 a. In this case one obtains a steady state solution with respect to the rotating reference frame and hence time integration is not needed. This simplifies the analysis considerably.

In case that one has to investigate e. g. the motion of a tire which comes into contact with a curb then a total Lagrangian description seems to be preferable, see Fig. 1 b. This is due to the fact that this process is not stationary. Here a full nonlinear

Figure 1: a) Rolling on plane   b) Rolling against obstacle.

2. Constitutive equations for contact interfaces. Due to the precision which is needed to resolve the mechanical behaviour in the contact interface, different approaches have been applied in the literature to model the mechanical behaviour in contact area.

Two main lines can be followed within the finite element method to impose contact conditions in normal direction. These are the formulation of the non–penetration condition as a purely geometrical constraint and the development of constitutive laws for the micro mechanical approach within the contact area. In this contribution we will restrict ourselves to the formulation of contact where the penalty method is applied to impose the inequality constraints.

For the tangential stick/slip motion we apply here as a constitutive relation for the frictional interface Coulomb's law. This model is the most simplest and hence is widely used in practical applications. However often Coulomb's law is not sufficient to model experimentally observed responses. For more discussion on this topic see e.g. [OM86] or [Wri95].

3. Weak form of contact contributions and overall solution strategies. The weak formulation of contact problems leads to variational inequalities, see [DL76] since contact conditions are represented as inequality constraints. Different possibilities exist for the numerical solution of these problems. Among them are the so called active set strategies which are applied in combination with Lagrangian multiplier or penalty techniques, see e.g. the text books of [Ber84] or [Lue84]; these methods are well known in optimization theory.

Most standard finite element codes which are able to handle contact problems use either the penalty or the Lagrangian multiplier method, for an overview and the mathematical framework, see e.g. [KO88]. The methods are designed to fulfill the constraint equations in normal direction in the contact interface. For the tangential part we need in general constitutive relations; associated techniques will be discussed later. A combination of the penalty and the Lagrangian multiplier technique leads to the so called augmented Lagrangian methods which try to combine the merits of both approaches. A general discussion of these techniques can be found in [GLT84] and with special attention also to inequality constraints in [Ber84]. However this technique requires an algorithmic treatment, the Uzawa method, which increases the total number of iterations. For applications of augmented Lagrangian techniques to contact problems within the finite element method, see e. g. [CA88], [SL92] or [ZWS95].

4. Discretization of the contact surfaces. When the discretization of contact surfaces is concerned one has to distinguish between the contact of two deformable bodies or the contact of a deformable body with a rigid obstacle. On a first glance it seems that the latter case is simply a special case of the first problem, which is true. But due to the fact that the surface description of a rigid obstacle can be given once and for all by the correct geometrical model this knowledge can be used within the discretization process. [HK90] have developed a formulation based on CAD–surfaces, [WP92] considered so called superquadrics to specify the geometry of contacting objects and [WI93] formulated the contact problem with splines. Especially in the case that non–steady rolling contact problems are considered there is need for a smooth interpolation as well for the deformed body as for the description of the rigid surface to avoid numerical noise stemming from a non-smooth interpolation. Hence special interpolation techniques which have C^1–continuity on the boundary are needed. These are just now under development, see e. g. [Pie97], [PC97], [TW99], [WKO00] and [WKOK99].

In case of the steady rolling a discretization can be performed in the standard way using all interpolation schemes known for large deformation contact including low order schemes. The most frequently used discretization is the so called node–to–segment approach. Here arbitrary sliding of a node over the entire contact area is allowed. Early implementations can be found in [HTK77] which have been developed for more and more general cases, [HGB85], [BC85] and [WVS90]. Now some finite element codes include also self–contact, see [HSS92].

5. Algorithms for the integration of constitutive equations in the contact area. Algorithms have only to be devloped for frictional contact since normal contact is governed by the constraints and hence no constitutive equation can be formulated for normal contact. for the frictional slip one has to solve an evolution equation which needs special algorithms. In early finite element applications often so–called "trial–and-error" algorithms have been applied which might not converge in some cases. More reliable methods are provided by the mathematical programming approach, [Kla86]. Another way which is now becoming more and more standard for numerical simulations involving friction is related to the possibility to recast the frictional interface laws in terms of non–associated plasticity. First formulations and applications in finite element analysis are found in [Fre76]. A theoretical basis was also provided by [MM78]. The major break through in terms of convergence behaviour and reliability of the solution algorithms came with the application of the return mapping schemes to frictional problems. Its application can be found in [Wri87] or [Gia89] for geometrically linear problems. This approach provides the possibility to develop algorithmic tangent

matrices which are needed to achieve quadratic convergence within Newton–type iterative schemes. Due to the non–associativity of the frictional slip these matrices are
non–symmetrical. For the case of large deformations associated formulations have been
developed in [JT88] for a regularized Coulomb friction law and in [WVS90] for different
frictional laws formulated in terms of non–associated plasticity.

6. Contact search algorithms. The search for the active set of contact constraints is not trivial in case of large deformations since a surface point of a body may
contact any portion of the surface of another body. Such point can even come into
contact with a part of the surface of its own body. Thus the search for the correct
contact location needs, depending on the problem, eventually considerable effort. An
implementation where each node of a surface is checked against each element surface in
the mesh is too exhaustive and thus computationally inefficient and refined algortihms
have to be constructed. This especially true when the contact of more then two bodies
has to be considered or when self–contact is possible. However this topic will be not
addressed here in detail.

7. Adaptive methods for rolling contact problems. Since numerical methods
for contact problems yield approximate solutions it is necessary to control the errors
inherited in the method. During the last ten years research activities have been focused
on adaptive techniques providing automatically a numerical model which is accurate
and reliable. The objective of adaptive techniques is to obtain a mesh which is optimal
in the sense that the computational costs involved are minimal under the constraint
that the error in the finite element solution is beyond a certain limit. Since the computational effort can be linked to the number of unknowns of the finite element mesh the
task is to find a mesh with minimum number of unknowns or nodes for a given error
tolerance. In general, adaptive methods rely on error indicators and error estimators
which can be computed *a priori* or *a posteriori*. For an overview over different techniques, see e.g. [Joh87] and references therein. Based on the error distribution a new
partially refined mesh can be constructed which yields a better approximate solution.
To obtain an optimal mesh in the sense of an equal solution quality it is desirable to
design the mesh such that the error contributions of the elements are equidistributed
over the mesh. During the last years a growing number of papers has been devoted to
this topic and applied to problems of solid and fluid mechanics, see e.g. [ZT89].

For rolling contact problems error indicators have been developed by [Nac95] and
[HW00]. The first author applied a special L_1–norm to measure the error in the contact
stresses. In [HW00] a residual based error estimator has been derived for linear elastic
bodies on the basis of the work by [CSW99] and then applied as an indicator to frictional
rolling contact problems with finite deformations.

2   Continuum solid mechanics and weak forms

The deformation of a solids is generally described the kinematical relations, the equations of balance and constitutive equations. This section summarizes the main equations which govern the deformation of solids. For a detailed treatment of this subject the reader should consult the literature, e. g. the standard book of [Eri67], [Mal69], [TN65], [TT60] or [Ogd84].

Here we will also state the general kinematical relations when another arbitrary reference configuration is introduced and specialize this for the case of rolling contact.

2.1   Kinematics

2.1.1   Motion and deformation gradient

The motion of body B is a temporally parametric series of configurations $\varphi_t\colon B \to \mathbb{E}^3$. For the position of the particle X at time $t \in \mathbb{R}^+$ we have

$$\mathbf{x} = \varphi_t\left(X\right) = \varphi\left(X, t\right). \tag{1}$$

This equation describes a curve in $\mathbb{E}^3$ for the particle X.
$\mathbf{X} = \varphi_0\left(X\right)$ defines the reference configuration of body B, where $\mathbf{X}$ is the position of particle X in this configuration. With (1) we have

$$\mathbf{x} = \varphi\left(\varphi_0^{-1}(\mathbf{X}), t\right). \tag{2}$$

For practical applications we do not need to differenciate between $\mathbf{X}$ and X. This simplifies the notation and we can write (2) as

$$\mathbf{x} = \varphi\left(\mathbf{X}, t\right), \tag{3}$$

where $\mathbf{X}$ depicts the position of particle X in the reference configuration B. With this, the positions $\mathbf{x}$ and $\mathbf{X}$ are described as vectors in $\mathbb{E}^3$ with respect to the origin $\mathbf{O}$, as shown in Fig. 2.

The equations of mechanics of continua can be formulated with respect to the deformed or the undeformed configuration of a body B. From the theoretical standpoint there is no difference whether the equations are referred to the current– or the reference configuration of the body. However one should consider implications due to physical modeling as in plasticity. When formulating numerical methods for continua considerable differences in efficiency can occur when either the equations are related to the spatial or the reference configuration. Thus we will define strain measures with respect to both configurations. Within this discussion we denote by small letters tensors which are referred to the current configuration $\varphi(B)$ and use capital letters for the reference configuration B.

To describe the deformation process locally we introduce the deformation gradient $\mathbf{F}$ which maps tangent vectors of the reference configuration to tangent vectors in the

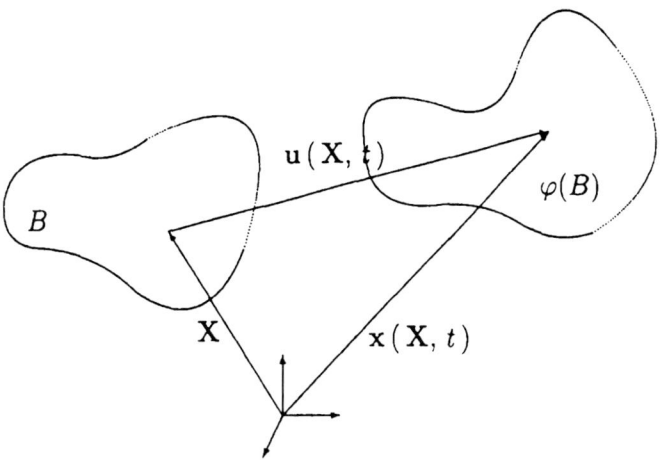

Figure 2: Configurations of body B.

spatial configuration. It is a tensor which associates to a material line element $d\mathbf{X}$ in B the line element $d\mathbf{x}$ in $\varphi(B)$:

$$d\mathbf{x} = \mathbf{F}\, d\mathbf{X} \tag{4}$$

The components of the deformation gradient follow from the direct notation $\mathbf{F} = \partial\mathbf{x}/\partial\mathbf{X}$ as partial derivatives $\partial x_i/\partial X_A = x_{i,A}$. With (3) we obtain

$$\mathbf{F} = \operatorname{Grad}\varphi(\mathbf{X}, t) = F_{iA}\,\mathbf{e}_i \otimes \mathbf{E}_A = \frac{\partial x_i}{\partial X_A}\,\mathbf{e}_i \otimes \mathbf{E}_A\,. \tag{5}$$

Since the gradient (5) is a linear operator the local transformation (4) is also linear. To preserve the continuous structure in B during the deformation the mapping (4) has to be one–to–one, i. e. $\mathbf{F}$ cannot be singular. This is equivalent to the condition

$$J = \det \mathbf{F} \neq 0\,, \tag{6}$$

where J defines the JACOBIAN determinant. Furthermore, to exclude self–penetration of the body, J has to be greater than 0. Thus its inverse exists which is denoted by $\mathbf{F}^{-1}$.

Once the deformation gradient $\mathbf{F}$ is known transformations of area and volume elements between B and $\varphi(B)$ can be derived. The transformation of area elements between B and $\varphi(B)$ is given by the formula due to NANSON (see e. g. [Ogd84], pp. 88)

$$d\mathbf{a} = \mathbf{n}\, da = J\,\mathbf{F}^{-T}\,\mathbf{N}\, dA = J\,\mathbf{F}^{-T}\, d\mathbf{A}\,. \tag{7}$$

In this equation $\mathbf{n}$ is the normal to the surface of $\varphi(B)$ and $\mathbf{N}$ denoted the normal to the surface of B. J is the JACOBI determinant defined in (6) and da respectively

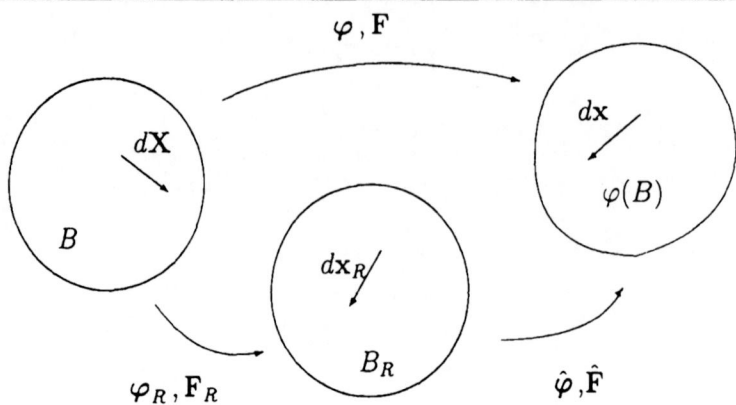

Figure 3: Reference frames to describe the deformation.

dA are the surface elements in the associated configuration. For the transformation of volume elements from the reference to the spatial configuration we have

$$dv = J \, dV \, . \tag{8}$$

With the introduction of a displacement vector $\mathbf{u}(\mathbf{X}, t)$ as the difference of position vectors of a point in the reference and the current configuration

$$\mathbf{u}(\mathbf{X}, t) = \boldsymbol{\varphi}(\mathbf{X}, t) - \mathbf{X} \tag{9}$$

we can write for the deformation gradient (5)

$$\mathbf{F} = \text{Grad}\,[\,\mathbf{X} + \mathbf{u}(\mathbf{X}, t)\,] = \mathbf{1} + \text{Grad}\,\mathbf{u} = \mathbf{1} + \mathbf{H} \, , \tag{10}$$

where $\mathbf{H} = \text{Grad}\,\mathbf{u}$ is the displacement gradient with respect to $\mathbf{X}$.

2.1.2 Special reference frames for rolling contact

In case of rolling contact it can be useful not to use the LAGRANGIAN description as given before, but to use a special rotating reference frame. From the continuum point of view, we will distinguish between three different configurations of the body, see Fig. 3.

B denotes the initial configuration of the body. With $\varphi(B)$ we describe the deformed configuration. Finally, $\varphi_R(B)$ is the rotating reference configuration which rotates with a given angular velocity ω_R. Thus we have for the deformation map

$$\varphi = \hat{\varphi} \otimes \varphi_R \tag{11}$$

where $\hat{\varphi}$ is the deformation relative to the rotation frame. For the line elements — needed later to define the strain measures — we have

$$dx = \mathbf{F}\, d\mathbf{X} = \hat{\mathbf{F}}\, d\mathbf{x}_R$$
$$d\mathbf{x}_R = \mathbf{F}_R\, d\mathbf{X} \tag{12}$$

with $\mathbf{F} = \dfrac{\partial \mathbf{x}}{\partial \mathbf{X}}$, $\hat{\mathbf{F}} = \dfrac{\partial \mathbf{x}}{\partial \mathbf{x}_R}$, $\mathbf{F}_R = \dfrac{\partial \mathbf{x}_R}{\partial \mathbf{X}}$. Hence

$$dx = \hat{\mathbf{F}}\, \mathbf{F}_R\, d\mathbf{X} \tag{13}$$

If the reference configuration $\varphi_R(B)$ is given by a pure spinning rotation then we can write

$$\mathbf{x}_R = \mathbf{R}\,\mathbf{X} \tag{14}$$

with the orthogonal Matrix $\mathbf{R}$ describing the rotation. Thus

$$dx = \hat{\mathbf{F}}\,\mathbf{R}\, d\mathbf{X} \tag{15}$$

Now the velocity is given by the material time derivative $\mathbf{v} = \dfrac{d\varphi}{dt}$. Using the special decomposition of the motion into the rotating frame φ_R and the deformation relative to this rotating reference configuration $\hat{\varphi}$, see (11) we obtain with $\mathbf{x} = \varphi(\mathbf{X},t) = \hat{\varphi}(\varphi_R(\mathbf{X},t),t)$ by the chain rule

$$\mathbf{v} = \frac{d\varphi}{dt} = \frac{\partial \hat{\varphi}}{\partial t} + \frac{\partial \hat{\varphi}}{\partial \varphi_R}\frac{d\varphi_R}{dt} \tag{16}$$

With (12) we can rewrite this equation and obtain

$$\mathbf{v} = \frac{d\varphi}{dt} = \frac{\partial \hat{\varphi}}{\partial t} + \hat{\mathbf{F}}\, \mathbf{v}_R \tag{17}$$

where the velocity $\mathbf{v}_R$ is defined by $\mathbf{v}_R = \dfrac{d\varphi_R}{dt}$. For the computation of the acceleration the chain rule has to be applied again to (16) which yields

$$\mathbf{a} = \frac{d^2\varphi}{dt^2} = \frac{\partial^2 \hat{\varphi}}{\partial t^2} + 2\frac{\partial^2 \hat{\varphi}}{\partial \varphi_R \partial t}\frac{d\varphi_R}{dt} + \frac{\partial^2 \hat{\varphi}}{\partial \varphi_R^2}\left[\frac{d\varphi_R}{dt}\right]^2 + \frac{\partial \hat{\varphi}}{\partial \varphi_R}\frac{d^2\varphi_R}{dt^2} \tag{18}$$

This equation can be shortened by using (12) which leads to

$$\mathbf{a} = \frac{\partial^2 \hat{\varphi}}{\partial t^2} + 2\dot{\hat{\mathbf{F}}}\,\mathbf{v}_R + \frac{\partial}{\partial \varphi_R}\hat{\mathbf{F}}\,\mathbf{v}_R^2 + \hat{\mathbf{F}}\,\mathbf{a}_R \tag{19}$$

where a_R is defined by $a_R = \dfrac{d^2\,\varphi_R}{dt^2}$. In case of a steady state spinning process $\hat{\varphi}$ is not explicitly time dependent which simplifies (16) and (18) considerably

$$v = \frac{\partial\hat{\varphi}}{\partial\varphi_R}\frac{d\varphi_R}{dt} = \hat{F}\,v_R \tag{20}$$

$$a = \frac{\partial^2\hat{\varphi}}{\partial\varphi_R^2}\left[\frac{d\varphi_R}{dt}\right]^2 + \frac{\partial\hat{\varphi}}{\partial\varphi_R}\frac{d^2\varphi_R}{dt^2} = \frac{\partial}{\partial\varphi_R}\hat{F}\,v_R^2 + \hat{F}\,a_R$$

For this case we assume that the reference frame has a constant angular velocity ω_R around a given axis. This defines a rigid body motion of the reference frame. Hence we obtain by using (14) for the velocity v_R in the reference frame with $X = R^{-1}x_R = R^T x_R$ and $\dot{R}\,R^T = \Omega_R$

$$v_R = \frac{\partial x_R}{\partial t} + \Omega_R\,x_R \tag{21}$$

The associated acceleration a_R is then computed from

$$a_R = \frac{\partial^2 x_r}{\partial t^2} + \dot{\Omega}_R\,x_R + \Omega_R\left(\frac{\partial x_R}{\partial t} + \Omega_R\,x_R\right) \tag{22}$$

To the skew symmetric tensor Ω_R we can associate an axial vector such that

$$v_R = \frac{\partial x_R}{\partial t} + \omega_R \times x_R \tag{23}$$

The first term disappears for a constant time independent rotation. Then combining (20) and (21) yields

$$v_R = \hat{F}\,\Omega_R x_R \tag{24}$$

Furthermore equations (20) and (22) lead then to the acceleration

$$a_R = \left(\frac{\partial}{\partial\varphi_R}\hat{F}\,\Omega_R\,x_R + \hat{F}\,\Omega_R\right)\Omega_R\,x_R\,. \tag{25}$$

2.1.3 Strain measures

In this section we describe different strain measures which will be applied later. One of the most common strain measures is the right CAUCHY–GREEN tensor C, which is refered to the initial configuration B

$$C = F^T\,F\,. \tag{26}$$

With respect to section 2.1.2 we have to investigate how the strain measures are affected by the introduced rotating reference configuration. For this purpose we apply (12) which gives $F = \hat{F}\,R$. This result can be used in (26) to compute the right CAUCHY GREEN strains

$$C = (\hat{F}\,R)^T\,\hat{F}\,R = R^T\,\hat{F}^T\,\hat{F}\,R = R^T\,\hat{C}\,R\,. \tag{27}$$

Thus by the choice of the rotating reference configuration the strains, $\hat{C}$, producing stresses exclusivly stem from the motion relative to the rotating frame.

Since the right CAUCHY GREEN strains are not zero at the initial state (there we have $F = 1 \Rightarrow C = 1$) it is convenient to introduce the GREEN–LAGRANGIAN strain tensor E which is refered to the initial configuration B

$$E = \frac{1}{2}(F^T F - 1) = \frac{1}{2}(C - 1). \tag{28}$$

Another interpretation of this strain measure follows from the fact that the CAUCHY GREEN tensor expresses the square of the infinitesimal line element dx via the material line element dX by $dx \cdot dx = dX \cdot C\, dX$. Thus the strain E is the difference of the square of the line elements in B und $\varphi(B)$.

For the rotated reference frame we obtain with (27)

$$E = \frac{1}{2}(R^T \hat{C} R - 1) = R^T \frac{1}{2}(\hat{C} - 1)R = R^T \hat{E} R. \tag{29}$$

which has the same structure as (27).

2.2   Balance laws

The partial differential equations which represent the local balance laws of continuum mechanics are summarized in this section. For a detailed derivation see e. g. ([Mal69], Chap. 5).

2.2.1   Balance of mass

The balance of mass m of a body is given by the relation

$$m = \int_B \rho_0\, dV = \int_{\varphi(B)} \rho\, dV = const. \tag{30}$$

where ρ_0 is the density in the initial configuration and ρ the density in the current configuration. Within the LAGRANGIAN description of a motion we can conclude assuming sufficient smoothness that $\rho_0 = J\rho$. This equation yields a relation for the volume elements in initial and current configuration

$$dv = \frac{\rho_0}{\rho}\, dV = J\, dV. \tag{31}$$

2.2.2   Local balance of momentum and moment of momentum

The local balance of momentum with respect to a volume element in the current configuration $\varphi(B)$ can be written as

$$\mathrm{div}\,\sigma + \rho\,\bar{b} = \rho\,\dot{v}, \quad \sigma_{ik,i} + \rho\,\bar{b}_k = \rho\,\dot{v}_k. \tag{32}$$

In this equation σ denotes the CAUCHY stress tensor. $\rho\bar{b}$ defines in (32) the volume or body force (e. g. due to gravitation). $\rho\dot{v}$ is the inertia force term, which can be neglected in the case of static analysis. Furthermore we have the CAUCHY theorem, which relates the stress vector t to the surface normal vector n by

$$t = \sigma^T n, \quad t_i = \sigma_{ik} n_i.$$

(33)

This relation has been stated here in direct notation and index and notation.

The local balance of angular of momentum yields in the absense of micropolar stresses, usually the case in non magnetic materials (see e.g. [TT60]),

$$\sigma = \sigma^T, \quad \sigma_{ik} = \sigma_{ki},$$

(34)

which dictats the symmetry of the CAUCHY stress tensor.

2.2.3   Transformation to the initial configuration, different stress tensors

Equations (32) and (34) are refered to the current configuration. Often one needs a formulation of these equations in quantities which are related to the initial configuration B. For this transformation, also often named *pull back*, we define more stress tensors, which follow from the equivalence of a force which is defined in B and $\varphi(B)$

$$\int_{\partial\varphi(B)} \sigma\, n\, da = \int_{\partial B} \sigma\, J\, F^{-T} N\, dA = \int_{\partial B} P\, N\, dA,$$

(35)

This relation defines the first PIOLA–KIRCHHOFF stress tensor P. We have the transformation

$$P = J\sigma F^{-T}, \quad P_{Ak} = J\sigma_{ik}(F_{iA})^{-1}$$

(36)

between the CAUCHY and the first PIOLA–KIRCHHOFF stresses, these mean the actual stresses in terms of the area of the initial configuration. Since in equation (36) the spatial quantity σ is only multiplied on one side by F P, it is a so-called two field tensor which one base vector lies in B and the other in $\varphi(B)$. After some manipulation we can transform now the local balance of momentum (32) to the initial configuration

$$\text{DIV}\, P + \rho_0\, \bar{b} = \rho_0\, \dot{v}$$

(37)

However when using (36) in the balance of angular of momentum (34) we see that the PIOLA–KIRCHHOFF stress tensor in general in nonsymmetric: $P\, F^T = F\, P^T$.

A symmetric stress tensor which is defined with regard to the reference configuration is the second PIOLA–KIRCHHOFF stress tensor, which follows from the complete *pull back* of CAUCHY stress tensor to the reference configuration B

$$
\begin{aligned}
S &= F^{-1}P = JF^{-1}\sigma F^{-T}, & (38)\\
S_{AB} &= (F_{Ai})^{-1}P_{Bi} = J(F_{Ai})^{-1}\sigma_{ik}(F_{kB})^{-1}. & (39)
\end{aligned}
$$

S does not represent an experimentally measurable stress. However, it is an essential stress measure that plays a prominent role in the constitutive theory. It is "work conjugated" (duality paired) with the GREEN–LAGRANGIAN strain tensor (28).

Besides the CAUCHY stress tensor σ, the KIRCHHOFF stress tensor τ is often employed, which is defined as the *push forward* of the second PIOLA–KIRCHHOFF stress tensor S to the current configuration

$$\tau = \mathbf{F} \mathbf{S} \mathbf{F}^T, \quad \tau = J \sigma. \tag{40}$$

2.3 Weak form of balance of momentum, variational principles

For the solution of boundary value problems stemming from the continuum we will employ numerical methods which base on variational formulations. Thus we need associated formulations which are given below.

The principle of virtual work is an equivalent formulation of the balance of momentum which often – due to its reduced regularity requirements – is called weak form of equilibrium. Since no constitutive equations enter a priori the weak form it is valid for all problem classes including plasticity, friction or non conservative loading. The derivation of the weak form starts from the local equilibrium equation (Div $\mathbf{P} + \rho_0 \bar{\mathbf{b}} = \rho_0 \dot{\mathbf{v}}$), which is multiplied by a vector valued function $\eta = \{\eta \mid \eta = 0 \text{ on } \partial B_u\}$ – often named virtual displacement or test function. Integration over the volume of the body under consideration yields

$$\int_B \text{Div} \, \mathbf{P} \cdot \eta \, dV + \int_B \rho_0 (\bar{\mathbf{b}} - \dot{\mathbf{v}}) \cdot \eta \, dV = 0. \tag{41}$$

Partial integration of the first term and usage of the divergence theorems leads with the boundary conditions to the weak form of

$$G(\varphi, \eta) = \int_B \mathbf{P} \cdot \text{Grad} \, \eta \, dV - \int_B \rho_0 (\bar{\mathbf{b}} - \dot{\mathbf{v}}) \cdot \eta \, dV - \int_{\partial B_\sigma} \bar{\mathbf{t}} \cdot \eta \, dA = 0. \tag{42}$$

The gradient of η can also be viewed at as a virtual variation $\delta \mathbf{F}$ of the deformation gradient

$$\delta \mathbf{F} = \frac{d}{d\epsilon} [\mathbf{F}(\mathbf{x} + \epsilon \eta)] \Big|_{\epsilon=0} \tag{43}$$

In (42) we can exchange the first PIOLA–KIRCHHOFF stress tensor with $\mathbf{P} = \mathbf{F} \mathbf{S}$ by the second PIOLA–KIRCHHOFF stress tensor and rewrite (42) as

$$G(\varphi, \eta) = \int_B \mathbf{F} \mathbf{S} \cdot \text{Grad} \, \eta \, dV - \int_B \rho_0 (\bar{\mathbf{b}} - \dot{\mathbf{v}}) \cdot \eta \, dV - \int_{\partial B_\sigma} \bar{\mathbf{t}} \cdot \eta \, dA = 0. \tag{44}$$

The first term in (44) denotes the virtual internal work or the stress divergence and the last two terms contain the virtual work of the external forces and the virtual work of the inertia term.

Again transformation of equation (44) with respect to the rotating reference configuration B_R yields

$$\int_B \mathbf{F}\mathbf{S} \cdot \operatorname{Grad} \eta \, dV \;=\; \int_{B_R} \hat{\mathbf{F}} \mathbf{R}\mathbf{S} \cdot \widehat{\operatorname{Grad} \eta \, \mathbf{R}} \, dV_R$$

$$\int_{B_R} \hat{\mathbf{F}} \hat{\mathbf{S}} \cdot \widehat{\operatorname{Grad} \boldsymbol{\eta}} \, dV_R \qquad (45)$$

where $\hat{\mathbf{S}}$ has been defined according to $\hat{\mathbf{S}} = \mathbf{R}\mathbf{S}\mathbf{R}^T$. Furthermore the inertia term
yields for a constant spinning motion

$$\int_B \rho_0 \, \dot{\mathbf{v}} \cdot \boldsymbol{\eta} \, dV \;=\; \int_{B_R} \rho_R \, \Omega_R^2 \, \mathbf{x}_R \cdot \boldsymbol{\eta} \, dV_R$$

$$- \int_{B_R} \rho_R \, (\hat{\mathbf{F}} \, \Omega_R \, \mathbf{x}_R) \cdot (\widehat{\operatorname{Grad} \eta \, \Omega_R \mathbf{x}_R}) \, dV_R \qquad (46)$$

where the first term denotes the body forces due to spinning. The second term is
associated with the inertia forces due to constant spinning.

Using (45) and (46) in (44) we obtain the weak form with respect to the rotating
frame

$$G(\hat{\boldsymbol{\varphi}}, \boldsymbol{\eta}) \;=\; - \int_{B_R} \rho_R \, (\hat{\mathbf{F}} \, \Omega_R \, \mathbf{x}_R) \cdot (\widehat{\operatorname{Grad} \eta \, \Omega_R \mathbf{x}_R}) \, dV_R + \int_{B_R} \hat{\mathbf{F}} \hat{\mathbf{S}} \cdot \widehat{\operatorname{Grad} \eta} \, dV_R$$

$$- \int_{B_R} \rho_R \, (\bar{\mathbf{b}} - \Omega_R^2 \, \mathbf{x}_R) \cdot \boldsymbol{\eta} \, dV_R - \int_{\partial B_{\sigma R}} \bar{\mathbf{t}} \cdot \boldsymbol{\eta} \, dA_R = 0 \,. \qquad (47)$$

2.4   Constitutive Equations

Since contact takes place at the the interface between bodies, the constitutive laws for
the bodies coming into contact which describe the material behaviour within the bodies
can be arbitrary and do not affect the main formulation of contact problems. However
it is clear that the physical properties of the surfaces of the bodies are influenced by
the general constitutive behaviour. Thus, to include a nonlinear constitutive equation
valid for large deformations, we discuss finite elasticity. Of course we can consider more
complicated constitutive relations which can also be of inelastic nature, but this is not
the aim of this book.

2.4.1   Hyperelastic response function

Throughout this section we briefly discuss hyperelastic constitutive relations, for more
detailed information, see e. g. [Ogd84]. These can be applied to describe the constitutive behaviour of e. g. rubber or foams. In case of small deformations these
constitutive equations reduce to the classical HOOKE's law of linear elasticity.

The constitutive equation or response function for the second PIOLA–KIRCHHOFF stress is in case of a hyperelastic material given by the partial derivative of the strain energy W function with respect to the right CAUCHY GREEN tensor, see e. g. [Ogd84],

$$S = 2 \frac{\partial W(\mathbf{C}, \mathbf{X})}{\partial \mathbf{C}}. \tag{48}$$

This response function represents a constitutive relation which fulfills the requirements of frame indifference and hence is objective. In case of a homogeneous material the strain energy W does not depend on $\mathbf{X}$. Here we will restricht ourselves to homogeneous isotropic materials. Then the strain energy function can be specialized and is represented by an isotropic tensor function

$$W(\mathbf{C}) = W(I_C, II_C, III_C). \tag{49}$$

The second PIOLA–KIRCHHOFF stresses follow now with (48) by using the chain rule

$$S = 2[(\frac{\partial W}{\partial I_C} + I_C \frac{\partial W}{\partial II_C}) \mathbf{1} - \frac{\partial W}{\partial II_C} \mathbf{C} + III_C \frac{\partial W}{\partial III_C} \mathbf{C}^{-1}]. \tag{50}$$

Within this equation the following results for the derivative of invariants with respect to tensors have been used

$$\frac{\partial I_C}{\partial \mathbf{C}} = \mathbf{1}, \quad \frac{\partial II_C}{\partial \mathbf{C}} = I_C \mathbf{1} - \mathbf{C}, \quad \frac{\partial III_C}{\partial \mathbf{C}} = III_C \mathbf{C}^{-1}. \tag{51}$$

For the special choice of the strain energy function W we obtain the simplest possible response function which is known as compressible NEO–HOOKIAN material. We choose

$$W(I_C, J) = g(J) + \frac{1}{2}\mu(I_C - 3). \tag{52}$$

For compressible materials function $g(J)$ in (52) has to be convex. Furthermore the following growth conditions must hold

$$\lim_{J \to +\infty} W \to \infty \quad \text{and} \quad \lim_{J \to 0} W \to \infty. \tag{53}$$

These conditions are equivalent with the conditions that the stress for a deformed body which volume goes to zero has to go to $-\infty$ and for a deformed body which volume goes to $+\infty$ also the stress has to go to $+\infty$. These growth conditions are fulfilled when the compressible part $g(J)$ is chosen, according to, [Cia88], as

$$g(J) = c(J^2 - 1) - d \ln J - \mu \ln, J \quad \text{with} \quad c > 0, d > 0 \tag{54}$$

The response function of the NEO–HOOKIAN material (52) follows now with (50) and yields for the second PIOLA–KIRCHHOFF stress tensor

$$S = \frac{\Lambda}{2}(J^2 - 1)\mathbf{C}^{-1} + \mu(\mathbf{1} - \mathbf{C}^{-1}) \tag{55}$$

where the constants c and d have been chosen as $c = \Lambda/4$ and $d = \Lambda/2$. The material constants Λ and μ are the LAMÉ constants which have to be determined by experiments.

Equation (55) can now be transformed to the rotated frame which will be applied for rolling contact. With $\hat{S} = R S R^T$, $C = R^T \hat{C} R$ and $C^{-1} = R^T \hat{C}^{-1} R$ we can write the (55) directly in terms of the stress $\hat{S}$ and the strain $\hat{C}$ as follows

$$\hat{S} = \frac{\Lambda}{2}(J^2 - 1)\hat{C}^{-1} + \mu(1 - \hat{C}^{-1}) \tag{56}$$

since the rotation tensor multiplies the tensors on both sides of (55) in the same way and thus cancels out.

2.4.2 Viscoelastic response function

Car tires, which are often investigated using rolling contact formulations, are made of rubber which does not only respond to deformations like a hyperelastic material but also shows inelastic behaviour which can be described by a viscoelastic material model. Here again one has to investigate whether a description using the rotated frame does change the constitutive equation or not, see above. Using a standard viscoelastic model, see [Chr80], we obtain for the second PIOLA–KIRCHHOFF stresses

$$S(t) = \bar{S}^e[E(t)] + \nu \int_{-\infty}^{t} e^{-\frac{t-s}{\tau}} \frac{\partial E}{\partial s} ds \tag{57}$$

Here t denotes the time, E is the GREEN strain tensor. $\bar{S}^e$ is the hyperelastic response function, see last section, and ν and τ are constitutive parameters. All quantities in (57) are referred to the initial configuration B. Transformation with respect to the rotating reference configuration yields with $\hat{S} = R S R^T$ and $E = R^T \hat{E} R$

$$\hat{S}(t) = \bar{S}^e[\hat{E}(t)] + \nu \int_{-\infty}^{t} e^{-\frac{t-s}{\tau}} \frac{d}{ds}[R(t) R^T(s)\hat{E}(s)R(s) R^T(t)] ds \tag{58}$$

It seems that in this model the rotation R depends on the complete motion and hence its history has to be known during the whole simulation. However the rotation only appears as $R(t) R^T(s)$ which means only the relative rotation between the two times t and s is involved. Since we assume steady state rotation the relative rotation is explicitly given by $e^{\Omega(t-s)}$. For a more detailed treatment of viscoelastic constitutive equations for rolling motions, see [TR94] or [GM99].

2.4.3 Incremental constitutive tensor

To derive the incremental constitutive tensor we have to compute the rate of the response function (48). Thus, the response function must be differentiated with respect to time. This will be here only performed for the hyperelastic constitutive equations. It leads to

$$\dot{S} = 2\frac{\partial^2 W}{\partial C\,\partial C}[\dot{C}],\tag{59}$$

and hence to an incremental relation between the rate of the second PIOLA–KIRCHHOFF stress tensor S and the right CAUCHY–GREEN tensor C. With the definition of a fourth order incremental constitutive tensor

$$\mathbb{C} = 4\frac{\partial^2 W}{\partial C\,\partial C}, \qquad \mathbb{C}_{ABCD} = 4\frac{\partial^2 W}{\partial C_{AB}\,\partial C_{CD}}\tag{60}$$

we obtain for (59)

$$\dot{S} = \mathbb{C}[\tfrac{1}{2}\dot{C}], \qquad \dot{S}_{AB} = \mathbb{C}_{ABCD}\tfrac{1}{2}\dot{C}_{CD}.\tag{61}$$

Thereafter we derive the incremental constitutive tensor for the constitutive equations (55). The response function (55) depends on the deformation via the inverse of the right CAUCHY–GREEN tensor and its determinant: $J = \sqrt{III_C}$. Thus for the computation of $\mathbb{C}$ using (60) the derivatives of J and C^{-1} with respect to C have to be computed.

The derivative of the JACOBIAN is given by

$$\frac{\partial J}{\partial C} = \frac{1}{2}J\,C^{-1}\tag{62}$$

The derivative of C^{-1} follows from relation, $\frac{\partial}{\partial C_{CD}}[C_{AM}C_{MB}^{-1}] = 0$, as

$$\frac{\partial C_{AB}^{-1}}{\partial C_{CD}} = -C_{AC}^{-1}C_{BD}^{-1}\tag{63}$$

Since C is symmetric we only need the symmetrical part of (63) and introduce the fourth order tensor $\mathbb{I}_{C^{-1}}$ which has the index notation

$$\mathbb{I}_{C^{-1}\,ABCD} = \frac{1}{2}\left(C_{AC}^{-1}C_{BD}^{-1} + C_{AD}^{-1}C_{BC}^{-1}\right)\tag{64}$$

With this preliminaries the constitutive tensor can be derived. After some algebraic manipulations we obtain

$$\begin{aligned}
\mathbb{C} &= \Lambda J^2\,C^{-1}\otimes C^{-1} + [2\mu - \Lambda(J^2-1)]\,\mathbb{I}_{C^{-1}},\\
\mathbb{C}_{ABCD} &= \Lambda J^2\,C_{AB}^{-1}C_{CD}^{-1} + [2\mu - \Lambda(J^2-1)]\,\mathbb{I}_{C^{-1}\,ABCD}.
\end{aligned}\tag{65}$$

Since the response function for the constitutive equation in the rotating frame, B_R, has the same form as the response function with respect to the initial configuration B we just have to exchange C by $\hat{C}$ in (65) to compute the associated incremental constitutive tensor in B_R.

2.5   Linearizations

Solution of nonlinear boundary value problems can in general only be obtained by approximate methods. Many of these methods, like the finite element method, base on the variational formulation of the field equations which is e. g. given by the weak form or principle of virtual work, hence equations (42) or (47) provide the starting point for a numerical method. For the solution of these nonlinear equations an iterative scheme has to be developed since the discretization of the weak form results in a nonlinear system of algebraic equations. Within many possible iterative algorithms NEWTON's method has been proven to be often the most efficient scheme since it exhibits quadratic convergence near the solution point. Within NEWTON's method a correction of the solution is achieved by the TAYLOR series expansion of the nonlinear equation set at a point where the approximated solution is already known. The necessary linearization can be computed with the aid of the directional derivative.

The linearization of the weak form is first derived with respect to the initial configuration which is basd on equation (42). We assume that the linearization is computed at a deformation state $\bar{\varphi}$ at which the body under investigation is in equilibrium. The linear part of the weak form is

$$L\,[\,G\,]_{\varphi=\bar{\varphi}} = G\,(\bar{\varphi}, \eta) + DG\,(\bar{\varphi}, \eta)\cdot\Delta\mathbf{u} \tag{66}$$

$G(\bar{\varphi}, \eta)$ is equal to (47), only φ is exchanged by the state $\bar{\varphi}$. The directional derivative of G, needed to compute the linearization, has only to be applied to the first term in (47) when the assumption of conservative loading is made

$$DG\,(\bar{\varphi}, \eta)\cdot\Delta\mathbf{u} = \int_{B} [DP(\bar{\varphi})\cdot\Delta\mathbf{u}]\cdot\mathrm{Grad}\,\eta\,dV\,, \tag{67}$$

all other terms do not depend on the deformation. The linearization of the first PIOLA–KIRCHHOFF stress tensor yields with $\mathbf{P} = \mathbf{F}\,\mathbf{S}$

$$DG\,(\bar{\varphi}, \eta)\cdot\Delta\mathbf{u} = \int_{B} \{\,\mathrm{Grad}\,\Delta\mathbf{u}\,\bar{\mathbf{S}} + \bar{\mathbf{F}}\,[DS(\bar{\varphi})\cdot\Delta\mathbf{u}]\,\}\cdot\mathrm{Grad}\,\eta\,dV\,. \tag{68}$$

Quantities labeled with a bar have to evaluated at $\bar{\varphi}$. For the linearization of the second PIOLA–KIRCHHOFF stresses equation (61) can be used, since linearization and time derivative are equivalent

$$DS(\bar{\varphi})\cdot\Delta\mathbf{u} = \bar{\mathbb{C}}\,[\,\Delta\bar{\mathbf{E}}\,]\,, \tag{69}$$

where the last term is the linearization of the GREEN–LAGRANGIAN strain tensor $\mathbf{E}$ at $\bar{\varphi}$

$$\Delta\bar{\mathbf{E}} = \frac{1}{2}\,(\bar{\mathbf{F}}^{T}\,\mathrm{Grad}\Delta\mathbf{u} + \mathrm{Grad}\Delta\mathbf{u}\,\bar{\mathbf{F}}) \tag{70}$$

which has the same structure as the variation $\delta\bar{E} = \frac{1}{2}(\bar{F}^T \operatorname{Grad} \eta + \operatorname{Grad}^T \eta \, \bar{F})$. The incremental elasticity tensor $\mathbb{C}$ which is evaluated with respect to the initial configuration is given with (60) by

$$\bar{\mathbb{C}} = 4 \left. \frac{\partial^2 W}{\partial C \, \partial C} \right|_{\varphi = \bar{\varphi}} \tag{71}$$

at $\bar{\varphi}$.

Inserting equation (71) in (68) completes the linearization

$$DG(\bar{\varphi}, \eta) \cdot \Delta\mathbf{u} = \int_B \{ \operatorname{Grad} \Delta\mathbf{u} \, \bar{S} + \bar{F}\bar{\mathbb{C}}[\Delta\bar{E}] \} \cdot \operatorname{Grad} \eta \, dV . \tag{72}$$

Note that also $\bar{\mathbb{C}}$ has to be computed at $\bar{\varphi}$. By making use of the trace operation and by considering symmetry of $\bar{\mathbb{C}}$ a compact form of (72) can be obtained

$$DG(\bar{\varphi}, \eta) \cdot \Delta\mathbf{u} = \int_B \{ \operatorname{Grad} \Delta\mathbf{u} \, \bar{S} \cdot \operatorname{Grad} \eta + \delta\bar{E} \cdot \bar{\mathbb{C}}[\Delta\bar{E}] \} \, dV \tag{73}$$

Note the symmetry of the linearization with respect to η and $\Delta\mathbf{u}$. The first term in (73) is the so called geometrical matrix or initial stress matrix. The second term contains the initial deformations which occur in the incremental constitutive tensor $\bar{\mathbb{C}}$.

The linearization of the weak form, defined in quantities of the rotated frame follows from (47). Since $\mathbf{R}$ and Ω_R are constant we obtain the same result as in (73. However there is one extra term stemming from the inertia forces, the first term in (47). The linearization of this term is however trivial since it is linear in the relevant deformation $\hat{\varphi}$. In total we derive for the linearization at the known state $\bar{\hat{\varphi}}$:

$$
\begin{aligned}
DG(\bar{\hat{\varphi}}, \eta) \cdot \Delta\mathbf{u} &= \int_{B_R} \{ \widehat{\operatorname{Grad} \Delta\mathbf{u}} \, \bar{\hat{S}} \cdot \widehat{\operatorname{Grad} \eta} + \delta\bar{\hat{E}} \cdot \bar{\hat{\mathbb{C}}}[\Delta\bar{\hat{E}}] \} \, dV_R \\
&\quad - \int_{B_R} \rho_R (\widehat{\operatorname{Grad}\Delta\mathbf{u} \, \Omega_R \times_R}) \cdot (\widehat{\operatorname{Grad} \eta \, \Omega_R \times_R}) \, dV_R \tag{74}
\end{aligned}
$$

With the last equations all relations, with respect to the initial and the current configuration, are known which have to applied within an iterative solution procedure, e. g. NEWTON's method. Thus the basis for the discretization using finite element method for nonlinear problems in solid mechanics is known.

3   Contact kinematics

Many technical contact problems involve large deformations. Thus we will formulate all contact relations for finite deformations and will look either at problems where one body is rolling over a given rigid surface or contact between two deformable bodies takes place. The first relations are used for steady state rolling contact whereas the second set of kinematical relations are used for non-steady contact formulations.

Figure 4: a) non–penetration b) penetration

The contact conditions are split into two main parts. The first is associated with the contact conditions in normal direction to the contact surface, denoted by Γ_c. These relations are also known as the non–penetration conditions since they provide a constraint which prevents the body from penetrating the contact surface. The second part of the contact conditions is related to the tangential contact and hence provides kinematical relations to establish either the stick constraint or the slip velocity on Γ_c.

3.1   Non–penetration condition

To formulate the non–penetration condition for rolling contact we follow the work of [CA88], [LS93] or [WM94] and define a minimum distance problem between the rolling object and the surface, see Fig. 4. The reference configuration which is used to define the contact conditions can be either the rotating frame or the initial configuration. In case when we have to distinguish both we will use the reference to B or B_R.

For a mathematical description of the problem it is useful to introduce convective coordinates $\boldsymbol{\xi} = (\xi^1, \xi^2)$ on the surface with which the rolling object is in contact, from now on called *master* surface. This leads to the definition of the master surface, described by the position vector: $\mathbf{X}_0 = \mathbf{X}_0(\xi^1, \xi^2)$. Now we can define for every point x on the deformed boundary of the rolling object the minimum distance problem

$$d = MIN\,\|\mathbf{x} - \mathbf{X}_0(\xi^1, \xi^2)\| \tag{75}$$

The solution of this problem provides the pair of convective coordinates, $\boldsymbol{\xi}$, which denotes the point on the master surface closest to point $\mathbf{x}$. $\boldsymbol{\xi}$ follows from the condition

$$\frac{d}{d\xi^\alpha}\,d = \frac{1}{\|\mathbf{x} - \hat{\mathbf{X}}_0(\xi^1, \xi^2)\|}[\mathbf{x} - \mathbf{X}_0(\xi^1, \xi^2)] \cdot \mathbf{X}_{0,\alpha}(\xi^1, \xi^2) = 0 \tag{76}$$

which is the closest point projection of point $\mathbf{x}$ onto the master surface. Since $\bar{\mathbf{X}}_{0,\alpha} = \mathbf{X}_{0,\alpha}(\bar{\boldsymbol{\xi}})$ are the tangent vectors to the convective coordinates of the master surface at the solution point, $\bar{\boldsymbol{\xi}}$, the vector $\mathbf{x} - \bar{\mathbf{X}}_0(\bar{\boldsymbol{\xi}})$ points in a direction normal to the master surface. Hence it can be used to define the non–penetration condition.

With

$$g_N = [\mathbf{x} - \bar{\mathbf{X}}_0(\bar{\xi})] \cdot \bar{\mathbf{N}}_0(\bar{\xi}) \tag{77}$$

we define the gap function where

$$\bar{\mathbf{N}}_0(\bar{\xi}) = \frac{\mathbf{x} - \bar{\mathbf{X}}_0(\bar{\xi})}{\|\mathbf{x} - \bar{\mathbf{X}}_0(\bar{\xi})\|} \quad \text{or}$$

$$= \frac{\bar{\mathbf{X}}_{0,1}(\bar{\xi}) \times \bar{\mathbf{X}}_{0,2}(\bar{\xi})}{\|\bar{\mathbf{X}}_{0,1}(\bar{\xi}) \times \bar{\mathbf{X}}_{0,2}(\bar{\xi})\|} \tag{78}$$

Both definitions can be used in (77) however the first is not well behaved for $g_N \longrightarrow 0$. Hence the second condition should be applied in (77). Function g_N then describes the state at the interface as follows

$$
\begin{aligned}
g_N &> 0 \quad \text{gap opening,} \\
g_N &= 0 \quad \text{perfect contact,} \\
g_N &< 0 \quad \text{penetration.}
\end{aligned}
$$

Thus contact is formulated by the inequality constraint

$$g_N \geq 0 \tag{79}$$

In case of a flat master surface which is often the case when rolling contact is considered we can simplify the representation of the master surface by cartesian coordinates. By defining the base vectors of the master surface as $\mathbf{E}_1$ and $\mathbf{E}_2$ we obtain for the normal $\mathbf{N}_0 = \mathbf{E}_3$. In this case the closest point projection (76) yields

$$[\mathbf{x} - \bar{\mathbf{X}}_0(X_1, X_2)] \cdot \mathbf{E}_\alpha = 0 \tag{80}$$

with the solution point $(\bar{X}_1, \bar{X}_2)$. Furthermore the gap is given by

$$g_N = [\mathbf{x} - \bar{\mathbf{X}}_0(\bar{X}_1, \bar{X}_2)] \cdot \mathbf{E}_3. \tag{81}$$

3.2 Tangential kinematical contact relations

The kinematical relations for the tangential motion in the contact area lead to the definition of the relative tangential velocity. This quantity can be obtained by the derivative of condition (76) with respect to time. This yields

$$(\mathbf{v} - \dot{\bar{\mathbf{X}}}_0) \cdot \bar{\mathbf{X}}_{0,\alpha} + (\mathbf{x} - \bar{\mathbf{X}}_0) \cdot \dot{\bar{\mathbf{X}}}_{0,\alpha} - \bar{\mathbf{X}}_{0,\beta} \dot{\xi}^\beta \cdot \bar{\mathbf{X}}_{0,\alpha} = 0 \tag{82}$$

Here $\mathbf{v}$ is the velocity associated to point x. With the metric $\bar{A}_{\alpha\beta} = \bar{\mathbf{X}}_{0,\alpha} \cdot \bar{\mathbf{X}}_{0,\beta}$ the components of the relative gap velocity in tangential direction are

$$\bar{A}_{\alpha\beta} \dot{\xi}^\beta = (\mathbf{v} - \bar{\mathbf{V}}_0) \cdot \bar{\mathbf{X}}_{0,\alpha} + (\mathbf{x} - \bar{\mathbf{X}}_0) \cdot \dot{\bar{\mathbf{X}}}_{0,\alpha} \tag{83}$$

where we have set: $\bar{\mathbf{V}}_0 = \dot{\bar{\mathbf{X}}}_0$.

In case of a flat master surface equation (83) simplifies with $\ddot{\bar{\mathbf{X}}}_{0,\alpha} = 0$, $\bar{\mathbf{X}}_{0,\alpha} = \mathbf{E}_\alpha$ and $\bar{A}_{\alpha\beta} = \delta_{\alpha\beta}$ to

$$\dot{X}_\alpha = (\mathbf{v} - \bar{\mathbf{V}}_0) \cdot \mathbf{E}_\alpha \tag{84}$$

where $\dot{X}_\alpha$ denotes the change in time of the projection point $(\bar{X}_1, \bar{X}_2)$ on the master surface and thus is the relative tangential velocity in the contact area. With the projection tensor $\mathbf{P}_\perp = [\mathbf{E}_\alpha \otimes \mathbf{E}_\alpha]$ we can reformulate (84) as

$$\mathbf{v}_\perp = \mathbf{P}_\perp (\mathbf{v} - \bar{\mathbf{V}}_0) \tag{85}$$

with $\mathbf{v}_\perp = \dot{X}_\alpha \mathbf{E}_\alpha$ being the tangential relative velocity vector in Γ_c.

With these relations we can now formulate the tangential contact conditions. The first is the non–slip or stick condition

$$\mathbf{v}_\perp = \mathbf{0} \tag{86}$$

and hence imposes a constraint on the relative tangential motion. It means that locally the rotating object is rolling and not sliding on the surface.

If the tangential forces exceed a certain limit in Γ_c then slip occurs. In that case the associated relative tangential velocity follows from a constitutive relation. Classically Coulomb's law is applied to determine the slip velocity, however also more complicated constitutive equations can be used to model the frictional behaviour in the contact interface. These constitutive equations will be considered in the next section.

Generally slip as well as stick can occur in the contact area. Thus we can subdivide Γ_c into: $\Gamma_c^{slip} \cup \Gamma_c^{stick} = \Gamma_c$. Within the numerical method applied to solve the rolling contact problem the stick and the slip area have to be computed.

In some cases it might be necessary in the numerical implementation of rolling contact, see e. g. [Nac93], to enforce the stick condition in a weak sense. Following [Nac93] one can use a least square fit

$$\int_{\Gamma_c^{stick}} v_\perp^2 \, d\Gamma \rightarrow MIN \implies \int_{\Gamma_c^{stick}} \mathbf{v}_\perp \cdot \delta\mathbf{v}_\perp d\Gamma = 0 \tag{87}$$

which leads with (84) and (86) to

$$\int_{\Gamma_c^{stick}} \mathbf{P}_\perp \delta\mathbf{v} \cdot \mathbf{P}_\perp (\mathbf{v} - \bar{\mathbf{V}}_0) \, d\Gamma = 0 \tag{88}$$

and hence yields a system of equations to determine the tangential components of $\mathbf{v}$ in the stick interface Γ_c^{stick} which fulfill the stick condition (86) in the weak sense

$$\int_{\Gamma_c^{stick}} \delta\mathbf{v} \cdot \mathbf{P}_\perp \mathbf{v} d\Gamma = \int_{\Gamma_c^{stick}} \delta\mathbf{v} \cdot \mathbf{P}_\perp \bar{\mathbf{V}}_0 \, d\Gamma . \tag{89}$$

Note that if slip occurs in a steady state computation of rolling contact then the dissipation due to the frictional forces in the slip zone has to be compensated by a moment around the spinning axis of the rolling body to preserve stationary motion.

3.2.1 Definition of creepage

For further reference we also will define another kinematical quantity which measures the creepage in the contact interface. Its definition is given by

$$s = \frac{\dot{X}_0 - \omega_R \times R}{\|\dot{X}_0\|}. \tag{90}$$

The creepage vector s can be decomposed in apart which is related to the flattening of the rolling body

$$s_F = \frac{\omega_R \times (r - R)}{\|\dot{X}_0\|} \tag{91}$$

and a partial slip due to rolling in the contact area

$$s_S = \frac{\dot{X}_0 - \omega_R \times r}{\|\dot{X}_0\|} \tag{92}$$

so that we have $s = s_F + s_S$.

3.3 Contact kinematics for two deformable bodies

Here we will look at problems where two or more bodies B^α approach each other during a finte deformation process which come into contact on parts of their boundaries denotes by Γ_c, see Fig. 5. We observe that two points, X^1 and X^2, in the initial configuration of the bodies which are distinct, can occupy the same position in the current configuration, $\varphi(X^2, t) = \varphi(X^1, t)$, within the deformation process.

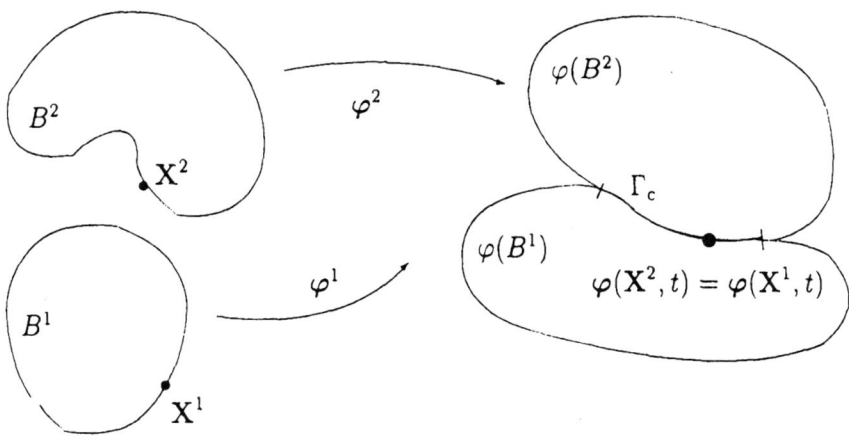

Figure 5: : Finite deformation contact.

In general two steps have to be performed to detect whether contact takes place or not. These are the global search for contact and the set–up of local kinematical relations which are needed to formulate the contact constraints. Here we will focus on the latter.

In a large deformation, continuum based formulation of contact kinematics the distance between the bodies being in contact is measured by minimization of the distance between a point on the boundary of one body and another point on the boundary of the other body. This yields a non–penetration condition see e.g. [CA88]. This non–penetration function plays also a significant role for the definition of the tangential velocity in the contact interface which is needed to formulate frictional problems, see e.g. [SL92], [WM92], [LS93].

Let us consider two elastic bodies B^α, $\alpha = 1, 2$, each of them occupying the bounded domain $\Omega^\alpha \subset R^3$. The boundary Γ^α of a body B^α consists of three parts: Γ^α_σ with prescribed surface loads, Γ^α_u with prescribed displacements and Γ^α_c where the two bodies B^1 and B^2 come into contact. In the contact area we have to formulate the constraint equations or the approach function for normal contact as well as the kinematical relations for the tangential contact.

3.3.1   Normal contact

Assume that two bodies come into contact. In that case the non–penetration condition can be developed according to Fig. 6

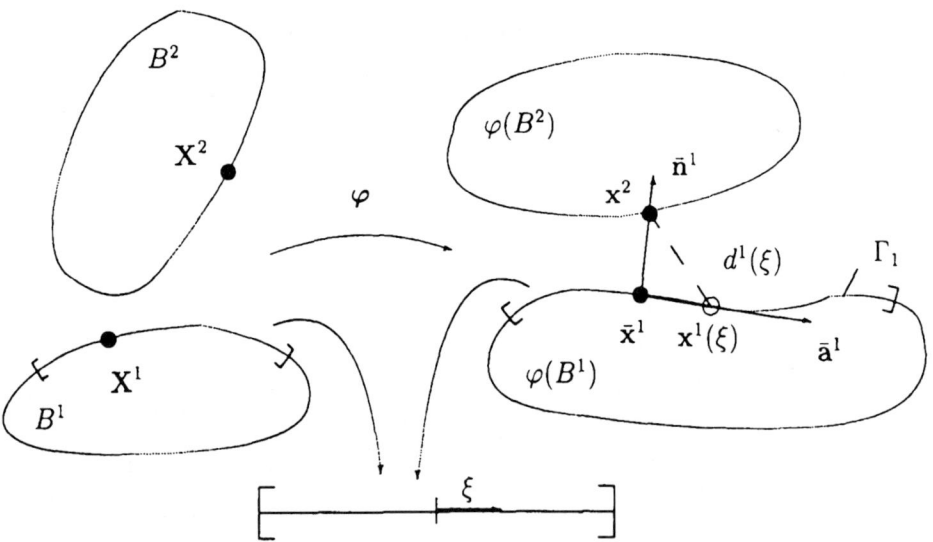

Figure 6: : a) Deformed configuration of bodies B^α,   b) Minimum distance

by a minimum distance problem between point x^2 and $\bar{x}^1$. x^α denote the coordinates of the current configuration $\varphi(B^\alpha)$ of body B^α: $x^\alpha = X^\alpha + u^\alpha$ where X^α is related to the initial configuration B^α and $\mathbf{u}^\alpha$ is the displacement field. The normal vector $\mathbf{n}^1$ is associated with body B^1. By assuming that the contact boundary describes, at least locally, a convex region, we can relate to every point $\mathbf{x}^2$ on Γ^2 a point $\bar{\mathbf{x}}^1 = \mathbf{x}^1(\bar{\boldsymbol{\xi}})$ on Γ^1 via the minimal distance problem

$$\|\mathbf{x}^2 - \bar{\mathbf{x}}^1\| = \min_{\mathbf{x}^1 \subseteq \Gamma^1} \|\mathbf{x}^2 - \mathbf{x}^1(\boldsymbol{\xi})\|, \tag{93}$$

see Fig 6 b which illustrates the two–dimensional case. $\boldsymbol{\xi} = (\xi^1, \xi^2)$ denotes the parametrization of the boundary Γ^1 via convective coordinates, see e.g. [WM94]. The parametrized surface Γ^1 is also called master surface in the following. (93) yields the condition

$$\frac{d}{d\xi^\alpha} \|\mathbf{x}^2 - \mathbf{x}^1(\xi^1, \xi^2)\| = \frac{\mathbf{x}^2 - \mathbf{x}^1(\xi^1, \xi^2)}{\|\mathbf{x}^2 - \mathbf{x}^1(\xi^1, \xi^2)\|} \cdot \mathbf{x}^1_{,\alpha}(\xi^1, \xi^2) = 0 \tag{94}$$

$\mathbf{x}^1_{,\alpha} = \mathbf{a}^1_\alpha$ is a vector tangent to the master surface. It has to be, at the solution point $\bar{\mathbf{x}}^1$ of (93), perpendicular to $\mathbf{x}^2 - \mathbf{x}^1(\xi^1, \xi^2)$ which thus is normal to the master surface, see Fig. 6 b.

Once the point $\bar{\mathbf{x}}^1$ is known, we can define the inequalitiy constraint of the non-penetration condition by introducing the gap function between the two materials $g_N = [\mathbf{x}^2 - \mathbf{x}^1(\bar{\boldsymbol{\xi}})] \cdot \mathbf{n}^1(\bar{\boldsymbol{\xi}})$. This function then describes the state at the interface as follows We can now state an inequalitiy constraint for the non–penetration condition in the

$g_N > 0$ gap opening,
$g_N = 0$ perfect contact,
$g_N < 0$ penetration.

interface

$$g_N = (\mathbf{x}^2 - \bar{\mathbf{x}}^1) \cdot \bar{\mathbf{n}}^1 \geq 0 \tag{95}$$

For some formulations which describe the contact it is useful to define a penetration function

$$\bar{g}_N = \begin{cases} (\mathbf{x}^2 - \bar{\mathbf{x}}^1) \cdot \bar{\mathbf{n}}^1 & \text{if } (\mathbf{x}^2 - \bar{\mathbf{x}}^1) \cdot \bar{\mathbf{n}}^1 < 0 \\ 0 & \text{otherwise}. \end{cases} \tag{96}$$

It defines the magnitude of penetration of one body into the other.

Function $\bar{g}_N$ and indicates a penetration of one body into the other and shows within an iterative solution process in which parts of Γ^α the constraint equations, preventing penetration, have to be activated. Thus (96) can be used to determine the contact area $\Gamma^\alpha_c \subseteq \Gamma^\alpha$.

3.3.2   Tangential contact

The tangential kinematical relations for contact have to represent as well stick as slip conditions. The stick condition means that no relative displacement is possible in tangential direction. Hence, by defining the relative tangential displacement we can formulate this condition in form of a constraint. In case of slip a tangential movement is possible, which is defined by the relative tangential velocity. The tangential relative displacement between two bodies is related to the change of the solution point $\bar{\xi} = (\bar{\xi}^1, \bar{\xi}^2)$ which has been obtained via the minimal distance problem (93) at time t. Thus

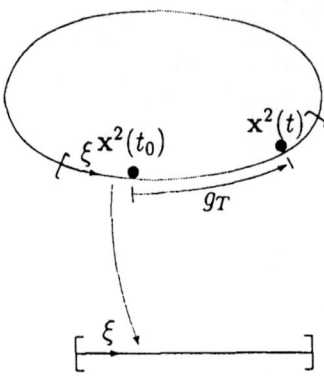

Figure 7: Relative tangential movement of point $\mathbf{x}^2$

first we will state tangential relative deformation of a point $\mathbf{x}^2$ related to material B^2 on the interfacial surface defined in terms of material B^1. With respect to Fig. 7 we obtain for the total slip between point $\mathbf{x}^2$ and the deformed master surface

$$g_T = \int_{t_0}^{t} \|\dot{\bar{\xi}}^\alpha \, \bar{\mathbf{a}}_\alpha^1\| \, dt \tag{97}$$

where t is the time, here used to parametrize the path of point $\mathbf{x}^2$. For the evaluation of this integral we have to compute the time derivative of ξ^α at the projection point $\bar{\mathbf{x}}^1$ of point x^2. This can be done by using (94). Thus we can compute the time derivative of ξ^α from the relation

$$[\mathbf{x}^2 - \bar{\mathbf{x}}^1] \cdot \bar{\mathbf{a}}_\alpha^1 = 0 \tag{98}$$

which is valid at the contact point since the difference $\mathbf{x}^2 - \bar{\mathbf{x}}^1$ is normal to the contact surface and $\bar{\mathbf{a}}_\alpha^1$ denotes the tangent vector to the surface Γ^1 at the minimal distance point, see Fig. 6 b. This yields the following result

$$\frac{d}{dt}[\mathbf{x}^2 - \bar{\mathbf{x}}^1(\bar{\xi}^1, \bar{\xi}^2)] \cdot \bar{\mathbf{a}}_\alpha^1 = [\mathbf{v}^2 - \bar{\mathbf{v}}^1 - \bar{\mathbf{a}}_\beta \, \dot{\bar{\xi}}^\beta] \cdot \bar{\mathbf{a}}_\alpha^1 + [\mathbf{x}^2 - \bar{\mathbf{x}}^1] \cdot \dot{\bar{\mathbf{a}}}_\alpha^1 = 0 \,. \tag{99}$$

With $\dot{\bar{\mathbf{a}}}_\alpha^1 = \bar{\mathbf{v}}_{,\alpha}^1 + \hat{\mathbf{x}}_{,\alpha\beta}^1 \, \dot{\bar{\xi}}^\beta$ we obtain $\dot{\bar{\xi}}^\beta$ from (99) which leads to the following system of equations

$$\bar{H}_{\alpha\beta} \, \dot{\bar{\xi}}^\beta = \bar{R}_\alpha \tag{100}$$

with

$$\bar{H}_{\alpha\beta} = [\bar{a}_{\alpha\beta} + g_N \bar{x}^1_{,\alpha\beta} \cdot \bar{n}^1]$$
$$\bar{R}_\alpha = [v^2 - \bar{v}^1] \cdot \bar{a}^1_\alpha + [x^2 - \bar{x}^1] \cdot \bar{v}^1_{,\alpha} \tag{101}$$
$$= [v^2 - \bar{v}^1] \cdot \bar{a}^1_\alpha + g_N \bar{n}^1 \cdot \bar{v}^1_{,\alpha}$$

A well known result from differential geometry introduces for $\bar{x}^1_{,\alpha\beta}(\bar{\xi}^1, \bar{\xi}^2) \cdot \bar{n}^1$ the curvature tensor $\bar{b}_{\alpha\beta}$. Thus we can rewrite $\bar{H}_{\alpha\beta} = [\bar{a}_{\alpha\beta} + g_N \bar{b}_{\alpha\beta}]$.

We define as the second important kinematical function the tangential relative velocity function on the current surface $\varphi^1(\Gamma^1_c)$ by setting

$$\mathcal{L}_v \, g_T := \dot{\bar{\xi}}^\alpha \, \bar{a}_\alpha \tag{102}$$

(102) determines the evolution of the tangential slip g_T which enters as a local kinematical the constitutive function for the contact tangential stress.

The stick condition is now given by the constraint

$$g_T = 0. \tag{103}$$

To obtain the slip amount related to the relative tangential velocity (102) we need a constitutive equation for the frictional behaviour in the contact interface.

REMARK

1. Note that the second terms on the right hand side of (100) depends on the penetration g_N. Thus in the case of a the strong enforcement of the non–penetration condition ($g_N = 0$) with Lagrangian multipliers this term vanishes. This yields

$$\dot{\bar{\xi}}^\beta = [v^2 - \bar{v}^1] \cdot \bar{a}^{1\beta} \tag{104}$$

Then the evolution $\mathcal{L}_v$ in (102) is given by the projection of the spatial velocities v^2 and $\hat{v}^1(\bar{\xi})$ evaluated at the contact points onto the tangential direction of the contact surface.

$$\mathcal{L}_v \, g_T = (\bar{a}^1_\alpha \otimes \bar{a}^{1\alpha})[v^2 - \bar{v}^1] \tag{105}$$

Since the unit tensor of the master durface is defined as $\mathbf{1} = \bar{a}^1_\alpha \otimes \bar{a}^{1\alpha} + \bar{n}^1 \otimes \bar{n}^1$ we can express (105) also by

$$\mathcal{L}_v \, g_T = \bar{P}[v^2 - \bar{v}^1] \quad \text{with} \quad \bar{P} = \mathbf{1} - \bar{n}^1 \otimes \bar{n}^1. \tag{106}$$

2. If we have a flat contact surface the curvature tensor $\bar{b}_{\alpha\beta}$ is zero.

For a penetration $g_N < 0$ we have to take into account the second term in (102) and the scaling factors $\bar{H}_{\alpha\beta}$, both consequences of the time dependence of $\bar{a}^1_\alpha$.

For the two–dimensional contact we can specify the result in (100) which then yields

$$\dot{\bar{\xi}} = \frac{1}{\bar{a}_{11} + g_N \bar{b}_{11}} \left\{ [\mathbf{v}^2 - \mathbf{v}^1(\bar{\xi})] \cdot \mathbf{x}^1,_\xi (\bar{\xi}) + g_N \, \bar{\mathbf{n}}^1 \cdot \mathbf{v}^1,_\xi (\bar{\xi}) \right\} \qquad (107)$$

where $\bar{a}_{11} = \mathbf{x}^1,_\xi (\bar{\xi}) \cdot \mathbf{x}^1,_\xi (\bar{\xi})$ describes the metric and $\bar{b}_{11} = \mathbf{x}^1,_{\xi\xi} (\bar{\xi}) \cdot \bar{\mathbf{n}}^1$ the curvature of the boundary. The vectors $\mathbf{v}^\alpha$ denote the velocities at $\mathbf{x}^\alpha$. Knowing the change of the coordinate $\bar{\xi}$ we can define the relative tangential velocity as

$$\mathcal{L}_v \, \mathbf{g}_T := \dot{\bar{\xi}} \, \mathbf{x}^1,_\xi (\bar{\xi}) . \qquad (108)$$

In the two–dimensional case we obtain for (97)

$$g_T = \int_{t_0}^{t} \|\dot{\bar{\xi}}^1 \, \bar{\mathbf{a}}^1_1\| \, dt = \int_{\xi_0}^{\xi} \|\bar{a}_{11}\| \, d\xi \qquad (109)$$

4   Weak forms, solution algorithms

For the formulation of the weak form we have to discuss only the additional terms due to contact in detail. The equations describing the behaviour of the bodies coming into contact do not change.

4.1   Weak form for Solids in Contact

For a numerical solution of the nonlinear boundary value problem summarized above we will use the finite element method. Thus we need the weak form of the local field equations. Due to the fact that the constraint condition (95) is represented by an inequality we obtain in general a variational inequality for the mechanical part. The general form can be written for the rolling contact with (47) as

$$-\int_{B_R} \rho_R \left(\dot{\mathbf{F}} \, \mathbf{\Omega}_R \mathbf{x}_R \right) \cdot (\widehat{\text{Grad}} \, \eta \, \mathbf{\Omega}_R \mathbf{x}_R) \, dV_R + \int_{B_R} \hat{\mathbf{F}} \, \hat{\mathbf{S}} \cdot \widehat{\text{Grad}} \, \eta \, dV_R$$

$$\geq \int_{B_R} \rho_R \left(\bar{\mathbf{b}} - \mathbf{\Omega}^2_R \mathbf{x}_R \right) \cdot \eta \, dV_R + \int_{\partial B_R} \bar{\mathbf{t}} \cdot \eta \, dA_R \qquad (110)$$

where the integration is performed with respect to the domain B_R occupied by the rolling body in the rotating reference configuration.

We now have to find the deformation $(\boldsymbol{\varphi}) \in \mathbf{K}$ such that (95) is fulfilled for all $(\boldsymbol{\eta}) \in \mathbf{K}$ with

$$\mathbf{K} = \{ (\boldsymbol{\eta}) \in \mathbf{V} \, | \, \boldsymbol{\eta} \cdot \mathbf{N}_0 \geq 0 \}, \qquad (111)$$

In case of finite elasticity the existence of the solution of (111) can be proved, see e.g. [Cia88] or [CHT92].

Algorithms for solving variational inequalities are given by mathematical programming, active set strategies or sequential quadratic programming methods, to name only a few. Each of these methods is well known from optimization theory, see e. g. [Lue84]. Applications to rolling contact can be found in [Kal90] or [Nac93].

Here we will concentrate on the active set strategies which are basis of the contact formulation in many existing finite element codes. Within this method the contact constraints are kept fixed during the iteration for equilibrium and thus can be treated like equality constraints. These can be introduced via Lagrangian multipliers or penalty terms. Within an active set strategy we can write the weak form as an equality since we know the active set within an incremental solution step. Then equation (110) can be recast in the following form

$$
\begin{aligned}
& -\int_{B_R} \rho_R \left(\hat{\mathbf{F}} \, \mathbf{\Omega}_R \, \mathbf{x}_R \right) \cdot \left(\widehat{\mathrm{Grad}\, \eta \, \mathbf{\Omega}_R \mathbf{x}_R} \right) dV_R + \int_{B_R} \hat{\mathbf{F}} \, \hat{\mathbf{S}} \cdot \widehat{\mathrm{Grad}\, \eta} \, dV_R \\
& -\int_{B_R} \rho_R \left(\bar{\mathbf{b}} - \mathbf{\Omega}_R^2 \, \mathbf{x}_R \right) \cdot \eta \, dV_R - \int_{\partial B_{\sigma R}} \bar{\mathbf{t}} \cdot \eta \, dA_R \qquad (112) \\
& + \text{"Contact Contributions"} = 0
\end{aligned}
$$

where the "contact contributions" are associated with the active constraint set Γ_c^{act}. $\eta \in V$ is the so called test function or virtual displacement which is zero at the boundary Γ_φ where the deformations are prescribed.

For two bodies being in contact we obtain the weak form of the interface by assuming that contact is active at the surface Γ_c^{act}. The formulation follows for two different cases as given below.

1. **Lagrangian multiplier method:**

$$
\int_{\Gamma_c^{act}} \left(\lambda_N \, \delta g_N + \lambda_T \cdot \delta \mathbf{g}_T \right) dA \qquad (113)
$$

   Here λ_N denotes the Lagrangian multiplier which can be identified as the contact pressure. δg_N is the variation of the normal gap. The term $\lambda_T \cdot \delta \mathbf{g}_T$ is associated with the tangential stick or slip motion and needs further discussion. In case of pure stick the relative tangential slip $\mathbf{g}_T$ is zero which yields a constraint equation from which λ_T follows as a reaction. In case of sliding the tangential stress vector $\mathbf{t}_T$ is determined by the constitutive law for frictional slip, see section 3.2, and thus we should write instead of $\lambda_T \cdot \delta \mathbf{g}_T \longrightarrow \mathbf{t}_T \cdot \delta \mathbf{g}_T$.

2. **Penalty method:** In this formulation a penalty term due to the constraint condition is added to the weak form (113). This means that once the constraint equation for g_N^- is violated

$$
\int_{\Gamma_c^{act}} \epsilon_N \, g_N^- \, \delta g_N^- \, dA, \quad \epsilon_N > 0 \qquad (114)
$$

has to be considered for normal contact. It can be shown, see e.g. [Lue84], that the solution of the Lagrangian multiplier method can be recovered from this formulation for $\epsilon_N \to \infty$, however high values of ϵ_N will lead to an ill-conditioned numerical problem. As in the Lagrangian multiplier method we have to distinguish between pure stick in the contact interface which yields a penalty term also for the tangential direction

$$\int_{\Gamma_c^{act}} (\epsilon_N \, \bar{g}_N \, \delta\bar{g}_N + \epsilon_T \, \mathbf{g}_T \cdot \delta\mathbf{g}_T) \, dA, \quad \epsilon_N > 0, \epsilon_T > 0 \tag{115}$$

and the slip condition which leads to

$$\int_{\Gamma_c^{act}} (\epsilon_N \, \bar{g}_N \, \delta\bar{g}_N + \mathbf{t}_T \cdot \delta\mathbf{g}_T) \, dA, \quad \epsilon_N > 0 \tag{116}$$

In the latter equation a frictional laws has to be applied.

In equations (113) to (116) the variation of the normal gap function g_N is needed which yields fortwo-body contact

$$\delta g_N = [\boldsymbol{\eta}^2 - \boldsymbol{\eta}^1(\bar{\xi}_1, \bar{\xi}_2)] \cdot \bar{\mathbf{n}}_1. \tag{117}$$

and for contact condition (77) between rolling object and rigid surface

$$\delta g_N = \boldsymbol{\eta} \cdot \bar{\mathbf{N}}_0 \tag{118}$$

Furthermore the variation of the tangential slip can be stated for two deformable bodies as

$$\delta\mathbf{g}_T = \delta\bar{\xi}^\alpha \, \bar{\mathbf{a}}_\alpha^1. \tag{119}$$

The variation of the tangential slip for the rolling problem on a flat surface yields with (85)

$$\delta\mathbf{g}_T = \delta\mathbf{v}_\perp = \mathbf{P}_\perp \, \delta\mathbf{v}, \tag{120}$$

REMARK:

1. Perturbed Lagrangian formulations can be used to combine both penalty and Lagrangian multiplier methods in a mixed formulation, see e.g. [Ode81] or [SWT85]. In this case the following functional

$$\Pi_p = \Pi + \int_{\Gamma_c} [\lambda_N \, g_N - \frac{1}{2 \, \epsilon_N} \lambda_N^2] \, d\Gamma \longrightarrow STAT \tag{121}$$

is defined for the frictionless case where Π denotes the total energy of the two bodies. The Lagrangian multiplier term is regularized by the second term in the integral which can be viewed as the complementary energy due to the Lagrangian multiplier. The variation leads to

$$\delta\Pi_p = \delta\Pi + \int_{\Gamma_c^{act}} [\lambda_N \delta g_N^- + \delta\lambda_N (g_N^- - \frac{1}{\epsilon_N}\lambda_N)] d\Gamma = 0 \qquad (122)$$

The first term is again associated with the Lagrangian multiplier formulation whereas the second term yields the "constitutive law": $\lambda_N = \epsilon_N g_N$ if evaluated locally. If we insert this result for λ_N in the first term of (122) we obtain the standard penalty formulation (116). Letting $\epsilon_N \longrightarrow \infty$ yields the classical Lagrangian multiplier method.

We note that equation (122) can also be a starting point for special mixed formulations (e.g. in finite element formulations when different interpolation functions are used for the Lagrange multiplier and the displacement field, see section 6).

2. A barrier method can also be applied to solve contact problems. In that case the associated variational formulation for the frictionless case is associated with the functional

$$\Pi_b = \Pi - \int_{\Gamma_c} \frac{\epsilon_N}{g_N} d\Gamma \longrightarrow STAT \qquad (123)$$

This technique has the advantage that all constraints can be always active. However if during the iteration process a penetration occurs this method does not converge Thus it has to be implemented with care.

3. A technique based on a new constraint functional which includes the penalty and the barrier formulation as limit cases has been developed lately and named methods of cross constraints, see [ZWS95]. Due to its construction the functional is also active when the gap function is open as in the barrier method, however a safe guard algorithm has not to be applied since the solution is not restricted to the feasable region.

4. A major problem associated with the numerical treatment of the penalty method is the ill–conditioning which arises when the penalty parameter ϵ_N assumes large values. A standard method to overcome the problem of ill–conditioning is based on the augmented Lagrangian technique, well known in optimization theory. This technique has been considered extensively within the context of incompressibility constraints [GLT84] and was also applied to contact problems, see [WST85] or [KO88] for frictionless contact. Recently this approach has been extended successfully also to large displacement contact problems including friction, see [AC91] or [LS93]. A formulation which accounts for micromechanical interface laws can be found in [WZ93].

The main idea is to combine the penalty method with Lagrangian multiplier methods. The augmented Lagrangian technique then yields an algorithm, known as UZAWA algorithm, where a Lagrangian multiplier $\bar{\lambda}_N$ is introduced and held constant during an iteration loop to solve the weak form which is nonlinear with respect to the deformation φ. This leads to the weak form

$$\delta\Pi + \int_{\Gamma_c} [\,\bar{\lambda}_N + \epsilon_N\, g_N\,)\,\delta g_N + \mathbf{t}_T \cdot \delta\mathbf{g}_T\,]\, dA = 0 \tag{124}$$

Since $\bar{\lambda}_N$ is unknown an update procedure for the Lagrangian multiplier has to be constructed within an iteration loop. The simplest update is: $\bar{\lambda}_{N_{new}} = \bar{\lambda}_{N_{old}} + \epsilon_N\, g_{N_{new}}$ which is only of first order accuracy. For other possibilities, see e.g. [Ber84] or in the context of finite element contact problems [AC91].

4.2   Local integration of constitutive equations

For the constitutive equations regarding the bodies coming in contact a large number of contributions can be found in the literature, hence this topic will not be discussed here. For elasticity integration is not needed, here a simple function evaluation of the elastic response function is sufficient. For the visco–elastic constitutive equation (58) a special integration procedure has to be applied, see [GM99]. For the integration of general inelastic constitutive equations, see [SH98]. Here we will only discuss the constitutive equations in the contact interface in detail.

For the normal contact a mere function evaluation –like for finite elasticity– can be used to obtain within the penalty method the contact pressure $p_N = \epsilon_N\, \bar{g}_N$.

The algorithmic update of the tangential stress $\mathbf{t}_{T\,n+1}$ and dissipation $\mathcal{D}^s_{n+1}$ is performed by the return algorithm based on an objective (backward Euler) integration of the evolution equation for the plastic slip, see e.g. [Wri87], [JT88], [Gia89], [WVS90].

The results can be summarized as follows: Integration of (102) gives the increment of the total slip within the time step Δt_{n+1}

$$\Delta\mathbf{g}_{T\,n+1} = (\,\bar{\xi}^\alpha_{n+1} - \bar{\xi}^\alpha_n\,)\,\bar{\mathbf{a}}_{\alpha\,n+1}. \tag{125}$$

The total slip has to be decomposed into an elastic, $\mathbf{g}^e$, and a slip, $\mathbf{g}^s$, part at time t_{n+1}

$$\mathbf{g}^e_{n+1} = \mathbf{g}_{n+1} - \mathbf{g}^s_{n+1} \tag{126}$$

Then we can compute the elastic trial state from the elastic or stick part

$$\mathbf{t}^{tr}_{T\,n+1} = c_T\,\mathbf{g}^e_{n+1} \tag{127}$$

and evaluate the COULOMB slip criterion $f_s = \|\,\mathbf{t}_T\,\| - \mu\,p_N \le 0$ at time t_{n+1}

$$
\begin{aligned}
\mathbf{t}^{tr}_{T\,n+1} &:= c_T\,(\,\mathbf{g}_{T\,n+1} - \mathbf{g}^s_{T\,n}\,) = \mathbf{t}_{T\,n} + c_T\,\Delta\mathbf{g}_{T\,n+1}, \\
f^{tr}_{s\,n+1} &:= \|\mathbf{t}^{tr}_{T\,n+1}\| - \mu\,p_{N\,n+1}.
\end{aligned}
$$

If this state is elastic ($f^{tr}_{s\,n+1} \le 0$) then no friction takes place and we have to use the elastic relation (127). In case that $f^{tr}_{s\,n+1} > 0$ we have to perform the return mapping.

Using the implicit Euler scheme, the slip rule

$$\mathcal{L}_v \, \mathbf{g}_T = \lambda \frac{\partial f_s}{\partial \mathbf{t}_T} \qquad (128)$$

yields

$$\mathbf{g}^s_{T\,n+1} = \mathbf{g}^s_{T\,n} + \lambda \, \mathbf{n}_{T\,n+1}$$
$$g_{v\,n+1} = g_{v\,n} + \lambda$$

With the standard arguments regarding the projection schemes, see e.g. [ST85], we obtain

$$\mathbf{t}_{T\,n+1} = \mathbf{t}^{tr}_{t\,n+1} - \lambda \, c_T \, \mathbf{n}_{T\,n+1},$$
$$\mathbf{n}_{T\,n+1} = \mathbf{n}^{tr}_{T\,n+1},$$

From the slip condition λ can be computed

$$f_s(\lambda) = \| \mathbf{t}^{tr}_{T\,n+1} \| - \mu \, p_{N\,n+1} - c_T \, \lambda = 0 \qquad (129)$$

which can be directly solved for λ

$$\lambda = \frac{1}{c_T} \left(\| \mathbf{t}^{tr}_{t\,n+1} \| - \mu \, p_{N\,n+1} \right) \qquad (130)$$

Knowing λ the stress update follows from (129) and the frictional slip from (129). Again we state the explicit results for Coulomb's model

$$\mathbf{t}_{T\,n+1} = \mu \, p_{N\,n+1} \, \mathbf{n}^{tr}_{T\,n+1},$$
$$\mathbf{g}^s_{T\,n+1} = \mathbf{g}^s_{T\,N} + \frac{1}{c_T} \left(\| \mathbf{t}^{tr}_{t\,n+1} \| - \mu \, p_{N\,n+1} \right) \mathbf{n}^{tr}_{T\,n+1}.$$

which completes the algorithm for the frictional interface law.

The dissipation due to the plastic slip is given by

$$\mathcal{D}^s_{n+1} = \begin{cases} 0 & \text{for } f^{trial}_{s\,n+1} \le 0 \\ \mathbf{t}_{T\,n+1} \cdot (\mathbf{t}^{trial}_{T\,n+1} - \mathbf{t}_{T\,n+1})/(c_T \, \Delta t_{n+1}) & \text{otherwise} \end{cases} \qquad (131)$$

5   Discretization of the continuum

The discretization of the domain contributions of the bodies being in contact is not the objective of this work. For a detailed treatment with respect to the finite element implementations of boundary-value-problems regarding large deformations, see e.g. [Cri91], [ZT91], or [Bat86] and references therein. However, in the next section the discretization of continua undergoing large strains using isoparametric elements is discussed briefly for completeness.

Within the finite element method we have different approximations. These are geometrical approximations of the domain B on which the boundary value problem is defined. Furthermore the associated fields, deformations or stresses, have to be approximated. Also the integrals are not evaluated exactly, since such as they are evaluated for the weak form, they have to be computed via numerical integration procedures. Collectively, these approximations are sources for errors inherent in the finite element method.

In this section a description of the interpolations, which are basis for a treatment using isoparametric elements, is given. Within this framework, we assume that the domain B is discretized by n_e finite elements, which leads to its geometrical approximation B^h:

$$\overline{B} \approx B^h = \overline{\bigcup_{e=1}^{n_e} \Omega_e} \,. \tag{132}$$

The configuration of one element is $\Omega_e \subset B^h$. ∂B^h denotes the boundary of the discretization B^h, which is in general also an approximation of the function describing the real boundary ∂B.

5.1 Isoparametric concept

The interpolation of given mechanical quantities within a finite element in B^h are given by the shape functions $N_I(\boldsymbol{\xi})$ defined on the reference element $\Omega_\square$, see Fig. 8. Thus, for every element Ω_e, there exists a transformation which relates the coordinates $\mathbf{X}_e = \mathbf{X}_e(\boldsymbol{\xi})$ to the coordinates $\boldsymbol{\xi}$ of the reference element $\Omega_\square$ by $\mathbf{X}_e = \sum_I N_I(\boldsymbol{\xi}) \mathbf{X}_I$. Hence all computations are performed with respect to the reference configuration. Only in very special cases the initial and current configuration of a finite element coincide. However this transformation is numerically easy to handle and allows the transformation of the reference element to arbitrary geometries. This feature leads to the fact that, in implementation of the method, there is almost no difference in the formulation of finite elements with respect to the current or the initial configuration.

Fig. 8 depicts the two possibilities to describe deformation in continuum mechanics using the isoparametric concept.

It can be seen easily that Fig. 8 is the discrete version of Fig. 2 where additionally we have now introduced the reference configuration $\Omega_\square$. The kinematical relations within one element are

$$F_e = j_e \, J_e^{-1} \quad \text{and} \quad J_e = \det F_e = \frac{\det j_e}{\det J_e}, \tag{133}$$

which show that the deformation gradient is uniquely defined by the isoparametric mapping of $\Omega_\square$ onto Ω_e in the initial configuration or onto $\varphi(\Omega_e)$ in the current configuration. In this equations tha gradients j_e and J_e are defined as follows

$$j_e = \text{Grad}_\xi \, \mathbf{x} = \frac{\partial \mathbf{x}}{\partial \boldsymbol{\xi}} = \sum_{I=1}^{n} N_{I,\xi}(\boldsymbol{\xi}) \, \mathbf{x}_I \otimes \mathbf{E}_\xi,$$

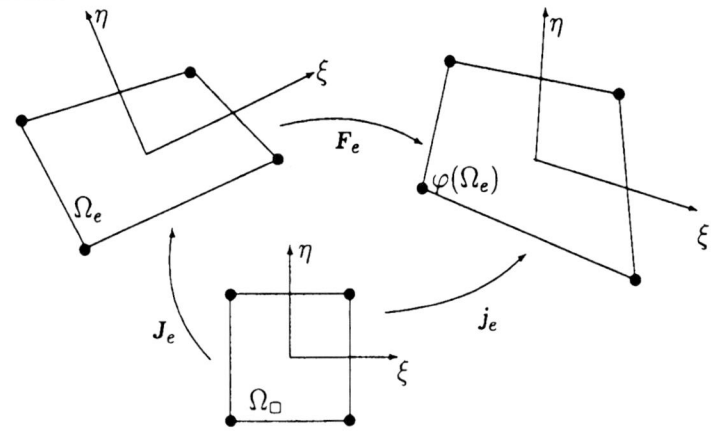

Figure 8: Isoparametric description of deformations

$$J_e = \mathrm{Grad}_\xi\, \mathbf{X} = \frac{\partial \mathbf{X}}{\partial \xi} = \sum_{I=1}^{n} N_{I,\xi}(\xi)\, \mathbf{x}_I \otimes \mathbf{E}_\xi \,. \tag{134}$$

Since the derivatives $N_{I,\xi}$ are scalar quantities, we can move them in front of the base vectors $\mathbf{E}_\xi$. This yields

$$j_e = \sum_{I=1}^{n} \mathbf{x}_I \otimes N_{I,\xi}(\xi)\, \mathbf{E}_\xi = \sum_{I=1}^{n} \mathbf{x}_I \otimes \nabla_\xi N_I \,,$$

$$J_e = \sum_{I=1}^{n} \mathbf{X}_I \otimes N_{I,\xi}(\xi)\, \mathbf{E}_\xi = \sum_{I=1}^{n} \mathbf{X}_I \otimes \nabla_\xi N_I \,. \tag{135}$$

$\nabla_\xi N_I$ is the gradient of the scalar function N_I with respect to the coordinates ξ.

  With this, it is simple to compute gradients with respect to the initial or the current configuration. For a vector field this reads as $\mathbf{u}_e$

$$\mathrm{Grad}\,\mathbf{u}_e = = \sum_{I=1}^{n} \mathbf{u}_I \otimes \nabla_X N_I \,,$$

$$\mathrm{grad}\,\mathbf{u}_e = = \sum_{I=1}^{n} \mathbf{u}_I \otimes \nabla_x N_I \,. \tag{136}$$

Analogous to the transformation of the derivatives between different configurations we obtain

$$\nabla_\xi N_I = J_e^T\, \nabla_X N_I \quad \text{and} \quad \nabla_\xi N_I = j_e^T\, \nabla_x N_I \,, \tag{137}$$

or the inverse relations

$$\nabla_X N_I = J_e^{-T}\, \nabla_\xi N_I \,, \quad \text{and} \quad \nabla_x N_I = j_e^{-T}\, \nabla_\xi N_I \,, \tag{138}$$

such that the gradient in (136) is completely defined in quantities which are defined in the reference configuration $\Omega_\square$ as

$$\text{Grad } \mathbf{u}_e \; = \; = \sum_{I=1}^{n} \mathbf{u}_I \otimes \mathbf{J}_e^{-T} \nabla_\xi N_I \,,$$

$$\text{grad } \mathbf{u}_e \; = \; = \sum_{I=1}^{n} \mathbf{u}_I \otimes \mathbf{j}_e^{-T} \nabla_\xi N_I \,. \tag{139}$$

The only difference in the formulation of both gradients in (139) lies in the exchange of the gradients $\mathbf{j}_e$ and $\mathbf{J}_e$, and therefore this approach is advantageous, especially for large deformation finite element formulations.

5.1.1   Isoparametric interpolation functions

Within the different possibilities to construct interpolation functions for isoparametric element, we will here follow the concept of the LAGRANGE interpolation, see e.g. [ZT89]. For a LAGRANGE polynominal of power $n-1$ we obtain in the one dimensional case

$$N_I(\xi) = \prod_{\substack{J=1 \\ J \neq I}}^{n} \frac{(\xi_J - \xi)}{(\xi_J - \xi_I)} \,. \tag{140}$$

For two– or three dimensional interpolations we choose a product formulation

$$N_J(\xi,\eta) = N_I(\xi)\, N_K(\eta) \quad \text{or} \quad N_J(\xi,\eta,\zeta) = N_I(\xi)\, N_K(\eta)\, N_L(\zeta) \,, \tag{141}$$

with $J = 1,\ldots n^{dim}$ and $I,K,L = 1,\ldots n$. (*dim* is the spatial dimension of the problem). The interpolation or shape functions are defined in the local coordinate system $\xi = \{\xi, \eta, \zeta\}$.

In the next section we will specify the isoparametric shape functions for three dimensional problems.

5.1.2   Two–dimensional shape functions

In the two–dimensional case quadrilateral and triangular finite elements have to be distinguished. Here we will discuss c^0–continuous shape functions which are linear as well as quadratic.

First triangular elements are considered. The simplest element with linear shape functions consists of three nodes, an element with quadratic interpolation needs 6 nodes to define the fields and geometry within an element. In Figure 9 the triangular element is depicted for the quadratic interpolation. For a linear element only the vertices 1 to 3 are necessary to define the interpolation. The element is shown in Fig. 9 in its

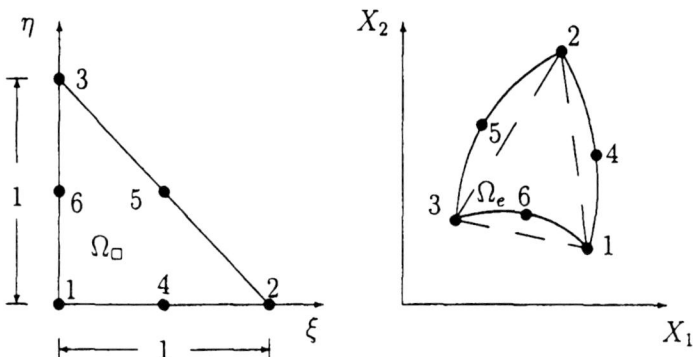

Figure 9: Three– and six node triangular element.

reference configuration Ω_0, denoted by the ξ-η-coordinates, and in its physical space denoted by X_1-X_2-coordinate system.

The shape functions for the linear case are defined by

$$N_1 = 1 - \xi - \eta \qquad N_2 = \xi \qquad N_3 = \eta \tag{142}$$

Here all partial derivatives with respect to ξ und η are constant.

The shape functions for the quadratical element are

$$\begin{aligned}
N_1 &= \lambda\,(2\lambda - 1), & N_4 &= 4\,\xi\,\lambda, \\
N_2 &= \xi\,(2\xi - 1), & N_5 &= 4\,\xi\,\eta, \\
N_3 &= \eta\,(2\eta - 1), & N_6 &= 4\,\eta\,\lambda,
\end{aligned} \tag{143}$$

with the abbreviation $\lambda = 1 - \xi - \eta$.

Next the shape functions for quadrilateral elements are defined. The simplest quadrilateral has 4 nodes. The associated interpolation for geometry and field variables is bi–linear.

$$N_I\,(\xi, \eta) = \frac{1}{2}\,(1 + \xi_I\,\xi)\,\frac{1}{2}\,(1 + \eta_I\,\eta). \tag{144}$$

where the coordinates ξ_I und η_I are associated with the vertices, see Fig. 10 on the reference element Ω_0.

$$\xi_1 = (-1, \; -1) \quad \xi_2 = (1, \; -1) \quad \xi_3 = (1, \; 1) \quad \xi_4 = (-1, \; 1). \tag{145}$$

The shape functions for the quadratic 9–node element follow again from the product formula (141) using quadratical interpolation. We obtain for the nodes, see Fig. 10,

- Vertices $(I = 1, 2, 3, 4)$:

$$N_I\,(\xi, \eta) = \frac{1}{4}\,(\xi^2 + \xi_I\,\xi)\,(\eta^2 + \eta_I\,\eta), \tag{146}$$

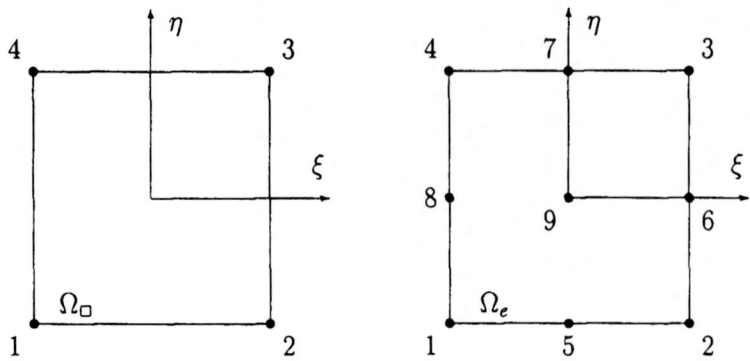

Figure 10: Isoparametric quadrilateral elements.

- Mid nodes $(I = 5, 6, 7, 8)$:

$$N_I(\xi, \eta) = \frac{1}{2}\xi_I^2(\xi^2 - \xi_I\,\xi)(1 - \eta^2) + \frac{1}{2}\eta_I^2(\eta^2 - \eta_I\,\eta)(1 - \xi^2). \qquad (147)$$

- and central node $(I = 9)$:

$$N_9(\xi, \eta) = (1 - \xi^2)(1 - \eta^2). \qquad (148)$$

It should be noted that this is not the only possibility to define these nine shape functions. Often a hierarchical formulation is used, see e. g. [ZT89].

The derivatives of the shape functions defined in the reference coordinates with respect to the coordinates in the physical space follow within the isoparametric concept by the chain rule

$$\frac{\partial u_e}{\partial X_\alpha} = \sum_{I=1}^{n} \frac{\partial N_I(\xi, \eta)}{\partial X_\alpha} u_I, \qquad (\alpha = 1, 2). \qquad (149)$$

Here the partial derivative of N_I with respect to X_α is computed according to (138)

$$\nabla_X N_I = \left\{ \begin{matrix} N_{I,1} \\ N_{I,2} \end{matrix} \right\} = J_e^{-T} \left\{ \begin{matrix} N_{I,\xi} \\ N_{I,\eta} \end{matrix} \right\}, \qquad (150)$$

with the JACOBI matrix J_e of an Ω_e element for the transformation between reference and initial configuration

$$J_e = \sum_{I=1}^{n} X_I \otimes \nabla_\xi N_I = \sum_{I=1}^{n} \left\{ \begin{matrix} X_{1I} \\ X_{2I} \end{matrix} \right\} \left\{ \begin{matrix} N_{I,\xi} \\ N_{I,\eta} \end{matrix} \right\}^T = \left[\begin{matrix} X_{1,\xi} & X_{1,\eta} \\ X_{2,\xi} & X_{2,\eta} \end{matrix} \right],$$

with $\quad X_{\alpha,\beta} = \sum_{I=1}^{n} N_{I,\beta} X_{\alpha I}. \qquad (151)$

This leads to an explicit from which allows to compute in (149) the derivatives with respect to $\mathbf{X}$

$$\begin{Bmatrix} N_{I,1} \\ N_{I,2} \end{Bmatrix} = \frac{1}{\det \boldsymbol{J}_e} \begin{bmatrix} X_{2,\eta} & -X_{2,\xi} \\ -X_{1,\eta} & X_{1,\xi} \end{bmatrix} \begin{Bmatrix} N_{I,\xi} \\ N_{I,\eta} \end{Bmatrix} . \tag{152}$$

5.1.3 Three dimensional shape functions

Finite elements for three–dimensional problems are either brick– or tetrahedron elements. Also here are isoparametric interpolations advantageous when arbitrary geometries have to be discretized. Besides bricks and tetrahedrons there are of course more elements possible, e. g. prismatic elements, which will not be discussed here. For general shape functions, see [DT85].

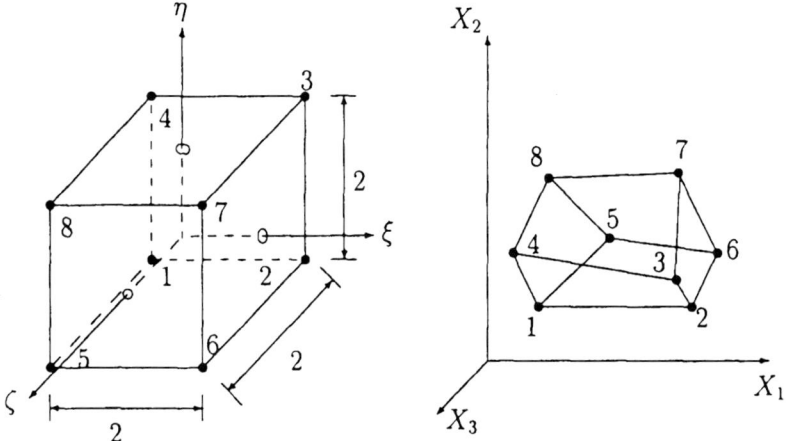

Figure 11: Isoparametric 8 node brick element.

For the three dimensional brick element, shown in Fig. 11, the following shape functions are used

$$N_I = \frac{1}{2}\left(1 + \xi_I \xi\right)\frac{1}{2}\left(1 + \eta_I \eta\right)\frac{1}{2}\left(1 + \phi_I \phi\right), \tag{153}$$

which result from the product formula (141). Fig. 11 depicts the associated element in its reference configuration, $\Omega_\square$, and the initial configuration, Ω_e. Quadratic elements can be designed with (141). This yields an interpolation with 27 nodes per element. However we will not give here the explicit representation which can be found in e. g. [DT85].

Shape functions for the tetrahedron elements can be developed analogously We obtain

- 4–node tetrahedron (linear interpolation)

$$N_1 = 1 - \xi - \eta - \zeta, \quad N_2 = \xi, \quad N_3 = \eta, \quad N_4 = \zeta. \tag{154}$$

- 10–nodes tetrahedron (quadratic interpolation)

$$
\begin{aligned}
&N_1 = \lambda\,(2\lambda-1), && N_6 = 4\xi\,\eta, \\
&N_2 = \xi\,(2\xi-1), && N_7 = 4\eta\,\lambda, \\
&N_3 = \eta\,(2\eta-1), && N_8 = 4\zeta\,\lambda, \\
&N_4 = \zeta\,(2\zeta-1), && N_9 = 4\xi\,\zeta, \\
&N_5 = 4\xi\,\lambda, && N_{10} = 4\eta\,\zeta,
\end{aligned}
\tag{155}
$$

with $\lambda = 1 - \xi - \eta - \zeta$

the local node numbers associated with these shape functions are depicted in Fig. 12.

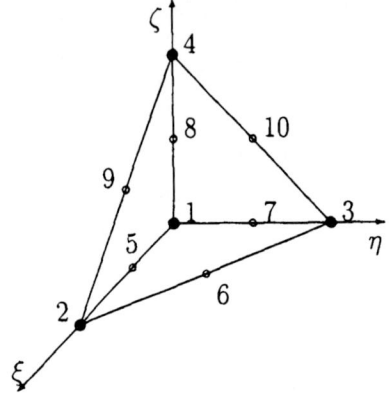

Figure 12: Isoparametric tetrahedron element, local node numbers.

The derivatives of the shape functions with respect to the coordinates of the initial or current configuration can be computed using (137). For the derivatives with respect to the coordinates of the initial configuration we have

$$
\nabla_X N_I = \left\{ \begin{array}{c} N_{I,1} \\ N_{I,2} \\ N_{I,3} \end{array} \right\} = J_e^{-T} \left\{ \begin{array}{c} N_{I,\xi} \\ N_{I,\eta} \\ N_{I,\zeta} \end{array} \right\}.
\tag{156}
$$

The JACOBI matrix J_e of element Ω_e, which is needed in this derivation, is given by (135) from

$$
J_e = \sum_{I=1}^{n} X_I \otimes \nabla_\xi N_I = \begin{bmatrix} X_{1,\xi} & X_{1,\eta} & X_{1,\zeta} \\ X_{2,\xi} & X_{2,\eta} & X_{2,\zeta} \\ X_{3,\xi} & X_{3,\eta} & X_{3,\zeta} \end{bmatrix}.
\tag{157}
$$

within this formula the components of J_e are computed from

$$
X_{m,k} = \sum_{I=1}^{n} N_{I,k}\,X_{m\,I},
$$

where the partial derivative with respect to k stands for a derivative with respect to ξ, η or ζ.

5.2   Discretization of the weak form

In general we can now apply the shape functions to decribe the interpolation for the geometry and the field variables within the weak forms. In this section we will do this in a brief form for the equations (44) and (47). Furthermore the linearizations of the weak forms are considered.

An isoparametric interpolation is chosen for each finite element Ω_e, which approximates the displacement field u and the geometry. The integrals of the weak form can then be written with the isoprametric interpolation as

$$\int_{B_R} (\dots)\, dV \approx \int_{B_{Rh}} (\dots)\, dV = \bigcup_{e=1}^{n_e} \int_{\Omega_e} (\dots)\, d\Omega = \bigcup_{e=1}^{n_e} \int_{\Omega_\square} (\dots)\, d\square \qquad (158)$$

The operator $\cup$ is introduced instead of a sum sign to denote the assembly process which has to be performed to obtain the set of nonlinear algebraic equations following from (158). The polynominal shape functions of the isoparametric interpolation ensures the fulfillment of the inter element continuity conditions as well as the fulfillment of the boundary conditions within the global system of equations. Since the assembly process is standard and well known it will not described in detail here, see e.g. [Bat82], [ZT89] or [GHSW99].

5.2.1   FE discretization of the weak form for rolling contact

The approximation of the weak form (47) requires the discretization of the virtual internal work $\int_B \mathbf{S} \cdot \delta\mathbf{E}\, dV$, of the inertia terms $\int_B \rho_0 \dot{\mathbf{v}} \cdot \boldsymbol{\eta}\, dV$ and of the volume– and surface loads $\int_B \rho_0\, \bar{\mathbf{b}} \cdot \boldsymbol{\eta}\, dV + \int_\Gamma \bar{\mathbf{t}} \cdot \boldsymbol{\eta}\, dA$. For the virtual internal work we need the variation of the GREEN–LAGRANGIAN strain tensors within the element Ω_e, see (158). With (136) one obtains

$$\delta\mathbf{E}_e = \frac{1}{2} \sum_{I=1}^{n} \left[\mathbf{F}_e^T \left(\boldsymbol{\eta}_I \otimes \nabla_X N_I \right) + \left(\nabla_X N_I \otimes \boldsymbol{\eta}_I \right) \mathbf{F}_e \right]. \qquad (159)$$

In this equation a finite element approximation of the deformation gradient (5) has to be applied which can be written with (136) within the element Ω_e as

$$\mathbf{F}_e = \sum_{K=1}^{n} \left(\mathbf{x}_K \otimes \nabla_X N_K \right). \qquad (160)$$

For the derivation of the matrix formulation needed within the computer implementation of finite elements index notation is necessary. This yields for (159)

$$\delta E_{e\, AB} = \frac{1}{2} \sum_{I=1}^{n} \left[F_{Ak}\, N_{I,B} + N_{I,A}\, F_{kB} \right] \eta_{k\, I} \qquad (161)$$

with components of the deformation gradienten $F_{kB} = \sum_{J=1}^{n} x_{kJ}\, N_{J,B}$.

Within the matrix formulation we can consider the symmetry of the GREEN–LAGRANGIAN strain tensor and its variation. Then it is possible to introduce instead of nine components for the three dimensional strain tensor only six components

$$\delta E_e = \begin{Bmatrix} \delta E_{11} \\ \delta E_{22} \\ \delta E_{33} \\ 2\,\delta E_{12} \\ 2\,\delta E_{23} \\ 2\,\delta E_{13} \end{Bmatrix}_e = \sum_{I=1}^{n} B_{LI}\,\eta_I, \tag{162}$$

which can be approximated as a sum over the element nodes I with the matrices

$$B_{LI} = \begin{bmatrix} F_{11}\,N_{I,1} & F_{21}\,N_{I,1} & F_{31}\,N_{I,1} \\ F_{12}\,N_{I,2} & F_{22}\,N_{I,2} & F_{32}\,N_{I,2} \\ F_{13}\,N_{I,3} & F_{23}\,N_{I,3} & F_{33}\,N_{I,3} \\ F_{11}\,N_{I,2} + F_{12}\,N_{I,1} & F_{21}\,N_{I,2} + F_{22}\,N_{I,1} & F_{31}\,N_{I,2} + F_{32}\,N_{I,1} \\ F_{12}\,N_{I,3} + F_{13}\,N_{I,2} & F_{22}\,N_{I,3} + F_{23}\,N_{I,2} & F_{32}\,N_{I,3} + F_{33}\,N_{I,2} \\ F_{11}\,N_{I,3} + F_{13}\,N_{I,1} & F_{21}\,N_{I,3} + F_{23}\,N_{I,1} & F_{31}\,N_{I,3} + F_{33}\,N_{I,1} \end{bmatrix} \tag{163}$$

The index L depicts in (162) that the matrix B_{LI} is linear in the displacements since we have $F_e = 1 + \mathrm{Grad}\,u_e$.

The stresses follow from the constitutive equation which will be specified in the associated sections. However we note, that the stresses have to be computed pointwise within the element and result e. g. in finite elasticity from a pure function evaluation of the response function. Since also the 2nd PIOLA–KIRCHHOFF stress tensor is symmetric, we need only its six independent components which yields the vector $S = \{\,S_{11}, S_{22}, S_{33}, S_{12}, S_{23}, S_{13}\,\}^T$. With these preliminaries, the virtual internal work can be written as

$$\int_{B_R} \delta E \cdot S \, dV \;=\; \bigcup_{e=1}^{n_e} \int_{\Omega_e} \delta E^T\, S_e \, d\Omega$$

$$= \bigcup_{e=1}^{n_e} \sum_{I=1}^{n} \eta_I^T \int_{\Omega_e} B_{LI}^T\, S_e \, d\Omega \tag{164}$$

$$= \bigcup_{e=1}^{n_e} \sum_{I=1}^{n} \eta_I^T \int_{\Omega_\square} B_{LI}^T\, S_e \, \det J_e \, d\square .$$

The last term in (165) reflects already the evaluation of the integrals with respect to the configuration of the isoparametric reference element. To shorten notation we introduce the vector

$$R_I\,(u_e) = \int_{\Omega_e} B_{LI}^T\, S_e \, d\Omega \tag{165}$$

and reformulate the virtual internal work

$$\int_{B_r} \delta \mathbf{E} \cdot \mathbf{S} \, dV = \bigcup_{e=1}^{n_e} \sum_{I=1}^{n} \boldsymbol{\eta}_I^T \mathbf{R}_I \left(\mathbf{u}_e \right) = \boldsymbol{\eta}^T \mathbf{R} \left(\mathbf{u} \right). \tag{166}$$

In this equation $\boldsymbol{\eta}$ is the test function of virtual displacement and $\mathbf{R}(\mathbf{u})$ is the stress divergence term also often called residual force vector which results from the assembly of all finite elements to the complete structure.

The loading terms are determined in an analogous way. After inserting the finite element approximations for the test function $\boldsymbol{\eta}$, it follows

$$\int_{B_R} \rho_0 \, \boldsymbol{\eta} \cdot \bar{\mathbf{b}} \, dV + \int_{\Gamma_{R\sigma}} \boldsymbol{\eta} \cdot \bar{\mathbf{t}} \, dA \;=\; \bigcup_{e=1}^{n_e} \sum_{I=1}^{n} \boldsymbol{\eta}_I^T \int_{\Omega_e} \rho_0 \, \bar{\mathbf{b}} \, N_I \, d\Omega$$
$$+ \bigcup_{r=1}^{n_r} \sum_{I=1}^{m} \boldsymbol{\eta}_I^T \int_{\Gamma_r} N_I \, \bar{\mathbf{t}} \, d\Gamma,$$

where n_r are the number of loaded element boundaries and Γ_l is the element surface of an element which is subjected to a surface load defined by the stress vector $\bar{\mathbf{t}}$, see Fig. 13. Observe that for the interpolation function of the surface loads we can use a function which is reduced by one dimension. Thus the surface loads in Fig. 13 which depicts a two dimensional body need as an approximation for the test function along the boundary a one dimensional function which are defined by m surface nodes (in Fig. 13 we have $m = 2$ nodes).

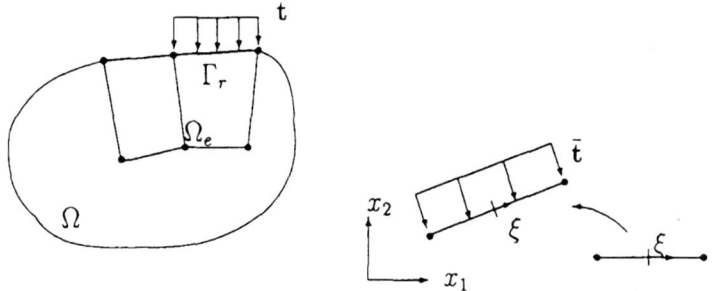

Figure 13: Discretization of surface loads.

Also here we can simplify by matrix notation and obtain with

$$\mathbf{P}_I = \int_{\Omega_e} N_I \, \rho \, \bar{\mathbf{b}} \, d\Omega \quad \text{und} \quad \mathbf{P}_I^\sigma = \int_{\Gamma_r} N_I \, \bar{\mathbf{t}} \, d\Gamma \tag{167}$$

the load vectors

$$\int_{B_R} \rho \boldsymbol{\eta} \cdot \bar{\mathbf{b}} \, dV + \int_{\Gamma_{R\sigma}} \boldsymbol{\eta} \cdot \bar{\mathbf{t}} \, dA = \bigcup_{e=1}^{n_e} \sum_{I=1}^{n} \boldsymbol{\eta}_I^T \mathbf{P}_I + \bigcup_{r=1}^{n_r} \sum_{I=1}^{n} \boldsymbol{\eta}_I^T \mathbf{P}_I^\sigma = \boldsymbol{\eta}^T \mathbf{P}. \tag{168}$$

The vector P contains all information with regard to the loads acting on the structure.

The matrix form of the terms due to the rotating reference coordinate system appearing in (47) yields

$$\int_{B_R} \rho_R \left(\hat{\mathbf{F}} \, \Omega_R \times_R \right) \cdot \left(\widehat{\mathrm{Grad}\, \eta} \, \Omega_R \times_R \right) dV_R$$

$$= \bigcup_{e=1}^{n_e} \int_{\Omega_e} (\Omega_R \times_R)^T \widehat{\mathrm{Grad}\eta}^T (\hat{\mathbf{F}}_e \Omega_R \times_R)\, d\Omega$$

$$= \bigcup_{e=1}^{n_e} \sum_{I=1}^{n} \eta_I^T \int_{\Omega_e} [(\Omega_R \times_R)^T \nabla N_I] (\hat{\mathbf{F}}_e \Omega_R \times_R)\, d\Omega \qquad (169)$$

$$= \bigcup_{e=1}^{n_e} \sum_{I=1}^{n} \eta_I^T \int_{\Omega_\square} [(\Omega_R \times_R)^T \nabla N_I] (\hat{\mathbf{F}}_e \Omega_R \times_R) \det J_e\, d\square .$$

Here we have made use of equation (136) which yields in matrix form $\sum_{I=1}^{n} \eta_I \nabla N_I$. Like in (165) we introduce the vector

$$\mathbf{Q}_I(u_e) = \int_{\Omega_e} [(\Omega_R \times_R)^T \nabla N_I] (\hat{\mathbf{F}}_e \Omega_R \times_R)\, d\Omega \qquad (170)$$

The matrix notation in (166), (168) and (170) yields now for the weak form (47)

$$\eta^T [\mathbf{R}(u) + \mathbf{Q}(u) - \mathbf{P}] = 0 . \qquad (171)$$

Due to the fact that the test function η is arbitrary, this leads to a nonlinear system of ordinary differential equations.

$$\mathbf{R}(u) + \mathbf{Q}(u) - \mathbf{P} = 0 \qquad \forall u \in \mathbb{R}^N . \qquad (172)$$

In (172) all quantities are evaluated with respect to the rotating frame B_R configuration. N is the total number of degree of freedoms which are contained in the unknown displacement vector u.

5.2.2 Linearization of the weak form

For an efficient solution of the nonlinear algebraic equation systems (172) NEWTON's method is applied which requires the linearization of (172). The linearization can be obtained by a direct discretization of the continuous formulation (73)

$$DG(\bar{\varphi}, \eta) \cdot \Delta u = \int_{B_R} \{ \mathrm{Grad}\, \Delta u\, \bar{\mathbf{S}} \cdot \mathrm{Grad}\, \eta + \delta \bar{\mathbf{E}} \cdot \bar{\mathbb{C}} [\Delta \bar{\mathbf{E}}] \}\, dV, . \qquad (173)$$

For the first term we obtain directly with

$$\mathrm{Grad}\,\Delta u^h \;=\; \sum_{K=1}^{n} \Delta u_K \otimes \nabla_X N_K \,,$$

$$\mathrm{Grad}\,\eta \;=\; \sum_{I=1}^{n} \eta_I \otimes \nabla_X N_I \tag{174}$$

the dicretization

$$\int_{B_R} \mathrm{Grad}\Delta u\,\bar{S} \cdot \mathrm{Grad}\,\eta\,dV = \bigcup_{e=1}^{n_e} \sum_{I=1}^{n} \sum_{K=1}^{n} \int_{\Omega_e} (\Delta u_K \otimes \nabla_X N_K)\,\bar{S}_e \cdot (\eta_I \otimes \nabla_X N_I)\,d\Omega\,,$$

which yields, with the rules for the dyadic and scalar products and with $\Delta u_K \cdot \eta_I = \eta_I^T \Delta u_K = \eta_I^T I \Delta u_K$,

$$\int_{B_R} \mathrm{Grad}\,\Delta u\,\bar{S} \cdot \mathrm{Grad}\,\eta\,dV = \bigcup_{e=1}^{n_e} \sum_{I=1}^{n} \sum_{K=1}^{n} \eta_I^T \int_{\Omega_e} \bar{G}_{IK}\,I\,d\Omega\,\Delta u_K \tag{175}$$

where the abbreviation

$$\bar{G}_{IK} = (\nabla_X N_I)^T\,\bar{S}_e\,\nabla_X N_K \tag{176}$$

has been used. The matrix form of the scalar product (176) can be derived if the gradients are described as vectors. This leads to

$$\bar{G}_{IK} = \begin{bmatrix} N_{I,1} & N_{I,2} & N_{I,3} \end{bmatrix} \begin{bmatrix} \bar{S}_{11} & \bar{S}_{12} & \bar{S}_{13} \\ \bar{S}_{21} & \bar{S}_{22} & \bar{S}_{23} \\ \bar{S}_{31} & \bar{S}_{32} & \bar{S}_{33} \end{bmatrix}_e \begin{Bmatrix} N_{K,1} \\ N_{K,2} \\ N_{K,3} \end{Bmatrix}\,. \tag{177}$$

Relation (175) is independent from the constitutive equation since only the stress at configuration $\bar{\varphi}$ has to be considered. Hence the matrix which is defined by (175) is often called initial stress matrix.

The second term in (73)

$$\int_{B_R} \delta\bar{E} \cdot \mathbb{C}[\Delta\bar{E}]\,dV$$

depends on the incremental constitutive tensor $\mathbb{C}$ which has to be evaluated at configuration $\bar{\varphi}$ and thus is directly connected to the constitutive equation. Since $\Delta\bar{E}$ has the same structure as $\delta\bar{E}$ we can write with (159)

$$\Delta E_e = \frac{1}{2} \sum_{I=1}^{n} \left[F_e^T\,(\Delta u_I \otimes \nabla_X N_I) + (\nabla_X N_I \otimes \Delta u_I)\,F_e \right]\,. \tag{178}$$

Now the matrix formulation follows with (163)

$$\Delta E_e = \sum_{I=1}^{n} B_{LI}\,\Delta u_I\,. \tag{179}$$

Introduction of this relation yields together with the incremental constitutive tensor $\bar{D}$

$$\int_{B_R} \delta\bar{\mathbf{E}} \cdot \bar{\mathbb{C}}\,[\Delta\bar{\mathbf{E}}]\,dV = \bigcup_{e=1}^{n_e} \sum_{I=1}^{n} \sum_{K=1}^{n} \boldsymbol{\eta}_I^T \int_{\Omega_e} \bar{B}_{LI}^T\,\bar{D}\,\bar{B}_{LK}\,d\Omega\,\Delta\mathbf{u}_K \qquad (180)$$

Thus we can summarize and obtain the discretization

$$\int_{B_R} \{\,\mathrm{Grad}\,\Delta\mathbf{u}\,\bar{\mathbf{S}} \cdot \mathrm{Grad}\,\boldsymbol{\eta} + \delta\bar{\mathbf{E}} \cdot \bar{\mathbb{C}}\,[\Delta\bar{\mathbf{E}}]\,\} \,dV = \bigcup_{e=1}^{n_e} \sum_{I=1}^{n} \sum_{K=1}^{n} \boldsymbol{\eta}_I^T\,\bar{K}_{T_{IK}}\Delta\mathbf{u}_K\,, \qquad (181)$$

Here matrix $\bar{K}_{IK}$ denotes the "tangent matrix" because it represents the tangent to the deformation at $\bar{\varphi}$

$$\bar{K}_{T_{IK}} = \int_{\Omega_e} \left[(\nabla_X N_I)^T\,\bar{\mathbf{S}}_e\,\nabla_X N_K + \bar{B}_{LI}^T\,\bar{D}\,\bar{B}_{LK} \right] d\Omega \qquad (182)$$

It is stated for the nodal combination I, K within a finite element Ω_e.

The contribution to the tangent matrix due to the terms referring to the rotating reference coordinate system appearing in (74) yields

$$\int_{B_R} \widetilde{\mathrm{Grad}\,\Delta\mathbf{u}}\,\bar{\mathbf{S}} \cdot \widetilde{\mathrm{Grad}\,\boldsymbol{\eta}}\,dV = \bigcup_{e=1}^{n_e} \sum_{I=1}^{n} \sum_{K=1}^{n} \boldsymbol{\eta}_I^T \int_{\Omega_e} \left[(\Omega_R\,\mathbf{x}_R)^T \nabla N_I\right] \left[\nabla N_K^T (\Omega_R\,\mathbf{x}_R)\right] d\Omega\,\Delta\mathbf{u}_K$$

$$(183)$$

It has to be added to (182).

In this notation the submatrix $\bar{K}_{T_{IK}}$ in (182) has the size $n_{dof} \times n_{dof}$, where n_{dof} is the number of degrees of freedom for one node within the finite element (in three-dimensional problems in continuum mechanics we have three degrees of freedom for each point, hence $n_{dof} = 3$). Indices I and K are nodes of an element and thus directly associated with the discretization. E.g. for a ten node tetrahedron we have $n = 10$, hence the total size of the tangent matrix $\bar{K}_{T_e}$ for one element is $(n \cdot n_{dof}) \times (n \cdot n_{dof}) = 30 \times 30$.

6   Discretization of contact contributions

The matrix formulation for the weak which is related to the continuum has been developed in the last section. Here we focus on the contact constraints. For reasons of simplicity we will restrict ourselves here to two dimensional formulations. Three dimensional contact discretizations can be found in e.g. [HGB85], [LS93] or [HC93].

When the discretization of surfaces in rolling contact is concerned one has to distinguish between between steady and non-steady processes and the contact of two deformable bodies or the contact of a deformable body with a rigid obstacle. On a first glance it seems that the latter case is simply a special case of the first problem, which is true. But due to the fact that the surface description of a rigid obstacle can be given once and for all by the correct geometrical model this knowledge can be used within the discretization process.

In the first applications of finite elements to contact problems of two deformable bodies only small changes in the geometry were assumed so that the geometrically linear theory could be applied. Then it is possible to incorporate the contact constraints on a purely nodal basis, see e.g. [FZ75]. Later also contact elements were developed which resulted from a degenerated solid element, see e.g. [SW79] or the textbook of [KO88]. A mathematical study of these classes of elements which also accounts for the correct integration rules can be found in [Ode81]. All of the above mentioned elements need a discretization in which the element nodes match each other in the contact interface. For the general case of nodes being arbitrary distributed along the possible contact interface between two bodies, which can occur when automatic meshing is used for two different bodies, [SWT85] developed a segment approach to discretize the contact interface.

For the general case of contact including large deformations finite element nodes on the contacting boundaries do not assume the same position, see Fig. 14 Then we define the active part of the contact interface Γ_c where $g_N \leq 0$ for the solution process. Right now most algorithms rely on discretizations which are based on nodes on the boundary of the bodies. Thus the algorithm has to find the active contact constraints denoted by $\mathcal{J}_A \in \mathcal{J}_C$, where $\mathcal{J}_C$ are all possible contact nodes. The most frequently used discretization is the so called node–to–segment approach. Here arbitrary sliding of a node over the entire contact area is allowed. Early implementations can be found in [HTK77] which have been developed for more and more general cases, [HGB85], [BC85] and [WVS90]. Special discretizazions for rolling contact can be found in e.g. [Nac93] and [Nac95].

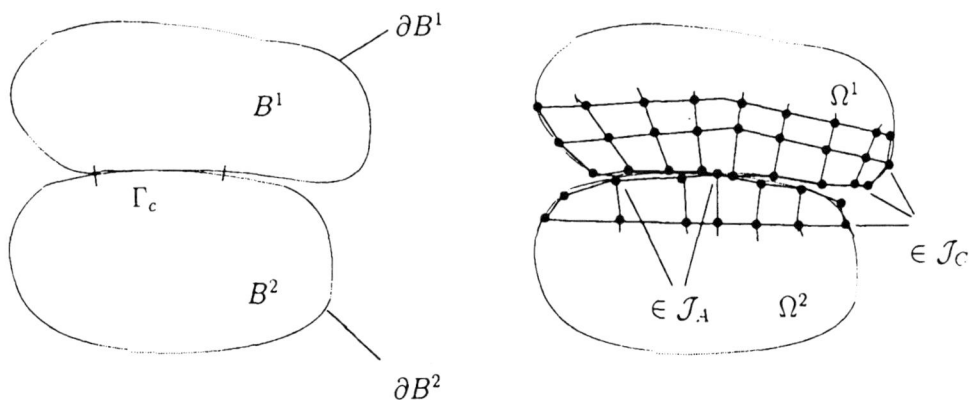

Figure 14: Discretization for large deformation contact.

The basic difference between the Lagrangian multiplier method (113) and the penalty approach (114) lies in the fact that the Lagrangian multiplier formulation is a mixed method which means that both variables λ_N and δg_N have to be discretized

$$\int_{\Gamma_c} \lambda_N \, \delta g_N \, d\Gamma \longrightarrow \int_{\Gamma_c^h} \lambda_N^h \, \delta g_{Nh} \, d\Gamma \tag{184}$$

with interpolations for λ_N^h and δg_{Nh}

$$\lambda_N^h = \sum_K M_K(\xi) \, \lambda_{NK} \quad \text{and} \quad \delta g_{Nh} = \sum_I N_I(\xi) \, \delta g_{NI} \tag{185}$$

Note that the interpolations have to be chosen in such a way that they fulfil the LBB condition since the Lagrangian multiplier metho is a mixed formulation, see e.g. [KO88].

Contrary, the penalty method needs only the discretization of the displacement variables

$$\int_{\Gamma_c} \epsilon_N \, g_{\bar{N}} \, \delta g_{\bar{N}} \, d\Gamma \longrightarrow \int_{\Gamma_c^h} \epsilon_N \, g_{Nh} \, \delta g_{Nh} \, d\Gamma \tag{186}$$

where usually the same interpolation is introduced for the gap function and its variation

$$g_{Nh} = \sum_I N_I(\xi) \, g_{NI} \quad \text{and} \quad \delta g_{Nh} = \sum_I N_I(\xi) \, \delta g_{NI} \tag{187}$$

In the following we will only discuss discretizations related to the penalty method. But also here one has to be careful when choosing the interpolation for the continuous contact. This follows from the fact that the penalty method is equivalent to a mixed method and hence the LBB condition plays again a role, see e.g. [KO88], for a detailed discussion of this matter.

6.1   Node–to–segment contact discretization

For the case of nonlinear deformations we will discuss the most simple element which is widely used in standard finite element codes. This discretization is named node–to–segment contact element and is widely used in nonlinear finite element simulations of contact problems. Due to its importance we like to consider this contact element in more detail. Assume that the discrete slave point (s) of body B^2 with coordinate x_s^2 comes into contact with the master segment (1)–(2) of body B^1 defined by the nodal coordinates x_1^1 and x_2^1, see Fig. 15. Then the kinematical relations can be directly computed using the equations stated in section 3. With linear interpolation for the master segment based on the introduction of the surface coordinate ξ along the master surface

$$\hat{x}^1(\xi) = x_1^1 + (x_2^1 - x_1^1) \, \xi \tag{188}$$

we derive that each segment has a constant tangent vector

$$\bar{a}_1^1 = \hat{x}^1(\xi),_1 = (x_2^1 - x_1^1) \tag{189}$$

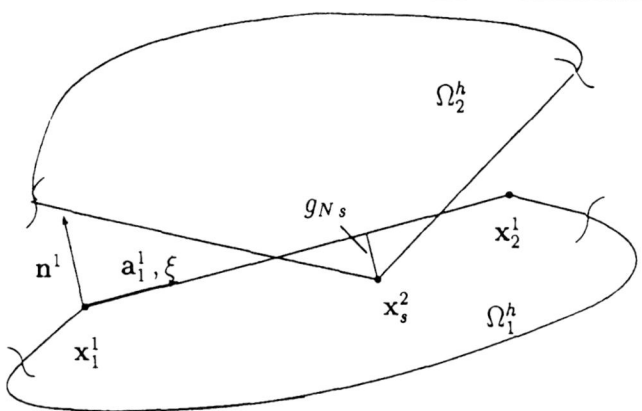

Figure 15: Node–to–segment contact element.

It is connected to an orthonormal base vector $\mathbf{a}_1^1$ by $\mathbf{a}_1^1 = \bar{\mathbf{a}}_1^1 / l$ with $l = \parallel \mathbf{x}_2^1 - \mathbf{x}_1^1 \parallel$ being the current length of the master segment. With the unit tangent vector $\mathbf{a}_1^1$ the unit normal to the segment (1)–(2) can be defined as $\mathbf{n}^1 = \mathbf{e}_3 \times \mathbf{a}_1^1$.

$\bar{\xi}$ and g_{Ns} are given by the solution of the minimal distance problem, i.e. by the projection of the slave node $\mathbf{x}_s$ in (s) onto the master segment (1)–(2)

$$\bar{\xi} = \frac{1}{l}(\mathbf{x}_s^2 - \mathbf{x}_1^1) \cdot \mathbf{a}_1^1 \qquad \text{and} \qquad g_{Ns} = [\mathbf{x}_s^2 - (1 - \bar{\xi})\mathbf{x}_1^1 - \bar{\xi}\mathbf{x}_2^1] \cdot \mathbf{n}^1. \qquad (190)$$

From these equations and the local continuous formulation (117) we compute directly the variation of the gap function δg_N on the straight master segment (1)–(2).

$$\delta g_{Ns} = [\boldsymbol{\eta}_s^2 - (1 - \bar{\xi})\boldsymbol{\eta}_1^1 - \bar{\xi}\boldsymbol{\eta}_2^1] \cdot \mathbf{n}^1. \qquad (191)$$

The local equation (119) yields the expression for $\delta\bar{\xi}$. With the interpolation for the variation $\hat{\boldsymbol{\eta}}^1(\xi) = \boldsymbol{\eta}_1^1 + \xi(\boldsymbol{\eta}_2^1 - \boldsymbol{\eta}_1^1)$ on the straight master segment (1)–(2) we specialize

$$\begin{aligned} \bar{H}_{\alpha\beta} &= (a_{\alpha\beta} + g_N b_{\alpha\beta}) \Longrightarrow \bar{H}_{11} = a_{11} = l^2 \\ \bar{R}_1 &= [\boldsymbol{\eta}^2 - \hat{\boldsymbol{\eta}}^1(\bar{\xi})] \cdot \bar{\mathbf{a}}_1^1 + g_{Ns}\bar{\mathbf{n}}^1 \cdot \hat{\boldsymbol{\eta}}_1^1(\bar{\xi})_{,\xi} \end{aligned} \qquad (192)$$

which leads to

$$\delta g_T = l\delta\bar{\xi} = [\boldsymbol{\eta}_s^2 - (1 - \bar{\xi})\boldsymbol{\eta}_1^1 - \bar{\xi}\boldsymbol{\eta}_2^1] \cdot \mathbf{a}_1^1 + \frac{g_{Ns}}{l}[\boldsymbol{\eta}_2^1 - \boldsymbol{\eta}_1^1] \cdot \mathbf{n}^1. \qquad (193)$$

Equations (191) and (193) characterize the main kinematical relations of the contact element in Figure 10.

In what follows we compute the contribution of the node–to–segment element to the weak form (114). The basic formulation for this discretization is analogous to the node–to–node element. Thus we assume that we know the normal force $P_{Ns} = p_{Ns} A_s$ and the tangential force $T_{Ts} = t_{Ts} A_s$ at the discrete contact point (s) of the contact element under consideration where A_s denotes the area of the contact element. Both forces, P_{Ns} and T_{Ts}, can be obtained from the constitutive relations discussed above. This leads to

$$\int_{\Gamma_c} (p_N\, \delta g_N + t_T\, \delta g_T)\, d\Gamma \longrightarrow \sum_{s=1}^{n_c} (P_{Ns}\, \delta g_{Ns} + T_{Ts}\, \delta g_{Ts}) \tag{194}$$

In practice we compute the normal force P_{Ns} from the penalty update $P_{Ns} = \epsilon\, g_{Ns}$ multiplied by the area of the contact element. For the tangential force T_{Ts} we have to perform an algorithmic update, see last section. Thus the contributions of one contact element takes the form

$$\delta g_{Ns}\, P_{Ns} + \delta g_{Ts}\, T_{Ts} \tag{195}$$

for the discrete contact point (s) with the mechanical relative variations analogous to (191) and (193). This equations can now be cast into a matrix formulation. For the normal part $(195)_1$ we set for the variation (191) of the penetration

$$\delta g_{Ns} = \boldsymbol{\eta}^T \mathbf{N}_s . \tag{196}$$

With the same notation we can express the variation (193) of the tangential gap

$$\delta g_{Ts} = \boldsymbol{\eta}^T \left(\mathbf{T}_s + \frac{g_{Ns}}{l}\, \mathbf{N}_{0s} \right) . \tag{197}$$

In these equations the following vectors have been used

$$\boldsymbol{\eta} = (\, \eta_s^2 \quad \eta_1^1 \quad \eta_2^1 \,)^T , \tag{198}$$

$$\mathbf{N}_s = \left\{ \begin{array}{c} \mathbf{n}^1 \\ -(1 - \bar{\xi})\, \mathbf{n}^1 \\ -\bar{\xi}\, \mathbf{n}^1 \end{array} \right\}_s , \qquad \mathbf{N}_{0s} = \left\{ \begin{array}{c} 0 \\ -\mathbf{n}^1 \\ \mathbf{n}^1 \end{array} \right\}_s , \tag{199}$$

and

$$\mathbf{T}_s = \left\{ \begin{array}{c} \mathbf{a}_1^1 \\ -(1 - \bar{\xi})\, \mathbf{a}_1^1 \\ -\bar{\xi}\, \mathbf{a}_1^1 \end{array} \right\}_s , \qquad \mathbf{T}_{0s} = \left\{ \begin{array}{c} 0 \\ -\mathbf{a}_1^1 \\ \mathbf{a}_1^1 \end{array} \right\}_s . \tag{200}$$

Thus the virtual mechanical work of the contact element can be written in the matrix formulation $\boldsymbol{\eta}^T \mathbf{G}_s^c$ with the contact element residual

$$\mathbf{G}_s^c = P_{Ns}\, \mathbf{N}_s + T_{Ts} \left(\mathbf{T}_s + \frac{g_{Ns}}{l} \mathbf{N}_{0s} \right) . \tag{201}$$

Due to this approach a pure displacement formulation of the contact problem is possible by expressing P_{Ns} through the penalty relation $P_{Ns} = \epsilon_N\, g_{Ns}$. This is in contrast to the Lagrangian multiplier technique, where $P_{Ns} = \lambda_{Ns}$. But we observe that this discretization can be applied to both methods.

Often a Newton–Raphson iteration is used to solve the global set of equations. Then the linearization of (201) is needed to achieve quadratic convergence near the solution point. The associated derivation is a little bit cumbersome and thus only the final results will be summarized for this discretization. Details of the frictionless case can be found in [WS85] and for contact including friction in [WVS90]. The tangent matrix for the normal contact is derived from the term $\delta g_{Ns} P_{Ns}$. Note that in (191) the change in $\bar{\xi}$ has to be considered as well as the change of the normal $\mathbf{n}^1$. For the penalty approach with $P_{Ns} = \epsilon_N g_{Ns}$ we obtain the tangent matrix

$$\mathbf{K}_{Ns}^c = \epsilon_N \left[\mathbf{N}_s \mathbf{N}_s^T - \frac{g_{Ns}}{l} \left(\mathbf{N}_{0s} \mathbf{T}_s^T + \mathbf{T}_s \mathbf{N}_{0s}^T + \frac{g_{Ns}}{l} \mathbf{N}_{0s} \mathbf{N}_{0s}^T \right) \right] \qquad (202)$$

The used matrices have been defined in (199) and (200). Note that in a geometrically linear case all terms vanish which are multiplied by g_{Ns}. This gives the simple matrix $\mathbf{K}_{Ns}^{Lc} = \epsilon_N \mathbf{N}_s \mathbf{N}_s^T$.

For the tangential contributions in the contact area we have to linearize the term $\delta g_{Ts} T_{Ts}$ which yields for the pure stick

$$\begin{aligned}
\mathbf{K}_{Ts}^c = c_T \Big\{ &(\mathbf{T}_s + \frac{g_{Ns}}{l} \mathbf{N}_{0s})(\mathbf{T}_s + \frac{g_{Ns}}{l} \mathbf{N}_{0s})^T \\
&+ \frac{g_{Ns}}{l} \Big[\mathbf{N}_{0s} \mathbf{N}_s^T + \mathbf{N}_s \mathbf{N}_{0s}^T - \mathbf{T}_{0s} \mathbf{T}_s^T - \mathbf{T}_s \mathbf{T}_{0s}^T \\
&- 2 \frac{g_{Ns}}{l} (\mathbf{N}_{0s} \mathbf{T}_{0s}^T + \mathbf{T}_{0s} \mathbf{N}_{0s}^T) \Big] \Big\}
\end{aligned} \qquad (203)$$

Also in this case all terms containing g_{Ns} disappear in a geometrically linear situation which yields $\mathbf{K}_{Ts}^{Lc} = c_T \mathbf{T}_s \mathbf{T}_s^T$. The case of frictional slip leads to an additional contribution in (203).

This expression is obtained together with the local update algorithm for the frictional contact, see (129). In the two dimensional case of the node–to–segement contact element as discussed above, the explicit matrix form results from the term $\delta g_{Ts\,n+1} T_{Ts\,n+1}$ and can be stated for a contacting node (s), with $\mathbf{K}_{Ts}^c$ as

$$\mathbf{K}_{Ts}^{Sc} = \mathbf{K}_{Ts}^c + \mu \epsilon_N \left(\mathbf{T}_s + \frac{g_{Ns}}{l} \mathbf{N}_{0s} \right) \mathbf{N}_s^T \qquad (204)$$

Note that this matrix is non-symmetric which corresponds to the non–associativity of Coulomb's frictional law.

6.2   Smooth Bezier interpolation for frictionless contact

Smooth interpolations in the contact interface can be applied especially for the case of rolling contact in the Lagrangian description. This leads to less noise in the response, since corners are omitted.

Note, that contact simulations based on implicit methods have often been limited due to the C^0 continuity, resulting in loss of convergence. The use of the explicit method was then the only possibility to handle such problems [FCM99]. To overcome the lack of a C^0 interpolation near the master nodes various formulations were developed.

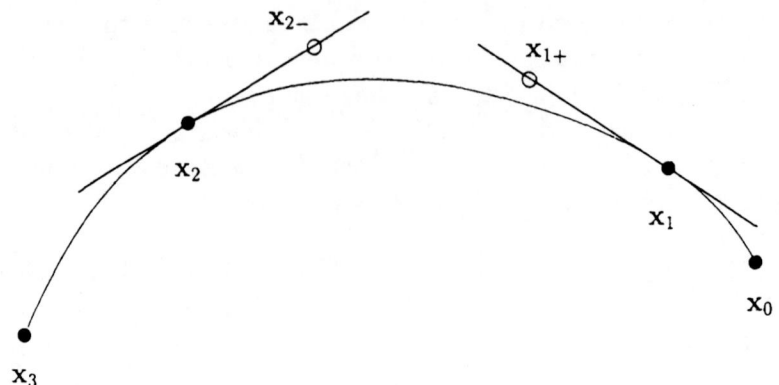

Figure 16: C^1 – continuous Bezier interpolation of contact surface.

Undefined or not uniquely defined normals can be locally treated as special cases like in [HC93] and [BC98]. Furthermore the normal can be averaged like in [PT92] and [WN99]. Also, as described in [LMM99], a continuous normal change with the normal vector that is not perpendicular to the contact surface, can be introduced. Sometimes as in [HA96] or [HK90], the master surface is supposed to be rigid. Hence various CAD C^1 surfaces can be defined, but then these surfaces cannot be deformed.

The simplest smooth interpolation is provided by C^1–continuous shape functions. This interpolation has been developed for frictionless contact in [Pie97] or [TW99]. Bezier polynomials which are introduced here to obtain a continuous normal field are cubic functions. As Hermitian functions these are defined by four points on the master surface, however in a different manner, see Figure 16. The Bezier interpolation for the segment described by node 1 and 2 yields

$$\mathbf{x}(\xi) = B_1(\xi)\mathbf{x}_1 + B_2(\xi)\mathbf{x}_{1+} + B_3(\xi)\mathbf{x}_{2-} + B_4(\xi)\mathbf{x}_2 \tag{205}$$

where the Bezier interpolation functions are defined as

$$
\begin{aligned}
B_1(\xi) &= \frac{1}{8}(1-\xi)^3 \\[4pt]
B_2(\xi) &= \frac{3}{8}(1-\xi)^2(1+\xi) \\[4pt]
B_3(\xi) &= \frac{3}{8}(1-\xi)(1+\xi)^2 \\[4pt]
B_4(\xi) &= \frac{1}{8}(1+\xi)^3
\end{aligned}
\tag{206}
$$

Observe that the interpolation lies in the convex hull spanned by the nodes $\mathbf{x}_1$, $\mathbf{x}_{1+}$, $\mathbf{x}_{2-}$ and $\mathbf{x}_2$.

Our main requirement for the interpolation is, that the tangent vectors of adjacent segments have to be equal to maintain C^1 continuity over segment boundaries. This condition can be applied to compute the interior points of the segment x_{1+}, x_{2-}. By defining the tangent vectors at nodes 1 and 2 as in the previous section we obtain

$$t_1 = \frac{\alpha}{2}(x_2 - x_0) \quad \text{and} \quad t_2 = \frac{\alpha}{2}(x_3 - x_1) \tag{207}$$

Now we take the derivative of (205) and evaluate this at the end points $\xi = -1$ and $\xi = +1$. By setting this equal to the tangent vectors t_α we obtain

$$x_{1+} = x_1 - \frac{\alpha}{2}(x_2 - x_0)$$

$$x_{2-} = x_2 + \frac{\alpha}{2}(x_3 - x_1) \tag{208}$$

Here α is a parameter which specifies how far nodes x_{1+} and x_{2-} are away from nodes x_1 and x_2, respectively. For different α the shape of the surface interpolation changes. In the limit for $\alpha \to 0$ we obtain an almost flat segment, however the corner region between adjacent segments is still C^1 continuous. Since the shape of the surface changes during the finite deformation process, α might be adapted within the calculation, see also the second numerical example. For standard applications a good choice for α is $\alpha = \frac{1}{3}$, see also [Pie97].

With (208) we can rewrite the interpolation (205). This leads to

$$x(\xi) = \sum_{i=0}^{3} \bar{B}_i(\xi) x_i \tag{209}$$

with

$$\bar{B}_0(\xi) = \frac{\alpha}{2} B_2(\xi)$$

$$\bar{B}_1(\xi) = B_1(\xi) + B_2(\xi) - \frac{\alpha}{2} B_3(\xi)$$

$$\bar{B}_2(\xi) = B_3(\xi) + B_4(\xi) - \frac{\alpha}{2} B_2(\xi) \tag{210}$$

$$\bar{B}_3(\xi) = \frac{\alpha}{2} B_3(\xi)$$

Now we compute the first and second derivative of $x(\xi)$ with respect to the surface coordinate ξ for later use

$$x_{,\xi}(\xi) = \sum_{i=0}^{3} \bar{B}_{i,\xi}(\xi) x_i$$

$$x_{,\xi\xi}(\xi) = \sum_{i=0}^{3} \bar{B}_{i,\xi\xi}(\xi) x_i \tag{211}$$

The expression for the variation of the gap using this interpolation is now

$$dg_{N_s} = \left[\delta x_s - \sum_{i=0}^{3} \bar{B}_i(\bar{\xi}) \delta x_i \right] \cdot \bar{n}^1 \tag{212}$$

which easily is expressed in matrix form as

$$dg_{N_s} = \delta\hat{\mathbf{x}}_s^T \mathbf{B}_n(\bar{\xi}) = \langle\, \delta\mathbf{x}_s^T\, , \delta\mathbf{x}_0^T\, , \delta\mathbf{x}_1^T\, , \delta\mathbf{x}_2^T\, , \delta\mathbf{x}_3^T\, \rangle \left\{ \begin{array}{c} \bar{\mathbf{n}}^1 \\ -\bar{B}_0(\bar{\xi})\,\bar{\mathbf{n}}^1 \\ -\bar{B}_1(\bar{\xi})\,\bar{\mathbf{n}}^1 \\ -\bar{B}_2(\bar{\xi})\,\bar{\mathbf{n}}^1 \\ -\bar{B}_3(\bar{\xi})\,\bar{\mathbf{n}}^1 \end{array} \right\} \tag{213}$$

Thus the residuum connected with the smooth Bezier contact formulation yields

$$d\,\Pi_\varepsilon^h = \sum_{s=1}^{n_c} \delta\hat{\mathbf{x}}_s^T \left[\, \varepsilon\, A_s\, g_{N_s}\, \mathbf{B}_n(\bar{\xi}) \,\right] \tag{214}$$

The linearization of the variation of the gap function can be derived from (117). For this we have to express $\delta\xi$ and $\Delta\xi$ in matrix form as well as $\delta\bar{\mathbf{x}}_{,\xi}^1$ and $\Delta\bar{\mathbf{x}}_{,\xi}^1$.

Let us first compute the variation of $\bar{\mathbf{x}}_{,\xi}^1$ which yields

$$\delta\bar{\mathbf{x}}_{,\xi}^1 = \delta\hat{\mathbf{x}}_s^T \mathbf{B}_{n,\xi}(\bar{\xi}) = \langle\, \delta\mathbf{x}_s^T\, , \delta\mathbf{x}_0^T\, , \delta\mathbf{x}_1^{T}\, , \delta\mathbf{x}_2^T\, , \delta\mathbf{x}_3^T\, \rangle \left\{ \begin{array}{c} 0 \\ \bar{B}_{0,\xi}(\bar{\xi})\,\bar{\mathbf{n}}^1 \\ \bar{B}_{1,\xi}(\bar{\xi})\,\bar{\mathbf{n}}^1 \\ \bar{B}_{2,\xi}(\bar{\xi})\,\bar{\mathbf{n}}^1 \\ \bar{B}_{3,\xi}(\bar{\xi})\,\bar{\mathbf{n}}^1 \end{array} \right\} \tag{215}$$

We obtain for the linearization

$$\begin{aligned} g_{N_s}\, d\bar{\mathbf{n}}^1 \cdot D\bar{\mathbf{n}}^1 &= \frac{g_{N_s}}{\|\,\bar{\mathbf{x}}_{,\xi}^1\,\|^2}\, \delta\bar{\mathbf{x}}_{,\xi}^1 \cdot \left[\bar{\mathbf{n}}^1 \otimes \bar{\mathbf{n}}^1\right] \Delta\bar{\mathbf{x}}_{,\xi}^1 \\ &= \delta\hat{\mathbf{x}}_s^T \left[\frac{g_{N_s}}{\|\,\bar{\mathbf{x}}_{,\xi}^1\,\|^2}\, \mathbf{B}_{n,\xi}(\bar{\xi})\, \mathbf{B}_{n,\xi}(\bar{\xi})^T \right] \Delta\hat{\mathbf{x}}_s \end{aligned} \tag{216}$$

Furthermore we define the matrix form of $(\delta\mathbf{x}_s^2 - \delta\bar{\mathbf{x}}^1) \cdot \bar{\mathbf{x}}_{,\xi}^1$ which is needed to compute $\delta\xi$

$$(\delta\mathbf{x}_s^2 - \delta\bar{\mathbf{x}}^1) \cdot \bar{\mathbf{x}}_{,\xi}^1 = \delta\hat{\mathbf{x}}_s^T \mathbf{B}_t(\bar{\xi}) = \langle\, \delta\mathbf{x}_s^T\, , \delta\mathbf{x}_0^T\, , \delta\mathbf{x}_1^T\, , \delta\mathbf{x}_2^T\, , \delta\mathbf{x}_3^T\, \rangle \left\{ \begin{array}{c} \bar{\mathbf{x}}_{,\xi}^1 \\ -\bar{B}_0(\bar{\xi})\,\bar{\mathbf{x}}_{,\xi}^1 \\ -\bar{B}_1(\bar{\xi})\,\bar{\mathbf{x}}_{,\xi}^1 \\ -\bar{B}_2(\bar{\xi})\,\bar{\mathbf{x}}_{,\xi}^1 \\ -\bar{B}_3(\bar{\xi})\,\bar{\mathbf{x}}_{,\xi}^1 \end{array} \right\} \tag{217}$$

The variation of the surface coordinate follows now with (217) and (215) in matrix notation

$$\delta\xi = \delta\hat{\mathbf{x}}_s^T \left[\, H_{\xi\xi}\,(\,\mathbf{B}_t(\bar{\xi}) + g_{N_s}\, \mathbf{B}_{n,\xi}\,)\,\right] = \delta\hat{\mathbf{x}}_s^T \mathbf{B}_\xi(\bar{\xi}) \tag{218}$$

where $H_{\xi\xi} = 1\,/\,(\,\bar{\mathbf{x}}_{,\xi}^1 \cdot \bar{\mathbf{x}}_{,\xi}^1 - g_{N_s}\,\bar{\mathbf{n}}^1 \cdot \bar{\mathbf{x}}_{,\xi\xi}^1\,)$.

The matrix form of the linearization of the gap function can be expressed with (216) and (218). Thus we obtain finally for the linearization of

$$
\begin{aligned}
\mathrm{Dd}\Pi_\varepsilon^h \;=\; &\sum_{s=1}^{n_c} \varepsilon\, A_s\, \delta\hat{\mathbf{x}}_s^T\, \big[\, \mathbf{B}_n(\bar{\xi})\, \mathbf{B}_n(\bar{\xi})^T \\
&-\, g_{N_s}\, \big(\, \mathbf{B}_{n,\xi}(\bar{\xi})\, \mathbf{B}_\xi(\bar{\xi})^T + \mathbf{B}_\xi(\bar{\xi})\, \mathbf{B}_{n,\xi}(\bar{\xi})^T + (\bar{\mathbf{x}}_{,\xi\xi}^1 \cdot \bar{\mathbf{n}}^1)\, \mathbf{B}_\xi(\bar{\xi})\, \mathbf{B}_\xi(\bar{\xi})^T \quad (219) \\
&-\, \frac{g_{N_s}}{\|\,\bar{\mathbf{x}}_{,\xi}^1\,\|^2}\, \mathbf{B}_{n,\xi}(\bar{\xi})\, \mathbf{B}_{n,\xi}(\bar{\xi})^T\,\big)\,\big]\, \Delta\hat{\mathbf{x}}_s
\end{aligned}
$$

which denotes the tangent matrix of the smooth Bezier contact formulation.

6.3 Smooth Bezier interpolation for frictional contact

To show that there is a certain variety in choosing the boundary discretization for continuous contact formulations we introduce for the frictional C^1-interpolation another discretization based only on to segments instead of the three used in the previous section, as described in [WKOK99]. We define two interpolating polynomials, see Fig. 17, by two *mid-nodes* (a point between two master surface nodes) and two tangent vectors. Mid-nodes $\mathbf{m}_{12}$ and $\mathbf{m}_{23}$ represent end-points of the polynomial while tangent vectors, $\mathbf{x}_2 - \mathbf{x}_1$ and $\mathbf{x}_3 - \mathbf{x}_2$ are defined by a line between master surface nodes. The so defined geometry, when also applied in the same way for the neighbouring segments, ensures C^1 continuity between adjacent contact segments and hence C^1 continuity on entire master contact surface. For each active contact segment, two interpolations are evaluated (in Fig. 17 this would be the interpolation defined by $\mathbf{m}_{12}$ and $\mathbf{m}_{23}$ and the interpolation defined by $\mathbf{m}_{23}$ and $\mathbf{m}_{34}$. That polynomial which has the minimum distance to the slave node $\mathbf{x}_s$ has to be chosen for the calculation of the contact residual and the associated tangent matrix. For simplicity, we suppose that the first polynomial defined by $\mathbf{m}_{12}$ and $\mathbf{m}_{23}$ is closer to the slave node, thus all vectors and matrices are described with respect to this interpolation. Evaluation of all quantities for the second interpolation is similar as for the first one.

According to Fig. 17, active segment nodes are represented by indices 2 and 3, while neighbouring segment nodes are represented by indices 1 and 4. The slave node $\mathbf{x}_s$ is represented in the following derivations by index 5. With this notation we can define the nodal displacement vector for a single two dimensional contact element as

$$
\mathbf{u} = \{u_{11}, u_{21}, u_{12}, u_{22}, u_{13}, u_{23}, u_{14}, u_{24}, u_{15}, u_{25}\}, \tag{220}
$$

where the first index describes the direction with respect to a cartesian coordinate system $\{\mathbf{e}_1, \mathbf{e}_2, \mathbf{e}_3\}$. The second index is the nodal number. With the nodal displacement vector we obtain a relation between the current configuration $\mathbf{x}$ and the reference configuration $\mathbf{X} = \mathbf{x}\,(t = 0)$ of a body, i.e.,

$$
\mathbf{x} = \mathbf{X} + \mathbf{u}. \tag{221}
$$

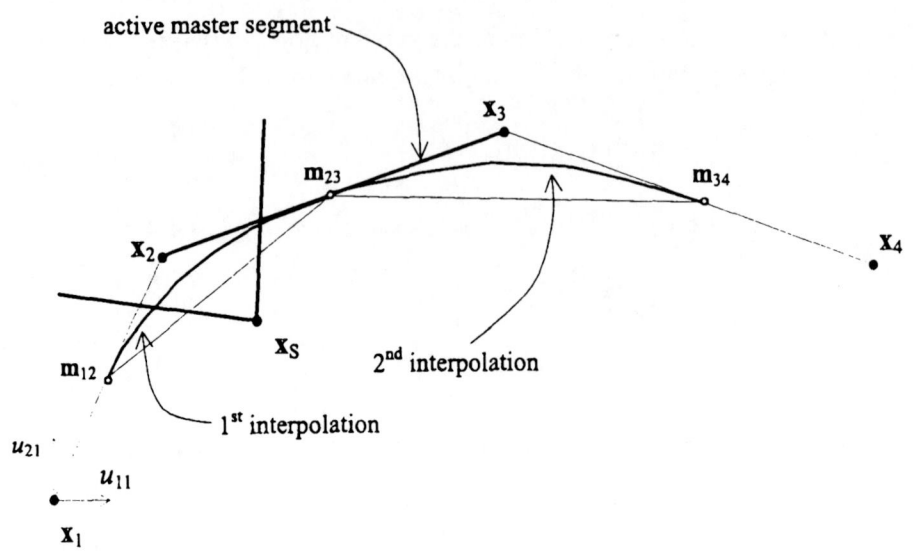

Figure 17: Description of the "active segment", nodal coordinates and nodal displacements

Now two C^1 cubic Bézier curves $\mathbf{x}(\xi)$ can be stated in terms of Bernstein polynomials:

$$B_i^m(\xi) = \binom{m}{i} \xi^i (1-\xi)^{m-i} \quad i=0,....m ,\tag{222}$$

where m is the order of the polynomial. For the case of third order cubic polynomial, the surface is given by

$$\mathbf{x}^1(\xi) = \mathbf{b}_0\, B_0^3(\xi) + \mathbf{b}_1\, B_1^3(\xi) + \mathbf{b}_2\, B_2^3(\xi) + \mathbf{b}_3\, B_3^3(\xi) ,\tag{223}$$

with the polynominals (207). Vectors $\mathbf{b}_i$ are the Bézier points that are defining Bézier polygon. The Bézier curve is always lying inside this polygon. This represents so called convex hull property of Bézier curves [Far93] which ensures numerical stability of the interpolation.

When this interpolation is applied to discretize the contact segments the end-points $\mathbf{b}_0$ and $\mathbf{b}_3$ are defined by the mid-nodes (see Fig. 18). The position of the Bézier points $\mathbf{b}_1$ and $\mathbf{b}_2$ has to be on the tangents $\mathbf{x}_2 - \mathbf{x}_1$ and $\mathbf{x}_3 - \mathbf{x}_2$, respectively. However the distance from the end points $\mathbf{b}_0$ and $\mathbf{b}_3$ is still arbitrary and can vary. The choice of this distance plays an important role in the description of rolling contact and will be discussed at the end of this section.

Let, because of simplicity, points $\mathbf{b}_1$ and $\mathbf{b}_2$ be defined by *quarter-nodes*, i.e. the points between nodes and mid-nodes. This choice ensures C^1, see [Far93].

As the cubic Bézier polynomial is defined explicitly, the path length given with (97) by

$$s\left(\overline{\xi}\right) = \int\limits_0^{\overline{\xi}} \sqrt{\left(\frac{\partial x_1^1\left(\xi\right)}{\partial \xi}\right)^2 + \left(\frac{\partial x_2^1\left(\xi\right)}{\partial \xi}\right)^2} \, d\xi, \tag{224}$$

where x_i^1 represents the i-th. member of the vector $\mathbf{x}^1\left(\xi\right)$. There is no explicit analytical solution of the integral (224), hence numerical Gauss integration has to be applied. In

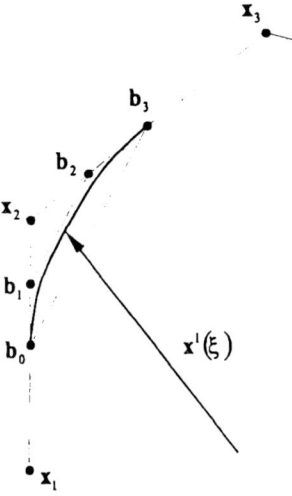

Figure 18: Bézier interpolation and Bézier points

case of rolling contact we have to consider circular cross sections which move on a given surface like wheels or tires. Smoothing of the contact discretization is then absolutely necessary since sliding and rolling of such bodies are significantly influenced by the contact surface geometry. The element based on the cubic Bézier interpolation shows a strong sensitivity with respect to the location of the the second $\mathbf{b}_1$ and the third $\mathbf{b}_2$ Bézier points (quarter-nodes, see Fig. 18). To obtain an optimal surface discretization we define for the moment the positions of these B ézier points as variable, see Fig. 19, using the parameter $\alpha \in [0, 1]$

$$\mathbf{b}_1 = \mathbf{b}_0 + \left(\mathbf{x}_2 - \mathbf{b}_0\right)\alpha \tag{225}$$
$$\mathbf{b}_2 = \mathbf{x}_2 + \left(\mathbf{b}_3 - \mathbf{x}_2\right)\left(1 - \alpha\right) \tag{226}$$

For a circular cross sections, when only a small number of elements is used, special care has to be taken when choosing α since otherwise the circular geometry might be very poorly represented. Different possibilities can be explored to obtain a better approximation of the geometry in case of the Bézier interpolation.

1. When all master segments have the same length, the parameter α can be computed from the requirement that the point $\mathbf{x}^1\left(\xi = 1/2\right)$ for a Bézier interpolation

should be the same as the point $\mathbf{x}^1$ ($\xi = 1/2$) for a Hermite interpolation. This results in parameter $\alpha = 2/3$ which is completely independent from the problem geometry. It is easy to see that in this case Bézier and Hermitian interpolation coincide. Note that this is only valid for the case of master segments with equal length.

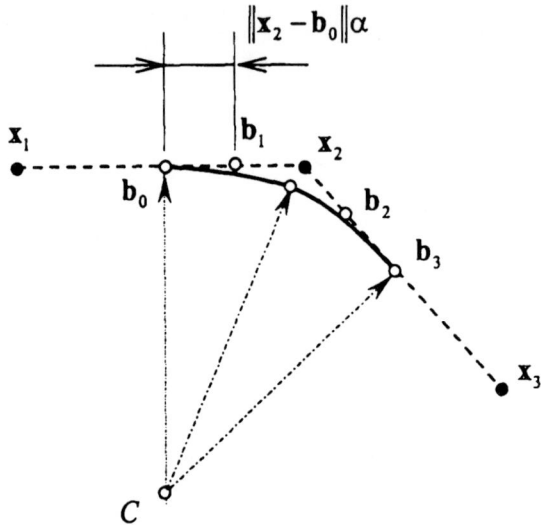

Figure 19: Definition of the parameter α for variable Bézier points

2. Another way of computing the parameter α is based on the requirement that the radius, defined as the distance from the circle center C to the point on the polynomial (Fig. 19), has to be the same for the points $\mathbf{x}$ ($\xi = 0$) and $\mathbf{x}$ ($\xi = 1/2$). This approximates the curvature of the interpolation pointwise. It yields a Bézier interpolation which is closer to the ideal arc shape then the interpolation with $\alpha = 2/3$. In this formulation, in comparison with the previous one, the parameter α depends on the geometry and has to be re-calculated for each different geometry description.

3. Finally one can approximate the curvature of the contact element by an average procedure. This means that the average curvature of the contact segment interpolation has to be equal to the radius of curvature $1/R$ given by the real geometry. This yields the integral

$$R \approx \frac{1}{S} \int_{\Gamma_e} \|\mathbf{x}^1_{,\xi\xi}\| \, d\Gamma \qquad (227)$$

where $x^1_{,\xi\xi}$ is the second derivative of the Bézier interpolation with respect to the path parameter ξ and S denotes the length of the segment. The evaluation of this integral using an interpolation with (225) and (226) yields for each contact segment a value for α.

The significance of smoothing the the circular geometry as well as the correct choice of parameter α, is demonstrated by means of examples in the next section.

Now we will derive the residual vector and the tangent matrix associated with the described contact element. In the case of pure stick, the residual vector is obtained by inserting (127). Considering the fact that the tangential stress vector is a function of the displacement $\mathbf{u}_{n+1}$ and the parameter $\bar\xi_{n+1}$ one has to use the chain rule to derive the expression for i-th component of the residual vector related to the displacement $u_{i_{n+1}}$. This yields

$$\left\{\Psi^{stick}_{n+1}\right\}_i = \varepsilon_N\, g_{N_{n+1}}\left[\frac{\partial g_{N_{n+1}}}{\partial u_{i_{n+1}}} + \frac{\partial g_{N_{n+1}}}{\partial\bar\xi_{n+1}}\frac{\partial\bar\xi_{n+1}}{\partial u_{i_{n+1}}}\right] +$$
$$+ t^{tr}_{T_{n+1}}\cdot\left[\frac{\partial g^{stick}_{T_{n+1}}}{\partial u_{i_{n+1}}} + \frac{\partial g^{stick}_{T_{n+1}}}{\partial\bar\xi_{n+1}}\frac{\partial\bar\xi_{n+1}}{\partial u_{i_{n+1}}}\right], \tag{228}$$

where index i has the range $i = 1,\ldots,10$. Note that one can show with the expression for g_N, (96), that $\partial g_N / \partial\xi$ is equal to zero and hence can be neglected in (228). However from a formal point of view, the latter result is not known a priori. Hence it was included when we used the Mathematica package SMS (Symbolic Mechanics System) [Kor97] for the automatic derivation of matrix formulae. The partial derivative of the path parameter with respect to the displacements is given by

$$\frac{\partial\bar\xi_{n+1}}{\partial u_{i_{n+1}}} = -\frac{\frac{\partial\left\|\mathbf{r}_{n+1}\left(\bar\xi_{n+1}\right)\right\|}{\partial u_{i_{n+1}}}}{\frac{\partial\left\|\mathbf{r}_{n+1}\left(\bar\xi_{n+1}\right)\right\|}{\partial\bar\xi_{n+1}}}. \tag{229}$$

The tangent matrix is obtained by the linearization of the residual vector

$$\left\{K^{stick}_{n+1}\right\}_{ij} = \frac{\partial\left\{\Psi^{stick}_{n+1}\right\}_i}{\partial u_{j_{n+1}}} + \frac{\partial\left\{\Psi^{stick}_{n+1}\right\}_i}{\partial\bar\xi_{n+1}}\frac{\partial\bar\xi_{n+1}}{\partial u_{j_{n+1}}}, \tag{230}$$

where index $j = 1,\ldots,10$. In the finite element code, when calculating the contact contribution of a single slave node, the residual vector is calculated for every iteration within the current time interval. Equation (228) can be re-formulated using equation for increment of the total tangential gap (??),

$$\left\{\Psi^{stick}_{n+1}\right\}_i = \varepsilon_N\, g_{N_{n+1}}\left[\frac{\partial g_{N_{n+1}}}{\partial u_{i_{n+1}}} + \frac{\partial g_{N_{n+1}}}{\partial\bar\xi_{n+1}}\frac{\partial\bar\xi_{n+1}}{\partial u_{i_{n+1}}}\right] +$$
$$+ t^{tr}_{T_{n+1}}\cdot\left\{\frac{\partial\left[\left(s_{n+1}\left(\bar\xi_{n+1}\right) - s_n\left(\bar\xi_n\right)\right)\mathbf{e}_{T_{n+1}}\left(\bar\xi_{n+1}\right)\right]}{\partial u_{i_{n+1}}} +\right.$$

$$+ \frac{\partial \left[\left(s_{n+1}\left(\bar{\xi}_{n+1}\right) - s_n\left(\bar{\xi}_n\right)\right) e_{T_{n+1}}\left(\bar{\xi}_{n+1}\right)\right]}{\partial \bar{\xi}_{n+1}} \frac{\partial \bar{\xi}_{n+1}}{\partial u_{i_{n+1}}} \right\}. \tag{231}$$

If sliding occurs, the residual vector is obtained by the inserting (131) into (116)

$$\left\{ \Psi_{n+1}^{slip} \right\}_i = \varepsilon_N \, g_{N_{n+1}} \left[\frac{\partial g_{N_{n+1}}}{\partial u_{i_{n+1}}} + \frac{\partial g_{N_{n+1}}}{\partial \bar{\xi}_{n+1}} \frac{\partial \bar{\xi}_{n+1}}{\partial u_{i_{n+1}}} \right] +$$

$$+ t_{T_{n+1}}^{slip} \cdot \left[\frac{\partial g_{T_{n+1}}^{slip}}{\partial u_{i_{n+1}}} + \frac{\partial g_{T_{n+1}}^{slip}}{\partial \bar{\xi}_{n+1}} \frac{\partial \bar{\xi}_{n+1}}{\partial u_{i_{n+1}}} \right], \tag{232}$$

while the tangent matrix is

$$\left\{ K_{n+1}^{slip} \right\}_{ij} = \frac{\partial \left\{ \Psi_{n+1}^{slip} \right\}_i}{\partial u_{j_{n+1}}} + \frac{\partial \left\{ \Psi_{n+1}^{slip} \right\}_i}{\partial \bar{\xi}_{n+1}} \frac{\partial \bar{\xi}_{n+1}}{\partial u_{j_{n+1}}} \tag{233}$$

For Coulomb's frictional law the residual vector for the slip case (131) can be re-formulated

$$\left\{ \Psi_{n+1}^{slip} \right\}_i = \varepsilon_N \, g_{N_{n+1}} \left[\frac{\partial g_{N_{n+1}}}{\partial u_{i_{n+1}}} + \frac{\partial g_{N_{n+1}}}{\partial \bar{\xi}_{n+1}} \frac{\partial \bar{\xi}_{n+1}}{\partial u_{i_{n+1}}} \right] +$$

$$+ \mu \, \varepsilon_N \, g_{N_{n+1}} \frac{t_{T_{n+1}}^{tr}}{\left\| t_{T_{n+1}}^{tr} \right\|} \cdot \left\{ \frac{\partial \left[\frac{1}{\varepsilon_T} \left(\left\| t_{T_{n+1}}^{tr} \right\| - \mu \, \varepsilon_N \, g_{N_{n+1}} \right) \frac{t_{T_{n+1}}^{tr}}{\left\| t_{T_{n+1}}^{tr} \right\|} \right]}{\partial u_{i_{n+1}}} + \right.$$

$$\left. + \frac{\partial \left[\frac{1}{\varepsilon_T} \left(\left\| t_{T_{n+1}}^{tr} \right\| - \mu \, \varepsilon_N \, g_{N_{n+1}} \right) \frac{t_{T_{n+1}}^{tr}}{\left\| t_{T_{n+1}}^{tr} \right\|} \right]}{\partial \bar{\xi}_{n+1}} \frac{\partial \bar{\xi}_{n+1}}{\partial u_{i_{n+1}}} \right\} \tag{234}$$

Equations (231), (230), (234) and (233) provide the basis for the automatic code generation using SMS.

6.3.1 Continuity of history variables between adjacent segments in the slip case

For both polynomial formulations described in the previous sections C^1 continuity of adjacent interpolations is ensured. In the frictionless case, where no history variables are needed, there is no influence on the residual vector and the tangent matrix if sliding occurs over several adjacent interpolations and segments. For frictional problems, where path length and traction vector are history variables, extra considerations are needed when sliding of the slave node over adjacent segments and interpolations occurs.

Let, according to Fig. 20, an interpolation which is defined between nodes x_1 and x_3 be called the *first* interpolation. Furthermore an interpolation, which is defined between x_2 and x_4 be called *second* interpolation. For both interpolations the path length is calculated via the integral from zero to $\bar{\xi}_{n+1}$. Let the path length for the *first* interpolation be denoted by the symbol $s_{n+1}^I \left(\bar{\xi}_{n+1} \right)$ and the path length for the *second* interpolation by the symbol $s_{n+1}^{II} \left(\bar{\xi}_{n+1} \right)$. These values are calculated according to equations (??) and (224) for each iteration and saved as variable which describes the history of the sliding path of the slave node.

The history variable $s_n \left(\bar{\xi}_n \right)$ that represents the path length for the last converged state is constant during the iterations within the same time increment. If, within the same active segment, the slave node slides from the *first* interpolation to the *second* or vice versa then the path length $s_{n+1} \left(\bar{\xi}_{n+1} \right)$ used in equations (129) and (231) has to be modified. The following possibilities have to be considered when updating the history variable describing the path length:

1. If sliding from the *first* to the *second* interpolation occurs then $s_{n+1} \left(\bar{\xi}_{n+1} \right)$ used for calculation of $\left(s_{n+1} \left(\bar{\xi}_{n+1} \right) - s_n \left(\bar{\xi}_n \right) \right)$ is modified according to

$$ s_{n+1} \left(\bar{\xi}_{n+1} \right) = s_{n+1}^{II} \left(\bar{\xi}_{n+1} \right) + s_{n+1}^I \left(\xi_{n+1} = 1 \right). \tag{235} $$

2. If, within the same active segment, the slave node slides from the *second* to the *first* interpolation then the path length $s_{n+1} \left(\bar{\xi}_{n+1} \right)$ has to be modified according to

$$ s_{n+1} \left(\bar{\xi}_{n+1} \right) = s_{n+1}^I \left(\bar{\xi}_{n+1} \right) - s_{n+1}^I \left(\xi_{n+1} = 1 \right). \tag{236} $$

3. If there is no sliding between interpolations of the same active segment than we set for the *first* interpolation

$$ s_{n+1} \left(\bar{\xi}_{n+1} \right) = s_{n+1}^I \left(\bar{\xi}_{n+1} \right), \tag{237} $$

   or for the *second* interpolation

$$ s_{n+1} \left(\bar{\xi}_{n+1} \right) = s_{n+1}^{II} \left(\bar{\xi}_{n+1} \right). \tag{238} $$

4. If the slave node slides between two adjacent segments, there is no need for a modification of the current path length $s_{n+1} \left(\bar{\xi}_{n+1} \right)$. Hence equations (237) and (238) are used. For such a case, procedures described by (235) and (236) could be activated unnecessarily. This can happen when e. g. a interpolation that was the *second* one for the active segment changes to the *first* interpolation, and vice versa. To suppress the use of procedures (235) and (236) for such a case, another additional history variable (switch) is introduced which monitors if an active master segment number has changed.

As mentioned before, history variables that are going to be used for the next time steps are not influenced by modifications (235) and (236). Detection of sliding between interpolations of the same active segment requires an additional history variable (switch) which is used to monitor if the interpolation number (*first* or *second*) has changed.

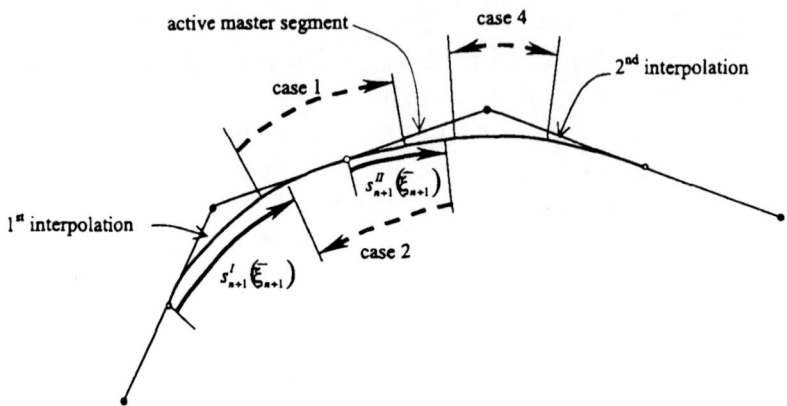

Figure 20: Description of the path length modification

7   Numerical examples

The first two examples are non-stationary processes which need a Lagrangian description for contact and a time integration algorithm. The third example then presents a stationary rolling process which can be handled by the ALE-formulation.

7.1   Dry cylindrical bearing

The rotation of a shaft in a dry cylindrical bearing is simulated. The shaft is modeled by an elastic cylinder which is pressed into the cylindrical bearing with an initial overlap and rotated for the angle $\varphi = 2\pi$, see Fig. 21. This example is similar to one described in [PC97].

The simulation is performed in two steps. First the elastic bearing of inner radius R_{T1} is brought into contact with the shaft of radius R_C by applying a radial displacement δ on the outer surface of the bearing (defined by the radii R_{T2}). This corresponds to the time interval $[0, 1]$ in Fig. 22. After this step, the displacements of the outer part of the bearing surface are fixed and the inner cylinder is rotated up to a final angle of $\varphi = 2\pi$. This part of the simulation correspondents to the time interval $[1, 2]$ in Fig. 22. The cylindrical shaft is modeled by assuming an elastic neo-hookean material with the bulk modulus $K = 1.75\ 10^5\ MPa$ and the shear modulus $G = 8.08\ 10^4\ MPa$. The bearing is also characterized by an elastic neo-hookean material with $K = 1.75\ MPa$

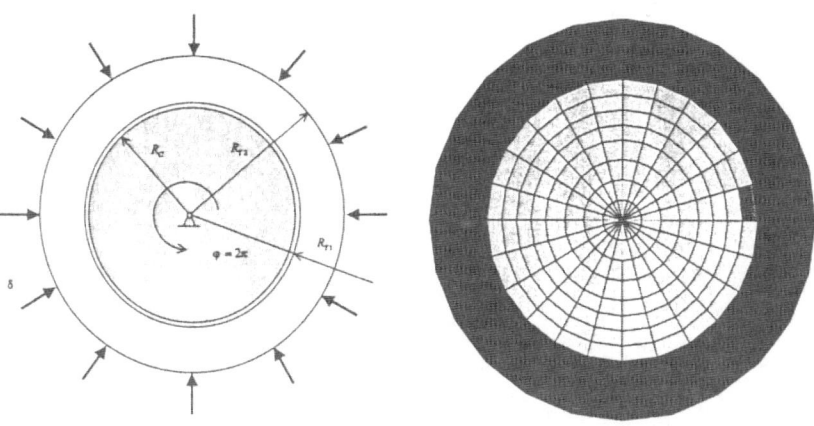

Figure 21: Problem description and the FE mesh

and $G = 0.81\,MPa$. The radius of the shaft is $R_C = 70\,mm$. The bearing is defined by the $R_{T1} = 70.1\,mm$ and $R_{T2} = 100\,mm$. Frictional behaviour is modeled by the Coulomb's law using as frictional coefficient $\mu = 0.3$.

The angular reaction (moment) is measured for the central node where the rotation is applied. Fig. 22 depicts the angular reaction for the straight node to segment (NTS) element, the smooth Bézier (BFRI), and the smooth Bézier contact element (BFRIalpha) with the parameter $\alpha = 2/3$. The surface of the shaft is defined to be the master surface. The applied displacement δ has to be adjusted to the different contact formulations This is due to the fact that the surface is not the same for the NTS and the smooth elements. Therefore the results of Bézier elements can be mutually compared, while NTS element can be only approximately compared with smooth elements. The displacement δ is chosen to be $\delta = 0.04\,mm$ for the NTS element and $\delta = 0.078\,mm$ for BFRI and BFRIalpha elements. From a simple analytical consideration we observe that

the angular reaction has to be constant. As can be observed in Fig. 22 the NTS element shows a non–smooth response stemming from the sudden normal changes between the contact elements and from the strong changes of the normal gap associated with the slave node according to the angle. The Bézier element (BFRI, $\alpha = 1/2$) also undergoes significant changes of the normal gap characterized by large but smooth variations in the angular reaction. These variations (waving) are caused by the geometry definition of Bézier element that is not following cylindrical surface as close as the Bézier element for $\alpha = 2/3$ do. As can be seen from Fig. 22, the BFRI element depict a very good, almost constant, solution for the angular reaction.

7.2   Simulation of a sliding and rolling wheel

In this section an example is presented that can be suggested as a benchmark test for the smooth 2D frictional contact elements since it analytical solution is known. A

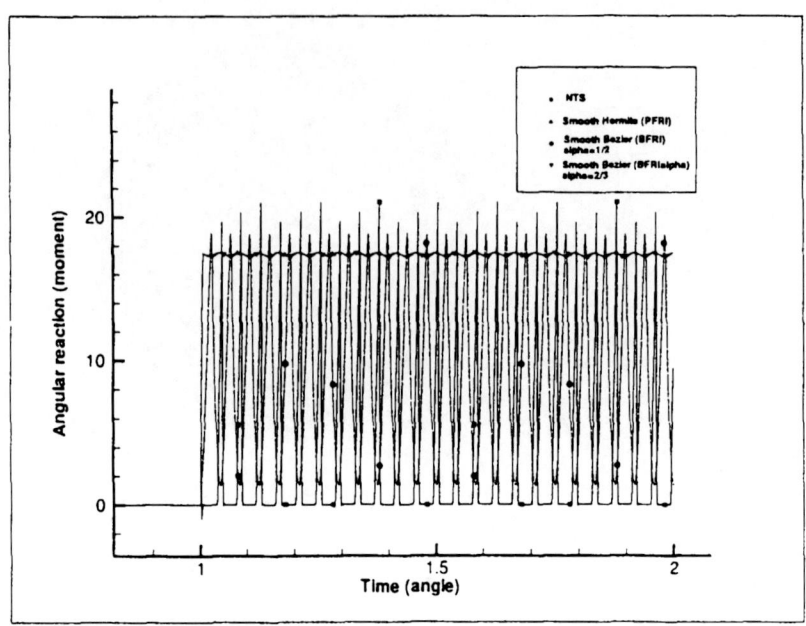

Figure 22: Comparison of the simple and the smooth frictional elements

wheel with radius r, density ρ, Coulomb friction coefficient μ, elasticity parameters is put on a flat surface under gravity loading. When an initial velocity v_0 for all points of the wheel is applied, the wheel first starts to slide without rolling. After a certain time the wheel starts to roll until finally pure rolling occurs, see Fig. 23. The analytical solution yields the time and distance after pure rolling starts

$$t = \frac{1}{3} \frac{v_0}{\mu \, g} \qquad l = \frac{5}{18} \frac{v_0^2}{\mu \, g}. \tag{239}$$

With a frictional coefficient $\mu = 0.3$ and gravity $g = 9.81 \ m/s$, the dependency of the length l according to the initial velocity is shown in Fig. 24. For an initial velocity of $v_0 = 1 \ m/s$ it follows from equation (239) that the length and the time, when pure rolling starts, is $l = 0.094 \ m$ and $t = 0.113 \ s$.

The FE simulation, see Fig. 25, is performed for a wheel with radius $r = 0.04 \ m$, density of steel $\rho = 7850 \ kg/m^3$, frictional coefficient $\mu = 0.3$ and elasticity parameters $K = 1.75 \ 10^5 \ MPa$, $G = 8.08 \ 10^4 \ MPa$. The dynamical problem is solved using the Newark method (Newmark parameters $\beta = 0.25$ and $\gamma = 0.5$). The diagrams in Fig.

26, Fig. 27 and Fig. 28 show horizontal velocities of the master nodes (wheel surface) and the central node. The time when pure rolling starts can be detected from the fact that some master nodes have horizontal velocity zero, while diagonally symmetric master nodes (plotted with the same gray scale) have velocity twice as big as the central node (dashed line).

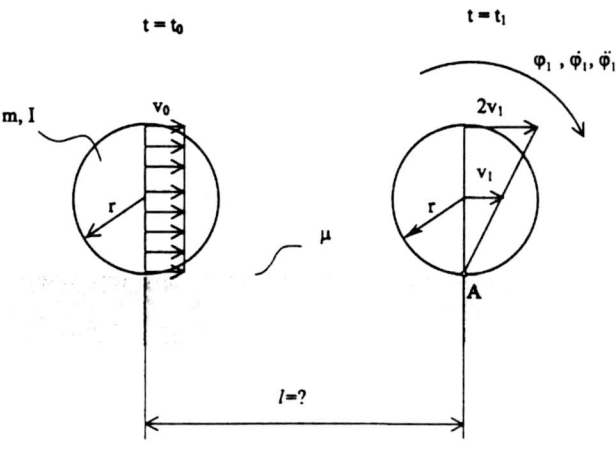

Figure 23: Description of the rolling wheel problem

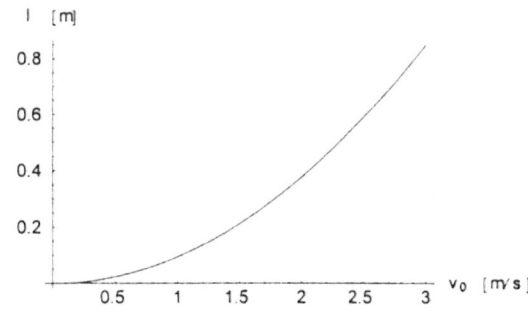

Figure 24: Relation between the initial velocity and the beginning of the pure rolling

Pure rolling for the straight NTS contact element occurred at $l = 0.124$ m. The overall behaviour is characterized by jumps and non-physical separations from the base surface, see Fig. 26. For all smooth formulations pure rolling started at $l = 0.099$ m. This represents a solution close to the analytical one. Figure 27 depicts the response for the smooth Bézier contact element with $\alpha = 1/2$ which is satisfactory since the circular geometry is not well represented by this choice of α. Figure 28 shows the response for the Bézier contact element with the parameter $\alpha = 0.65292$ calculated from requirement of the same radii for x $(\xi = 0)$ and x $(\xi = 1/2)$. As can be seen from Fig. 26 to Fig. 28, the straight contact element discretization leads to errouneous results whereas the smooth interpolation with $\alpha = 0.65292$ yields better results, closed to the analytical solution.

The above derived interpolations have been implemented in the finite element anal-

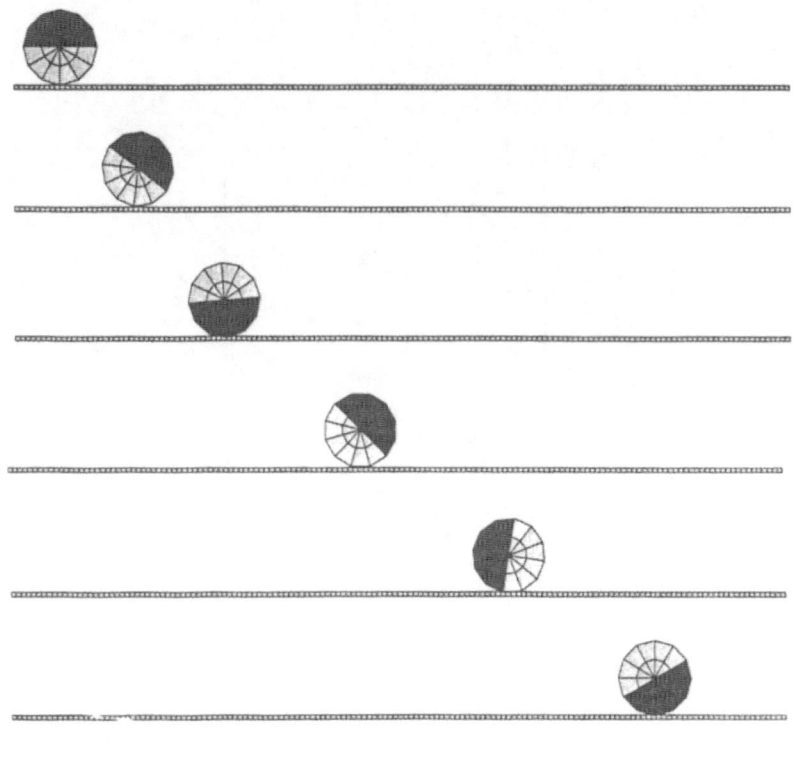

Figure 25: FE simulation of dynamical motion with contact

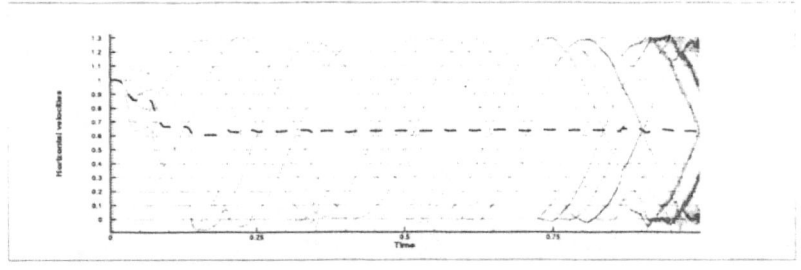

Figure 26: Straight (NTS) node–to–segment frictional contact element

ysis program FEAP, see [ZT89]. To show the performance of the smooth contact discretizations, we make a comparison with the classical node–to–segment contact formulations with straight segments. All derivations of residuum and tangent matrices for the frictionless case have been given above.

The presented examples demonstrate the importance of contact surface smoothing. Sudden normal changes, that are common for the standard node–to–segment contact

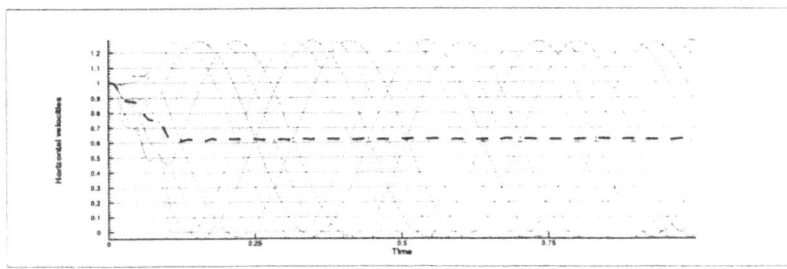

Figure 27: Smooth Bézier ($\alpha = 1/2$) frictional contact element (BFRI)

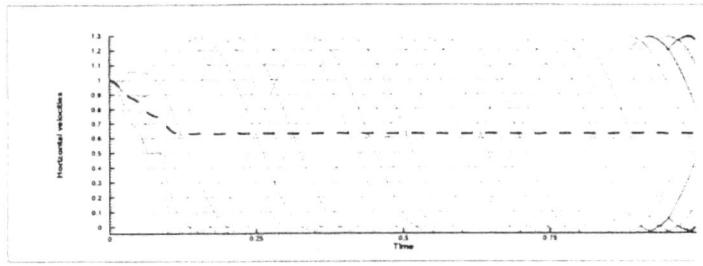

Figure 28: Smooth Bézier frictional contact element, $\alpha = 0.65292$

formulations, introduce significant errors in the finite element discretization of contact phenomena. With this C^1 continuity is always satisfied. It results into quadratic rate of convergence and thus a robust FE contact element. Use of Bézier interpolation yields short and compact element code, security against curve looping, but also the possibility of defining sharp corners. The latter can easily be defined by collapsing Bézier points near the sharp master corner into the master node of the sharp corner

7.3   The sheet/plate rolling simulation

The advantage of the smooth contact elements can be used for the sheet/plate metal forming simulations. Particularly when deformations and stresses of roller are not of primary interest, smoothing of the cylindrical roller, defied as the master surface, enables a reduction of the required number of elements needed for discretization of the roller. In Fig. 29 is illustrated an example of the rolling simulation where the displacement is applied on the horizontal plate. The fixed steel roller ⌀140 mm can rotate around its axis. The displacement is applied on the rubber plate 478x85 mm supported on its lower edge. The roller is modeled as the elastic, neo-hookean, finite strain material characterized with the bulk modulus of $K = 1.75 \cdot 10^5 \ MPa$ and the shear modulus of $G = 8.08 \cdot 10^4 \ MPa$. The plate is modeled as the elastic, neo-hookean,

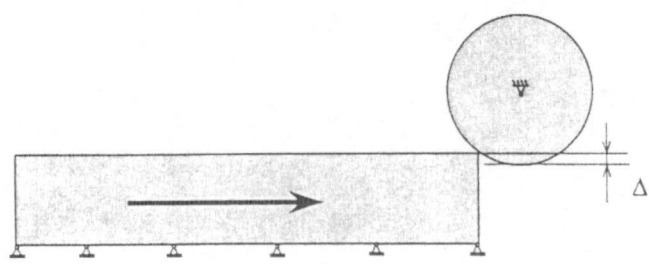

Figure 29: Description of sheet/plate forming-problem

finite strain material characterized with the bulk modulus of $K = 55.6\ MPa$ and the shear modulus of $G = 3.4\ MPa$. The cylinder is discretized by the relatively coarse mesh, as shown in Fig. 30, the overlapping is approximately $\Delta \approx 5\ mm$. There is

Figure 30: FE-mesh of sheet/plate forming

also a difference in the master surface geometry between the simple and the smooth node-to-segment elements. As the reactions for the central cylinder node are measured, there would be a difference in the reaction due to the differences in the mater surface geometry. The contact is modeled using: the simple node-to-segment (NTS), and the Bézier node-to-segment contact element with the parameter $\alpha = 0.65444$. This parameter is evaluated from the requirement for the same distance between the circle center and the points $^{n+1}\mathbf{x}\,(^{n+1}\xi = 0)$ and $^{n+1}\mathbf{x}\,(^{n+1}\xi = 1/2)$. Fig. 31 and Fig. 32 show the horizontal and the vertical reaction for the central node.

The horizontal reaction shows clearly that the continuous contact discretization yields better results since the normal reaction has to be zero in this example This fact is only reproduced by the smooth discretization, see Fig. 31. The difference in the reactions in Fig. 32 appears due to the differences in the master surface geometry, i.e., the nodal element geometry is the same for the both cases. Therefore, when modeling such a problem one has to take into account an adjustment of discretized geometry in order to obtain model similar to the non-discretized problem. Figures 33 to Fig. 34 depict contours of the vertical stresses for the smoothed and non-smoothed

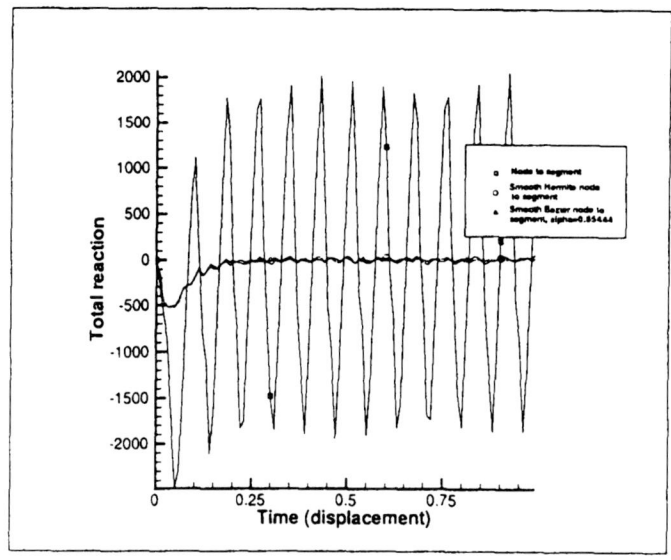

Figure 31: Horizontal reactions for the central node

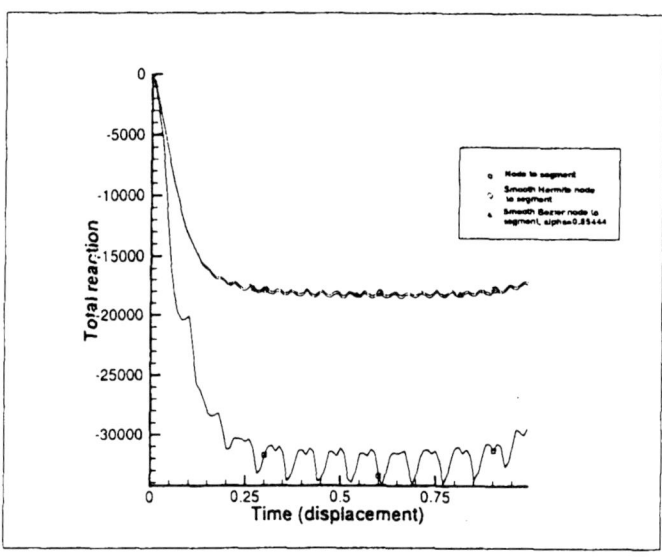

Figure 32: Vertical reactions for the central node

approach. For the simple NTS element (see Fig. 33), significant difference in the vertical stresses appears between the two characteristic positions of the roller (the positions "A" and "B"). For the same characteristic positions of the roller there is no significant difference in the vertical stresses when the smooth polynomial elements

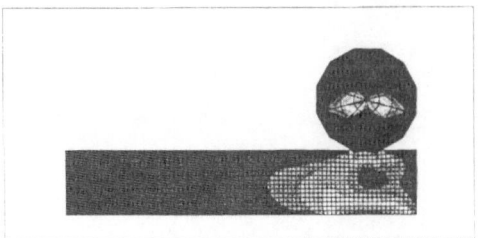 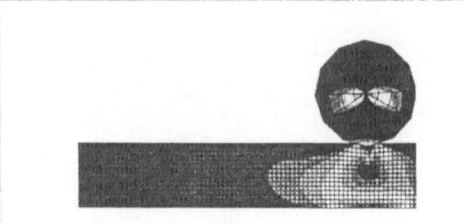

Figure 33: Vertical stresses for the simple NTS contact element - positions A and B

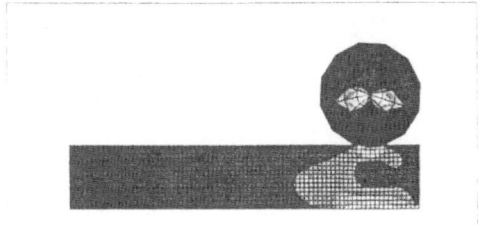 

Figure 34: Vertical stresses for the smooth contact element - positions A and B

are used (see Fig. 34). As the plotting facilities have not been developed for the symbolically generated polynomial elements, overlapping of the master surface near the master node in Fig. 34 in reality does not appear.

7.4 Rolling contact of a ring

In this example we discuss rolling of a cylinder on a flat rigid foundation. The formulation used is the one for steady state rolling. The ring is made of a hyperelastic material with the Lame constants $\mu = 20$ and $\Lambda = 80$. Its inner radius is $R_i = 1$ and the outer radius is $R_a = 2$. The angular velocity of the rolling cylinder is $\omega = 10$. The cylinder is pressed down on the surface by prescribed displacement of $\delta_V = 0.2$. To obtain pure rolling in the contact interface the tangential velocity of the rigid foundation is prescribed as $V_0 = \omega (R_a - \delta_V)$. First we look at the case of frictionless contact. The first mesh with the stress distributions are shown in Fig. 36 and 37. The adaptive method then produces the error distribution for this mesh depicted in Fig.38. Finally Fig. 39 and 40 show the stress components computed with the refined mesh. Note that in frictionless rolling contact we have almost the same result like in a static analysis, the only difference follows form the loading term due to the rotation.

Fig. 36 First mesh: σ_{yy}

Fig. 37 First mesh: τ_{xy}

Fig. 38 Error distribution

Fig. 39 Second mesh: σ_{yy}

Fig. 40 Second mesh: τ_{xy}

The same procedure can be carried out for rolling contact with friction in the contact interface. Here we have used Coulomb's law with a frictional constant $\mu_F = 0.3$. Again the stress distribution is shown for the initial mesh in Fig. 41 and 42. Figure 43 contains information about the error of the finite element computation using the first mesh. Due to that the mesh is adapted. The results for two stress components are depicted in Fig. 44 and Fig. 45 for the adapted mesh. Due to friction the stress distribution is no longer symmetric.

Fig. 41 First mesh: σ_{yy}

Fig. 42 First mesh: τ_{xy}

Fig. 43 Error distribution

Fig. 44 Second mesh: σ_{yy}

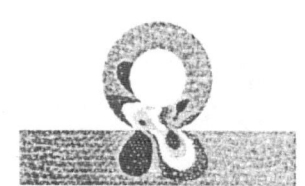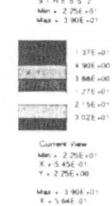

Fig. 45 Second mesh τ_{xy}

References

[AC91]   P. Alart and A. Curnier. A mixed formulation for frictional contact problems prone to newton like solution methods. *Computer Methods in Applied Mechanics and Engineering*, 92:353–375, 1991.

[Bat80]   R. C. Batra. Quasistatic indentation of a rubber–covered roll by a rigid roll. *International Journal for Numerical Methods in Engineering*, 17:1823–1833, 1980.

[Bat82]   K. J. Bathe. *Finite Element Procedures in Engineering Analysis*. Prentice-Hall, New Jersey, 1982.

[Bat86]   K. J. Bathe. *Finite-Elemente-Methoden, Matrizen und lineare Algebra. Die Methode der finiten Elemente. L"osung von Gleichgewichtsbedingungen und Bewegungsgleichungen; Deutsche "Ubersetzung von P. Zimmermann.* Springer-Verlag, Berlin–Heidelberg–New York, 1986.

[BC85]   K. J. Bathe and A. B. Chaudhary. A solution method for planar and axisymmetric contact problems. *International Journal for Numerical Methods in Engineering*, 21:65–88, 1985.

[BC98]   E. Bittencourt and G. J. Creus. Finite element analysis of three-dimensional contact and impact in large deformation problems. *Computers and Structures*, 69:219–234, 1998.

[Ber84]   D. P. Bertsekas. *Constrained Optimization and Lagrange Multiplier Methods.* Academic Press, New York, 1984.

[CA88]   A. Curnier and P. Alart. A generalized newton method for contact problems with friction. *J. Mec. Theor. Appl.*, 7:67–82, 1988.

[Chr80]   R. M. Christensen. A nonlinear theory of viscoelastocity for application to elastomers. *Journal of Applied Mechanics*, 47:762–768, 1980.

[CHT92]   A. Curnier, Q. C. He, and J. J. Telega. Formulation of unilateral contact between two elastic bodies undergoing finite deformation. *C. R. Acad. Sci. Paris*, 314:1–6, 1992.

[Cia88]   P. G. Ciarlet. *Mathematical Elasticity I: Three-dimensional Elasticity.* North-Holland, Amsterdam, 1988.

[Cri91]   M. A. Crisfield. *Non–linear Finite Element Analysis of Solids and Structures*, volume 1. J. Wiley, Chichester, 1991.

[CSW99]   C. Carstensen, O. Scherf, and P. Wriggers. Adaptive finite elements for elastic bodies in contact. *SIAM Journal Scientific Computing*, 20:1605–1626, 1999.

[DL76]   G. Duvaut and J. L. Lions. *Inequalities in Mechanics and Physics.* Springer Verlag, Berlin, 1976.

[DT85]   G. Dhatt and G. Touzot. *The Finite Element Method Displayed.* J. Wiley, Chichester, 1985.

[Eri67]   A.C. Eringen. *Mechanics of Continua.* J. Wiley & Sons, New York, London, Sidney, 1967.

[Far93]   G. Farin. *Curves and Surfaces for Computer Aided Geometric Design. A Practical Guide.* Academic Press, Boston, third edition, 1993.

[FCM99]   L. Fourment, J. L. Chenot, and K. Mocellin. Numerical formulations and algorithms for solving contact problems in metal forming simulation. *International Journal for Numerical Methods in Engineering*, 46:1435–1463, 1999.

[Fre76]   B. Fredriksson. Finite element solution of surface nonlinearities in structural mechanics with special emphasis to contact and fracture mechanics problems. *Computers and Structures*, 6:281–290, 1976.

[FZ75]   A. Francavilla and O. C. Zienkiewicz. A note on numerical computation of elastic contact problems. *International Journal for Numerical Methods in Engineering*, 9:913–924, 1975.

[GHSW99] D. Gross, W. Hauger, W. Schnell, and P. Wriggers. *Technische Mechanik 4*. Springer, Berlin, dritte edition, 1999.

[Gia89]   A. E. Giannokopoulos. The return mapping method for the integration of friction constitutive relations. *Computers and Structures*, 32:157–168, 1989.

[GLT84]   R. Glowinski and P. Le Tallec. Finite element analysis in nonlinear incompressible elasticity. In *Finite Element, Vol. V: Special Problems in Solid Mechanics*. Prentice–Hall, Englewood Cliffs, New Jersey, 1984.

[GM99]   S. Govindjee and P. A. Mihalic. Viscoelastic constitutive relations for the steady spinning of a cylinder. *International Journal for Numerical Methods in Engineering*, 1999.

[HA96]   A. Heege and P. Alart. A frictional contact element for strongly curved contact problems. *International Journal for Numerical Methods in Engineering*, 39:165–184, 1996.

[HC93]   J.-H. Heegaard and A. Curnier. An augmented lagrangian method for discrete large–slip contact problems. *International Journal for Numerical Methods in Engineering*, 36:569–593, 1993.

[HGB85]   J. O. Hallquist, G. L. Goudreau, and D. J. Benson. Sliding interfaces with contact–impact in large–scale lagrangian computations. *Computer Methods in Applied Mechanics and Engineering*, 51:107–137, 1985.

[HK90]   E. Hansson and A. Klarbring. Rigid contact modelled by cad surface. *Engineering Computations*, 7:344–348, 1990.

[HSS92]   J. O. Hallquist, K. Schweizerhof, and D. Stillman. Efficiency refinements of contact strategies and algorithms in explicit fe programming. In D. R. J. Owen, E. Hinton, and E. Onate, editors, *Proceedings of COMPLAS III*, pages 359–384. Pineridge Press, 1992.

[HTK77]   T. R. J. Hughes, R. L. Taylor, and W. Kanoknukulchai. A finite element
           method for large displacement contact and impact problems. In K. J.
           Bathe, editor, *Formulations and Computational Algorithms in FE Analy-
           sis*, pages 468–495, Boston, 1977. MIT–Press.

[HW00]   G. D. Hu and P. Wriggers. On the adaptive finite element method
           of steady–state rolling contact for hyperelasticity in finite deformations.
           Computer Methods in Applied Mechanics and Engineering, 2000.

[Joh87]   C. Johnson. *Numerical solution of partial differential equations by the
           finite element method.* Cambridge University Press, 1987.

[JT88]   W. Ju and R. L. Taylor. A perturbed lagrangian formulation for the
           finite element solution of nonlinear frictional contact problems. *Journal of
           Theoretical and Applied Mechanics*, 7:1–14, 1988.

[Kal90]   J. J.. Kalker. *Three-dimensional Elastic Bodies in Rolling Contact.* Kluwer
           Academic Publishers, Dordrecht, 1990.

[Kla86]   A. Klarbring. A mathematical programming approach to three-
           dimensional contact problems with friction. *Computer Methods in Applied
           Mechanics and Engineering*, 58:175–200, 1986.

[KO88]   N. Kikuchi and J. T. Oden. *Contact Problems in Elasticity: A Study of
           Variational Inequalities and Finite Element Methods.* SIAM, Philadelphia,
           1988.

[Kor97]   J. Korelc. Automatic generation of finite-element code by simultaneous
           optimization of expressions. *Theoretical Computer Science*, 187:231–248,
           1997.

[LMM99]   W. N. Liu, G. Meschke, and H. A. Mang. A note on the algorithmic
           stabilization of 2d contact analyses. In L. Gaul and C. A. Brebbia, ed-
           itors, *Computational Methods in Contact Mechanics IV*, pages 231–240,
           Southhampton, 1999. Wessex Institute.

[LS93]   T. A. Laursen and J. C. Simo. A continuum–based finite element formu-
           lation for the implicit solution of multibody, large deformation frictional
           contact problems. *International Journal for Numerical Methods in Engi-
           neering*, 36:3451–3485, 1993.

[Lue84]   D. G. Luenberger. *Linear and Nonlinear Programming.* Addison–Wesley
           Publishing Company, second edition, 1984.

[Mal69]   L. E. Malvern. *Introduction to the Mechanics of a Continuous Medium.*
           Prentice-Hall, Inc., Englewood Cliffs, 1969.

[MM78]   R. Michalowski and Z. Mroz. Associated and non–associated sliding rules in contact friction problems. *Archives of Mechanics*, 30:259–276, 1978.

[Nac93]   U. Nackenhorst. On the finite element analysis of steady state rolling contact. In M. H. Aliabadi and C. A. Brebbia, editors, *Contact Mechanics*, pages 53–60, Southampton, 1993. Computational Mechanics Publications.

[Nac95]   U. Nackenhorst. An adaptive finite element method to analyse contact problems. In M. H. Aliabadi and C. Alessandri, editors, *Contact Mechanics II*, Southampton, 1995. Computational Mechanics Publications.

[Ode81]   J. T. Oden. Exterior penalty methods for contact problems in elasticity. In W. Wunderlich, E. Stein, and K. J. Bathe, editors, *Nonlinear Finite Element Analysis in Structural Mechanics*, Berlin, 1981. Springer.

[Ogd84]   R. W. Ogden. *Non-Linear Elastic Deformations*. Ellis Horwood und John Wiley, Chichester, 1984.

[OL86]   J. T. Oden and T. L. Lin. On the general rolling contact problem for finite deformations of a viscoelastic cylinder. *Computer Methods in Applied Mechanics and Engineering*, 52:297–367, 1986.

[OM86]   J. T. Oden and J. A. C. Martins. Models and computational methods for dynamic friction phenomena. *Computer Methods in Applied Mechanics and Engineering*, 52:527–634, 1986.

[PC97]   G. Pietrzak and A. Curnier. Continuum mechanics modeling and augmented lagrangian formulation of multibody, large deformation frictional contact problems. In D. R.J. Owen, E. Hinton, and E. Onate, editors, *Proceedings of COMPLAS 5*, pages 878–883, Barcelona, 1997. CIMNE.

[Pie97]   G. Pietrzak. Continuum mechanics modelling and augmented lagrangian formulation of large deformation frictional contact problems. Technical Report 1656, EPFL, Lausanne, 1997.

[PT92]   P. Papadopoulos and R. L. Taylor. A mixed formulation for the finite element solution of contact problems. *Computer Methods in Applied Mechanics and Engineering*, 94:373–389, 1992.

[PZ84]   J. Padovan and I. Zeid. Finite element analysis of steadily moving contact fields. *Computers and Structures*, 2:111–200, 1984.

[SH98]   J. C. Simo and T. J. R. Hughes. *Computational Inelasticity*. Springer, New York, Berlin, 1998.

[SL92]   J. C. Simo and T. A. Laursen. An augmented lagrangian treatment of contact problems involving friction. *Computers and Structures*, 42:97–116, 1992.

[ST85]   J. C. Simo and R. L. Taylor. Consistent tangent operators for rate-independent elastoplasticity. *Computer Methods in Applied Mechanics and Engineering*, 48:101–118, 1985.

[SW79]   J. T. Stadter and R. O. Weiss. Analysis of contact through finite element gaps. *Computers and Structures*, 10:867–873, 1979.

[SWT85]   J. C. Simo, P. Wriggers, and R. L. Taylor. A perturbed lagrangian formulation for the finite element solution of contact problems. *Computer Methods in Applied Mechanics and Engineering*, 50:163–180, 1985.

[TN65]   C. Truesdell and W. Noll. The nonlinear field theories of mechanics. In S. Flügge, editor, *Handbuch der Physik III/3*. Springer, Berlin, Heidelberg, Wien, 1965.

[TR94]   P. Le Tallec and C. Rahier. Numerical methods of steady rolling for non-linear viscoelastic structures in finite deformations. *International Journal for Numerical Methods in Engineering*, 37:1159–1186, 1994.

[TT60]   C. Truesdell and R. Toupin. The classical field theories. In *Handbuch der Physik III/1*. Springer, Berlin, Heidelberg, Wien, 1960.

[TW99]   R. L. Taylor and P. Wriggers. Smooth surface discretization for large deformation frictionless contact. Technical report, University of California, Berkeley, February 1999. Report No. UCB/SEMM-99-04.

[WI93]   P. Wriggers and M. Imhof. On the treatment of nonlinear unilateral contact problems. *Ing. Archiv*, 63:116–129, 1993.

[WKO00]   P. Wriggers and L. Krstulovic-Opara. On smooth finite element discretization for frictional contact problems. *Zeitschrift für angewandte Mathematik und Mechanik*, 2000.

[WKOK99]   P. Wriggers, L. Krstulovic-Opara, and J. Korelc. Development of 2d smooth polynomial frictional contact element based on a symbolic approach. In Wunderlich, Stein, Ramm, and Wriggers, editors, *Proceedings of ECCM*, München, 1999.

[WM92]   P. Wriggers and C. Miehe. On the treatment of contact contraints within coupled thermomechanical analysis. In D. Besdo and E. Stein, editors, *Proc. of EUROMECH, Finite Inelastic Deformations*, Berlin, 1992. Springer.

[WM94] P. Wriggers and C. Miehe. Contact constraints within coupled thermome-chanical analysis - a finite element model. *Computer Methods in Applied Mechanics and Engineering*, 113:301–319, 1994.

[WN99] S. P. Wang and E. Nakamachi. The inside-outside contact search algorithm for finite element analysis. *International Journal for Numerical Methods in Engineering*, 40:3665–3685, 1999.

[WP92] J. R. Williams and A. P. Pentland. Superquadrics and modal dynamics for discrete elements in interactive design. *Engineering Computations*, 9:115–127, 1992.

[Wri87] P. Wriggers. On consistent tangent matrices for frictional contact prob-lems. In G.N. Pande and J. Middleton, editors, *Proceedings of NUMETA 87*, Dordrecht, 1987. M. Nijhoff Publishers.

[Wri95] P. Wriggers. Finite element algorithms for contact problems. *Archive of Computational Methods in Engineering*, 2:1–49, 1995.

[WS85] P. Wriggers and J.C. Simo. A note on tangent stiffnesses for fully nonlinear contact problems. *Communications in Applied Numerical Methods*, 1:199–203, 1985.

[WST85] P. Wriggers, J.C. Simo, and R.L. Taylor. Penalty and augmented la-grangian formulations for contact problems. In J. Middleton and G.N. Pande, editors, *Proceedings of NUMETA Conference*, Rotterdam, 1985. Balkema.

[WVS90] P. Wriggers, T. Vu Van, and E. Stein. Finite-element-formulation of large deformation impact- contact -problems with friction. *Computers and Structures*, 37:319–333, 1990.

[WZ93] P. Wriggers and G. Zavarise. On the application of augmented lagrangian techniques for nonlinear constitutive laws in contact interfaces. *Communications in Applied Numerical Methods*, 9:815–824, 1993.

[ZT89] O. C. Zienkiewicz and R. L. Taylor. *The Finite Element Method, 4rd Ed.*, volume 1. McGraw Hill, London, 1989.

[ZT91] O. C. Zienkiewicz and R. L. Taylor. *The Finite Element Method, 4rd Ed.*, volume 2. McGraw Hill, London, 1991.

[ZWS95] G. Zavarise, P. Wriggers, and B. A. Schrefler. On augmented lagrangian algorithms for thermomechanical contact problems with friction. *International Journal for Numerical Methods in Engineering*, 38:2929–2949, 1995.

PLASTIC DEFORMATION IN ROLLING CONTACT

K.L. Johnson
Cambridge University, Cambridge, UK

LECTURE 1: ONSET of PLASTIC YIELD

It will have been evident from the analysis of rolling contact of *elastic* bodies that the contact zone is one of high stress concentration. With metallic solids, high loads would be expected to lead to plastic yielding and permanent deformation in the highly stressed zone, which is a precursor to failure by wear and fatigue. This group of four lectures will be devoted to the mechanics of inelastic deformation in rolling contact. Before considering rolling contact, the relevant basic principles of the theory of plasticity will be reviewed and applied to the stationary contact of curved bodies.

1. Elements of plastic behaviour.

1.1 Simple tension and compression

The characteristic behaviour of a ductile material loaded and unloaded in tension and compression is well known, and is illustrated in Fig.1.1. The response in tension and compression is linear and elastic up to a yield stress σ_y, followed by plastic deformation and strain hardening. When the strains become large a distinction must be made between the nominal stress (load/initial cross-section area) and true or Cauchy stress (load/current area) and also between the nominal strain (extension or compression/original length) and the logarithmic strain ($\log_e\{$current length / original length$\}$). For an isotropic material, whose properties are independent of direction, the yield strengths in tension and compression have equal magnitudes and the curves of true stress and logarithmic strain are anti-symmetrical. However for most of the topics covered in these lectures the strains will be assumed to be sufficiently small for the difference between true and nominal stress and strain to be negligible.

During unloading the initial response is elastic with the same modulus E. When followed by reverse loading many materials yield at a stress of lower magnitude than in the initial loading, followed by strain hardening at a steeper rate. This is known as the 'Bauchinger effect' and is due to locked-in stresses in the micro structure of the material.

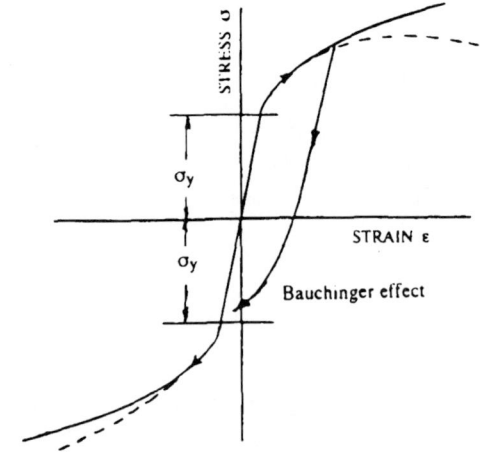

Fig.1.1 ———True stress-log. strain; – – –Nominal stress-nominal strain

For the purpose of carrying out calculations in the plastic range it is valuable to have simple idealised models of this behaviour. A hierarchy of such models, chosen to represent different aspects of the above behaviour, is presented in Fig. 1.2.

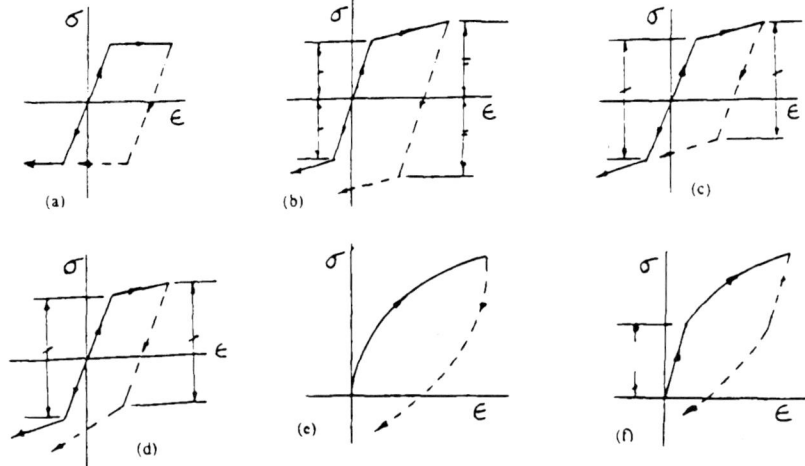

Fig. 1.2   ———— Loading;   – – – Unloading

(a) *Elastic-perfectly plastic* : Equal yield in tension and compression; zero strain hardening.

(b) *Isotropic hardening*: Equal strain hardening in tension and compression.

(c) *Linear kinematic hardening*: Constant range between yield in tension and compression; equal strain hardening rates.

(d) *Non-linear kinematic hardening*: Constant range of yield stress; unequal hardening rates.

(e) *Power-law solid*: Zero elastic response; $\sigma = C\, \varepsilon^n$.

(f) *Elastic-power law hardening*: $\sigma = E\, \varepsilon$, $\varepsilon \le \sigma_y / E$; $\sigma = \sigma_y (E\, \varepsilon / \sigma_y)^n$, $\varepsilon \ge \sigma_y / E$.

1.2 Complex stress

The general state of stress at a point in a homogeneous solid is a second order tensor having six components; three direct stresses: σ_{xx}, σ_{yy}, σ_{zz} and three shear stresses: σ_{xy}, σ_{yz}, σ_{zx}, as shown in Fig. 1.3a and expressed by the symmetric matrix:

$$\sigma_{ij} = \begin{bmatrix} \sigma_\alpha & \sigma_{xy} & \sigma_{xz} \\ \sigma_{xy} & \sigma_{yy} & \sigma_{yz} \\ \sigma_{xz} & \sigma_{yz} & \sigma_{zz} \end{bmatrix} \qquad (1.1)$$

By a suitable choice of axes the shear stresses vanish, as shown in Fig. 1.3b, to give:

$$\sigma_{ij} = \begin{bmatrix} \sigma_1 & 0 & 0 \\ 0 & \sigma_2 & 0 \\ 0 & 0 & \sigma_3 \end{bmatrix}$$ (1.2)

where the *principal stresses* σ_1, σ_2 and σ_3 and the *principal axes* n_1, n_2 and n_3 are given by the eigenvalues and eigenvectors respectively of matrix (1.1).

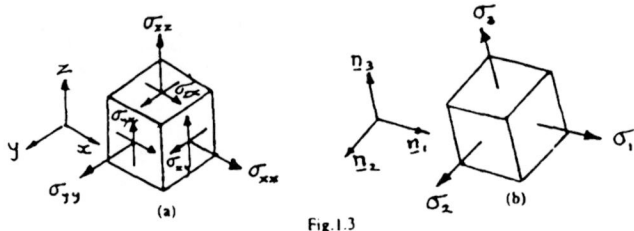

Fig.1.3

Since the state of stress at a point in an elastic body is uniquely determined by the principal stresses, the *yield criterion* i.e. the stress state at the onset of plastic yield can be expressed:

$$f\{\sigma_1, \sigma_2, \sigma_3\} = K$$ (1.3)

where K is a constant for the material. Any combination of principal stresses can be resolved into a hydrostatic stress $\bar{\sigma} = (1/3)(\sigma_1 + \sigma_2 + \sigma_3)$ together with three shear stresses $(1/3)(\sigma_1 - \sigma_2)$, $(1/3)(\sigma_2 - \sigma_3)$. and $(1/3)(\sigma_3 - \sigma_1)$, as shown in Fig.1.4.

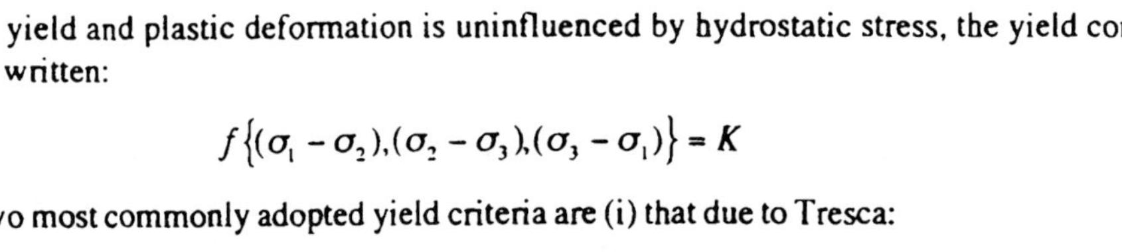

Fig.1.4

Since yield and plastic deformation is uninfluenced by hydrostatic stress, the yield condition (3) can be written:

$$f\{(\sigma_1 - \sigma_2),(\sigma_2 - \sigma_3),(\sigma_3 - \sigma_1)\} = K$$ (1.4)

The two most commonly adopted yield criteria are (i) that due to Tresca:

$$\text{max. of } |\sigma_1 - \sigma_2|, |\sigma_2 - \sigma_3|, |\sigma_3 - \sigma_1| = \sigma_y = 2k$$ (1.5)

and (ii) that due to von Mises:

$$(\sigma_1 - \sigma_2)^2 + (\sigma_2 - \sigma_3)^2 + (\sigma_3 - \sigma_1)^2 = 2\sigma_y^2 = 6k^2$$ (1.6)

where σ_y is the yield stress in simple tension ($\sigma_1 = \sigma_y$, $\sigma_2 = \sigma_3 = 0$) and k is the yield stress in simple shear ($\sigma_1 = -\sigma_2 = k, \sigma_3 = 0$). For a von Mises material we now define an *equivalent stress* σ_e and *equivalent plastic strain* increment $d\varepsilon_e$ by:

$$\sigma_e = [(1/2)\{(\sigma_1 - \sigma_2)^2 + (\sigma_2 - \sigma_3)^2 + (\sigma_3 - \sigma_1)^2\}]^{1/2} \qquad (1.7)$$

$$d\varepsilon_e^p = [(2/9)\{(d\varepsilon_1^p - d\varepsilon_2^p)^2 + (d\varepsilon_2^p - d\varepsilon_3^p)^2 + (d\varepsilon_3^p - d\varepsilon_1^p)^2\}]^{1/2} \qquad (8)$$

The strain hardening rate c is then given by:

$$c = d\sigma_e / dd\varepsilon_e^p \qquad (9)$$

A revealing representation of the yield criteria is provided by a hexagonal plot of the principal stresses in the so called π--plane, as shown in Fig.1.5. Clearly the hydrostatic component of stress, which comprises three equal vectors at $120°$, is represented by a point at the origin O. In this plot it may be shown that the Tresca criterion (1.5) is represented by a regular hexagon and the von Mises criterion (1.6) by the circumscribing circle of radius σ_y, as shown in Fig.1.5. For an isotropic material, invariance of yield stress with direction of the principal axes requires that the yield locus in the π-plane should have $15°$ symmetry which is seen to be satisfied by both the criteria shown. During elastic deformation the states of stress lie within, or in the limit, on the yield locus; during plastic deformation the stress state lies on the locus. It is apparent from this figure that the difference between the von Mises and Tresca criteria is never more than 15%.

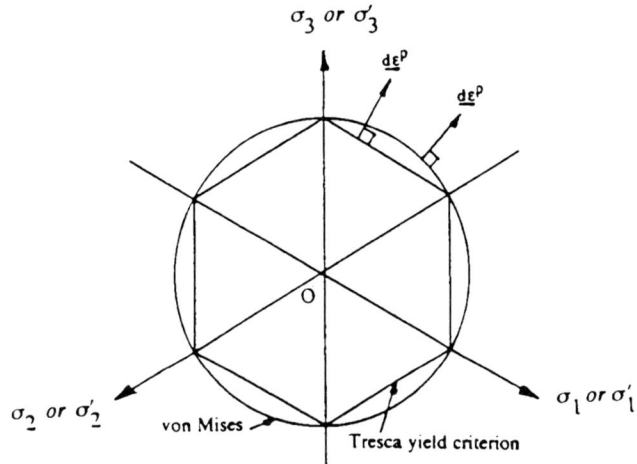

Fig1.5 The π-plane in principal stress space.

Provided that the material is inherently stable, i.e. requires a positive work input to cause an increment of plastic flow, it follows that the plastic strain increment vector $\underline{d\varepsilon}^p$ is in a direction *normal* to the yield locus at the current stress state (see Fig.1.5). This enables a *flow rule* to be written such that during plastic deformation:

$$\frac{d\varepsilon_1^p}{\sigma_1'} = \frac{d\varepsilon_2^p}{\sigma_2'} = \frac{d\varepsilon_3^p}{\sigma_3'} = \lambda \qquad (1.10a)$$

alternatively

$$\frac{d\varepsilon_\alpha^p}{\sigma_\alpha'} = \frac{d\varepsilon_w^p}{\sigma_{yy}'} = \frac{d\varepsilon_{zz}^p}{\sigma_{zz}'} = \frac{d\varepsilon_{xy}^p}{\sigma_{xy}} = \frac{d\varepsilon_{yz}^p}{\sigma_{yz}} = \frac{d\varepsilon_{zx}^p}{\sigma} = \lambda \qquad (1.10b)$$

where λ is an arbitrary constant and the *deviatoric stress* $\sigma' = \sigma - \bar{\sigma}$.

We can now generalise the models of plastic behaviour itemised in Fig.1.2 to states of complex stress (see Fig.1.6).

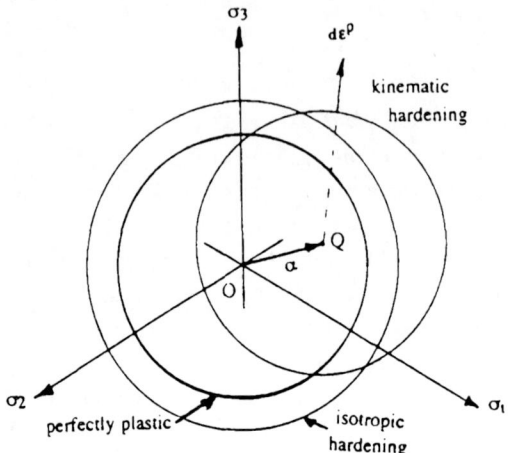

Fig.1.6   Simple hardening models in the π-plane.

(a) *Elastic-perfectly plastic* : During plastic deformation the yield locus remains fixed in stress space, without change of size or shape.

(b) *Isotropic hardening*: The yield locus expands radially.

(c) *Linear kinematic hardening*: The yield locus moves in stress space without change in size or shape. The hardening rate is independent of the position of the locus.

(e) *Non-linear kinematic hardening*: The hardening rate depends on the stress state in relation to the position of the locus.

(f) *Power law hardening* (isotropic):

$$\sigma_e = C\varepsilon_e^n \tag{1.11}$$

consistent with the flow rule of equation (10).

(g) *Elastic-hardening*: See textbooks in Plasticity.

3.   Onset of yield in Hertz contact

3.1 Frictionless line contact (plane strain):

Solid bodies in rolling contact have curved profiles. In the absence of friction forces at the interface the stresses and deformation under a normal load P are given by the Hertz theory. For simplicity we shall first discuss the case of *line contact* in which a state of plane strain exists.

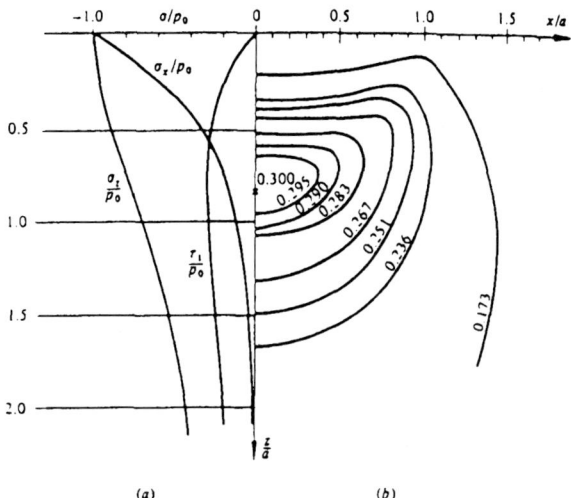

Fig. 1.7 Contact of cylinders: (a) subsurface stresses along the axis of
symmetry, (b) contours of principal shear stress τ_1.

An elastic cylinder and plane in frictionless contact are shown in Fig.1.7. The half-width of contact a and the contact pressure $p(x)$ under a load P are given by the Hertz relationships:

$$a = \left(\frac{4PR}{\pi E *} \right)^{1/2} \quad \text{and} \quad p(x) = p_o \sqrt{1-(x/a)^2} \tag{1.12}$$

where $p_o = 2P/\pi a$ and $E* = [(1-v_1^2)/E_1 + (1-v_2^2)/E_2]^{-1}$. The variation of stresses σ_{xx} and σ_{zz} along the z-axis are shown in Fig.1.7a and are given by:

$$\sigma_{xx} = -(p_o/a)\{(a^2 - 2z^2)(a^2 + z^2)^{-1/2} - 2z\} \tag{1.13a}$$

$$\sigma_{zz} = -(p_o/a)(a^2 + z^2)^{-1/2} \tag{1.13b}$$

and
$$\sigma_{yy} = v(\sigma_{xx} + \sigma_{zz}) \tag{1.13c}$$

Since Oz is an axis of symmetry, the shear stress σ_{zx} is zero and the principal stresses are: $\sigma_1 = \sigma_{xx}$, $\sigma_3 = \sigma_{zz}$, $\sigma_2 = v (\sigma_{xx} + \sigma_{zz})$. Consider first the mid-point of the surface O ($x = z = 0$), at which $\sigma_1 = \sigma_3 = -p_o$, $\sigma_2 = -2v\, p_o$. Substituting these values into the Tresca condition (1.5), yield occurs when p_o reaches the value $\sigma_y / (1-2v)$. Even if p_o exceeds this value the plastic strain which occurs is a small fraction of the yield strain σ_y/E, which in general can be neglected. With an incompressible material, whose Poisson's ratio $v = 1/2$, the stress state at O is completely hydrostatic and no yielding would be expected. Turning to conditions beneath the surface, the Tresca criterion (1.5) gives:

$$\left| \sigma_1 - \sigma_3 \right| = \left| \sigma_{xx} - \sigma_{zz} \right| = \sigma_y = 2k \tag{1.14}$$

It is simple to show from the expressions (13) that the principal shear stress τ_1 $(= (1/2)(\sigma_1 - \sigma_2))$ has a maximum value of 0.30 p_o at a depth $z = 0.78$ a. Contours of $\tau_1(x,z)$ throughout the field are shown in Fig.1.7b, confirming that the value at $(0,0.78)$ is the true maximum. At this point, for reasonable values of Poisson's ratio, σ_2 is the intermediate principal stress. So that we can state that, by the Tresca criterion, yield will initiate at the point $(0,0.78)$ when

$$(p_o)_y = 1.67\ \sigma_y = 3.3k \tag{1.15}$$

Inserting the expressions for the principal stresses into the von Mises criterion (1.6) gives:

$$(p_o)_y = 1.79\ \sigma_y = 3.1k \tag{1.16}$$

which, as expected, is not very different.

1.2 Frictionless point contacts

The onset of yield in point contacts is found in the same way. The stresses on the z-axis of an axi-symmetric Hertz contact are given by Johnson 1985 (p.62 and Fig.4.3). They vary in a similar way to those shown for line contact in Fig.1.7. The maximum principal shear stress ($v = 0.3$) has a maximum of 0.31p_o at a depth $z = 0.48$ a so that, by Tresca, yield initiates at:

$$(p_o)_y = 1.60\ \sigma_y = 3.2\ k \tag{1.17}$$

and by von Mises:

$$(p_o)_y = 1.60\ \sigma_y = 2.8\ k \tag{1.18}$$

In this last case (1.18), it follows from the Hertz relationships that the load to initiate yield is given by:

$$P_y = 8.3\ R^2\ \sigma_y{}^3\ /\ E^{*2} \tag{1.19}$$

which demonstrates the importance of yield stress or hardness in supporting contact loads without plastic deformation.

Bodies of general shape will have an elliptical contact area of semi-axes a and b. The magnitude and location of $(\tau_1)_{max}$ are shown in Table 1.1 to vary only slightly over the complete range of eccentricity of the ellipse.

b/a	0	0.2	0.4	0.6	0.8	1.0
z/b	0.785	0.745	0.665	0.590	0.530	0.480
$(\tau_1)_{max}/p_o$	0.300	0.322	0.325	0.323	0.317	0.310

Table 1.1

If the load is increased above that for first yield plastic deformation spreads from the point of initiation. It is evident from the stress contours in Fig.1.6b that, to begin with, plastic deformation is confined to a small enclave below the surface; the plastic strains will the be restricted to an elastic order of magnitude. Only after an appreciable increase in load will then plastic zone expand to the free surface outside the contact and enable unconstrained flow to take place. When an elastic sphere indents an elastic-perfectly plastic half-space, the variation of mean

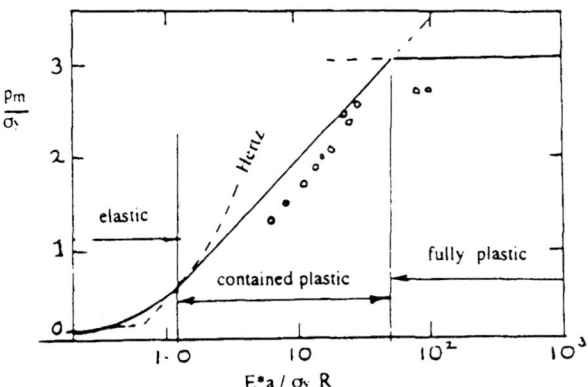

Fig.1.8 Plastic indentation by an elastic sphere of radius R

indentation pressure with the size of an axi-symmetrical indentation is shown in Fig.1.8. Three regimes are revealed: (i) *elastic* : when the contact pressure is less than that given by (1.17 or 18) (ii) *contained plastic*: when the plastic zone lies below the surface, and (iii) *fully plastic*: when unrestricted plastic flow can take place. In zone (i) the mean pressure p_m = (2/3) p_0, where p_0 is given by Hertz. In zone (ii) an approximate expression for the mean contact pressure has been proposed by Johnson (1985):

$$\frac{p_m}{\sigma_y} \cong \frac{2}{3}\left\{L7 + \ln\left(\frac{Ea}{3\sigma_y R}\right)\right\}$$ (1.20)

In zone (iii), which is equivalent to a Brinell hardness test, the mean pressure or hardness is approximately given by: p_m = 3 σ_y. The size of the indentation a is expressed by the non-dimensional parameter $(Ea / \sigma_y R)$, which may be interpreted as the ratio of the strain imparted by the indenter (a/R) to the yield strain (σ_y/E) of the plastically deforming material.

1.3 Sliding contacts with friction

We shall now investigate the onset of yield in sliding contact, or combined rolling and sliding, as between a pair of gear teeth, and restrict the analysis to the plane-strain situation of line contact (Fig.1.9). With similar materials in contact, the contact width a and pressure distribution p(x) are given by Hertz (equation 12). With dissimilar materials a and p(x) are influenced by friction, but the effect is small and will be neglected. During sliding the frictional traction is assume to vary as:

$$q(x) = \mu\, p(x) = \mu\, p_o\sqrt{1-(x/a)^2}$$ (1.21)

The stresses within the solid are now due to the combined effect of the pressure p(x) and the frictional traction q(x). Computed contours of principal shear stress $\tau_1(x,z)$ are plotted in Fig.1.9a. Compared with Fig.1.7b, the contours are no longer symmetrical about the z-axis, the maximum value is slightly increased and located slightly closer to the surface. However the situation at the surface is completely changed. The stress state due to the normal pressure is

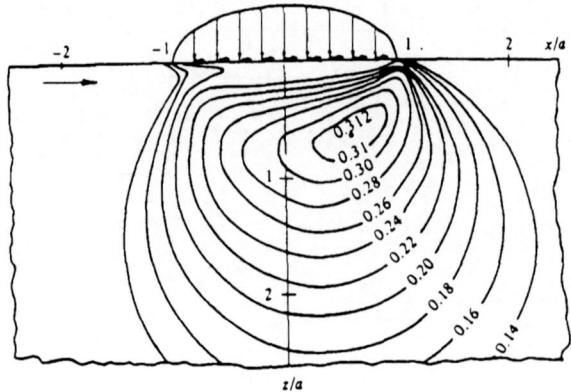

Fig.1.9a Contours of the principal shear stress τ_1 beneath a sliding
contact, $Q_x = 0.2P$.

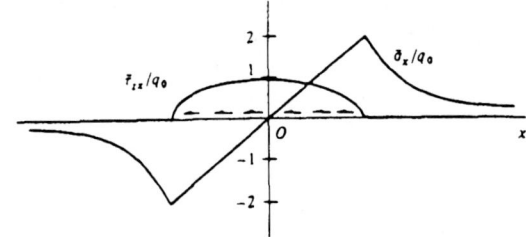

Fig.1.9b Surface stresses due to frictional traction $q = q_0(1 - x^2/a^2)^{1/2}$.

effectively hydrostatic at all points in the contact, so that there is no tendency to yield (except for
the negligible plastic flow in the lateral direction). Due to the frictional traction q(x) the stresses are:

$$\sigma_{xx}(x,0) = -2\mu x / a$$
$$\sigma_{yy}(x,0) = -2v\mu x / a$$
$$\sigma_{zz}(x,0) = 0 \qquad\qquad | \quad -a \le x \le a \qquad\qquad (1.22)$$
$$\sigma_{zx}(x,0) = -q(x) = -\mu\, p_o \sqrt{1 - (x/a)^2}$$

as shown in Fig.1.9b. The principal shear stress $\tau_1(x,0)$ in the plane of deformation is given by:

$$\tau_1(x,0) = (1/2)(\sigma_3 - \sigma_1) = (1/2)\sqrt{(\sigma_{zz} - \sigma_{xx})^2 + 4\sigma_{zx}^2}$$
$$= \mu\, p_o \qquad\qquad (1.23)$$

This result shows that yield initiates simultaneously at all points in the contact surface when $\tau_1 = k$,

i.e. when $(p_0)_y / k = \mu$ (1.24)

Comparing with equation (1.15), we see that yield at the surface will precede that beneath the
surface if the coefficient of friction µ exceeds about 0.3. *This result has practical importance in
relation to wear and surface degradation in sliding/rolling contact.*

References

Calladine,C.R. (1969) *Engineering Plasticity*, Pergamo

Johnson,K.L. (1985) *Contact Mechanics*, C.U.P.

Lubliner,J. (1990) *Plasticity Theory*, Macmillan.

LECTURE 2: SHAKEDOWN in ROLLING CONTACT

1.   Response of elastic-plastic structures to cyclic loads

Shakedown principles were developed to examine the steady-state response of elastic-plastic structures to repeated cyclic loads. They have two notable advantages: (i) they address the steady cyclic state directly without having to follow the plastic loading history from the start and (ii) they rely on *elastic* distributions of stress, which are more readily obtained than elastic-plastic stresses. This situation is appropriate to rolling contacts, in which the surface material in the majority of applications is subjected to many cycles of load. It is the steady cyclic state which controls failure rather than the initial transient.

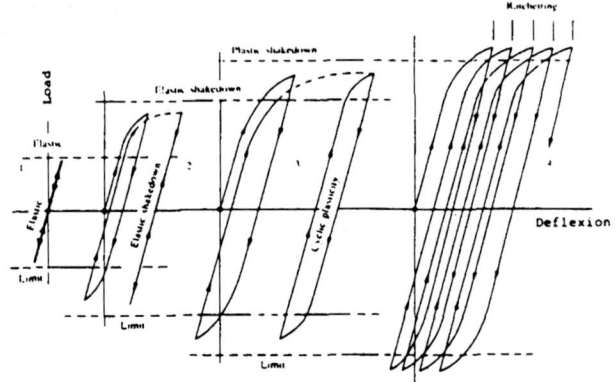

Fig.2.1.   Structural response to cyclic loading:

The response of a rate-independent elastic-plastic structure to cyclic loads is illustrated in Fig.2.1. Four regimes of behaviour can be identified:

(i) *Perfectly elastic*: If the loads are sufficiently small for no element of the structure to reach its elastic limit, the response will be perfectly elastic and reversible throughout.

(ii) *Elastic shakedown::* In this regime the elastic limit is reached in the first few cycles, but the steady-state is entirely elastic. The maximum load for which elastic shakedown can be achieved is known as the *shakedown limit.* .

(iii) *Cyclic plasticity* (or plastic shakedown): In this regime the steady-state consists of a closed cycle of plastic deformation.

(iv) *Ratchetting* (or incremental collapse): In this regime the structure accumulates increments of uni-directional plastic strain, leading to collapse.

Since a structure which is subjected to steady cyclic plasticity is likely in due course to fail by fatigue, and one which accumulates plastic strain by ratchetting is likely to fail by ductile fracture, the elastic shakedown limit provides a rational design criterion for cyclically loaded structures.

Three possible consequences of plastic deformation in the early cycles influence shakedown in the steady state: (i) residual stresses; (ii) strain hardening; (iii) geometry changes.

Generally, though not exclusively, residual stresses are protective in that they make subsequent yielding less likely and hence promote elastic shakedown.

The effect of strain hardening depends on the form it takes (see Fig.1.6). Isotropic hardening (a poor model of real material behaviour) causes elastic shakedown in one cycle of load. and is of no further concern here. Linear kinematic hardening leads either to elastic shakedown or cyclic plasticity. To model ratchetting behaviour it is necessary to adopt some form of non-linear kinematic hardening. A perfectly plastic material (zero hardening), on the other hand, is capable of displaying the behaviour of all four regimes.

2.   Shakedown theorems

The tools of the trade for the study of shakedown of ductile structures are the shakedown theorems of the theory of plasticity. For perfectly plastic solids (no hardening) these comprise the statical theorem due to Melan (1938) and the kinematical theorem due to Koiter (1956). The statical theorem has been extended by Ponter (1976) and Mandel (1976) to cover kinematically hardening materials. The theorems are stated below. For their derivation and proof, see the original papers or plasticity texts (e.g. Lubliner 1990).

2.1 Statical (lower bound) theorem

'If any system of self-equilibrating residual stresses $\rho*$ can be found which, in combination with the stresses σ due to a repeated load P, do not exceed yield at any time. then elastic shakedown will take place'. The maximum load which, together with the 'true' distribution of residual stress just reaches yield, is referred to as the 'shakedown limit'. Any other distribution of residual stress will give a *lower bound* to the shakedown limit.

1.2 Kinematical (upper bound) theorem

'If any kinematically acceptable mechanism of plastic collapse can be found in which the rate of doing work by the elastic stresses due to the load exceeds the rate of plastic deformation. then incremental collapse will take place'. The ratio of the work done by the elastic stresses to the plastic dissipation has a maximum in the 'true' mechanism of collapse, so any other mechanism gives an *upper bound* to the shakedown limit.

1.3 Kinematic hardening theorem

This theorem is based on the fact that a kinematically hardening material can achieve shakedown partly by developing residual stresses ρ, but also by a displacement of the yield locus ('back stress') α. The sum of $\rho + \alpha$ can be thought of as an effective residual stress $\rho*$,which can be used in the statical theorem to find a lower bound to the shakedown limit. For any element of material $\rho*$ must remain constant through the loading cycle. Unlike the 'true' residual stress ρ, however, it does not need to satisfy the equilibrium equations of the continuum and cannot be separated into its components ρ and α.

In the application of these theorems any acceptable yield criterion, e.g. Tresca or von Mises can be used.

2. Application to plane rolling line contacts

2.1 Frictionless (free rolling) contact - static approach

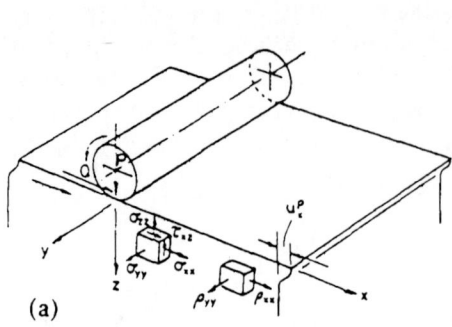

(a)

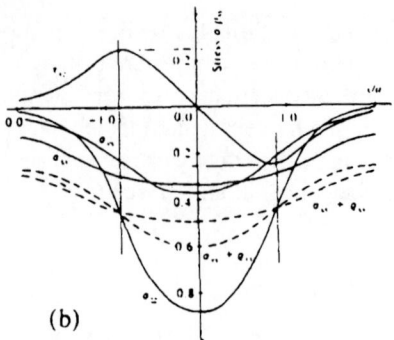

(b)

Fig.2.2a. Rolling/sliding contact of a cylinder with an elastic-plastic half-space. Contact stresses σ_{xx}, σ_{yy}, σ_{zz} and τ_{zx} which exceed the elastic limit introduce residual stresses ρ_{xx} and ρ_{yy} (compressive).

Fig.2.2b. Elastic stresses in frictionless line contact. The addition of residual stresses ρ_{xx} and ρ_{yy} leads to shakedown provided $(\tau_{zx})_{max} \leq k$.

An elastic cylinder of radius R in rolling line contact with an elastic-perfectly plastic half-space is shown in Fig.2.2a. In the first instance the frictional traction force Q will be taken to be zero. The elastic contact stresses $\sigma_{xx}(x)$, $\sigma_{yy}(x)$, $\sigma_{zz}(x)$ and $\tau_{zx}(x)$ at a depth z = 0.5 a due to the load P per unit length are shown by the full lines in Fig.2.2b. The semi-contact width and maximum contact pressure are given by Hertz:

$$a = (4\,PR\,/\,\pi E^*)^{1/2} \qquad \text{and} \qquad p_0 = (PE^*/\,\pi P)^{1/2} \tag{2.1}$$

If the elastic limit of the half-space is exceeded residual stresses will be introduced. If, as expected, the surface of the half-space remains plane after rolling, the residual stresses will be independent of x and y. Assuming conditions of plane strain, $\rho_{xy} = \rho_{yz} = 0$, and for equilibrium with the free surface $\rho_{zz} = \rho_{zx} = 0$. We are left, as shown in Fig.2.2a, with $\rho_{xx}(z)$ and $\rho_{yy}(z)$, which vary with depth z only.

To apply Melan's theorem we choose the residual stresses $\rho_{xx}(z)$ and $\rho_{yy}(z)$ to promote shakedown, i.e. to avoid yield. $\rho_{yy}(z)$ can be chosen to ensure that $(\sigma_{yy} + \rho_{yy})$ is the intermediate principal stress. Then for yield not to be exceeded by the Tresca criterion:

$$(\sigma_1 - \sigma_3)^2 = \{(\sigma_{xx} + \rho_{xx}) - \sigma_{zz}\}^2 + 4\,\tau_{zx}^2 \leq 4\,k^2 \tag{2.2}$$

This condition can just be satisfied with $(\tau_{zx})_{max} = k$, by choosing $\rho_{xx} = \sigma_{zz} - \sigma_{xx}$. In the elastic contact stress field $(\tau_{zx})_{max} = 0.25\,p_0$ at $x = \pm 0.87a$, z = 0.5a, which gives a lower bound to the shakedown limit:

$$p_o^s \geq 4k \tag{2.3}$$

The same result is obtained by the von Mises criterion, by choosing $\sigma_{xx} + p_{xx} = \sigma_{yy} + p_{yy} = \sigma_{zz}$, as shown by the broken lines in Fig.2.2b.

2.2 Frictionless (free rolling) contact - kinematic approach

To apply the kinematical theorem we have to select a mechanism whereby increments of plastic strain can be continuously accumulated. The simplest mechanism would seem to be shear along a plane at a depth h below the surface through an incremental plastic displacement Δu^P. The work done by the elastic stresses is $(\tau_{zx} \Delta u^P)$ and the plastic dissipation is $(k \Delta u^P)$. By Koiter's theorem, incremental collapse (ratchetting) will occur if $\tau_{zx} \Delta u^P \geq k \Delta u^P$. Thus the optimum upper bound to the shakedown limit is obtained by choosing h = 0.5 a at which τ_{zx} has its maximum value 0.25 p_0. Then:

$$p_o^s \leq 4k \tag{2.4}$$

Comparing (2.4) with (2.3) reveals that lower and upper bounds are equal, so that $p_0{}^s$ = 4k is the exact value of the elastic shakedown limit for a perfectly plastic solid. In practice a small amount of strain hardening spreads the shear plane into a narrow shear band, as revealed by experiments.

2.3 Rolling with sliding line contact

When sliding accompanies rolling a frictional traction is exerted at the interface given by:

$$q(x) = \mu \, p(x) = \mu \, p_0\{1 - (x/a)^2\}^{1/2} \tag{2.5}$$

The shakedown limit by the Tresca criterion for a perfectly plastic material is still given by $(\tau_{zx})_{max}$ = k but, as discussed in Lecture I, Section 1.3, the effect of the frictional traction (2.5) is to give rise to two peaks of τ_{zx}, one at the surface and one sub-surface (see Fig.1.9). The magnitude and location of the sub-surface peak as a function of the coefficient of friction μ has been computed by Johnson & Jefferis (1963). At the surface $(\tau_{zx})_{max}$ = $\mu \, p_0$. The shakedown limit found by using the von Mises yield criterion is very similar.

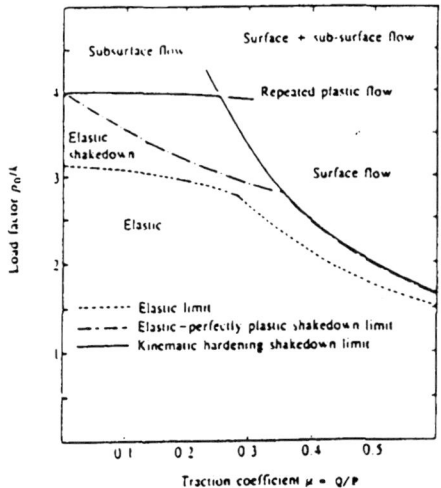

Fig.2.3. Shakedown map in line contact for a perfectly plastic and a kinematically hardening material

Using the von Mises data a *shakedown map* of load factor $(p_0{}^s/k)$ against μ has been constructed in Fig.2.3. The map shows both the variation in the initial yield stress and the shakedown limit, indicating the increase in load carrying capacity due to the introduction of residual stresses. In the frictionless case this represents an increase in stress from $\sigma_y = 3.1$ k (eq. 1.16) to 4 k (eq. 2.3), i.e. 29% which corresponds to an increase in load of 66%. The curves show a discontinuity at $\mu \approx 0.3$ when the critical stress move from sub-surface to surface.

2.4   Rolling of line contacts with partial slip

In a rolling contact where the traction coefficient Q/P is less than the coefficient of limiting friction μ, partial slip occurs at the contact interface. With a line contact in plane strain Carter (1926) showed that in a strip of width c at the leading edge of the contact the surfaces roll without slip, where c is given by:

$$c/a = \{1 - (Q/\mu P)\}^{1/2} \tag{2.6}$$

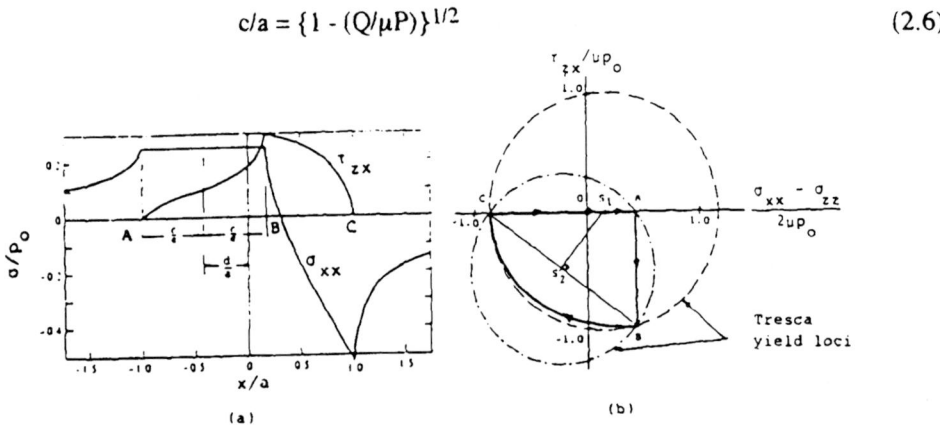

Fig.2.4.   Tractive rolling contact stresses in line contact with partial slip (Q/P < μ).
(a) Variation with position in the contact.   (b) Trajectory in stress space.
— — — —   Yield locus for perfectly plastic solid.
— · — · —   Yield locus for kinematic hardening.

The distributions at the surface (z = 0) of the elastic stress components σ_{xx} and τ_{zx} ($\sigma_{zz} = 0$) are shown in Fig.2.4a [see notes on elastic rolling contact]. These stresses are plotted in stress space having co-ordinates $(\sigma_{xx} - \sigma_{zz})/\mu p_0$ and $\tau_{zx}/\mu p_0$ in Fig.2.4b, to give the stress trajectory OABCO. We note that the Tresca yield locus (2.2) maps into this plane as a circle of radius $k/\mu p_0$ and centre at the point $(-\sigma_{xx}/\mu p_0, 0)$. The shakedown limit for a perfectly plastic solid is then obtained by finding the smallest circle having its centre on the horizontal axis which circumscribes the stress trajectory as shown. If the radius of this circle is r_{min}, we have $k/\mu p_0 = r_{min}$, so that

$$p_0{}^s = 1/\mu r_{min} \tag{2.7}$$

For a material which strain hardens kinematically the yield locus is free to move anywhere in stress space as shown in Fig.1.6. This enables the trajectory to be circumscribed by a circle of much smaller radius, with a consequent increase in the shakedown limit. Shakedown maps for rolling line contact with partial slip, so found, are presented in Fig.2.5 for both perfectly elastic and kinematically harden materials.

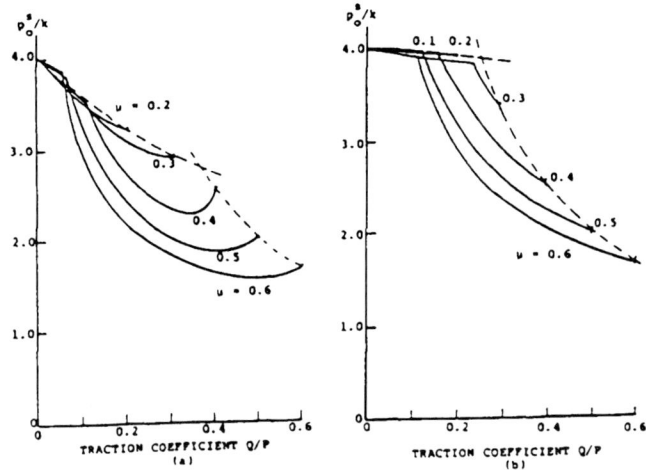

Fig.2.5.   Shakedown maps for rolling line contact with partial slip $(Q/P < \mu)$.
(a) Elastic-perfectly plastic solid.   (b) Kinematically hardening solid.

As the traction force approaches limiting friction, the stress distribution approaches that of complete sliding [eq.(1.22)] and the trajectory in the stress space of Fig.2.4b becomes a semi-circle, centre O. The perfectly plastic and kinematic hardening yield loci then become identical circles of radius k and centre O, so that for high friction, when yield is governed by surface stresses, the shakedown and elastic limits by Tresca are identical. For low friction when yield is governed by sub-surface stresses, it is the *range* between the maximum and minimum values of τ_{zx} which governs the shakedown limit. The shakedown limits for a kinematic hardening material have been added to Fig.2.3, which shows that, with low values of μ when yield is governed by the sub-surface stress, the shakedown limit is enhanced by hardening. With high values of μ, when yield occurs at the surface, the shakedown limit is not improved by hardening.

3.   Application to rolling point contacts

A 'point contact' between elastic bodies, loaded by a normal force P, develops an elliptical (or circular) contact area with a contact pressure given by:

$$p(x,y) = p_0\{1 - (x/a)^2 - (y/b)^2\}^{1/2} \tag{2.8}$$

where a and b are semi-axes of the ellipse.

Determination of the shakedown limits for point contacts is much more difficult than for line contacts. All six components of residual stress can, and indeed do, exist. The situation is illustrated in Fig.2.6. Any tendency for the material of the rolling track to shear ahead of its surroundings is resisted by residual shears ρ_{xy}, ρ_{yz} and ρ_{zx} as shown. By neglecting these residual stresses and by assuming that shakedown is governed by $(\tau_{zx})_{max}$, as in the case of line contact, a lower bound to the elastic shakedown limit was found by Hills & Ashelby (1982), both with and without frictional traction. By admitting the residual shear system shown in Fig.2.6, a somewhat improved lower bound was obtained by Hearle (1984), which is displayed in the point contact shakedown map by curve D in Fig.2.8.

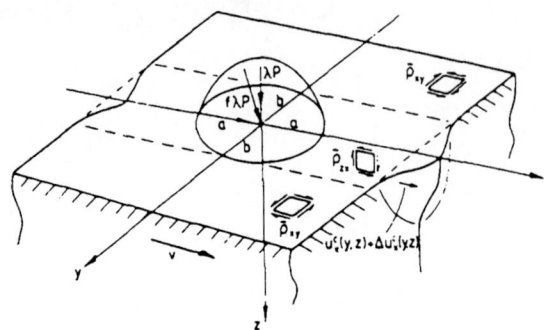

Fig.2.6.   Mode of deformation in point contact : displacement of a surface segment $u'_x(y,z)$.

To investigate if and when ratchetting might take place appeal was made to the kinematical theorem [Ponter et al. (1985)]. When, for example, a hard ball rolls on an elastic plastic half-space plastic deformation gives rise to a permanent shallow groove, which has the effect of reducing the contact stress and thereby promoting shakedown. To avoid this complication, the problem was transformed into that of a Hertz pressure distribution (2.8) traversing the surface of an elastic-perfectly plastic half-space.

The chosen mechanism of incremental collapse (ratchetting) is shown in Fig.2.7. Shear is assumed to occur on a curved surface s, so that a sliver of material in the rolling track moves forward by an increment in displacement δ. The shear stress $\hat{\tau}(s)$ on this plane is given by:

$$\hat{\tau}(s) = \{\tau_{xy}^2 + \tau_{zx}^2\}^{1/2} \tag{2.9}$$

where τ_{xy} and τ_{zx} are given by the Hertz elastic theory.

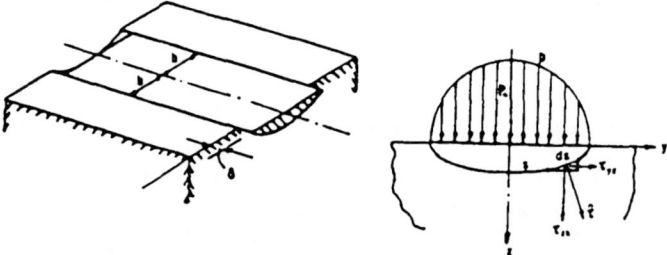

Fig.2.7.   Application of Kolter's kinematical theorem to shakedown of a point contact (a) mechanism of incremental collapse: (b) stress contour.

We define $\hat{\tau}_m(s)$ as the maximum value of $\hat{\tau}(s)$ encountered by an element of material on the shear surface during a cycle of load. The work done by these elastic stresses in an increment of displacement is thus:

$$p_o \delta \int_o^L (\hat{\tau}_m / p_o) ds$$

while the plastic dissipation in the same increment is:

$$kL\delta$$

Thus, by the kinematical theorem, ratchetting would be expected if

$$\frac{p_o}{L}\int_o^L (\hat{\tau}_m / p_o)ds \geq k \tag{2.10}$$

To obtain the 'best' upper bound to the ratchetting threshold the shape and location of the shear surface should be chosen to minimise the integral on the left-hand side of (2.10). This has been done by a numerical optimising procedure to give the upper bound to the ratchetting limit shown by curve B in the shakedown map in Fig.2.8.

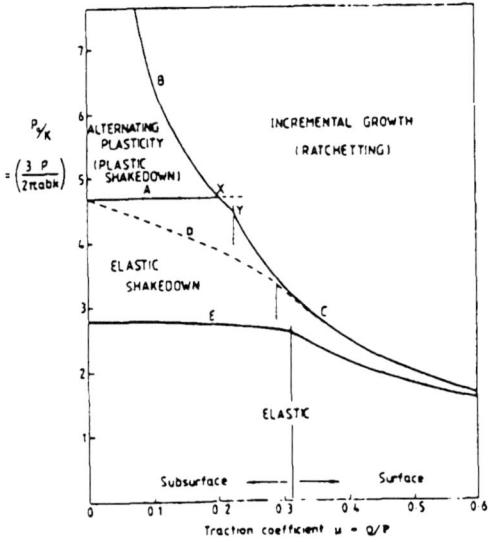

Fig.2.8.   Shakedown map for point contacts.   Curve E: elastic limit: A: upper bound to elastic shakedown limit: D: lower bound to elastic shakedown limit: B-C: upper bound to ratchetting limit (by Koiter's theorem).

It turns out, particularly at low values of the traction coefficient, that the load factors (p_o/k) approaching the ratchetting threshold give rise to stresses which exceed yield in a fully contained enclave beneath the centre of the contact. In such cases Ponter argues that, since the load is insufficient to allow ratchetting, residual stresses $\bar{\rho} = -(1/2)\{\hat{\tau}(x)_{max} - \hat{\tau}(x)_{min}\}$ will develop in those enclaves such that the plastic deformation take the form of closed cycles i.e. 'plastic shakedown'. Loads for which the resultant stress $(\hat{\tau} + \bar{\rho})$ does not reach yield give an upper bound to the elastic shakedown limit (curve A in Fig.2.8).

References

Hearle,A.D. (1984) *PhD Dissertation*, Cambridge University.

Hills,D.A. & Ashelby,D.W. (1982) *Wear*, **7 5**, 221.

Johnson,K.L. (1992) *J.Mech.A/Solids*, **1 1**, 155.

Johnson & Jefferis (1963) *Proc.I.Mech.E. Symp. on Rolling Contact Fatigue*, London.

Koiter,W.T. (1956) *Konikl. Ned. Ak. Wetenschap*, **B 59**, 24.

Mandel,J. (1976) *Mech.Res.Comm.* **3**, 251,483.

Melan,E. (1938) *Ing.-Arch.* **9, 116.**

Ponter,A.R.S. (1985) *J.Mech.Phys.Solids*, **3 3**, 339.

LECTURE 3 : RATCHETTING

1. Introduction

While the elastic shakedown limit provides a rational and effective design criterion for the trouble-free working of rolling contacts, it is becoming clear that cyclic plastic deformation and ratchetting play an important role in their failure by fatigue and wear. To understand the mechanism of these failures and to make rational estimates of fatigue life and wear rates, therefore, it is necessary to be able to make reliable estimates of ratchetting rates under prescribed cyclic loading. This question will be addressed in this lecture.

Research into ratchetting in rolling contact was stimulated by the experiments of Crook (1957) who sectioned two loaded rollers into which soft radial pins had been inserted. After rolling the pins acquired a permanent plastic shear deformation as shown in Fig.3.1. The frictional traction at the interface was neglgibly small; only necessary to overcome the bearing friction in the driven roller. The direction of the shear was such that the surface moved ahead of the core in the direction of rotation of each roller, leading to the description *forward flow*. In a repeat of Crook's experiment, Merwin (1962) showed that the plastic strain accumulated steadily over many cycles of load (revs. of the rollers), demonstrating that the phenomenon was a classic example of ratchetting in an elastic-plastic component [see Fig.3.2]

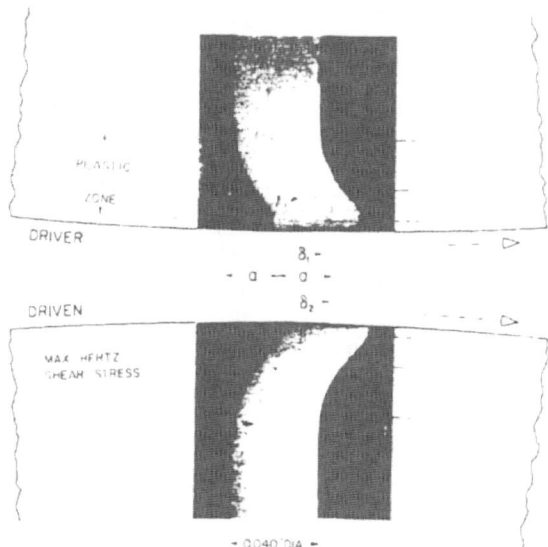

Fig.3.1 Forward flow in freely rotating discs shown by deformed pins, [from Crook (1957).

Ratchetting in rolling contact has different manifestations. In free rolling, or when the traction coefficient is small, plastic shear accumulates beneath the surface as seen in Fig.3.1. However, under high traction, the plastic flow is located in a thin layer at the surface. The situation pictured in Fig.3.1 is one in which the bulk material of the rollers exceeds the shakedown limit. A different situation may occur, notably in sliding contacts, in which the bulk stresses are within the

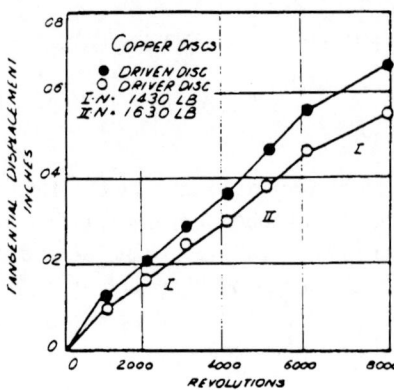

Fig.3.2 Measurements of forward flow in rolling discs from Merwin (1962).
(a) copper, (b) duralumin.

elastic limit, but the surface irregularities - asperities - in which the stress is higher, are subjected to plastic flow and ratchetting on a microscopic scale. Finally most analyses of ratchetting have been confined to plane strain deformation in line contact, even though most practical situations involve three-dimensional point contacts.

When this problem first arose the numerical finite element method for elastic-plastic problems was virtually undeveloped. The prospect was daunting. Rolling contact has to proceed step-by-step from some assumed initial state through a sufficient distance for a steady state to be reached in which the stresses do not change with time. This process has then to be repeated to reach a cyclic steady-state in which the residual stresses at the end of a load cycle are identical with those at the start. It is not surprising, therefore, that ways have been sought to short circuit this procedure.

In order to model ratchetting in rolling contact it is first necessary to choose an appropriate model of the material. The elastic-perfectly plastic model (Fig.1.2a & Fig.1.6) is the simplest and most commonly used. To include the effects of strain hardening is not so straightforward. Isotropic hardening (Fig.1.2b & Fig.1.6) leads to shakedown in one cycle; linear kinematic hardening (Fig.1.2c &Fig.1.6) always leads to cyclic plasticity. To model ratchetting it is necessary to include hardening in a non-linear way, i.e. for the hardening rate to be a function of the displacement α of the yield locus from its initial position (Fig.1.2d & Fig.1.6).

2.   Elastic-perfectly plastic material; subsurface deformation.

Reference to the shakedown map in Fig.2.3 indicates that plastic deformation should be dominant beneath the surface in free rolling ($Q = 0$) and with low traction coefficients ($Q < 0.3\ P$). The first plane strain calculations of residual stresses and ratchetting rate in a freely rolling contact were made by Merwin & Johnson (1963). On the grounds that the plastically deforming zone is small and fully contained beneath the surface, it was assumed that the total strain field (elastic + plastic) is approximately the same as an elastic field under the same load. The stresses were the computed from the Prandtl-Reuss stress-strain equations for an elastic-perfectly plastic solid. This calculation does not provide stresses which satisfy the equilibrium equations. However *overall* equilibrium was achieved at the completion of the calculation by relaxing the residual stresses ρ_{zz}

and $(\tau_{zx})^r$. The calculation then yields the residual shear strain $(\gamma_{zx})^r =$ and the non-zero residual stresses $\rho_{xx}(z)$ and $\rho_{yy}(z)$. The displacement of the surface relative to the core ('forward flow') for each cycle of load is then found by integrating the residual shear strain with depth,

i.e.
$$\delta = \int_0^\infty (\gamma_{zx})^r \, dz \qquad\qquad (3.1)$$

This displacement is plotted non-dimensionally as a function of load factor (p_0/k) in Fig.3.3, where p_0 is the max. Hertz pressure and k is the shear yield strength of the solid. The residual stresses $\rho_{xx}(z)$ and $\rho_{yy}(z)$ are shown in Fig.3.4.

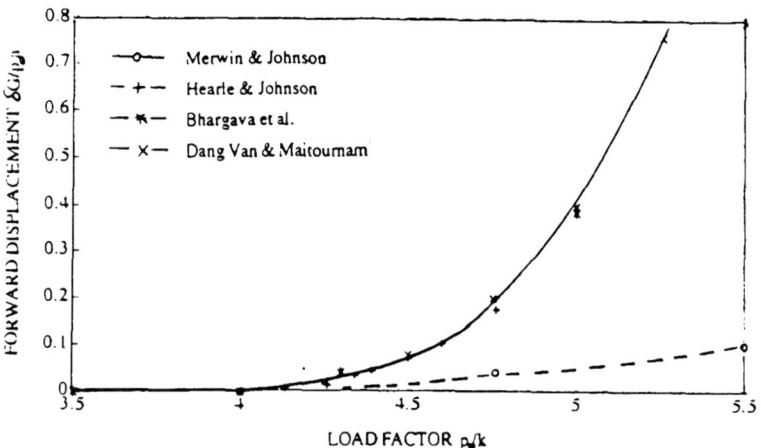

Fig.3.3 Calculated forward flow in free rolling; elastic-perfectly plastic.

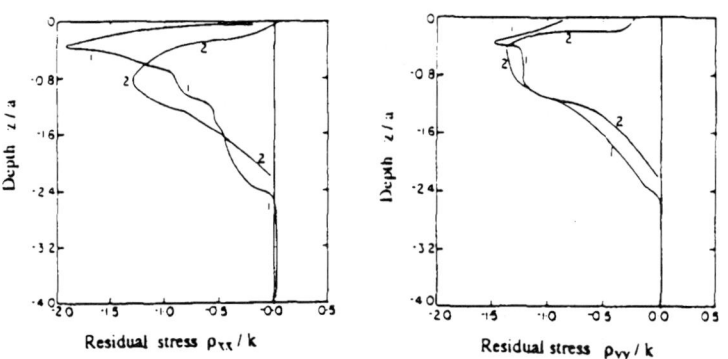

Fig.3.4 Variation of residual stress with depth. (1) Bhargava et al. (2) Merwin & Johnson.

A completely different technique of calculating forward flow was proposed by Hearle & Johnson (1987), which has the advantage of approaching the steady cyclic state directly. It was assumed that plastic deformation is restricted to shear on planes parallel to the surface, i.e. to the strain component γ_{zx}. This deformation was modelled by continuous distributions of glide dislocations on a series of parallel planes, as shown in Fig.3.5. The form, magnitude and extent of these distributions of dislocations is determined by the necessity of maintaining the net value of the shear stress $\tau_{zx} = k$ in the plastically deforming region (Fig.3.5a). Interaction between the zones of positive and negative shear stress at entry and exit results in an asymmetric distribution of stress and strain, such that the forward shear in the exit zone exceeds the backward shear in the entry zone. The difference comprises the net forward flow. It is reflected in the difference in size of the plastic zones shown in Fig.3.5b. Computations of the forward displacement δ by Hearle & Johnson are added to Fig.3.3.

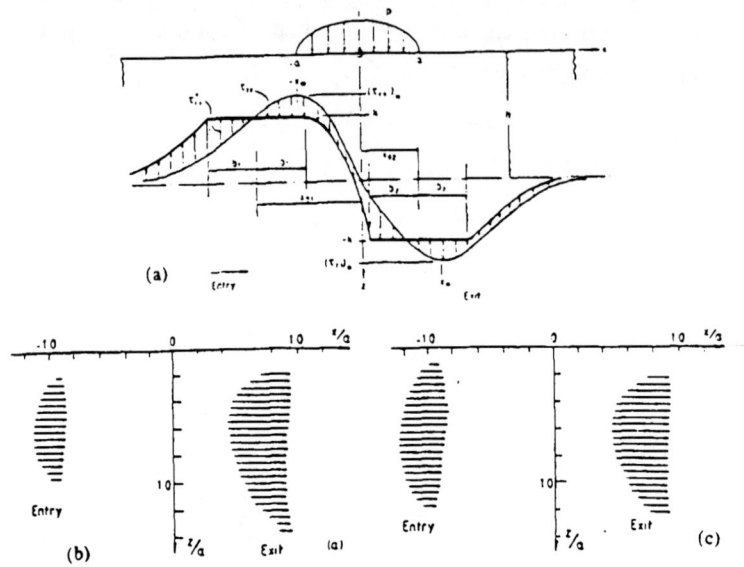

Fig.3.5 Distributed dislocations [Hearle & Johnson]. (a) Stress distribution on a glide plane.
(b) Plastic zone. Q / P = 0. (c) Plastic zone, Q / P = - 0.05.

The dislocation technique is equally adaptable to rolling with traction. The *elastic* distribution of shear stress τ_{zx} now has unequal magnitudes at entry and exit, which influences the relative size of the two plastic zones (Fig.3.5c) and hence the relative magnitude of the plastic shear strains. A sufficiently large negative tractive force reverses the direction of the net flow, giving a *backward* displacement. The non-dimensional displacement $(\delta G/p_oa)$ at a load factor $(p_o/k) = 4.76$ is plotted as a function of traction coefficient (Q/P) in Fig.3.6.

Meanwhile several researchers were investigating forward flow in elastic-plastic rolling contact by the finite-element method. Bhargava et al.(1985) computed forward displacement and residual stresses after 4 cycles of load, which are added to Fig.3.3 & 3.4. F-E results were obtained for both free and tractive rolling by Murakami et al.(1990) and by Dang Van & Maitournam (1993) which are included in Fig.3.3 and Fig.3.6.

Consider first Fig.3.3 which displays the different calculations of forward displacement (ratchetting rate) for a perfectly plastic material in free rolling. Ratchetting vanishes, as expected, at a load factor less than the shakedown limit ($p_o/k = 4$). It rises rapidly when the load factor exceeds about 4.7. Good agreement is found between the approximate calculations of Hearle & Johnson and the F-E analyses of Bhargava et al. and Dang Van & Maitournam. The predictions of Merwin

& Johnson are out of line and much lower. Evidently approximating the deformation field to that in elastic contact introduces serious errors and should be discarded.

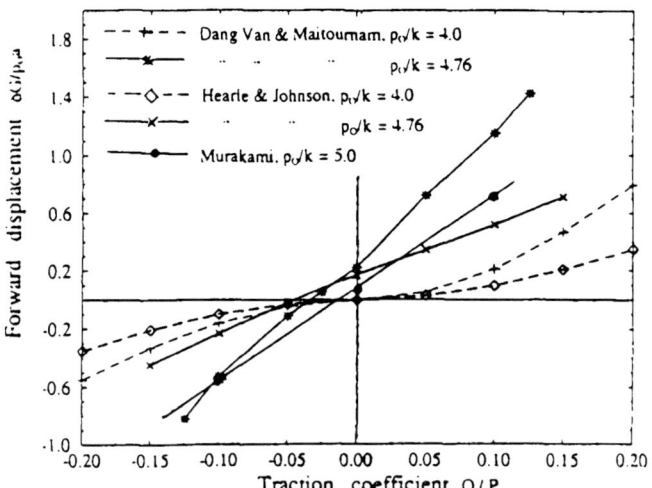

Fig.3.6. Forward flow in tractive rolling: elastic-perfectly plastic.

The various calculations of ratchetting rate in tractive rolling are presented in Fig.3.6. The rate varies roughly in proportion to the traction coefficient Q/P, with a bias towards forward displacement in free rolling (Q/P = 0), but here the different calculations are inconsistent. No satisfactory explanation has, so far, been found for the inconsistency or which result is the more correct.

The calculations of ratchetting rate in free rolling in Fig.3.3 are compared with experimental measurements in Fig.3.7. These measurements gave early support for the original calculations by Merwin & Johnson (curve A); it was only when the more reliable calculations became available (curve B) that a discrepancy between theory and experiment became apparent. This might have been due, in part, to the difficulty of reproducing plane strain conditions in the experiment, but a more like cause was the effect of strain hardening in the experimental material. Calculations for a hardening material are described in the next section.

General consistency is found between the calculations of residual stress, including those by Merwin & Johnson., as indicated in Fig.3.4. Both components of residual stress ρ_{xx} and ρ_{yy} are compressive in the plastically deforming zone beneath the surface, with values of order of the shear yield stress k. Note that they cannot exceed the yield stress in compression.

When the traction coefficient exceeds 0.3, yield initiates at the surface. Assuming the ratio of the tangential traction q to the normal pressure p is a constant coefficient of sliding friction μ, equation (1.24) demonstrates that all points in the contact zone reach yield simultaneously. With an elastic-perfectly plastic material this leads to a singular situation. The shear traction q cannot increase further, so that an increase in contact pressure p causes an automatic reduction in μ. Plastic deformation take the form of a vanishingly thin surface layer shearing through an indeterminate strain.

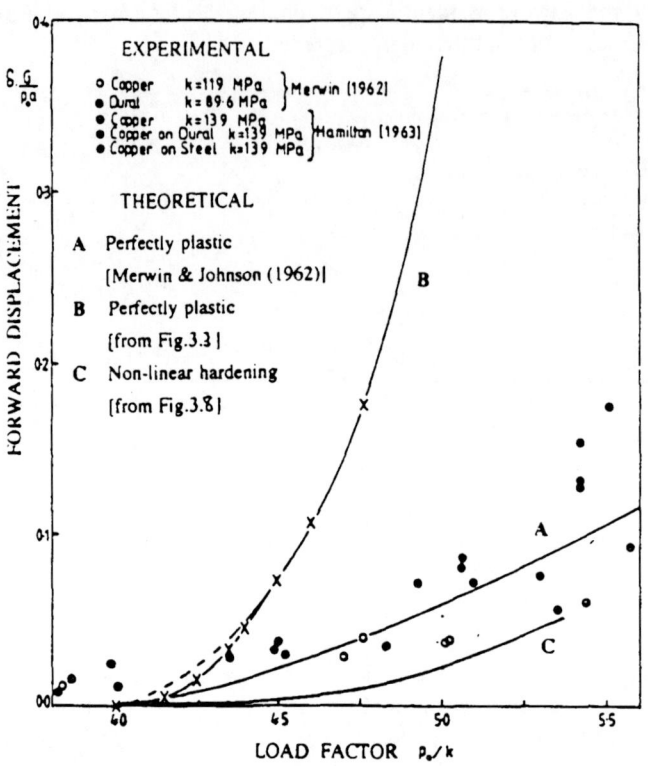

Fig.3.7 Comparison of theories of forward flow (ratchetting rate) with experiment .

3.   Non-linear kinematic hardening model

As discussed earlier, to reproduce ratchetting in a hardening material a *non-linear* hardening model is necessary. Bower (1989) adapted the Armstrong & Frederick (1966) for this purpose. For a material which displays a constant ratchetting rate, the model may be expressed as follows. A displaced von Mises yield locus in deviatoric stress space is shown in the figure, where the deviatoric stress $\sigma' = \sigma - \bar{\sigma}$ and shift in the centre of the yield locus $\alpha' = \alpha - \bar{\alpha}$.

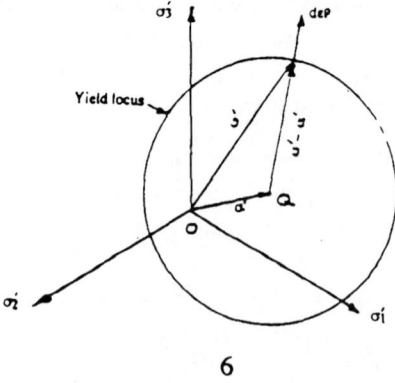

6

The locus may then be expressed by

$$f = \{(3/2)(\sigma - \alpha):(\sigma - \alpha')\}^{1/2} - \sqrt{3}k = 0 \qquad (3.2)$$

Plastic flow follows the normality rule:

$$d\varepsilon^P = d\lambda \frac{\partial f}{\partial \sigma} = \frac{\sqrt{3}}{2} d\lambda \frac{\sigma - \alpha'}{k} \qquad (3.3)$$

where $d\varepsilon^P$ is an increment in plastic strain and $d\lambda$ is its modulus:

$$d\lambda = \{(2/3)(d\varepsilon^P : d\varepsilon^P)\}^{1/2} \qquad (3.4)$$

The hardening characteristics are determined by the relationship between the displacement of the yield locus and the plastic strains. Here we assume:

$$d\alpha = (2/3)c\, d\varepsilon^P - \gamma\, \alpha\, d\lambda \qquad (3.5)$$

where the constants c and γ govern the hardening and ratchet rates respectively.

4. Hardening material: free rolling

Bower & Johnson (1989) applied the above hardening model to calculate the ratchetting rate in free rolling (Q=0) by the 'distributed dislocation' method of Hearle & Johnson. Plastic shear is still assumed to take place on parallel planes (Fig.3.5) but, instead of a constant yield stress k, the stress τ_{zx} is related to the plastic strain ε_{zx}^P by equations (3.2-5). The parameters of the model: k, c, and γ were determined by independent cyclic tests on hard drawn copper specimens. The calculated forward displacement of the surface is plotted as a function of the load parameter p_0/k in Fig.3.8. These results are also displayed in Fig.3.7 (curve C), where they are compared with similar calculations for a perfectly plastic material (curve **B**). The large effect of a modest amount of hardening on the ratchetting rate is clearly shown. Further, the calculated rate with hardening is much closer to the measurements by Merwin and Hamilton.

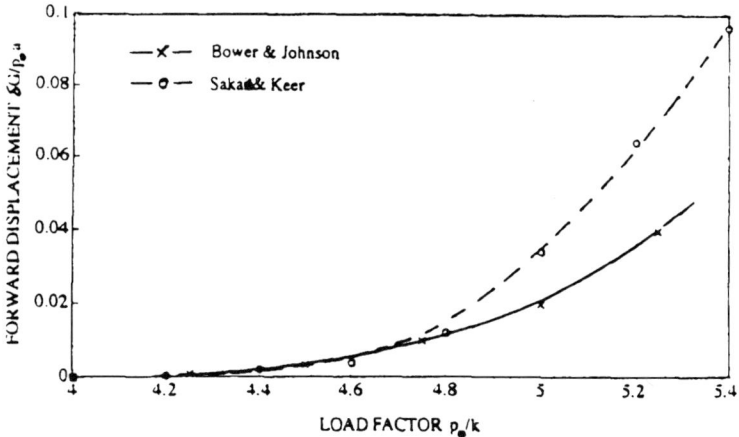

Fig.3.8. Forward flow in free rolling: non-linear kinematic hardening.

More recently Sakae & Keer (1997) have adapted a F-E method of analysing cyclic loading of elastic-plastic structures due to Zarka & Casier (1979), which iterates on the steady cyclic state, and thereby obviates the need for a step-by-step approach. Their calculations of forward displacement for the same material parameters used by Bower & Johnson are added to Fig3.8. They compare well with Bower & Johnson at light loads, but diverge at higher loads. This might be due to the restriction in the dislocation method to the single component of stress τ_{zx}, which becomes increasingly approximate as the size of the contact region enlarges.

Sakae & Keers' analysis also provides values for the residual stresses ρ_{xx} and ρ_{yy}. They were found to be distributed similarly to those for a perfectly plastic material (Fig.3.4) but reduced in magnitude by about 20%.

5. Hardening material: high friction

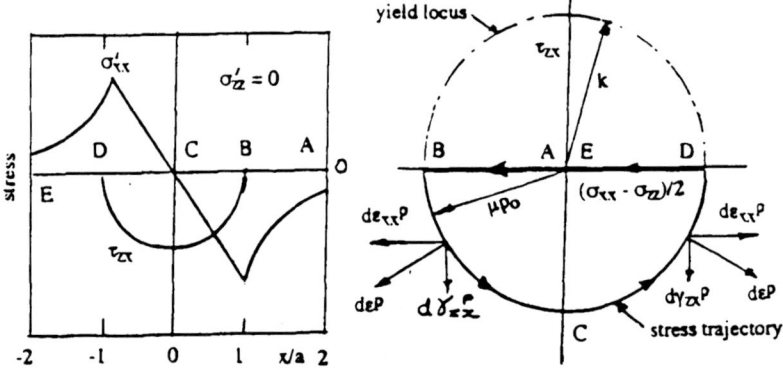

Fig.3.9 Deviatoric stresses at the surface in rolling with sliding.

Combined rolling and sliding with high friction ($\mu > 0.3$) results in yield at the surface. In contrast with a perfectly plastic material, the ability to strain harden permits the formation of a plastically deforming surface layer of finite thickness. Bower & Johnson assumed that the stresses in the layer remained Hertzian (i.e. elastic) as shown for the surface in Fig.3.9, together with an in-plane residual stress ρ_{xx}. When plotted in stress space, with axes ($\sigma_{xx} - \sigma_{zz}$)/2, τ_{zx}, the variation in stress through the cycle maps into the semi-circular trajectory ABCDE. The Tresca yield locus in plane strain is also circular in this space, so that when the radius of the trajectory μp_0 equals that of the yield locus k, yield occurs thoughout the contact area BCD. The plastic strain increments act normal to the yield locus, as shown. The longitudinal component $d\varepsilon_{xx}^P$ is of opposite sign at entry and exit. The residual stress ρ_{xx} was found by iteration such that the net extension through the cycle $\Delta\varepsilon_{xx}^P = 0$. The shear component $d\gamma_{zx}^P$ is uni-directional throughout the cycle and is responsible for the ratchetting effect.

The plastic strains consistent with these stresses were found from the hardening equations (3.2-5). Calculating the ratchetting strain increment through the thickness of this layer enables the displacement of the surface to be found by equation (3.1). The ratchetting rates so found are shown in Fig.3.10, where they are compared with a few measurements made by Bower in a rolling contact disc machine. It was found that ratchetting rates at different values of the traction coefficient μ, could be correlated onto a single curve when plotted to a base of μ (p_0/k). In the high friction regime the shakedown limit by the Tresca criterion is given by $p_0^s = k / \mu$, hence the parameter $\mu (p_0/k) = p_0/p_0^s$ for all values of μ $(\mu > 0.3)$.

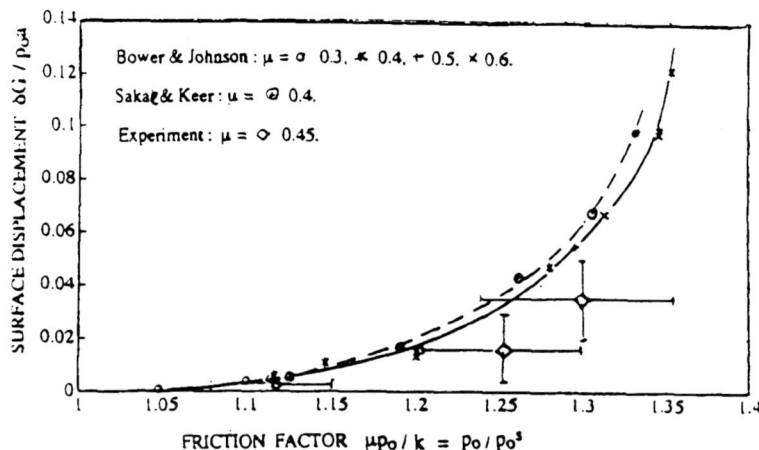

Fig.3.10 Surface flow with high friction ($\mu > 0.3$): non-linear kinematic hardening.

Sakae & Keer also used their F-E method to calculate surface flow, with results also shown in Fig.3.10. The agreement with the approximate calculations of Bower and Johnson is much closer in the high friction case than in free rolling.

References

Armstrong,P.J. & Frederick,C.O. (1966), *CEGB Rept* . Rd/B/N 737.

Bhargava,V. Hahn,G.T. & Rubin,C.A. (1985), *ASME J.Appl.Mech.* **5 2**, 75.

Bower,A.F. (1989), *J.Mech.Phys.Solids*, **3 7**,455.

Bower,A.F. & Johnson,K.L. (1989), *J.Mech.Phys.Solids*, **37**, 471.

Crook,A.W. (1957), *Proc. Inst. Mech. Engrs.* **171**, 187.

Dang Van,K & Maitournam,M.H. (1993), *J.Mech.Phys.Solids*, **41**, 1691.

Hearle,A.D. & Johnson,K.L. (1987), *ASME, J.Appl.Mech.* **5 4**, 1.

Merwin,J.E. (1962), *PhD Dissertation*, Cambridge University.

Merwin,J.E. & Johnson,K.L. (1963), *Proc.Inst.Mech.Engrs.* **173**, 667.

Murakami,Y. Sakae,C. Ichimaru,K. & Morita,T. (1990), *Proc Japanese Int.Tribology Conf.* Nagoya, p.563.

Sakae,C. & Keer,L.M. (1997), *J.Mech.Phys.Solids*, **45**, 1577.

LECTURE 4: FAILURE OF METALLIC ROLLING CONTACTS

1. Modes of failure

We shall be concerned in this lecture with the failure of rolling contacts in which the material can be modelled as an elastic-plastic solid, i.e. metals, most commonly steel. The most common modes of failure which are driven by contact stress may be briefly summarised as follows.

Wear is characterised by the steady progressive removal of material from the surface. It may be categorised as *mild*, in which the rate of removal is slow and tolerable, or *severe*, which rapidly renders the component inoperable.

Fatigue in rolling contact is very characteristic. Material is removed in the form of relatively large pieces leaving pits in the initially smooth surface, which has lead to this mode of failure being referred to as *pitting*.

Scuffing is a mode of failure brought about by failure of the lubricant to maintain low friction in a combined rolling and sliding contact. It is strongly influenced by the temperature in the contact and is characterised by a sudden change from very mild to very severe wear of the surfaces.

Corrugation refers to the progressive development of regular ripples on rolling surfaces which were initially flat of having a random roughness [corrugation will be the subject of separate lectures].

Competing modes of failure

An engineering component can generally fail in different ways depending on the design, material and operating conditions. Such ways are often referred to as *competing modes of failure*. As an example we shall consider involute gear teeth which undergo a relative motion of rolling and sliding. Under the action of the contact stress the tooth surfaces can fail by wear, pitting or scuffing.

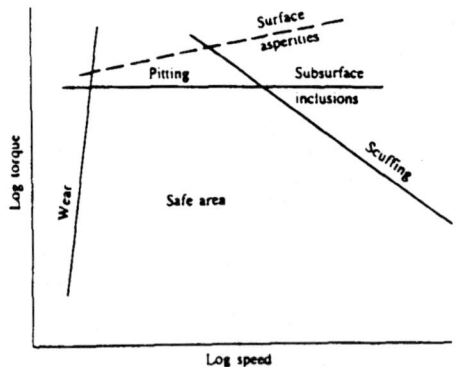

Fig.4.1   Failure map for involute gears, from Blok (1954).

A failure map, originally proposed by Blok (1954), is presented in Fig.4.1, in terms of the operational variables: torque and speed. Wear only occurs at very low speed when lubrication is ineffective. Pitting is principally governed by stress, i.e. torque. Scuffing is precipitated by temperature rise in the contact, which can be shown to be proportional to $(torque)^{3/4}$ x $(speed)^{1/2}$. A line of constant temperature rise thus has a slope of -2/3, as shown.

Technological developments can make significant changes to the safe operating capacity of the system. The introduction of case-hardened gear teeth raised the pitting boundary beyond the intersection of the wear and scuffing boundaries, thereby eliminating pitting as a possible failure mode. However the introduction of extreme pressure additives to lubricating oils pushed back the scuffing boundary such that pitting was reintroduced as a mode of failure, but at a higher torque than before.

Plastic deformation as a source of failure

In this lecture we shall discuss the modes of failure of rolling contacts in which plastic deformation plays an important part: wear and fatigue (pitting). All real surfaces are rough, even those finished by grinding and polishing, which ensures that the real area of contact at the tips of the asperities is much less than the apparent area, calculated by the Hertz theory. Thus it is necessary to distinguish between two different situations. In the first, the bulk contact stresses, calculated by applying the Hertz theory to the geometry of the contact neglecting surface roughness, exceed the elastic limit. In the second, the bulk contact stresses lie within the elastic limit, and plastic deformation is restricted to individual asperities. As we shall see, it is possible to apply the mechanics of plastic deformation described in the previous lectures to both these situations.

2. Wear

Except for situations of extreme contact stress, the process of wear is assumed to be associated with the deformation of the surface asperities. In modelling this process it is usual to assume that one surface is much harder than the other, so that material is only removed from the softer surface. This simplification is not as restrictive as it might appear, since a few percent difference in hardness is observed to cause a large bias in the wear rate towards the softer surface.

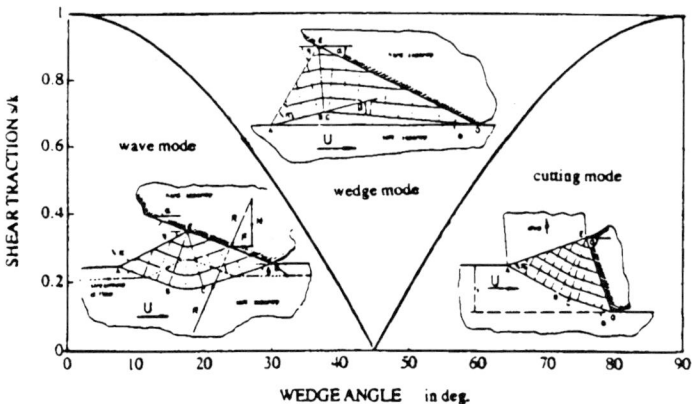

Fig.4.2   Three modes of deformation in ploughing by a rigid wedge, from Challen & Oxley (1979).

A revealing model of the mechanics of metallic wear was proposed by Challen & Oxley (1979) using slip-line theory for *rigid-perfectly plastic solids*. The action of single hard asperity is represented by a rigid 2-dimensional wedge ploughing a soft surface as shown in Fig.4.2a. Three modes of deformation are identified: (i) wave motion, (ii) ploughing, (iii) cutting. The conditions leading to the different modes are a function of the wedge angle α and the frictional shear traction s (normalised by the shear yield stress k), as shown in Fig.4.2b. The cutting mode corresponds to *abrasive wear* by hard angular particles or debris; the ploughing mode corresponds to prow or wedge formation as described by Hutchings (1992); the wave motion is a form of *ratchetting* as discussed in the previous lecture.

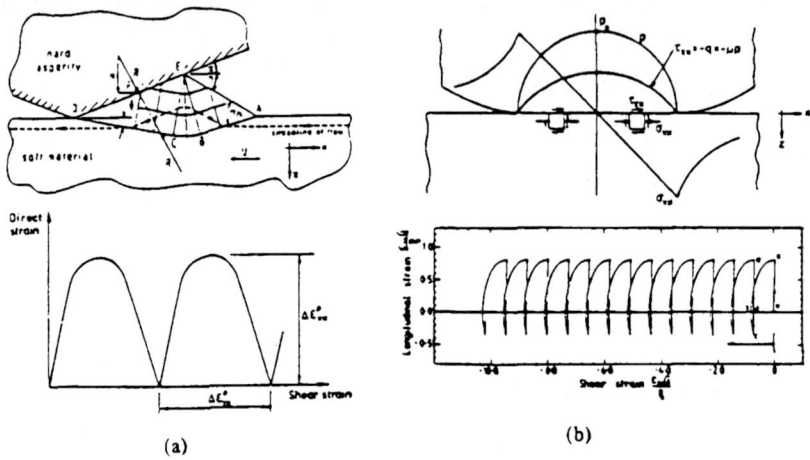

(a)   (b)

Fig.4.3. Ratchetting cycles of plastic strain in sliding contact. (a) Rigid-plastic; (b) elastic-plastic.

The cycle of longitudinal and shear plastic strain during repeated passages of the wedge in the wave mode is shown in Fig.4.3a. It may be seen to comprise a reversing component of extensional strain ε_{xx}^{P} and a uni-directional (ratchetting) strain γ_{zx}^{P}. The defect of this model, particularly at small wedge angles, lies in the exclusion of elastic strain in the rigid-plastic idealisation. This shortcoming is overcome in the elastic-plastic analysis of ratchetting by Bower & Johnson (1989), in which a 2-dimensional sliding asperity is represented by a rigid cylinder. The cycle of total strains ε_{xx} and γ_{zx} is shown in Fig.4.3b. It is similar to the rigid-plastic model, except that the magnitude of the plastic strains is much smaller.

The question now arises: what is the relationship between the plastic strain cycles shown in Fig.4.3 and the rupture necessary to produce detached wear particles. If the reversing strain were acting alone it would be expected that the number of cycles to failure would be given by the Coffin-Manson relationship:

$$N_f = (2C/\Delta\varepsilon_f)^{1/\alpha} \tag{4.1}$$

where N_f is the number of cycles to failure, $(\Delta\varepsilon_f)$ is the range of reversing (fatigue) strain, C is a constant of magnitude comparable with the fracture strain in monotonic loading and the index $n \approx 0.5$. In the circumstances where a ratchetting strain accompanies a fatigue strain, as in Fig.4.3, ductile failure in shear might occur when the accumulated uni-directional strain reaches a critical value ε_c, whose value is again related to the fracture strain in monotonic loading. The number of cycles to failure in this case N_r would then be given by:

$$N_r = \varepsilon_c / \Delta\varepsilon_r \qquad\qquad (4.2)$$

where $\Delta\varepsilon_r$ is the increment of ratchetting strain per cycle. In a graph of $\log\Delta\varepsilon$ against $\log N$, equations (4.1) and 4.2) plot as straight lines of slope $-(1/n)$ and -1 respectively (Fig.4.4). Kapoor (1994) extracted from the literature data on cyclic loading tests in which both reversing and ratchetting strains were measured. The hypothesis was then made that the two modes of failure: low cycle fatigue (LCF) or ratchetting (RF) are competitive in the sense that the actual failure mode is that which gives the lowest number of cycles to failure by equations (4.1) and (4.2). The results are shown in Fig.4.4. In all cases except those in which the ratchetting strain $\Delta\varepsilon_r$ was zero or small, the cycles to failure followed the RF line (eq.4.2).

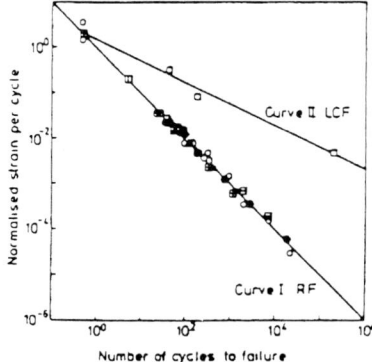

Fig.4.4   Cyclic strain rupture tests.
Competing modes of failure:
Low cycle fatigue (LCF); ratchetting failure (RF).

Fig.4.5   Cumulative plastic shear at the surface of pearlitic rail
steel after 17500 cycles, showing ductile cracks and
wear flakes, from Tyfour (1995).

These results suggest a mechanism of material removal which is referred to as *ratchetting wear*. This process is illustrated by the micrograph in Fig.4.5, which shows a section through a disc of pearlitic railway rail steel which has been run in a rolling contact disc machine under conditions of rolling with sliding. Severe plastic shear has accumulated in a surface layer about 0.5 mm thick. Microcracks are clearly seen towards the surface of the layer, aligned with the direction of the accumulated plastic shear. Where these cracks intersect the surface thin slivers of material become detached and form plate-like wear particles. It should be noted that the plastic strain at which the cracks appear is much in excess of that at which rupture occurs in a shear test at atmospheric pressure. This is due to the high hydrostatic pressure beneath a concentrated contact which increases ductility.

The process of ratchetting wear outlined above suggests the possibly of calculating wear rates from first principles. From the results of Bower & Johnson (see Fig.3.10) the increment of ratchetting per cycle $\Delta\varepsilon^P$ can be expressed:

$$\frac{\Delta \varepsilon_r E^*}{p_o} = f\left\{\frac{\mu\, p_p}{k}\right\} = f\left\{\frac{p_o}{p_o^s}\right\}$$ (4.3)

Where $p_o^s = k/\mu$ is the shakedown limit in the high friction regime. Assuming that the normal compression of the contact is given by Hertz,

$$\frac{p_o}{E^*} = \frac{1}{2}\frac{a}{R} = \frac{1}{2}\sqrt{\frac{\delta}{R}}$$ (4.4)

where a is the semi-contact width, R is the radius of the sliding cylinder and δ is the elastic compression of the surface. From equations (4.3) and 4.4) we get

$$\frac{\Delta\varepsilon_r}{\sqrt{\delta/R}} = f\left\{\frac{\mu\, E^*}{k}\sqrt{\frac{\delta}{R}}\right\} = f\left\{\frac{E^*}{p_o^s}\sqrt{\frac{\delta}{R}}\right\}$$ (4.5)

This expression refers to a single hard asperity sliding over a elastic-plastic plane surface. Kapoor et al.(1994) have considered the situation of a hard sliding surface with many spherical or cylindrical asperities. In this case the compression δ in equation (4.5) becomes the r.m.s. roughness σ and 1/R the mean curvature κ, so that the parameter on the R.H. side of equation (4.5) becomes:

$$\frac{E^*}{p_o^s}\sqrt{\sigma\kappa} = \Psi_s$$ (4.6)

where the parameter Ψ_s is a form of plasticity index [see Greenwood & Williamson (1966)], which governs wear in the ratchetting mode.

The mechanism of ratchetting wear discussed above, based on the ratchetting analysis of Bower & Johnson (1989), refers to the high friction condition where plastic flow is concentrated in a surface layer. An alternative approach, appropriate to lubricated contacts ($\mu < 0.2$), has been developed by Kapoor & Johnson (1994) and Kapoor et al.(1996). Here the shear deformation is sub-surface and wear occurs by the extrusion of very thin films [see Akagaki & Kato (1987)].

3. Rolling contact fatigue (pitting)

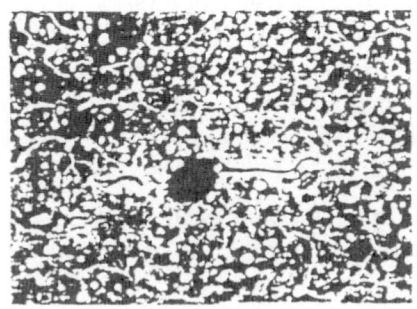 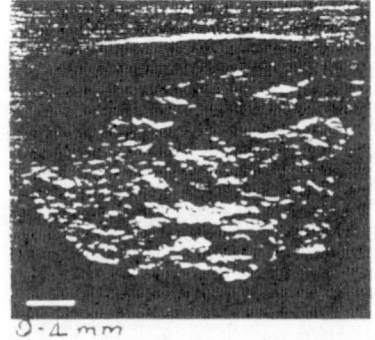

Fig.4.6 Pitting failure of roller bearing. (a) Sub-surface MnS inclusion, 5 μm dia with micocracks (b) Fatigue spall.

A typical fatigue failure can be divided into two principle phases: (i) *initiation*, in which a micro-crack is generated by cyclic plastic deformation, usually located at a point of high stress concentration or at a defect in the material, and (ii) *propagation* in which the crack extends at an increasing rate until complete rupture occurs. This is also true of fatigue in rolling contact. Following a period of initiation, a crack propagates beneath the surface until a piece falls out, leaving a distinct pit in the surface.

Forty years ago Johnson (1953) showed that rolling contact fatigue failures were prevalent when the Hertz contact pressure p_0 exceeded about half the shakedown limit p_0^s, indicating the presence of initiation sites of local stress concentration. These sites were identified as (i) non-metallic inclusions located beneath the surface at the depth of the maximum reversing shear stress (about 0.6 a), (ii) surface asperities whose height d exceeds the thickness of the lubricant film h (i.e. $\Lambda = h/d <$ 1), (iii) scratches of dents in the surface due to particulate contamination or bad handling. The morphology of the pits are different depending on whether the initiation site is beneath (i) or at the surface (ii) and (iii). A typical subsurface pit or 'spall', commonly found in rolling contact bearings, is shown in Fig.4.6. The fatigue crack, which forms the bottom of the spall can propagate considerable distance beneath and parallel to the surface before breaking out to leave a large, shallow pit. The development of a typical surface initiated pit is shown in Fig.4.7. It has the form of a concoidal sea shell. The crack propagates into the solid at an acute angle (about 15°) from an initiation site on the surface, to a depth of the maximum Hertz shear stress, and then breaks off or curves up to the surface to leave a pit. The appearance on the surface is a crack of V or arrow shape, pointing in the direction of motion of the surface. When sliding accompanies rolling such pits are almost entirely confined to the slower moving surface. The large fraction of the total number of cycles to failure before a crack is visible indicates that the initiation phase occupies most of the life. In circumstances where the bulk contact stress is low, but the surfaces are rough and poorly lubricated, micro-pits covering the surface can develop on the asperity scale. The mechanism of formation appears to be similar.

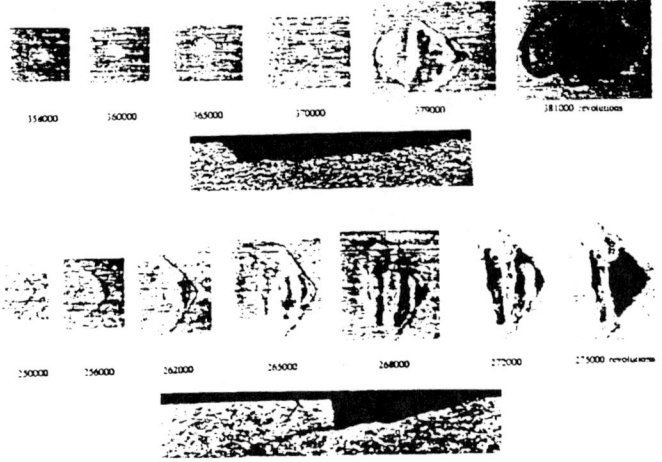

Fig.4.7   Pitting failures of gear steel in rolling & sliding, initiating at a microcrack in the slower moving surface.

Surface and sub-surface initiated cracks are competing modes of failure (see Fig.4.7). Prior to 1950, sub-surface spalls predominated in rolling contact bearings and were responsible for early failures. The introduction of vacuum de-gassed bearing steel greatly reduced the number and size of sub-surface inclusions and changed the predominant mode of failure to that initiated by surface asperities. The life is then sensitive to the lubricant film thickness, as shown.

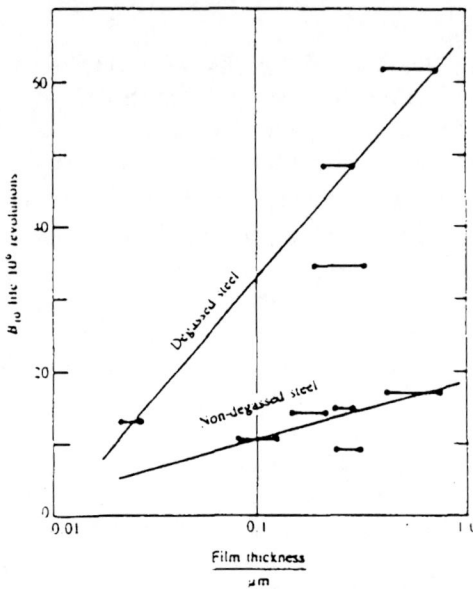

Fig.4.8   Competitive effect of sub-surface inclusions and surface roughness on ball bearing fatigue life.

It was pointed out by Way (1935) that surface initiated pits in rolling contact only develop on the slower surface and only in the presence of a liquid: oil or water. He proposed that these observations could be explained by fluid being forced into the cracks by the contact pressure. Fatigue cracks are normally propagated in Mode I by tensile stress. Mode II propagation only appears possible if the crack faces are held apart to eliminate crack face friction. But tensile stresses are generally small or absent in a Hertz contact, which makes Way's hypothesis attractive. It has been examined using fracture mechanics by Bower (1988), Murakami et al.(1994) and others.

Bower modelled an inclined plane-strain crack in an elastic half-space by distributed dislocations: a Mode I (tensile) crack by 'climb' dislocations and a Mode II (shear) crack by 'glide' dislocations, as shown in Fig.4.9. Fluid can influence the crack in three ways: (i) by reducing crack face friction, (ii) by transmitting the contact pressure to the tip of the crack, as suggested by Way, and (iii) by closing the mouth of the crack and trapping fluid within it. Bower found that both mechanisms (ii) and (iii) generated Mode I stress intensity factors at the tip of the crack, thus tending to propagate the crack, but only mechanism (iii) favoured propagation on the slower surface only. In the case of the faster surface, the cracks are inclined in the opposite direction and fluid is squeezed out from tip to mouth by the motion of the load.

Murakami et al. used a different numerical technique: the 'body force' method, in which doublets of normal and tangential concentrated forces P and Q, as shown in Fig.4.9, were distributed along the line of the crack. This technique has the advantage of handling 3-dimensional cracks. It was possible to follow the spread of the crack surface and to calculate the angle of V shape. They concluded that the crack propagated in Mode II (shear) initially and subsequently in Mode I by the action of trapped fluid.

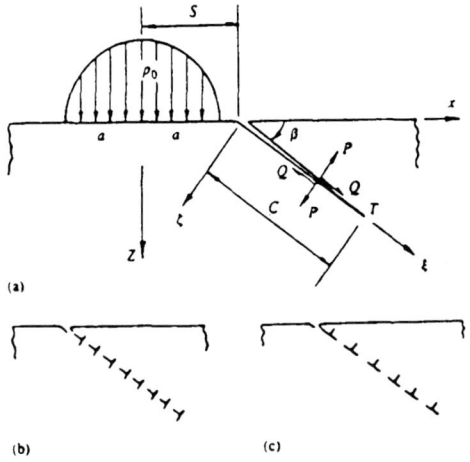

(a)

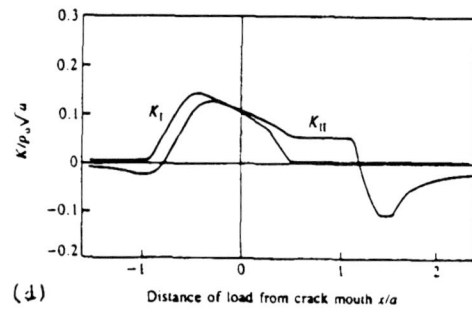

(d)   Distance of load from crack mouth x/a

Fig.4.9   (a) Inclined crack in an elastic half-space, showing force doublets P-P and Q-Q.

(b) Modelling Mode I by pile dislocations.

(c) Modelling Mode II by glide dislocations.

(d) Mode I and Mode II stress intensity factors.

The process of rolling contact fatigue now seems clear in its main features. Micro-cracks are initiated in a surface layer by the process of plastic ratchetting and ductile shear fracture. This may be on the bulk scale, as shown in Fig.4.5, or on the asperity scale. During rolling with sliding, they are inclined in opposite directions on the two surfaces but, even in free rolling, the forward direction predominates [see Lecture 3]. On the slower surface they are inclined in the direction to facilitate penetration and entrapment of the fluid, and so to propagate into fully developed pits.

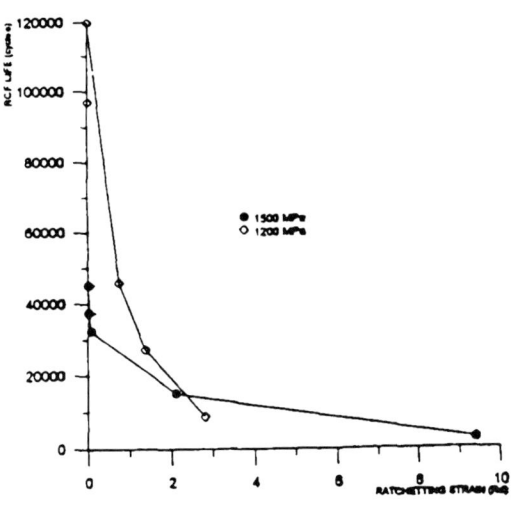

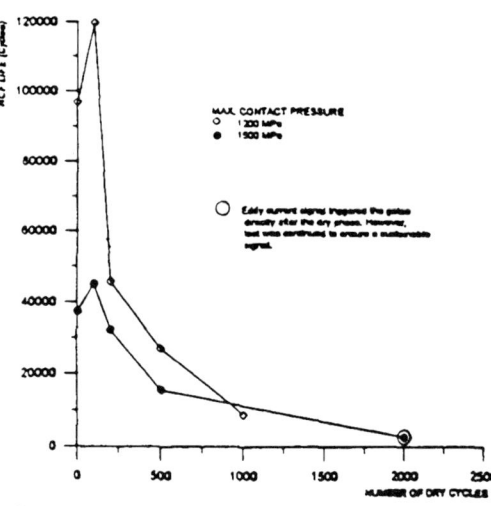

Fig.4.10 (a)   RCF life as a function of total ratchetting strain   (b)   Effect of initial dry cycling on the subsequent RCF life.

Striking support for this view has been provided by Tyfour et al.(1996), who ran rolling contact fatigue tests on pearlitic rail steel lubricated by water, which were preceded by varying number of cycles running dry. Following a short transient phase in which the coefficient of friction starts at about 0.12, when running dry it rises to a steady value of about 0.45, whereas running wet it rises only slightly above 0.12. The high friction when running dry increases the ratchetting rate and the development of the inclined surface cracks as shown in Fig.4.5. The accumulated plastic shear strain, shown in Fig.4.10a, was observed to increase in proportion to the number of dry cycles. The deleterious effect on fatigue life is clearly shown in Fig.4.10b.

The fact that the inclination of the initiating cracks is determined by the motion of the surface and the direction of sliding suggests that a reversal of rotation in mid-test should halt the process and that there should follow a quiescent period while the plastic shear at the surface is reversed. Such an experiment was carried out by Tyfour & Beynon (1994), in which the direction of rolling was reversed periodically after an interval of C_r cycles and the life (number of cycles to failure L_r) measured. In uni-directional rolling the life was L_u. The relative life (L_r / L_u) is plotted against the 'reversal factor' (C_r / L_u) in Fig.4.11. Reversed rolling of any period is seen to increase the life; The maximum effect, almost doubling the life, is when the reversal period is about 1/3 of the uni-directional life.

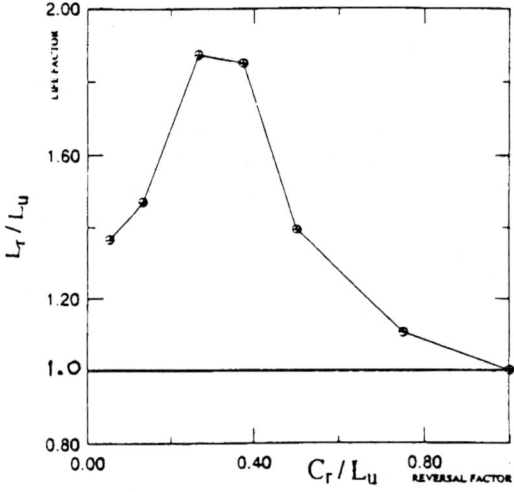

Fig.4.11 *Effect of rolling direction reversal on RCF life*

References

Akagaki,T & Kato,K. (1987), *Wear*,**117**, 179.

Blok,H (1954), *Proc.Inst.Mech.engrs. Conf. on Gearing*, London, p.114.

Bower, A.F. (1988), *ASME J.Trib.* **110**, 704.

Bower,A.F. & Johnson,K.L. (1989), *J.Mech.Phys.Solids*, **37**, 471.

Challen,J.M. & Oxley,P.L.B. (1979) *Wear*,**111**, 275.

Greenwood,J.A. & Williamson, (1966), *Proc.Roy.Soc.London*, **A295**, 300.

Hutchings,I.M. *Tribology*, Arnold 1992.

Johnson,K.L. (1953), *Inst.Mech.Engrs. Symp. on Rolling Contact Fatigue*, London, p.155.

Kapoor,A. Williams,J.A. & Johnson,K.L. (1994), *Wear*, **175**, 81.

Kapoor,A & Johnson,K.L. (1994), *Proc.Roy.Soc.London*, **A445**, 367.

Kapoor,A Johnson,K.L. & Williams,J.A. (1996), *Wear*, **200**, 38.

Murakami,Y Sakai,C. & Ichimura,K (1994), *STLE Trib.Trans.* **37**, 445.

Tyfour,W.R. *PhD Dissertation*, Leicester University, 1995.

Tyfour,W.R. Beynon,J.H. & Kapoor,A. (1996), *Wear*, **197**, 255.

Tyfour,W.R. & Beynon,J.H. (1994), *Trib.Int.* **27**, 275.

Way,S (1935), *ASME J.Appl.Mech.* **51**, 49.

NON-STEADY STATE ROLLING CONTACT AND CORRUGATIONS

K. Knothe
Technical University of Berlin, Berlin, Germany

1 Introduction to problems of rail corrugation [1]

Phenomenology and classification of rail corrugation. Field observations. Metallographic investigations. General model of short pitch corrugation formation. Feedback loop between transient dynamics and long-term wear. Similar phenomena on wheel surfaces (out-of-round wheels).

1.1 Definition of non-steady state contact mechanics

What does non-steady state contact mechanics mean? Let us assume that an observer is moving together with the wheel along the rail, looking at wheel/rail contact stresses. Only in very few cases do the observed contact stresses remain constant with time, e.g. for a wheel running in a central position on a straight, undisturbed track or for steady state curving. In most cases the observed contact stresses vary with time. If we restrict our investigations to harmonic variations of irregularities, normal forces or other contact parameters, the wavelength of the harmonic variation can be considered as a measure of non-steady state behaviour. As long as the wavelength is large compared to the contact diameter in rolling direction (say by a factor of twenty), a particle moving through the contact patch does not take note of varying contact conditions. *This situation shall be called quasi-steady state.* However, if the wavelength of the variation becomes smaller than 10, the particle takes note of these variations. *This is the non-steady state situation.* Mathematically, in such a non-steady state situation partial derivatives with respect to time $(\partial f/\partial t)$ have to be taken into account, as will be shown in Section 2.

1.2 Classification of rail corrugation. Feedback loop

Rail corrugation, that are quasi-periodic irregularities on the rail tread, are found throughout the world on rail surfaces. Different types and different wavelengths exist. The type we are interested in is called short pitch corrugation. It is shown in Fig. 1.

The most common wavelengths of short pitch corrugation are in the range of 20 – 100 mm. Serious problems with short pitch rail corrugation have been observed at BR and DB on straight or slightly curved track for high speed traffic. Such problems include extremely high noise levels and damage of track components, especially rail fastening systems.

As a typical example for the distribution of corrugation wavelengths the results of measurements for a DB high speed line (200 km/h) are given in Fig. 2. The most frequent wavelengths are near 40 mm. The maximum recorded depth of 180 μm would not be acceptable nowadays because of the increase of noise (roaring rails).

---

[1] The section is mainly based on Frederick and Bugden (1983) and on Baumann (1998).

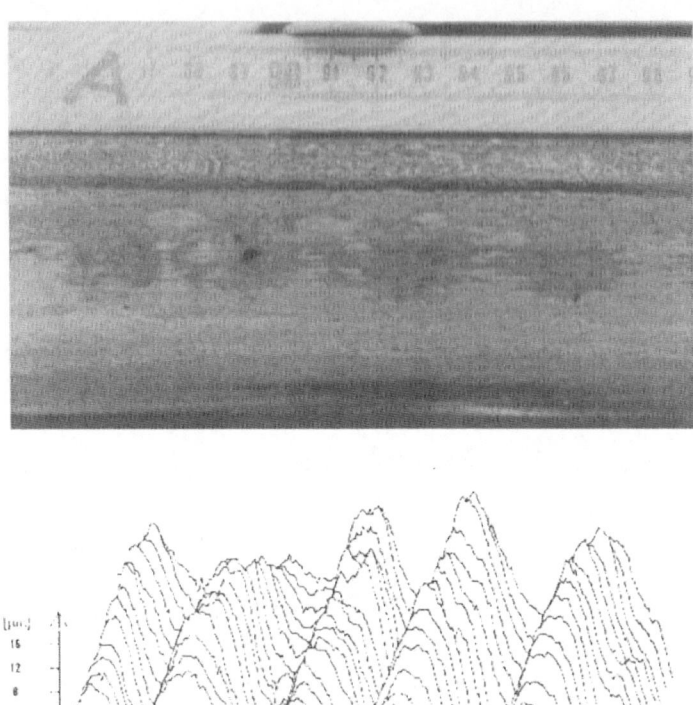

Figure 1. Typical corrugated rail sample (S 54) with profilogram from Baumann (1998)

This type of rail corrugation is not the only one which can be observed. Grassie and Kalousek (1993) have tried to provide a classification of different types of rail corrugation which is shown in a slightly modified version in Table 1. Short pitch rail corrugation, which has already been mentioned, is given in row 6. There are two other types shown in row 4 and 5 where wavelengths in the range of 20 to 80 cm can also be observed. It seems possible to summarize row 5 and 6 to a single row, but row 4 is different. This will become clear later.

We shall now look at the two categories which have been used by Grassie and Kalousek for the classification of rail corrugations:

- A wavelength fixing mechanism and
- a damage mechanism.

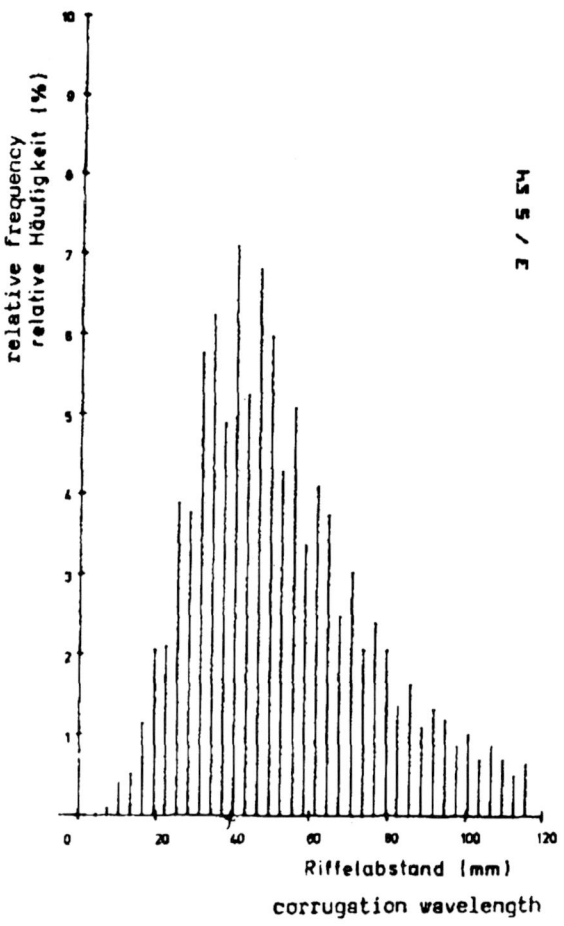

Figure 2. Relative frequency of different corrugation wavelengths measured over 3 kilometers (from Widmayer, 1983)

Both mechanisms together form a feedback loop, see Fig. 3

According to Grassie the damage mechanism of short pitch corrugation is wear of troughs mainly from longitudinal slip whereas the wavelength-fixing mechanism is unknown. Both, *system dynamics* (contact mechanics and structural mechanics) as well as *long-term behaviour* have to be considered for the investigation of the wavelengths-fixing mechanism.

System dynamics

- The wavelengths of short pitch rail corrugation are between 2 and 10 cm and the contact diameter in rolling direction is approximately 1 cm, so non-steady state *contact mechanics* is mandatory. This will be discussed in detail in Section 2.
- *Structural mechanics* of all components also have to be modeled correctly. Considering vehicle speed between 20 and 70 m/s a model of wheel and rail in the frequency range

	Type	Wavelength (mm)	Wavelength-fixing mechanism	Damage mechanism
1	Heavy haul corrugation	200-3000	P2-resonance	Plastic flow in troughs
2	Light rail corrugation	500-1500	P2-resonance	Plastic bending
3	Contact fatigue corrugation	150-450	P2-resonance	Rolling contact fatigue
4	Rutting corrugation	50 (trams) 200 (RATP) 150-450 (FAST)	Torsional resonance of wheel-set. Peak of vertical dynamic force. Stick slip.	Wear of troughs from longitudinal oscillation.
5	Short pitch corrugation (type 1) Booted sleepers	40-60 (RATP) 51-57 (Baltimore)	Sleeper resonance Flexural resonance of wheel-set	Wear of troughs. from lateral oscillation. Plastic flow of peaks
6	Short pitch corrugation (type 2) Roaring rails	25-80	Unknown	Wear of troughs (mainly from longitudinal slip)

Table 1. Classification of corrugation on rails (slightly modifies from Grassie and Kalousek (1993). References can be found there.)

between 200 Hz and 3000 Hz must be available, which is at least qualitatively reliable. This will be discussed in Section 3.

Long-term behaviour (damage mechanism) The description of long-term behaviour (called damage mechanism in Grassie and Kalousek, 1993) is even more complicated. Therefore phenomenological and metallurgical aspects will now be considered.

1.3   Phenomenological observations

Observations on the development of rail corrugations by BR are shown in Fig 4. Two different types of rails have been used for the observations. "Ground rails" means, that newly laid rails were first ground, and "unground rails" were used as they came from the

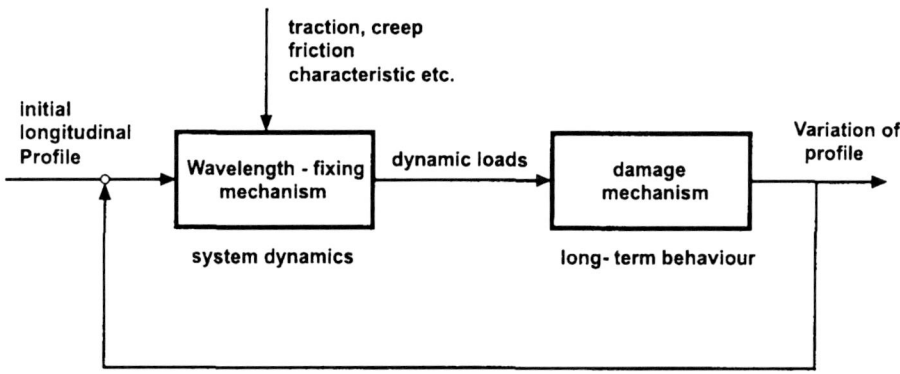

Figure 3. Components of a general corrugation mechanism

rolling mill. It can be seen that for ground rails the initial peak-to-trough height in the wavelength range of interest was approximately 10 μm, whereas for unground rails the initial height was nearly 40 μm.

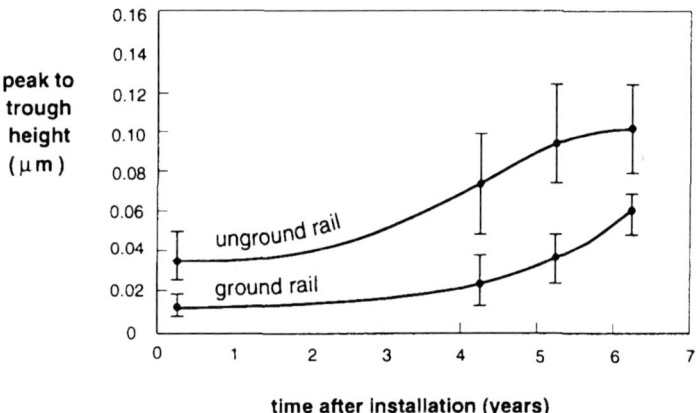

Figure 4. Development of rail corrugation at Rugby comparing ground and unground rails (taken from Frederick and Bugden, 1983)

It is useful to distinguish between three stages of corrugation. Before corrugation starts, there is a period of at least one year for ground rails (much shorter for unground rails) where no corrugations appear. This is a kind of running-in process. The most relevant feature of this period is, that a white, shiny running band is forming on the rail tread called WEL (white etching layer or white layer). The thickness of this WEL is 5 or at most 10 μm. The unground rail after a few months begins to corrugate, which means that quasi-periodic pattern are forming on the rail tread. We have called this stage *corrugation initiation*, see Fig. 5. It is followed by a stage of more rapid *corrugation growth*. During this stage on the one hand the WEL is growing on the corrugation peak and may become up to 100 μm thick. On the other hand the white layer is very often worn away in the

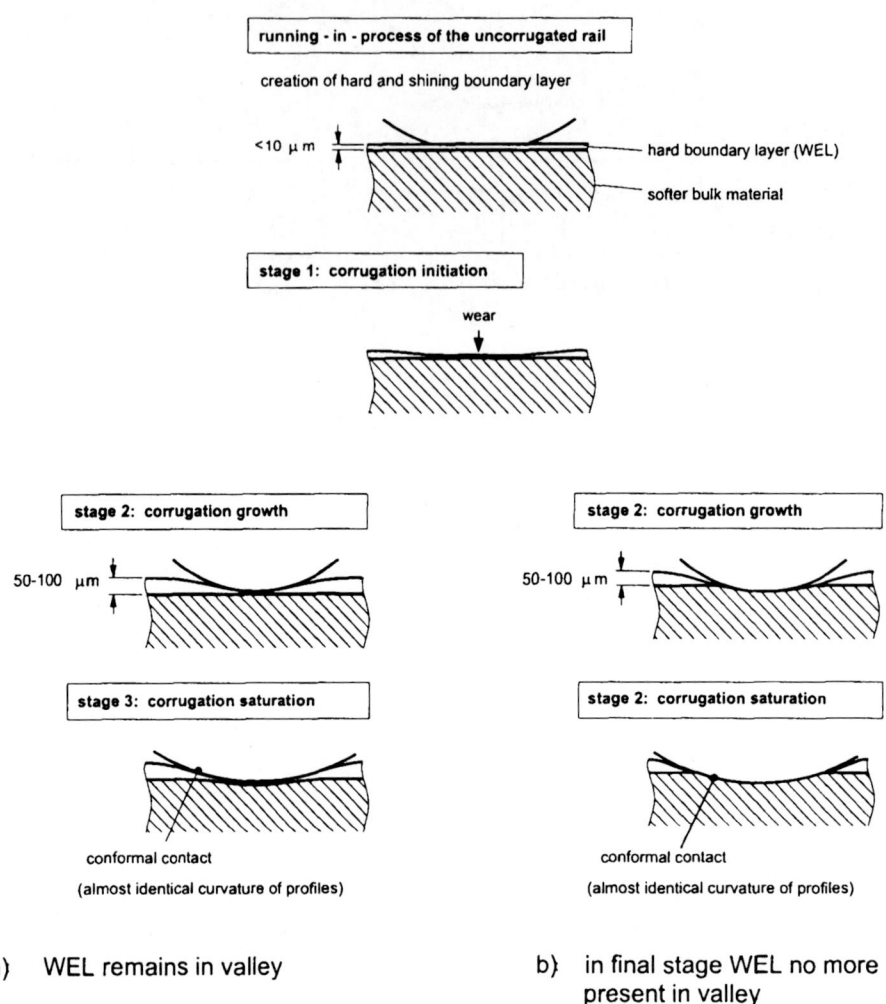

Figure 5. Different stages of rail corrugations (from Baumann et al., 1996b)

trough. As soon as this happens, corrugations have their typical appearance: a sequence of bright, shiny peaks and dark, grey troughs (see Fig. 1). The process is closed by a *corrugation saturation* stage.

The final height of saturated corrugation mainly depends on the corrugation wavelength. Assuming a pure sinusoidal irregularity the corrugation height can be estimated by simple, geometric considerations: The wheel must fit into the trough as otherwise wear would no longer be possible. The maximum height would be 90 μm for a wavelength of 30 cm or 160 μm for a wavelength of 40 mm. At DB corrugations are ground if the mean height is more than 80 μm in order to avoid unacceptable noise.

Measurements of Baumann (1998) have shown that the peaks and the troughs are not only distinguished by their appearance but also by their micro-structure. The R_a-values of micro-roughness on the peaks and in the troughs are $R_a = 0.23\mu m$ and $R_a = 0.57\mu m$ respectively. Troughs are more than twice as rough as peaks.

We have tacitly assumed that wear is the most important damage mechanism. This is evident for a uniform reduction of the height of the rail head. If plastic deformation were responsible then a plastic flow of material to the flanges and the formation of "noses" should be observed, which is only observed for extremely high loads. However, plastic deformation may be relevant for corrugated rails.

1.4   Metallurgical observations

Figs. 6 and 7 show micro-sections near the corrugated peak and at the trough of a corrugation. In both situations plastic deformations can be observed. Near the peak the plastic flow seems to be much more pronounced. As already discussed, the WEL layer is only observed near the peak, whereas in the troughs, see Fig. 7, mostly no WEL remains.

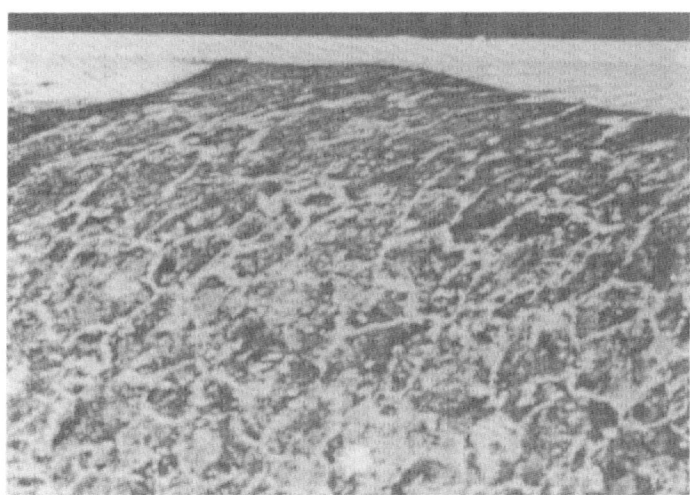

Figure 6. Micro-section showing plastic deformation beneath the peak of a corrugated rail (from Frederick and Bugden, 1983)

In Fig. 8 we tried to describe the composition of the surface layers schematically. At depths of 3 to 4 mm the pearlitic bulk material seems to be nearly undeformed. Near the WEL an intensive plastic deformation can be observed. Immediately below the WEL the deformation of the pearlite is so strong that the ferritic matrix contains particles of broken cementite. Coming into the white layer it seems that cementite gradually disappears. The WEL itself can be substructured. In the case which is shown in Fig. 8 the lower part of the WEL still contains particles which probably are remainders of cementite whereas the upper part is nearly homogeneous. It is quite surprising that there is always a sharp boundary between the WEL and the deformed bulk material.

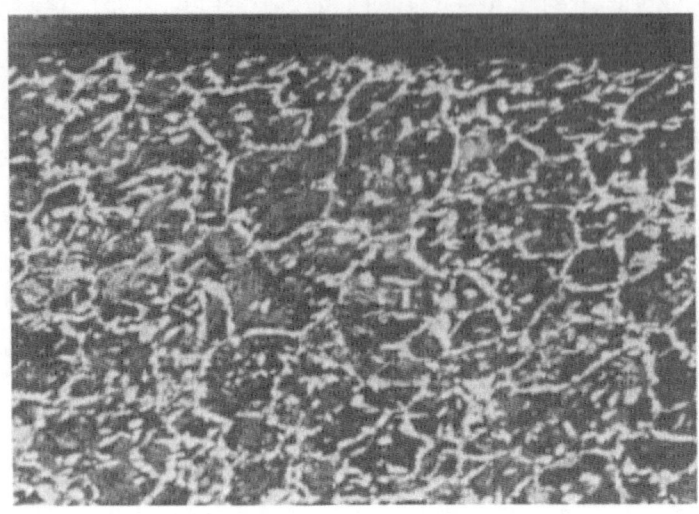

Figure 7. Microsection showing plastic deformation at a trough (from Frederick and Bugden, 1983)

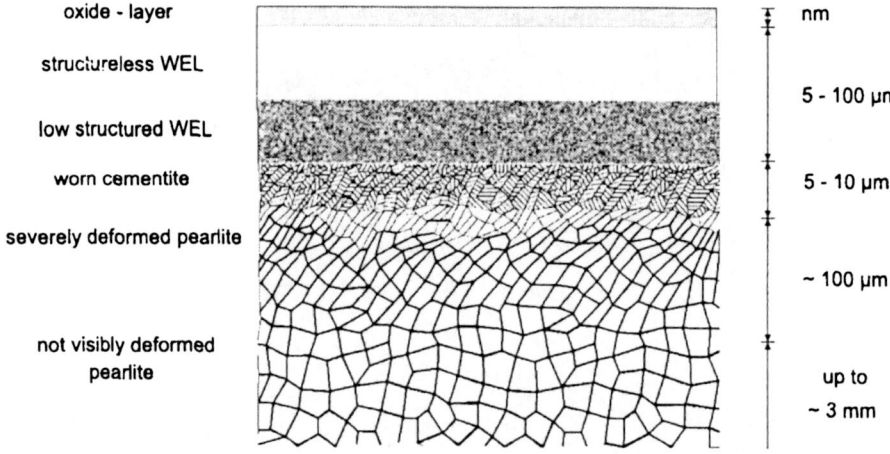

Figure 8. Structure of layers below the rail surface (from Baumann et al., 1996b)

One of the properties of WEL is that it is much harder than the pearlitic bulk material or the plastically deformed material. Hardness measurements have been performed e.g. by Baumann (1998) , the results of which are given in Fig. 9. The maximum hardness values immediately below the corrugation peak are more than four times higher than the hardness of the bulk material. Only a very small increase in hardness is found at the corrugation trough. Due to this increase in hardness the WEL has a higher wear resistance. Measurements by Baumann indicate that the wear resistance at the corrugation peak is at least twice the wear resistance at the trough.

There is no doubt, however, that even surfaces with a continuous running band of WEL continue to lose material. The annual rate of this loss exceeds the thickness of the WEL. This implies that WEL is continuously being newly formed from the bulk material.

There is a second consequence of the increase in hardness. The WEL is very brittle so that micro-cracks can often be observed, especially on heavily corrugated rails, see Fig. 10. The WEL material no longer has any ductility. If a net of micro-cracks has formed, then small parts of WEL may break off. In these situations WEL can be pressed down into the softer pearlitic material causing severe plastic flow to the surface. Fig. 11. No observation has be found up to now that such defects are healed by new formation of WEL. It can therefore be assumed that there is a continuous flow of softer material to the surface, where it is worn away.

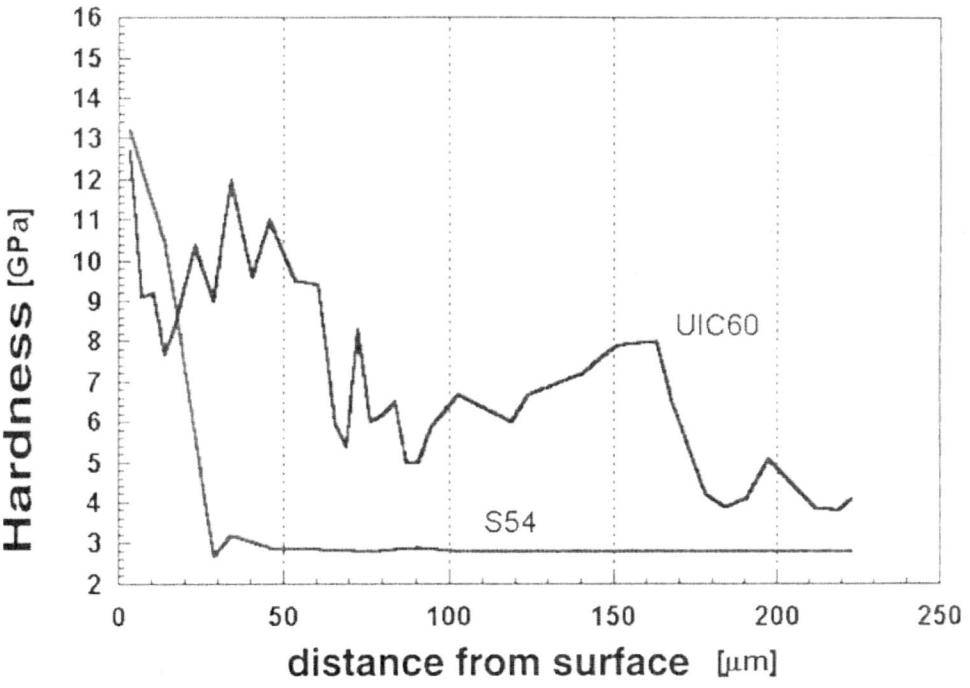

Figure 9. Hardness-depth profile on white layers (Baumann et al., 1996b)

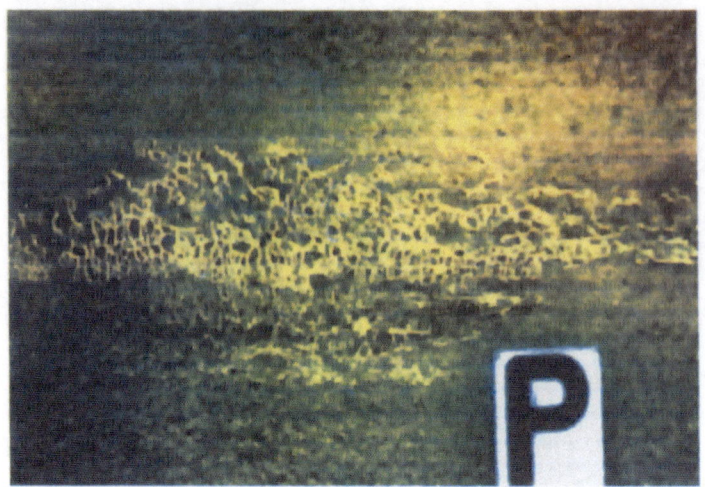

Figure 10. Patterns of microcracks on the rail peak of a heavily corrugated rail from Baumann et al. (1996b)

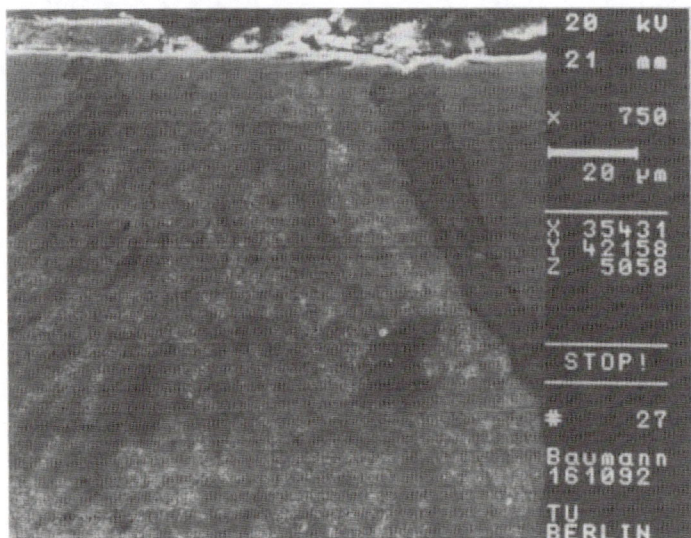

Figure 11. Flow of softer bulk material through a defect of white layer due to severe plastic deformation (from Baumann, 1998)

1.5 Additional phenomenological observations

If corrugations of a certain height have developed then grinding is the only remedy. It is, however, not as helpful as grinding of newly laid rails. Fig. 12 shows a careful measurement of BR. Before grinding a corrugation height of more than 100 μm was measured. After grinding the remaining irregularities over 27 cm distance are less than 10 μm. Four month later a regrowth of corrugation is observed where peaks and troughs appear at nearly the same position. Possible reasons are (1) variable material properties, (2) variable residual stresses or (3) fixation of corrugation pattern due to system dynamics. Probably these three effects are acting together. If the track structure remains unchanged, then no remedy exists for (3). To avoid variable material properties after grinding, corrugated rails are often ground twice or even three times so that white layers and variable material properties completely disappear. The situation is even more complicated with variable residual stresses. A ground rail may be completely smooth, but quasi-periodic irregularities may appear after ten wheels running over the rail due to relaxation of residual stresses.

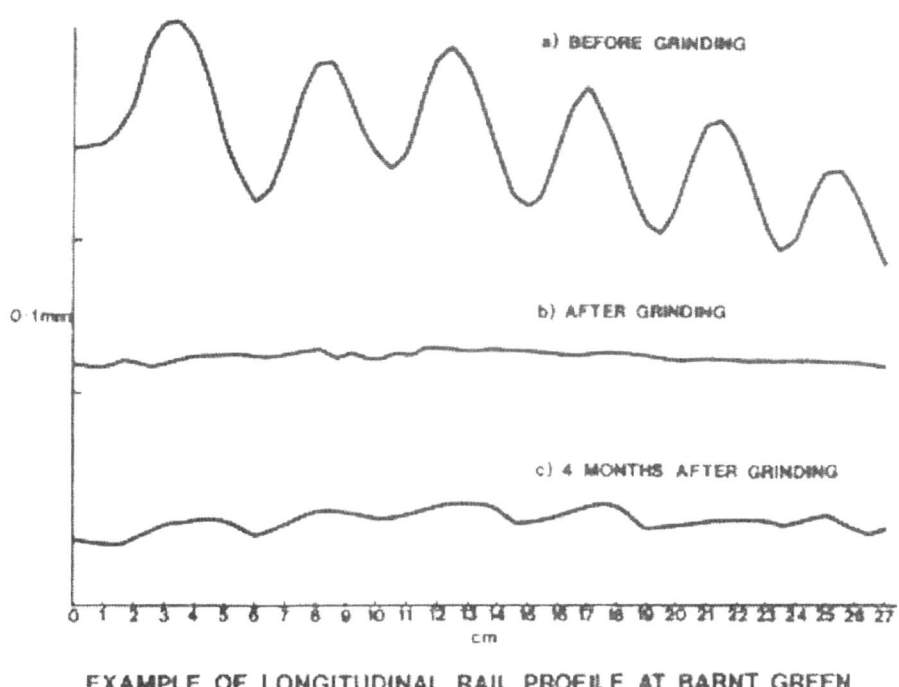

EXAMPLE OF LONGITUDINAL RAIL PROFILE AT BARNT GREEN

Figure 12. Regrowth of corrugation following rail grinding (from Frederick and Bugden, 1983)

1.6 Interpretation and conclusions

The most important questions from a metallurgical point of view are: What type of material is WEL? How is WEL formed?

It has been assumed by several authors that WEL is martensite which is formed through a typical temperature induced martensite transformation. Several arguments have been formulated by Baumann et al. (1996a) and Baumann (1998) against this hypothesis: The hardness of WEL is considerably higher than the hardness of typical martensite which has been obtained by a temperature induced transformation, and the austenitization temperature necessary for such a temperature induced martensite transformation, even considering the influence of high contact pressure, would be approximately 600 °C. To get such a temperature creepages of 10 % or even more would be necessary (Knothe and Liebelt, 1995). This is possible for controlled driving units (Lang and Roth, 1993), it is however unlikely for normal running.

According to Baumann it seems to be likely that a stress-induced transformation is responsible for WEL formation. Baumann has shown that material like WEL is obtained in a powder mill where the global temperature was not higher than 300 °C. This does not mean that temperature is irrelevant. First the stress induced transformation is intensified by an increase in temperature and secondly the temperature increase to 200 or 300 °C results in additional thermally induced stresses. Probably temperature increase in the surface layer is responsible for the sharp boundary between white layer and bulk material.

The conclusions which can be drawn from these phenomenological and metallurgical observations are as follows:

– If one is interested in an analysis including the final saturation stage of corrugation then a completely non-linear analysis of short term dynamics and long term behaviour is unavoidable.
– The damage mechanism should include not only wear but also shake-down effects, plastic deformation, work hardening and formation of WEL due to temperature or stress induced transformation. Up to now nobody has done such an analysis. For certain aspects like formation of white layers not even the physical laws are known.
– An investigation of this type mainly would be of theoretical interest. From a practical point of view one is mainly interested in the initiating state of corrugations. It is assumed that the initial irregularities are sufficiently small so that a purely linear analysis is possible. It has to be checked during the analysis that this assumption is always fulfilled.
– Additionally it is assumed that a running band of white layer remains and the corrugation height remains so small that plastic deformation has not be taken into account. Then wear is the dominant damage mechanism.

1.7   Corrugations on running surfaces of wheels

We restrict our attention to circumferential geometric irregularities of the wheel surface with wavelengths between 1200 and 300 mm, also known as out-of-round or polygonal wheels, which were one of the causes of the Eschede disaster (Böhmer et al., 2000). The irregularities of out-of-round wheels are not as periodic as short pitch corrugations. The damage mechanism seems to be the same as for short pitch corrugation on rails, namely wear. However, no agreement exists between experts concerning the wavelength fixing mechanism. There are at least four hypotheses:

1. **Feedback loop between structural dynamics and wear**. The candidates for wavelength fixing mechanisms are either the P2-resonance of the track or the first bending mode of the wheel-set axle, or perhaps both (Morys and Kuntze, 1997 and Morys, 1998).

2. **Hardness irregularities** of the wheel surface result in irregular wear resistance. As a consequence initial surface irregularities can occur (Mombrei and Rode, 1998).
3. Due to **incorrect reprofiling** of out-of-round wheels small periodic irregularities remain after reprofiling (Mombrei and Rode, 1998)
4. Insufficient **dynamic balancing** of wheel-sets can result in small fluctuations of contact forces and thus in polygonalization even if the original surface is geometrically smooth (Meinke and Szolc, 1995 and Meinke and Morys, 1998).

Probably not only one mechanism is active but different mechanisms are acting together. For none of these mechanisms non-steady state contact mechanics has to be considered as the wavelength always is much higher than the dimensions of the contact patch.

2   Linear model of non-steady state contact mechanics [2]

Basic ideas and basic equations of non-steady state contact mechanics. Frequency domain and time domain representation. Numerical results. Interpretation.

2.1   Basic ideas

When is non-steady state contact mechanics used? It has already been discussed that for harmonic motion the characterizations "steady state" and "non-steady state" contact mechanics depend on the wavelength of the motion and the contact diameter. One can consider steady state contact mechanics if the wavelength of the motion is considerably higher, say by a factor of 10, than the contact length, see Fig. 13:

$$\text{steady state:} \quad \frac{2a}{L} = \frac{2a\,f}{v} < \frac{1}{10}.$$

Typical examples shall be considered assuming a vehicle speed of $v = 80$km/h and a contact diameter $2a = 1$cm. L is the wavelength and f the frequency.

vertical car-body resonance ($f = 1$ Hz)   a/L = 1/4400,
wheel-set hunting ($L > 5$ m)   a/L < 1/1000,
short pitch corrugation ($L \simeq 3$ cm)   a/L = 1/6,
pinned-pinned-resonance ($f = 1100$ Hz) a/L = 1/4,
rolling noise (e.g. $f = 2200$ Hz)   a/L = 1/2.

It is therefore evident that steady state contact mechanics is sufficient for vertical and lateral vehicle dynamics.

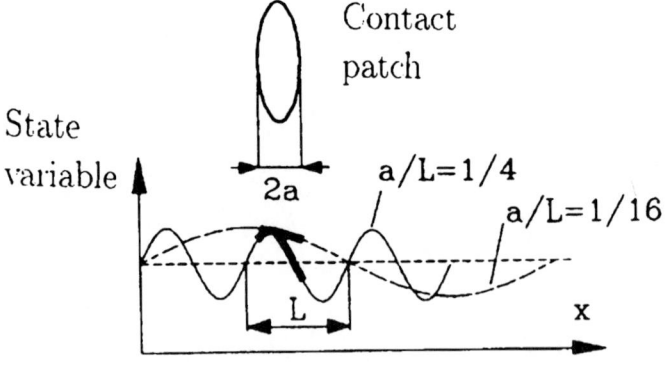

Figure 13. Characterization of steady state and non-steady state contact mechanics by relative contact length (Groß-Thebing, 1993)

---

[2] The paper is mainly based on Groß-Thebing (1993)

Linearization with respect to a reference state When performing non-steady state contact mechanical calculations, we shall restrict ourselves to a completely linear analysis. Non-linear effects only can be dealt with in a time-step integration procedure. Such an integration procedure is extremely time consuming so that extensive parameter variations are nearly impossible.

As shown in Fig. 14 two calculation steps are necessary. Each state variable is divided into a constant reference value and a varying disturbance with respect to this reference value. First the reference values are obtained by a non-linear, steady state contact mechanical analysis. In a second step the disturbances are calculated by a linear but non-steady state analysis.

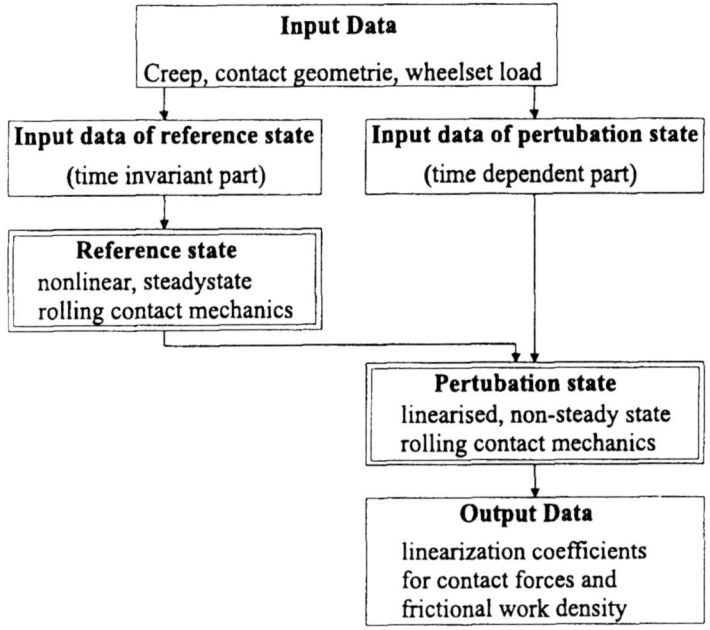

Figure 14. Concept of linearization of non-steady state contact mechanics with respect to a non-linear, steady state reference state (Groß-Thebing, 1993)

Assumptions It is useful to consider first all the basic assumptions of such a contact model:

A1 Linearity assumption: It is assumed that displacements and strains are sufficiently small, that the material of both bodies is linear and elastic and that the relative velocities or creepages in the contact patch are also sufficiently small.

A2 Homogenity assumption: The material is homogeneous and isotropic.

A3 Equality assumption: Wheel and rail are of the same material.

A4 Smoothness assumption: The surfaces are completely smooth.

A5 Half-space assumption: The dimensions of the contact patch are small compared with characteristic dimensions of the bodies near the contact point (e.g. radius of curvature) so that both bodies can be considered as half-spaces.

A6 **Second order polynomial assumption**: The surfaces of both bodies can be described by second order polynomials.

A7 **Coulomb assumption**: Coulomb's law with a constant coefficient of friction can be used.

A8 **Inertia assumption**: The rolling speed is small compared with the propagation speed of Raleigh waves so that inertia effects can be neglected during contact mechanical analysis.

A9 The rolling direction coincides with the direction of one of the axes of the contact ellipse.

A10 The **separability assumption** is discussed in 2.2.

2.2 Development of a linearized contact model

Contact model The idea of a *contact model* will be used in a special way, see Fig. 15. We assume that the contact between wheel and rail can be described similar as a spring and a damper. With the assumptions which have been made e.g. the vertical contact between wheel and rail can be modeled as a spring (Hertzian contact spring).

The model which is shown in Fig. 15 of course is more complicated. Input data of the contact model, given on the left side are displacement and velocities of wheel and rail in the contact point and irregularities of wheel and rail surface. One should be a little more precise: There are two rigid frames on the wheel and the rail structure which are connected to both bodies near the contact point. Within the frame, contact mechanical deformations take place, outside the frame structural dynamics are prevalent. It is assumed that both aspects can be separated (**A10: Separability assumption**).

Outputs are shown on the right hand side: forces and moments acting at the contact point between wheel and rail. If linearization is possible, then there is a linear relation between forces and moments on the one hand and displacements, velocities and irregularities on the other hand.

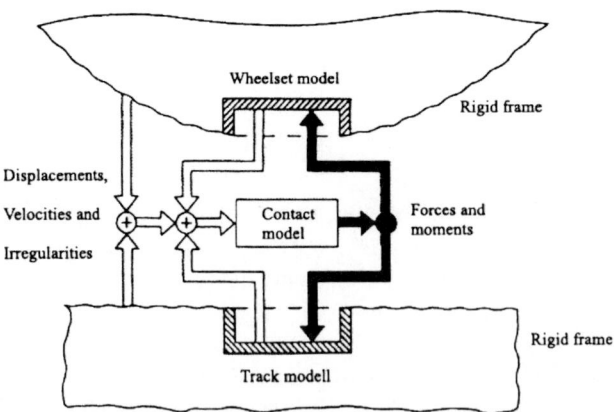

Figure 15. Contact model (Knothe, 1998)

2.3   Basic equations of non-steady state rolling contact

Reference state and superimposed time varying state All variables consist of a reference state and a time varying part. Assuming harmonic motion, a variable $r(t)$ can be expressed

$$r(t) = r_0 + \Delta r \, \cos(\Omega t + \varphi). \tag{1}$$

It is often convenient to use complex notation

$$r(t) = r_0 + \frac{1}{2} \left(\Delta \hat{r} \, e^{\lambda t} + \Delta \hat{\bar{r}} \, e^{\bar{\lambda} t} \right), \tag{2}$$

where overlining denotes a conjugated complex expression and
$\lambda = \alpha + i\omega,$ for free vibrations or
$\lambda = i\Omega,$ for forced vibrations.
Instead of time dependent variables, locally dependent variables can be used

$$r(x) = r_0 + \frac{1}{2} \left(\Delta \hat{r} \, e^{\lambda_x x} + \Delta \hat{\bar{r}} \, e^{\bar{\lambda}_x x} \right), \tag{3}$$

where now

$$\lambda_x = \frac{2i\pi V}{L}. \tag{4}$$

We always shall use an abbreviated notation, where

$$\Delta r(x) = \Delta \hat{r} \, e^{\lambda_x \bar{r}} \quad \text{or} \quad \Delta r(t) = \Delta \hat{r} \, e^{\lambda t}. \tag{5}$$

Co-ordinate systems In Fig. 16 the co-ordinate systems are shown for the case where there is no subdivision into a reference state and a time-variant state. Two different co-ordinate systems are used describing particles of body 1 (wheel: ξ_1, η_1) and body 2 (rail: ξ_2, η_2) moving through the contact patch. The transformation between the moving contact co-ordinate systems and the spatially fixed co-ordinate system is

$$x(t) = V_m(t - t_0) + \xi(t), \quad y = \eta \quad \text{and} \quad z = \zeta. \tag{6}$$

The situation is more complicated if there is a time variant state which is superimposed on the reference state as explained in Fig. 17. We assume that the radii of the contact ellipse are time dependent variables,

$$a(t) = a_0 + \Delta \hat{a} \, e^{\lambda t},$$
$$b(t) = b_0 + \Delta \hat{b} \, e^{\lambda t}. \tag{7}$$

If the same co-ordinate system (ξ, η), see Fig. 17a, is used not only for the reference state but also for the different time dependent states then trouble has to be expected as small variations Δa and Δb cannot be seized sufficiently accurately unless an extremely

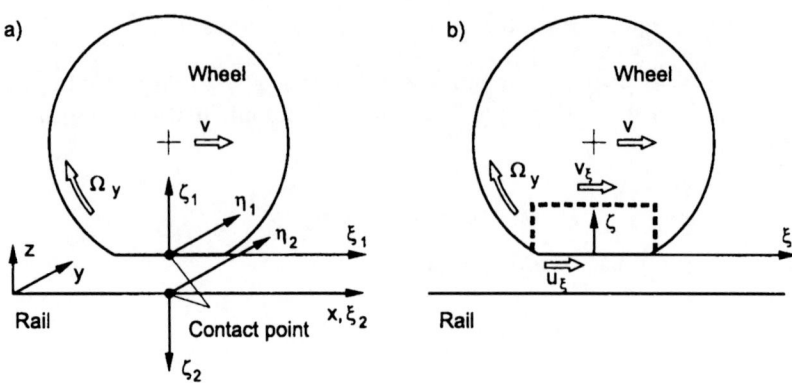

Figure 16. Co-ordinate systems (a) and kinematic relations (b) (from Groß-Thebing, 1993)

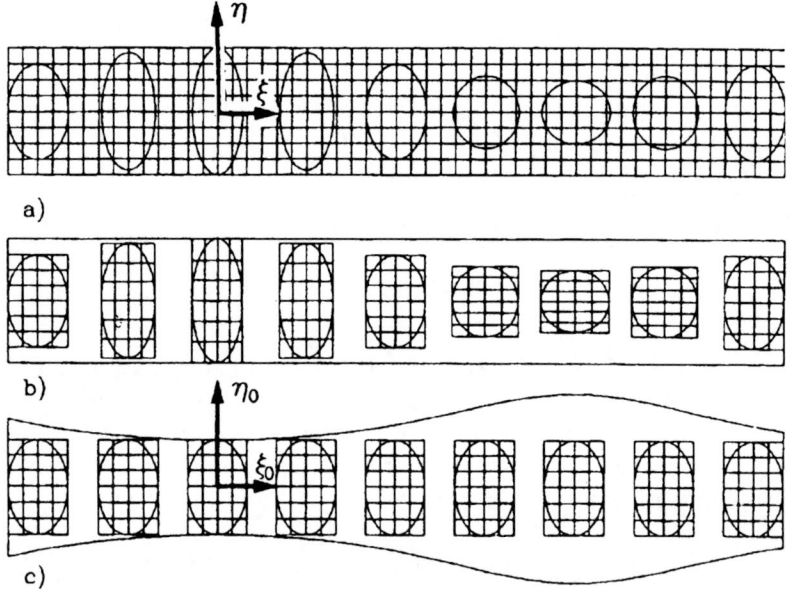

Figure 17. Mapping of time-variant contact ellipses on the contact ellipse of the steady state problem (Groß-Thebing, 1993)

fine discretization is used. We therefore introduce different co-ordinate systems for the different time variant states, see Fig 17b. This is a kind of mapping. The fluctuating contact ellipses are always mapped onto the contact ellipse of the reference state. In the image domain the radii of the mapped ellipses remain constant, now the scale of the co-ordinate system is varying, see Fig. 17c.

The transformation between both co-ordinate systems now reads

$$\xi_0(t)/a_0 = \xi(t)/a(t), \tag{8}$$

$$\eta_0(t)/b_0 = \eta/b(t). \tag{9}$$

Introducing the assumed law for the time depending variables $a(t)$ and $b(t)$, see eq. (7), gives

$$\xi(t) = \xi_0(t)\left(1 + \frac{\Delta\hat{a}}{a_0}e^{\lambda t}\right),$$

$$\eta = \eta_0(t)\left(1 + \frac{\Delta\hat{b}}{b_0}e^{\lambda t}\right). \tag{10}$$

The total derivative $\frac{d}{dt}$ has to be formulated as

$$\frac{d}{dt} = \frac{\partial}{\partial t} + \frac{\partial}{\partial\xi_0}\frac{d\xi_0}{dt}. \tag{11}$$

with a time dependent expression $d\xi_0/dt$. Using eq. (8) and taking into account that due to eq. (6) $d\xi/dt = -V$, we obtain

$$\frac{d\xi_0}{dt} = -V\left[1 + \left(1 + \frac{\lambda\xi_0}{V}\right)\frac{\Delta\hat{a}}{a_0}e^{\lambda t}\right]. \tag{12}$$

Kinematic relations The core of the basic equations are the kinematic relations. First the equations without superimposed time-dependent terms are considered. The velocities of two contacting particles of body 1 and 2 and the relative velocity are

$$\mathbf{v}_i(\xi_i,\eta_i) = \frac{d}{dt}\left\{\begin{array}{c}\xi_i(t)\\\eta_i\end{array}\right\} + \frac{d\mathbf{u}_i(\xi_i(t),\eta_i)}{dt}, \qquad \mathbf{v}(\xi,\eta) = \frac{d}{dt}\left\{\begin{array}{c}\xi_1(t)-\xi_2(t)\\\eta_1(t)-\eta_2(t)\end{array}\right\} + \frac{d\mathbf{u}(\xi,\eta)}{dt}, \tag{13}$$

where

$$\mathbf{v} = \mathbf{v}_1 - \mathbf{v}_2 \quad\text{and}\quad \mathbf{u} = \mathbf{u}_1 - \mathbf{u}_2. \tag{14}$$

Considering

$$\frac{d\mathbf{u}(\xi,\eta)}{dt} = \frac{\partial\mathbf{u}(\xi,\eta)}{\partial t} + \frac{\partial\mathbf{u}(\xi,\eta)}{\partial\xi}\frac{\partial\xi}{\partial t} = \frac{\partial\mathbf{u}(\xi,\eta)}{\partial t} - V_m\frac{\partial\mathbf{u}(\xi,\eta)}{\partial\xi}. \tag{15}$$

the following kinematic equation is obtained:

$$\frac{\mathbf{v}(\xi(t),\eta,t)}{V_m} = \left\{\begin{array}{c}\nu_\xi(t) - \nu_\zeta(t)\eta\\\nu_\eta(t) + \nu_\zeta(t)\xi(t)\end{array}\right\} - \frac{\partial\mathbf{u}(\xi(t),\eta,t)}{\partial\xi} + \frac{1}{V_m}\frac{\partial\mathbf{u}(\xi(t),\eta)}{\partial t}. \tag{16}$$

The expression on the left hand side contains local creepages. The first expression on the right hand side is the rigid body creep, the second is the steady state creep due to elastic deformations and the last one is non-steady state creep. For steady state rolling contact $\partial u(\xi(t), \eta, t)/\partial t = 0$, therefore this expression can be omitted.

A more tedious analysis has to be performed for our situation where there is a reference state (index 0) and a superimposed time variable state. One has to substitute $\mathbf{v}(\xi, \eta), \mathbf{u}(\xi, \eta), \xi, \eta$ with $\mathbf{v}(\xi_0, \eta_0) + \Delta\hat{\mathbf{v}}(\xi_0, \eta_0), \mathbf{u}(\xi_0, \eta_0) + \Delta\hat{\mathbf{u}}(\xi_0, \eta_0), \xi_0, \eta_0$ respectively and additionally eq. (12) has to be introduced. This tedious transformation is necessary as the whole numerical analysis shall be performed in the (ξ_0, η_0) coordinate system.

It makes no sense to go through this derivation step by step, see Groß-Thebing (1993). Only the final result is given, which can be divided into the kinematic equations for the reference state and the kinematic equations for the time variant state, where higher order terms are neglected:

$$\frac{\mathbf{v}_0(\xi_0, \eta_0)}{V_m} = \left\{ \begin{matrix} \nu_{\xi 0} - \nu_{\zeta 0}\eta_0 \\ \nu_{\eta 0} + \nu_{\zeta 0}\xi_0 \end{matrix} \right\} - \frac{\partial \mathbf{u}_0(\xi_0, \eta_0)}{\partial \xi_0}, \tag{17}$$

$$\frac{\Delta\hat{\mathbf{v}}(\xi_0, \eta_0)}{V_m} = \left\{ \begin{matrix} \Delta\hat{\nu}_\xi - \Delta\hat{\nu}_\zeta\eta_0 \\ \Delta\hat{\nu}_\eta + \Delta\hat{\nu}_\zeta\xi_0 \end{matrix} \right\} + \nu_{\zeta 0} \left\{ \begin{matrix} -\eta_0 \frac{\Delta\hat{b}}{b_0} \\ \xi_0 \frac{\Delta\hat{a}}{a_0} \end{matrix} \right\} - \frac{\partial \Delta\hat{\mathbf{u}}(\xi_0, \eta_0)}{\partial \xi_0} + \frac{\lambda}{V_m}\Delta\hat{\mathbf{u}}(\xi_0, \eta_0)$$

$$+ \left(1 - \xi_0\frac{\lambda}{V_m}\right) \frac{\Delta\hat{a}}{a_0} \frac{\partial \mathbf{u}_0(\xi_0, \eta_0)}{\partial \xi_0}. \tag{18}$$

Of course eq. (17) is the same as (16), taking into account that $\partial u(\xi, \eta)/\partial t = 0$ in eq. (16), whereas eq. (18) contains several additional terms mainly due to the variation $\Delta\hat{a}$ and $\Delta\hat{b}$ of the contact radii.

If eq. (17) has been solved, $\partial \mathbf{u}_0(\xi_0, \eta_0)/\partial \xi_0$ has to be substituted into eq. (18) to obtain the following version of the kinematic equation which contains only the local reference creepages and no other state variables of the reference state:

$$\frac{\Delta\hat{\mathbf{v}}(\xi_0, \eta_0)}{V_m} = \left\{ \begin{matrix} \Delta\hat{\nu}_\xi - \Delta\hat{\nu}_\zeta\eta_0 \\ \Delta\hat{\nu}_\eta + \Delta\hat{\nu}_\zeta\xi_0 \end{matrix} \right\} + \nu_{\zeta 0} \left\{ \begin{matrix} -\eta_0 \frac{\Delta\hat{b}}{b_0} \\ \xi_0 \frac{\Delta\hat{a}}{a_0} \end{matrix} \right\} - \frac{\partial \Delta\hat{\mathbf{u}}(\xi_0, \eta_0)}{\partial \xi_0} + \frac{\lambda}{V_m}\Delta\hat{\mathbf{u}}(\xi_0, \eta_0)$$

$$+ \left(1 - \xi_0\frac{\lambda}{V_m}\right) \frac{\Delta\hat{a}}{a_0} \left(\left\{ \begin{matrix} \Delta\hat{\nu}_\xi - \Delta\hat{\nu}_\zeta\eta_0 \\ \Delta\hat{\nu}_\eta + \Delta\hat{\nu}_\zeta\xi_0 \end{matrix} \right\} - \left\{ \begin{matrix} \Delta\hat{\nu}_\xi - \Delta\hat{\nu}_\zeta\eta_0 \\ \Delta\hat{\nu}_\eta + \Delta\hat{\nu}_\zeta\xi_0 \end{matrix} \right\} \right). \tag{19}$$

It is customary to use only displacements $\mathbf{u}_0(\xi_0, \eta_0)$ so that the strains $\partial \mathbf{u}_0(\xi_0, \eta_0)/\partial t$ have to be substituted. Therefore eq. (17) as well as eq. (19) have to be integrated. In the adhesion area the integrations is simple as

$$\mathbf{v}_0(\xi_0, \eta_0) = 0 \quad \text{and} \quad \Delta\hat{\mathbf{v}}(\xi_0, \eta_0) = 0 \quad \text{in the adhesion zone.}$$

In the sliding area $\mathbf{u}_0(\xi_0, \eta_0)$ and $\Delta\hat{\mathbf{u}}(\xi_0, \eta_0)$ have to be assumed. The simplest way is to approximate them as constant within each element of the contact patch and then to perform the integration.

Constitutive equations Not only local creepages but also tangential stresses are assumed to be constant within each element j. The displacements $\mathbf{u}(\xi_i, \eta_i)$ in the midpoint of an element of the discretized contact patch are obtained as

$$\mathbf{u}(\xi_i, \eta_i) = \sum_{j=1}^{N} \int_{\xi_j^u}^{\xi_j^o} \int_{\eta_j^u}^{\eta_j^o} \mathbf{G}(\xi_j - \xi^*, \eta_j - \eta^*) \mathbf{q}(\xi_j, \eta_j) \, d\xi^* \, d\eta^*. \tag{20}$$

Matrix $\mathbf{G}$ contains the Boussinesq-Cerutti functions and can be integrated. Again it has to be considered that in the non-steady state case there are varying element boundaries. The result for the displacements $\mathbf{u}_{0i}$ and $\Delta\hat{\mathbf{u}}_i$ of an element i due to stresses $\mathbf{q}_{0j}$ and $\Delta\hat{\mathbf{q}}_j$ can be understood best in the abbreviated matrix form:

$$\mathbf{u}_{0i} = \sum_{j=1}^{I} \mathbf{F}_{0ij} \mathbf{q}_{0j}, \tag{21}$$

$$\Delta\hat{\mathbf{u}}_i = \sum_{j=1}^{I} \mathbf{F}_{0ij} \Delta\hat{\mathbf{q}}_j + \sum_{j=1}^{I} \mathbf{F}_{aij} \mathbf{q}_{0j} \frac{\Delta\hat{a}}{a_0} + \sum_{j=1}^{I} \mathbf{F}_{bij} \mathbf{q}_{0j} \frac{\Delta\hat{b}}{b_0}. \tag{22}$$

Frictional law In the sliding area Coulomb's frictional law for the reference state can be written as

$$|\mathbf{q}_0(\xi, \eta)| = \mu p_{max0} \sqrt{1 - \frac{\xi_0^2}{a_0} - \frac{\eta_0^2}{b_0}} \tag{23}$$

$$\mathbf{c}_0^T(\xi, \eta) \, \mathbf{v}_0(\xi, \eta) = 0, \tag{24}$$

where $\mathbf{c}_0^T = \{-q_{\eta 0}, q_{\xi 0}\}$. Eq. (24) is used to simplify the linearization with respect to the reference state. For the disturbances one gets

$$\mathbf{q}_0^T(\xi, \eta) \Delta\hat{\mathbf{q}}(\xi, \eta) = |\mathbf{q}_0(\xi, \eta)|^2 \frac{\Delta\hat{p}_{max}}{p_{max0}}, \tag{25}$$

$$\mathbf{c}_0^T \Delta\hat{\mathbf{v}}(\xi, \eta) + \mathbf{d}_0^T \Delta\hat{\mathbf{q}}(\xi, \eta) = 0, \tag{26}$$

where additionally $\mathbf{d}_0^T = \{v_{\eta 0}, -v_{\xi 0}\}$. After some transformations the following linear relation is obtained:

$$\Delta\hat{\mathbf{q}}(\xi, \eta) = \frac{1}{\mathbf{q}_0^T \mathbf{v}_0(\xi, \eta)} \left(|\mathbf{q}_0(\xi, \eta)|^2 \mathbf{v}_0(\xi, \eta) \frac{\Delta\hat{p}_{max}}{p_{max0}} + \mathbf{c}_0 \mathbf{c}_0^T \Delta\hat{\mathbf{v}} \right). \tag{27}$$

What has not been considered up to now is that the boundary between adhesion and sliding area may vary too. This is a non-linear effect which is disregarded.

Tangential forces and spin moment For the calculation of the resultant creep forces, the tangential stresses, which are constant within each element, have to be integrated. A similar equation is obtained for the spin moment. For the reference state the resultant vector reads

$$
\left\{ \begin{array}{c} T_\xi \\ T_\eta \\ M_\varsigma \end{array} \right\} = \int \int\limits_A \left\{ \begin{array}{c} q_\xi(\xi,\eta) \\ q_\eta(\xi,\eta) \\ q_\eta(\xi,\eta)\xi - q_\xi(\xi,\eta)\eta \end{array} \right\} d\xi \, d\eta = \sum_{i=1}^{I} \left\{ \begin{array}{c} q_{\xi 0i} \\ q_{\eta 0i} \\ q_{\eta 0i}\xi_{0i} - q_{\xi 0i}\eta_{0i} \end{array} \right\} \Delta\xi_{01}\Delta\eta_{0i}.
$$

$$(28)$$

2.4   Solution of the linearized basic equations

As all basic equations of linearized non-steady state rolling contact mechanics have been formulated and a discretized contact patch exists, in Groß-Thebing (1993) the following discretized state variables were introduced:

  – $\Delta\hat{q}_i$, the amplitudes of the tangential stresses in the adhesion zone and
  – $\Delta\hat{v}_i$, the amplitudes of the sliding velocities in the sliding area.

The analysis gives no problems. It is quite similar to the analysis which can be performed with CONTACT. The main difference is that the analysis is completely linear so that no iteration is necessary.

If harmonic oscillation is assumed, e.g. due to harmonic short wavelength irregularities, then

$$
\lambda = i\Omega = 2\pi i \frac{V}{L}.
$$

$$(29)$$

One obtains the following result:

$$
\left\{ \begin{array}{c} \Delta\hat{T}_\xi \\ \Delta\hat{T}_\eta \\ \Delta\hat{M}_\varsigma \end{array} \right\} = \left[\begin{array}{ccc} \partial T_\xi/\partial v_\xi & \partial T_\xi/\partial v_\eta & \partial T_\xi/\partial v_\varsigma \\ \partial T_\eta/\partial v_\xi & \partial T_\eta/\partial v_\eta & \partial T_\eta/\partial v_\varsigma \\ \partial M_\varsigma/\partial v_\xi & \partial M_\varsigma/\partial v_\eta & \partial M_\varsigma/\partial v_\varsigma \end{array} \right] \left\{ \begin{array}{c} \Delta\hat{v}_\xi \\ \Delta\hat{v}_\eta \\ \Delta\hat{v}_\varsigma \end{array} \right\}
$$

$$
+ \left[\begin{array}{ccc} \partial T_\xi/\partial a & \partial T_\xi/\partial b & \partial T_\xi/\partial p_{max} \\ \partial T_\eta/\partial a & \partial T_\eta/\partial b & \partial T_\eta/\partial p_{max} \\ \partial M_\varsigma/\partial a & \partial M_\varsigma/\partial b & \partial M_\varsigma/\partial p_{max} \end{array} \right] \left\{ \begin{array}{c} \Delta\hat{a} \\ \Delta\hat{b} \\ \Delta\hat{p}_{max}. \end{array} \right\}
$$

$$(30)$$

Comparing eq. (30) with Kalker's linear, steady state creep-force/creep matrix

$$
\left\{ \begin{array}{c} \Delta T_\xi \\ \Delta T_\eta \\ \Delta M_\varsigma \end{array} \right\} = G a_0 b_0 \left[\begin{array}{ccc} C_{11} & 0 & 0 \\ 0 & C_{22} & \sqrt{a_0 b_0}\,C_{23} \\ 0 & -\sqrt{a_0 b_0}\,C_{23} & a_0 b_0\,C_{33} \end{array} \right] \left\{ \begin{array}{c} \Delta v_\xi \\ \Delta v_\eta \\ \Delta v_\varsigma \end{array} \right\}
$$

$$(31)$$

two basic differences are found:

- Both matrices of eq. (30) are complex whereas Kalker's matrix is not. The reason is that eq. (30) describes non-steady state processes.
- There are two matrices in eq. (30) compared with only one matrix in eq. (31). The reason is that Kalker's matrix has been obtained by linearizing with respect to a zero reference state. Otherwise a second matrix would also have been obtained.

2.5   Numerical results of non-steady state contact mechanics

Harmonic variations with respect to a reference state The simplest way to present the results of the linear contact model is to use frequency responses, either in the complex plane or as amplitude and phase. The details of the reference state are shown in Table 2:

	Notation	Units	
Young's modulus	E	N/mm^2	2.1×10^5
Poisson's number	σ		0.25
Coefficient of friction	μ		0.4
Radius in rolling direction	a	mm	9.6
Radius in lateral direction	b	mm	14.4
Longitudinal creep	$\nu_{\xi 0}$	%	0.1
Lateral creep	$\nu_{\eta 0}$	%	0
Spin creep	$\nu_{\zeta 0}$	1/m	0
Normal force	N_0	kN	50
Longitudinal creep force	$T_{\xi 0}$	kN	-9.8
Lateral creep force	$T_{\eta 0}$	kN	0
Spin moment	$M_{\zeta 0}$	Nm	0
Proportion of sliding area		%	29

Table 2. Data of the reference state

All frequency responses are normalised with respect to the creep coefficients for the steady state case. Therefore all frequency amplitudes take the value "1" for $a/L = 0$. The frequency response for the longitudinal creep is shown in Fig. 18. All other frequency responses of eq. (30) look similar.

For the complex frequency response the L/a values are given as parameters. It can be seen that even for $L/a = 20$, a relevant phase angle has been obtained whereas the amplitude is nearly the same as for $L/a \to \infty$.

Time-domain models The results of the last subsection are restricted to a frequency domain analysis. If the excitation is transient it can be of interest to have time domain

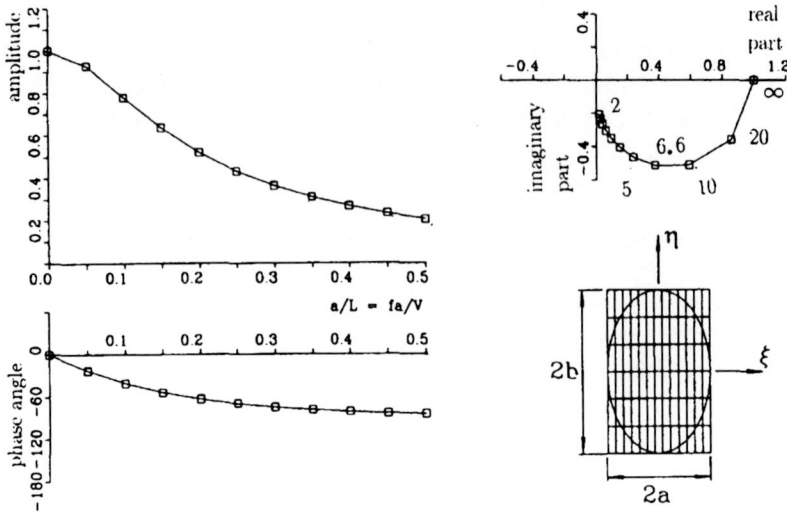

Figure 18. Frequency response for the normalised longitudinal creep force depending on a harmonically varying longitudinal creep ($\nu_{\xi ref} = 0.1\%$)

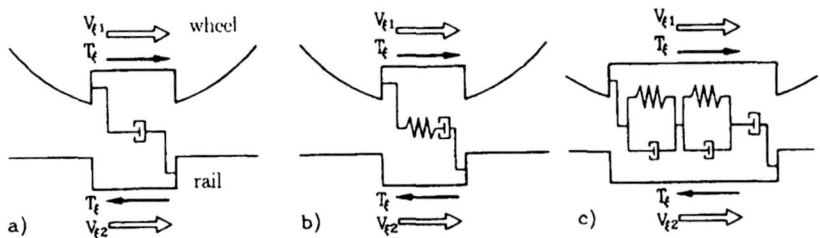

Figure 19. Model of steady state rolling contact (**a**), simplest model of non-steady state rolling contact (**b**) and advanced model (**c**)

models instead of frequency responses. How such models can be obtained shall be explained starting with the complex frequency response of Fig. 18. A simple approximation of the frequency response $\partial \hat{T}_\xi / \partial \hat{\nu}_\xi$ is given by the following equation:

$$\frac{\partial \hat{T}_\xi}{\partial \hat{\nu}_\xi} = \frac{Z_0}{1 + i4\pi \frac{a}{L} N_1}. \tag{32}$$

The whole equation is premultiplied with the denominator term and instead of the partial derivative ∂ we introduce Δ:

$$\left(1 + i4\pi \frac{a}{L} N_1\right) \Delta \hat{T} = Z_0 \Delta \hat{\nu}_\xi. \tag{33}$$

This equation is equivalent to the following ordinary differential equation:

$$\Delta T(t) + N_1 \frac{2a}{V} \frac{d\Delta T(t)}{dt} = Z_0 \Delta \nu_\xi(t). \tag{34}$$

The equivalence of eqs. (33) and (34) can be seen immediately by introducing

$$\Delta T(t) = \hat{\Delta T} e^{i2\pi \frac{V}{a} \frac{a}{L} t} \quad \text{and} \quad \Delta \hat{\nu}_\xi(t) = \hat{\Delta \nu}_\xi e^{i2\pi \frac{V}{a} \frac{a}{L} t}.$$

Now, the mechanical interpretation of the simplest model of non-steady state rolling contact mechanics is evident, see Fig. 19. Whereas linearized steady state rolling contact can be modelled as a simple damper, non-steady state rolling contact has to be modelled in the simplest case as a spring and a damper in series.

In most cases, even for more complicated frequency responses, it is possible first to approximate the frequency responses by broken rational polynomials, to perform a partial fraction decomposition and to interpret the results as mechanical models, see e.g. Wu (1997).

2.6 Linearized wear analysis

It has already been mentioned in Section 1 that wear is the only damage mechanism which is considered in the long term model. Details of such an analysis are again found in Groß-Thebing (1993). The main assumption is (see also Section 4) that wear (that is the mass Δm of material which is worn away), is proportional to the frictional work,

$$\Delta m \propto W_F.$$

Looking at a particle (x, y) of the rail surface, the frictional work density $w_F(x, y)$ for a single wheel running over this particle is obtained by integrating the frictional power density p_F:

$$w_F(x, y) = \int_{t_{lead}}^{t_{trail}} p_F(\xi(t), \eta, t) \, dt, \tag{35}$$

where t_{lead} is when the particle is entering the contact patch at the leading edge and t_{trail} is when it leaves the patch. The integration again has to be performed in the co-ordinate system (ξ_0, β_0). For details see Groß-Thebing (1993).

The frictional power density p_F has to be divided into a constant part (reference state) (p_{F0}) and a harmonically varying part $\Delta \hat{p}_F$. As the frictional power density is a product of the tangential stresses and the local relative velocity, we obtain

$$\begin{aligned} p_F = \mathbf{q}^T \mathbf{v} &= \{\mathbf{q}_0 + \Delta \mathbf{q}\}^T \{\mathbf{v}_0 + \Delta \mathbf{v}\} \\ &\simeq \mathbf{q}_0^T \Delta \mathbf{v} + \Delta \mathbf{q}^T \mathbf{v}_0 \\ &= p_{F0} + \Delta p_F. \end{aligned} \tag{36}$$

Therefore a linear perturbation part Δp_F only exists if the reference creep is not zero.

If one is only interested in a harmonic variation of the height due to wear, $w_F(x,y)$ also has to be integrated between $\eta = -b_0$ and $\eta = +b_0$. The final result reads

$$\Delta \hat{W}_F = \left\{ \frac{\partial W_F}{\partial v_\xi}, \frac{\partial W_F}{\partial v_\eta}, \frac{\partial W_F}{\partial v_\zeta} \right\} \left\{ \begin{array}{c} \Delta \hat{v}_\xi \\ \Delta \hat{v}_\xi \\ \Delta \hat{v}_\xi \end{array} \right\} + \left\{ \frac{\partial W_F}{\partial A}, \frac{\partial W_F}{\partial B}, \frac{\partial W_F}{\partial d} \right\} \left\{ \begin{array}{c} \Delta \hat{A} \\ \Delta \hat{B} \\ \Delta \hat{d} \end{array} \right\}. \quad (37)$$

What has been obtained again are complex linearization coefficients. They indicate, how the amplitude $\Delta \hat{W}_F$ of the frictional work density is influenced by the amplitudes of the harmonically varying input data, e.g. $\Delta \hat{v}_\xi$.

Two typical examples are given in Fig. 20. The data of the reference state are the same as in Table 2. The linearization coefficients which have been used for normalization can be found in the subsequent table:

Notation	Units	creep values		
Lateral reference creep $v_{\eta 0}$	%	0.15	0.3	0.45
Sliding part of contact patch	%	32	71	100
$\partial W_0/\partial v_\eta$	$(kJ/m^2)/\%$	-0.13	-0.18	-0.14
$\partial W_0/\partial d$	$(kJm^2)/mm$	0.4	-12.4	-47.1

2.7 Excitation model

Finally we will consider the excitation mechanism. In reality the wheel is rolling on the rail and either the wheel or the rail surface have irregularities, see Fig. 21a. The exact treatment of this situation is complicated, as one has to consider discrete support and the possibility of wave propagation in the rail. Therefore in most cases a simplified model is used: instead of a moving wheel-set a moving irregularity is introduced (Fig. 21b). A band which contains all the irregularities is pulled through the wheel and the rail. The wheel is allowed to rotate and to perform small displacements with respect to the bogie or the inertia frame. Therefore it is possible to take into account gyroscopic effects of the rotating wheel-set. Different support stiffness of the track due to different positions of the wheel on the rail also can be dealt with. Contact mechanics remain unaffected by this simplified excitation model. What is neglected are effects due to wave propagation in the rail and parametric excitation due to discrete sleeper support.

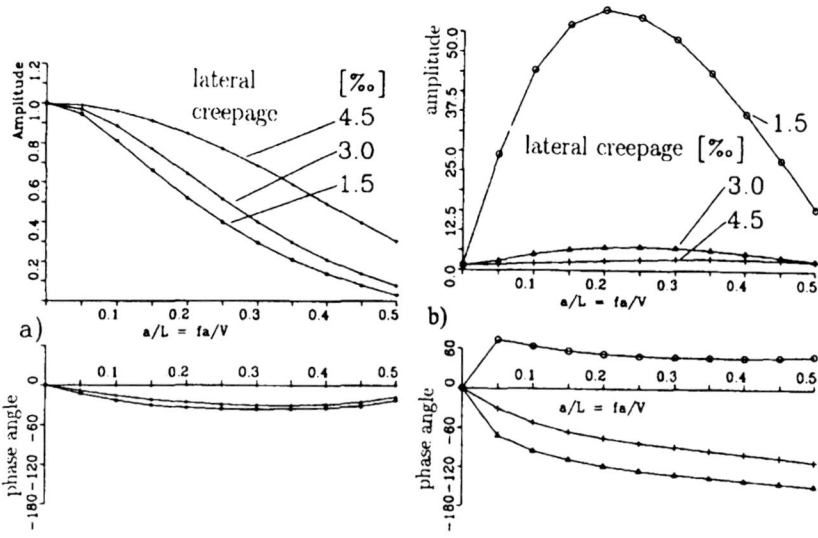

Figure 20. Frequency responses of frictional work for height variation due to a fluctuation of lateral creepage and due to a fluctuation of elastic penetration, taken from Groß-Thebing (1993)

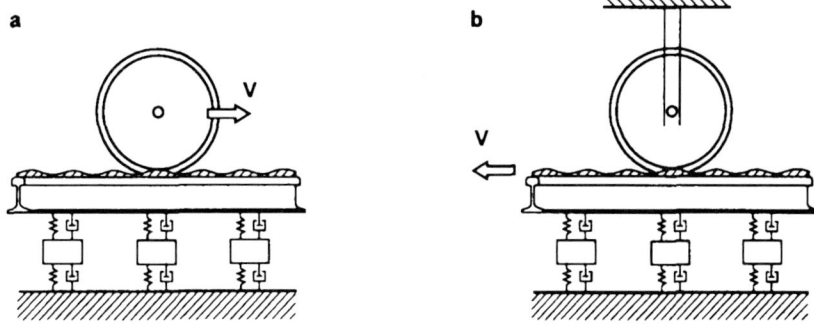

Figure 21. Excitation model. Model of moving wheel-set (a), model of moving excitation (b)

3   Structural mechanics of wheel and rail wheel-set [3]

Modelling and dynamic analysis of wheel and rail in the high frequency range. Direct and cross receptances of wheel and rail. Comparison with measurements.

3.1   Concept of structural dynamics

As in contact mechanics, the main assumption is that the mechanical models are completely linear. Therefore an analysis of structural dynamics of vehicle and track is possible in the frequency domain. As an input the harmonically fluctuating contact forces and probable moments acting on the wheel and the rail are used (black arrows in Fig. 15). The results are fluctuation displacements and slopes at the contact point (open arrows in Fig. 15). As in Section 2 we only consider complex amplitudes of these values. The results can be written as

$$\Delta \hat{u}_r = \mathbf{F}_r(i\Omega)\,\Delta \hat{q}_r, \quad \text{and} \quad \Delta \hat{u}_w = \mathbf{F}_w(i\Omega)\,\Delta \hat{q}_w. \tag{38}$$

The two matrices $\mathbf{F}_r$ and $\mathbf{F}_w$ are called receptance matrices or dynamic flexibility matrices of the rail and the wheel-set. They depend on the exciting frequency $\Omega = 2\pi f$. The expressions on the diagonal are so-called *direct receptances* whereas the out-of diagonal expressions are *cross receptances*. The main task of this section is to provide algorithms which are able to determine these receptances.

Track receptances in general are more important for corrugation initiation than wheel receptances, as different types of wheels with different dynamic behaviour are rolling over the same track. In 3.2 the dynamic behaviour of a wheel-set will be discussed briefly, followed by a detailled discussion of track receptances in 3.3

3.2   Wheelset dynamics

Mechanical model of the vehicle   In most cases only the wheel-set is important for vibrations which initiate rail corrugation. Bogie and car-body vibrations at higher frequencies are largely isolated from the wheelset.

Different models exist for wheelsets, see e.g. Knothe and Grassie (1993). In the mechanical model which has been used in the analysis the axle is modelled as a Timoshenko beam, the wheels (including the wheel rim) are modelled by plate elements in stretching and bending (Hempelmann and Knothe, 1989) or by shell elements (Kose, 1998), the disc brakes are considered as rigid bodies, see Fig. 22. For the numerical analysis Fourier series in circumferential direction are used to take advantage of the rotational symmetry.

Numerical results of wheelset receptances   As a result of wheelset analysis the lateral direct receptances of a DB wheelset type 88 is shown in Fig. 23. The results are given for a resting wheelset and for a vehicle speed of 216 km/h corresponding to

---

[3] The section is mainly based on Hempelmann and Knothe (1989), on Ripke and Knothe (1991) and on Kose (1998)

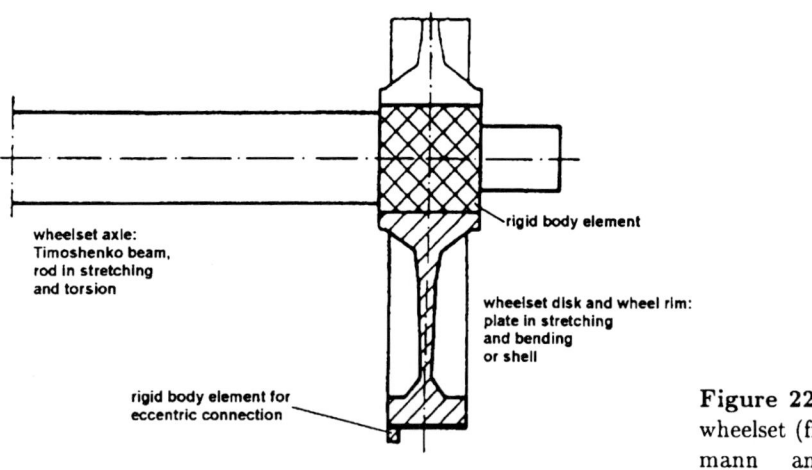

Figure 22. Model of a wheelset (from Hempelmann and Knothe, 1989)

$\Omega = 120$ rad/s. The splitting up of the resonance peaks which can be observed e.g. near 1000 Hz is a consequence of gyroscopic effects. The corresponding mode is a bending mode of the wheel disc with three nodal lines.

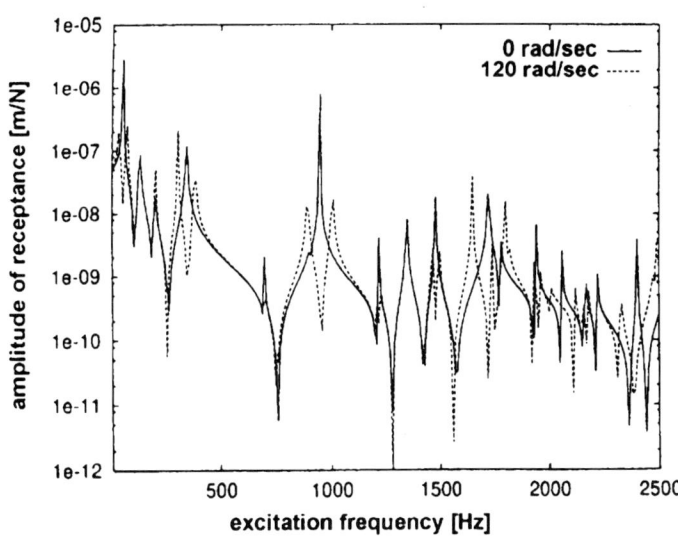

Figure 23. Lateral direct receptance of a DB wheelset type 88 (from Kose, 1998) from $v = 0$ and $v = 216$ km/h corresponding to $\Omega = 120$ rad/s.

3.3 Track dynamics

Mechanical model of the track The track is an infinite, discretely supported structure, consisting of rail, pad, sleeper, ballast und foundation. A mechanical model model
of a track section, which is usually used for corrugation initiation analysis, is shown as
Fig. 24.

- For vertical vibrations the rail is modelled as a Timoshenko beam (see Fig. 25a).
  For lateral and torsional analysis rail head and rail foot are modelled as independent
  beams, connected by a plate as a model for the web.
- The pad between rail foot and sleeper is modelled as spring and damper in parallel
  in three directions. If necessary rotational spring and dampers can also be used
- For reasons of simplicity the sleeper is modelled as a rigid body. It would also be
  possible without difficulty to model the sleeper as a Timoshenko-beam.
- Ballast and halfspace foundation are considered together and modelled as visco-elastic
  foundation.

The main problem is to get appropriate parameters for the pad and for the visco-elastic
foundation below the sleeper.

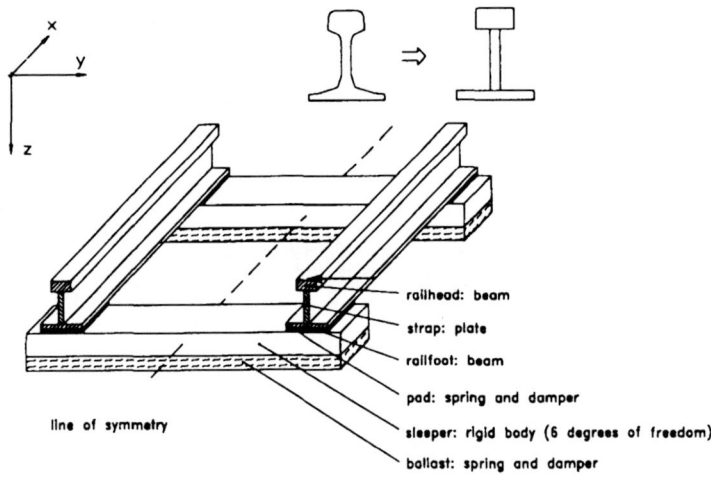

Figure 24. Model of a track section (Ripke and Knothe, 1991)

Derivation of track receptances The derivation of track receptances will be explained for the example shown in Fig. 26. An infinite, discretely supported track is excited
midspan by a harmonically varying force $P\,e^{i\Omega t}$.

   The track is subdivided into sections (see Fig. 27a), each of these sections is further
subdivided into two Timoshenko beam sections and a special section for the supporting

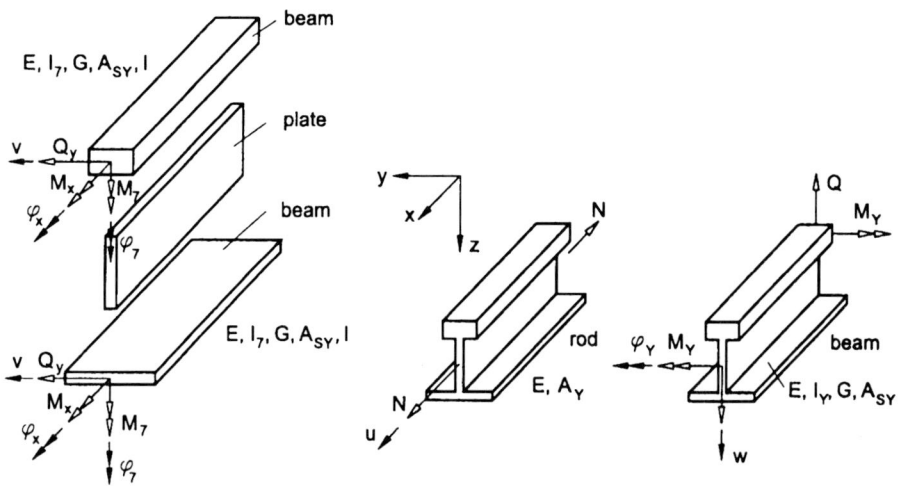

Figure 25. Rail model for lateral dynamics (**a**) and vertical-longitudinal dynamics (**b**) (Ripke and Knothe, 1991)

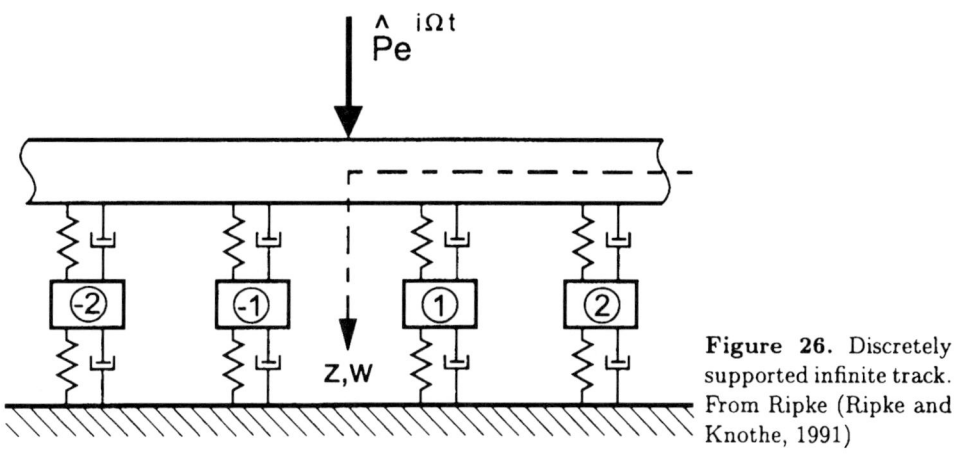

Figure 26. Discretely supported infinite track. From Ripke (Ripke and Knothe, 1991)

point (Fig. 27b). The state variables of the beam are collected in a vector $\Delta\mathbf{x}(x,t)^T = \{-w(x,t), \beta(x,t), M(x,t), Q(x,t)\}$. For harmonically fluctuating state variables we can write

$$\Delta\mathbf{x}(x,t)^T = \Delta\hat{\mathbf{x}}(x)\, e^{i\Omega t}. \tag{39}$$

As the track is subdivided into sections of the same kind, for each of theses sections we are interested in the following relation

$$\Delta\hat{\mathbf{x}}_i = \mathbf{T}(i\Omega)\Delta\hat{\mathbf{x}}_{i-1}, \tag{40}$$

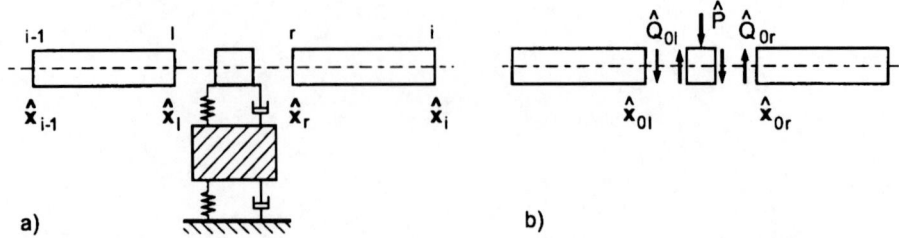

Figure 27. Subdivided track section and point of excitation (from Gasch and Knothe, 1989)

where $\mathbf{T}(i\Omega)$ is the so-called *transfer matrix*. For the two beam elements of Fig. 27b, a transfer matrix for beam elements can be used:

$$\Delta\hat{\mathbf{x}}_i = \mathbf{T}_{beam}(i\Omega)\,\Delta\hat{\mathbf{x}}_l \qquad \text{and} \qquad \Delta\hat{\mathbf{x}}_r = \mathbf{T}_{beam}(i\Omega)\,\Delta\hat{\mathbf{x}}_{i-1}. \tag{41}$$

The transfer matrix of the beam section is obtained from the stiffness and damping matrices of a beam element. This is straightforward and shall not be discussed here. The transfer matrix of the support element is given by the following equation:

$$\left\{\begin{array}{c} -\Delta\hat{w} \\ \Delta\hat{\beta} \\ \Delta\hat{M} \\ \Delta\hat{Q} \end{array}\right\}_r = \begin{bmatrix} 1 & 0 & 0 & 0 \\ 0 & 1 & 0 & 0 \\ 0 & 0 & 1 & 0 \\ -k_{supp} & 0 & 0 & 0 \end{bmatrix} \left\{\begin{array}{c} -\Delta\hat{w} \\ \Delta\hat{\beta} \\ \Delta\hat{M} \\ \Delta\hat{Q} \end{array}\right\} \qquad \text{where} \quad k_{supp} = \frac{(k_b + i\Omega c_b)(kp + i\Omega c_p - m_s\Omega^2)}{k_b + k_p + i\Omega(c_b + c_p) - m_s\Omega^2}$$

$$\tag{42}$$

or abbreviated

$$\Delta\hat{\mathbf{x}}_r = \mathbf{T}_{supp}(i\Omega)\,\Delta\hat{\mathbf{x}}_l. \tag{43}$$

Finally the transfer matrix of the track section we are looking for is obtained:

$$\Delta\hat{\mathbf{x}}_i = \mathbf{T}_{beam}\!\left(\frac{l}{2}, i\Omega\right)\mathbf{T}_{supp}(i\Omega)\,\mathbf{T}_{beam}\!\left(\frac{l}{2}i\Omega\right)\Delta\hat{\mathbf{x}}_{i-1} = \mathbf{T}_{sect}(i\Omega)\,\Delta\hat{\mathbf{x}}_{i-1}. \tag{44}$$

Let 0 be the point of midspan excitation. The state vector $\mathbf{x}_{or}$ then is transfered to a point j by the following equation:

$$\mathbf{x}_j = \underbrace{[\mathbf{T}\,\mathbf{T}\,\mathbf{T}\ldots\mathbf{T}]}_{j\ \text{times}}\mathbf{x}_{or} = \mathbf{T}^j\,\mathbf{x}_{or}. \tag{45}$$

As there is always some damping in the system, the state vector must have completely died away at infinity. This shall be taken into account by considering a single section j with the transfer matrix $\mathbf{T}$. For this transfer matrix first the eigenvalue problem

$$\mathbf{T}(\Omega)\,\varphi_i = \lambda_i\,\varphi_i \tag{46}$$

has to be solved. The meaning of eq. (46) is that an eigenvector, transferred through the matrix section, has only to be multiplied by a proportionality factor λ, which is generally complex. Now the state vector $\mathbf{x}_{j-1}$ at the beginning of the section and the state vector $\mathbf{x}_j$ at the end of the section can be superposed from the eigenvectors φ_i of the transfer matrix $\mathbf{T}$,

$$\mathbf{x}_{j-1} = \sum_{i=1}^{4} \varphi_i q_i \quad \text{and} \quad \mathbf{x}_j = \sum_{i=1}^{4} \lambda_i \varphi_i q_i. \tag{47}$$

The amplitudes of two of the eigenvalues, say λ_1 and λ_2 are less than 1. The amplitudes of the two remaining eigenvalues $\lambda_3 = 1/\lambda_1$ and $\lambda_4 = 1/\lambda_2$ are greater than 1. Therefore a state vector which is transferred to $+\infty$ can only consist of φ_1 and φ_2 as otherwise the state vector would not die away. Therefore

$$\mathbf{x}_{0r} = \begin{bmatrix} \varphi_1 & \varphi_2 \end{bmatrix} \begin{Bmatrix} q_1 \\ q_2 \end{Bmatrix}_{0r} \quad \text{and}$$

$$\mathbf{x}_{0l} = \begin{bmatrix} \varphi_3 & \varphi_4 \end{bmatrix} \begin{Bmatrix} q_3 \\ q_4 \end{Bmatrix}_{0l}. \tag{48}$$

Now the boundary conditions at point 0 can be formulated:

$$\mathbf{x}_{0l} - \mathbf{x}_{0r} = \begin{Bmatrix} 0 \\ 0 \\ 0 \\ P \end{Bmatrix}. \tag{49}$$

Introducing eqs. (48) into eq. (49) an algebraic system of equations is obtained

$$\begin{bmatrix} \varphi_1 & \varphi_2 & \varphi_3 & \varphi_4 \end{bmatrix} \begin{Bmatrix} -q_1 \\ -q_2 \\ q_3 \\ q_4 \end{Bmatrix} = \begin{Bmatrix} 0 \\ 0 \\ 0 \\ P \end{Bmatrix}, \tag{50}$$

which can be solved without difficulty.

The same algorithm can be used not only for the determination of lateral receptances of a rail on discrete supports but also for extremely refined rail models, as always only an eigenvalue problem for one section has to be solved.

Numerical results The vertical direct receptances (dynamic flexibilities) of the track are shown in Fig. 28. The receptance for an excitation midspan between two sleepers is shown by the thin line whereas the thick line shows an excitation near the sleeper. For the excitation midspan three resonances can be seen at 140 Hz, 460 Hz and 1070 Hz. At 140 Hz rail and sleeper are vibrating in phase whereas at 460 Hz they are nearly out of phase. The third resonance peak at 1070 Hz is only found for an excitation at midspan. For this frequency the rail is vibrating nearly sinusoidally with nodes at the sleepers. This standing wave is called pinned-pinned-mode (ppm). As it is extremely difficult to

excite this mode when the load is near the sleeper, the corresponding receptance curve
has a pronounced minimum (antiresonance) near the ppm-frequency.

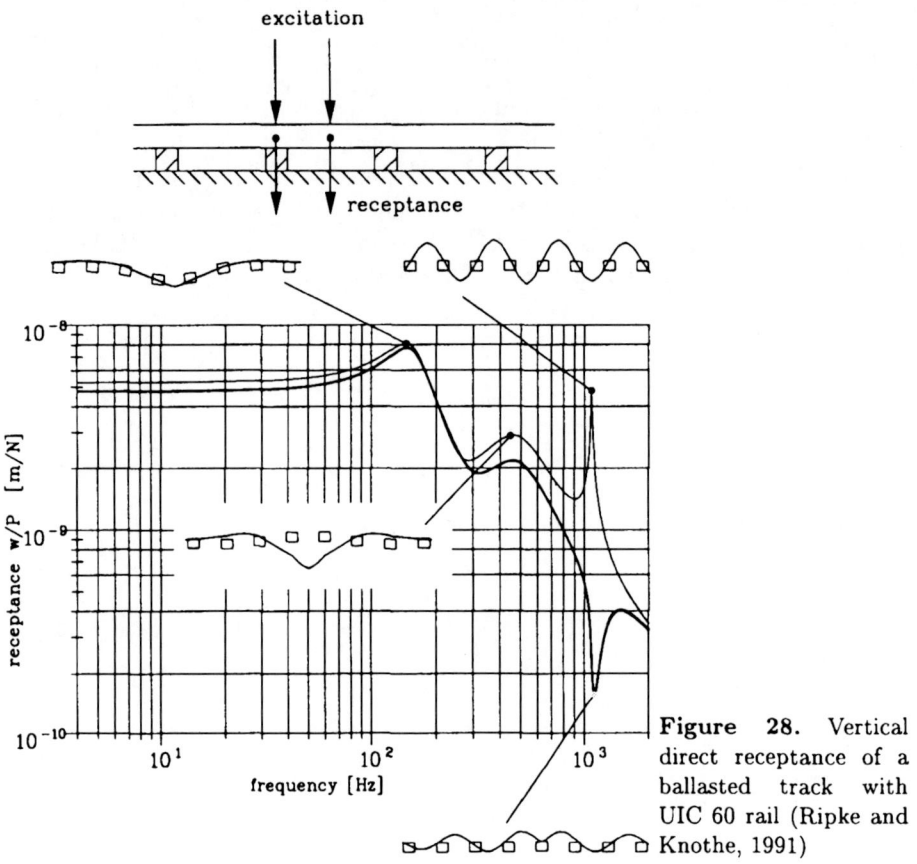

Figure 28. Vertical direct receptance of a ballasted track with UIC 60 rail (Ripke and Knothe, 1991)

The situation is similar for lateral excitation, see Fig. 29. However, as the rail, the
pad, and the ballast are softer laterally, the three resonances for excitation at midspan are
shifted to lower frequencies (1st resonance: 60 Hz, 2nd resonance: 160 Hz, 3rd resonance
pinned-pinned-mode: 560 Hz). As the rail has been modelled laterally as a double-beam,
more resonances and antiresonances are found in the frequency range up to 2000 Hz. An
additional pronounced peak for excitation at midspan is found at 1820 Hz. The vibration
modes at these frequencies are given in Fig. 30.

Comparison with measurements Static measurements were already carried out in
the last century by Schwedler (1882). Dynamic measurements already can be found in
Koch (1932), but they were not very reliable. Receptance behaviour for frequencies up
to 1500 Hz has been measured by Grassie et al. (1982), see Fig. 31, and compared
with theoretical results. As with all other measurements this comparison is not fully

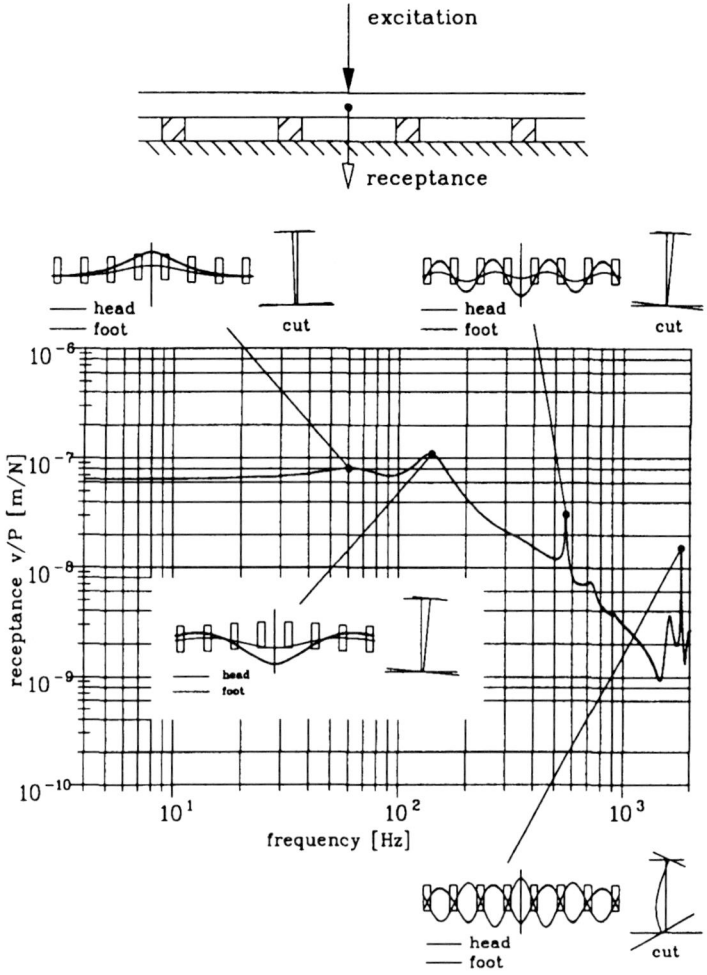

Figure 29. Lateral direct receptance of a ballasted track with UIC 60 rail (from Ripke and Knothe, 1991)

satisfactory, as the measured data usually are used to adapt calculated and measured results by an appropriate choice of pad and ballast parameters. Looking at the vibration modes in Fig. 28, it is evident that the ballast mainly influences the first resonance whereas the pad influences the second one. As a linear model with spring and damper in parallel has been used for pad and ballast/underground it is clear that the parameters of such a model are only valid near the corresponding resonance.

Grassie has used a harmonic excitation in order to get receptances. Nowadays mainly calibrated hammers are used. A result of such measurements compared with numerical results (again the pad and ballast values have been adapted) is shown in Fig. 32.

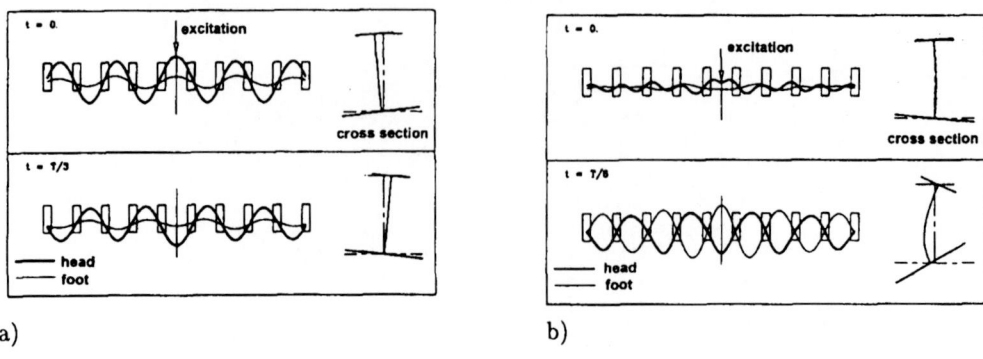

a)   b)

Figure 30. Lateral pinned-pinned-modes at 560 and 1820 Hz (Ripke and Knothe, 1991)

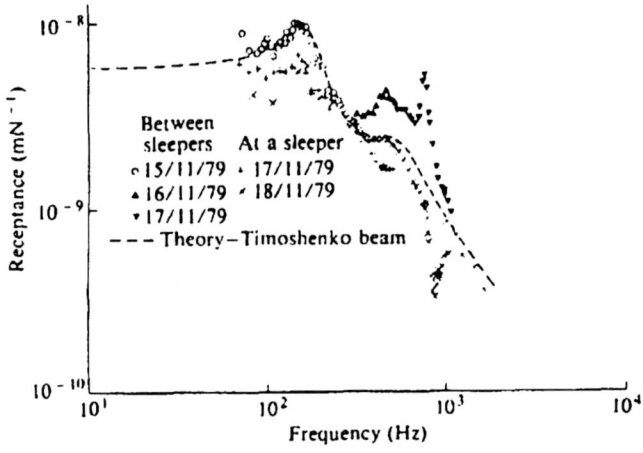

Figure 31. Vertical direct receptances measured by Grassie et al. (1982) and comparison with numerical results

Reliability and limits of validity   The track receptances are as important for the analysis of corrugation initiation as the non-steady state contact mechanics. It is therefore important to consider in some detail the reliability and the limits of validity from a theoretical point of view.

– **Reliability of track data:** Concerning *rail* and *sleeper* there is now problem. Geometric and material data are available for both components and are reliable. *Ballast* and *subgrade* data cause much more problems, however track dynamics is mainly influenced by the ballast in the frequency range up to 200 Hz. The frequency range which is of interest for corrugation initiation is usually higher, between 300 and 2000 Hz.
   The main problem is reliable pad data. According to Knothe and Grassie (1993) which contains most of the available data till 1993, pad stiffnesses can vary between 50 and 2000 kN/mm. Concerning pad damping the situation is even more complicated. Usually the pad is modelled as a linear, viscoelastic component. Using the

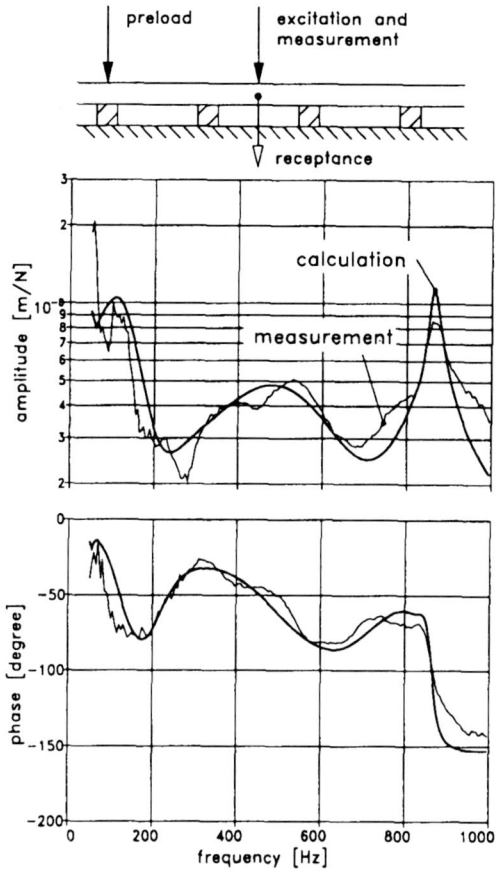

Figure 32. Measured vertical direct receptances (from Ripke, 1992)

available data in literature, it has been shown in Hempelmann (1994) that stiffness and damping of rail pad data found in literature are approximately proportional:

$$c_{p,track} = 1.3 \times 10^{-3} \, k_{p,track} \, [\text{s}]. \tag{51}$$

As already mentioned, pad data from measurements and therefore eq. (51) are only valid near the corresponding resonance frequency. Linear, viscoelastic damping can be also described as a complex, dynamic, frequency dependent stiffness, where the real part is constant and the imaginary part is proportional to the exciting frequency:

$$k_{p,dyn} = k_p + i\Omega c_p. \tag{52}$$

Recent laboratory experiments (e.g. Knothe et al., 1997) have shown that for a preloaded rail pad not only the real part but also the imaginary part of the complex stiffness does not depend on Ω:

$$k_{p,dyn,lab} = k_p \left(1 + i\eta\right). \tag{53}$$

This is equivalent to the fact that the parameter of viscous damping decreases with increasing frequency. Combining eqs. (51), (52) and (53) one gets the following relation for an approximate, viscoelastic damping, which can be used if only k_p is available

$$c_p \simeq \frac{c_{p,track}}{k_{p,track}} \frac{\Omega}{\Omega_{res}} k_p, \qquad \text{where} \qquad \frac{c_{p,track}}{k_{p,track}} \simeq 1.3 \times 10^{-3}. \tag{54}$$

It has to be mentioned briefly that the pad stiffness also depends on the preload. For higher loads the stiffness is also higher.

Very little data is available on lateral pad stiffness. In lateral direction the pad is generally softer by a factor of 4 to 20, depending on the geometric shape of the pad.

– **To what extent do track receptances depend on the rail model:**
  The rail model which has been used is a Timoshenko beam vertically and a double beam with additional torsional rigidity laterally. In order to test whether this model is sufficient up to 3 kHz it is necessary to use two- or threedimensional components in order to model the rail. Models of this kind have been used to determine the dispersion relations of the rail by Scholl (1987) and Knothe et al. (1994). A detailed analysis of rail receptances using plate elements has been performed by Umlauf (1994). Of course, the pad has not been modelled as a spring and a damper but more realistic as a viscoelastic foundation. The more realistic rail model is now much more complicated, however the same algorithm can be used as in 3.3.
  The amplitudes of vertical and lateral receptances for an excitation at midspan are shown in Figs. 33a and 33b. For the vertical direct receptance the results are very similar. The main difference is that due to the more realistic pad modelling the resonance peak of the pinned-pinned-mode at 1070 Hz is less pronounced. Greater deviations are found for the lateral direct receptance (Fig. 33b).

– **Must moving loads be considered?**
  All receptances have been obtained for resting, harmonically varying loads. This is not correct as the train is moving on the track and thus the fluctuating loads are moving too. A method to consider harmonically varying, moving loads on a discretely supported beam has been proposed by Jezequel (1981). Based on this theory a detailed analysis for a UIC 60 rail can be found in Ilias and Knothe (1992). As a main result in Fig. 34 the amplitude of the vertical direct receptance of a track is shown when the fluctuating load is just midspan between two sleepers.

Evidently the receptances now depend on the vehicle speed. For low frequencies, however, the dependency is negligible. For a static load ($\Omega = 0$) a maximum exists at 440 m/s. This *resonance speed* is far beyond realistic vehicle speeds. The only important deviation for realistic vehicle speeds is found near the pinned-pinned-mode. The resonance peak is split up into two peaks. This corresponds to the splitting-up of wheel eigenfrequencies due to gyroskopic effects. It has been shown by Sibaei that corrugation growth rate is not influenced very much by this splitting-up.

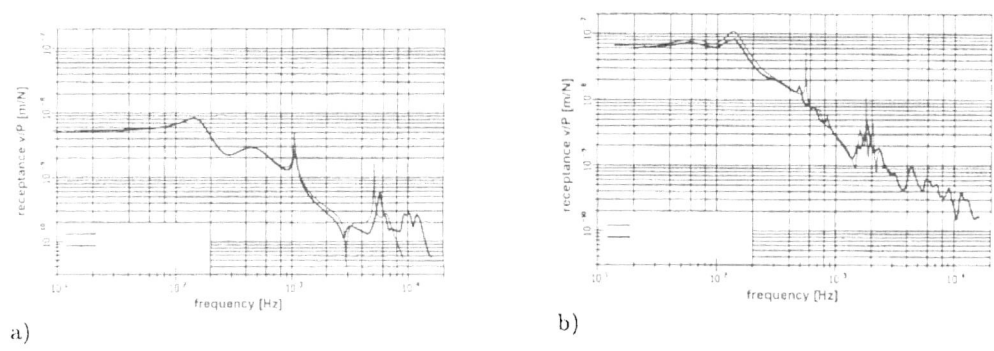

a)   b)

Figure 33. Vertical (**a**) and lateral direct receptances (**b**)of a track using a simple beam model (thin line) or a sophisticated plate model for the rail (from Umlauf, 1994)

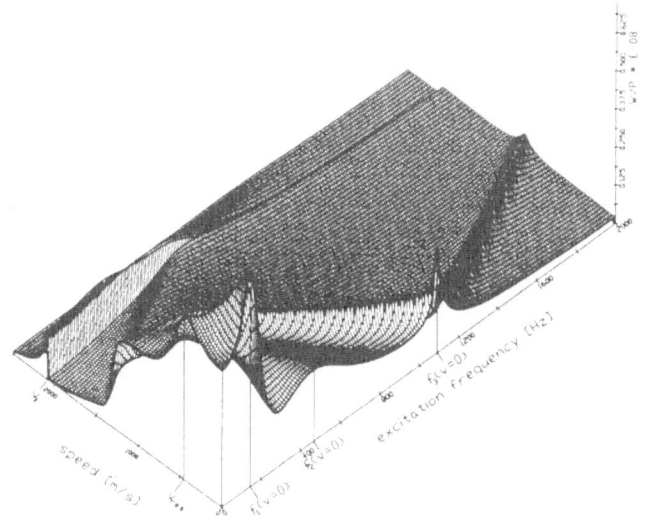

Figure 34. Direct receptance of a moving, harmonically varying load (from Ilias and Knothe, 1992

3.4   Conclusions

Vertical and lateral receptances of the track and the wheelset are available now in the frequency range between 50 and 2500 Hz. The main open problem is to get reliable data of pad stiffness and damping. Both receptances can be tied together with the the frequency responces of non-steady state contact mechanics to form a model high-frequency dynamics of the interacting system (Fig. 35). Inputs of this interaction model are fluctuating profile irregularities. Outputs are fluctuating contact parameters (e.g. forces, creepages, contact radii). As the model has been linearized with respect to a preloaded reference state, the data of this reference state also enters as input data. What is mainly missed are relations of contact geometry and kinematics. They will be provided in the next section.

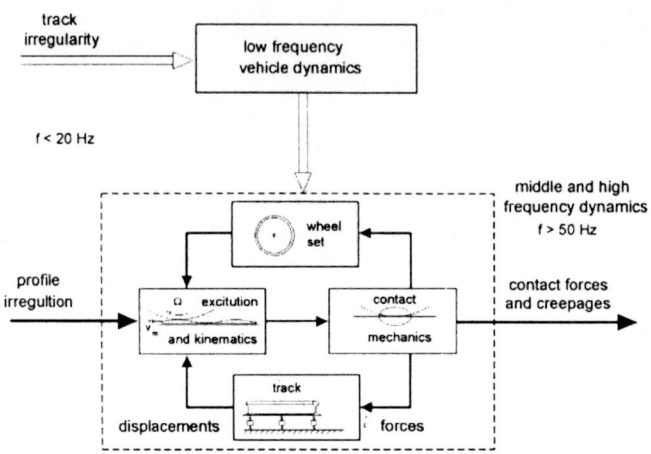

Figure 35. Feedback loop of linearized wheelset-track dynamics (Müller, 1998)

4   Linear model of short term dynamics and long term behaviour [4]

Complete linear model of high frequency analysis. Filter effects. Linearization with respect to a reference state. Stability analysis. Modelling of long term wear mechanisms. Results in principle. Discussion of other possible long-term mechanisms.

4.1   Basic concept

The basic concept of a feed-back loop between high frequency, short term dynamics and long term behaviour has already been demonstrated in Fig. 3. It is explained in Fig. 36 in some more detail. Small irregularities are always present on the rail surface. Therefore an excitation of wheel and rail vibrations is unavoidable. This results in fluctuating contact forces and creepages and therefore also in fluctuating frictional power. Assuming wear to be proportional to frictional work, we also get fluctuating wear along the rail surface which has to be superimposed to the initial irregularities. The process then starts again. For millions of wheels running over the rail, irregularities in a certain wavelength range may be intensified whereas others decrease.

A model of this type was probably first used to investigate corrugations on bearing rings of printing cylinders (Engl et al., 1983). For the investigation of rail corrugations it was proposed by Frederick (1987) and in parallel by Valdivia (1987 and 1988). Later on the theory was refined by Frederick and Sinclair (1992), by Hempelmann (1994), Bhaskar and Johnson et al. (1997a and 1997b) and by Müller (1998). The basic idea also has been used e.g. by Tassilly and Vincent (1991).

For the general model of Fig. 36 linearity is generally not necessary. For the numerical analysis, however, linearity is the basic assumption as otherwise a time-consuming numerical integration would be unavoidable, see e.g. Ilias (1998/99).

4.2   Linear model of high frequency dynamics

High frequency dynamics for vertical track irregularities   We first restrict ourselves to the simplest case and only consider sinusoidal vertical track irregularities as excitation mechanism, see Hempelmann (1994). Vibration of wheel and rail are only possible in vertical and lateral direction. As linearity has been assumed the analysis can be performed in the frequency domain. The system of equations which is obtained for the amplitudes of fluctuating state variables is given as eq. (55). The symbol ˆ, denoting a complex amplitude of harmonically fluctuating state variables, is always omitted.

---

[4] The section is mainly based on Hempelmann (1994) and on Müller (1998)

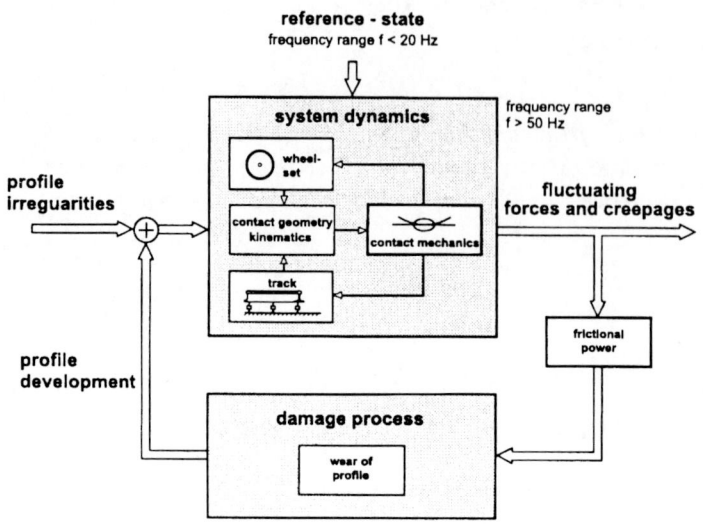

Figure 36. Feed-back loop between high frequency dynamics and long-term wear behaviour (Hempelmann, 1994)

$$
\begin{bmatrix}
-1 & 0 & 0 & \frac{\partial p_0}{\partial A} & \frac{\partial p_0}{\partial d} & 0 & 0 & 0 & 0 & 0 \\
0 & -1 & 0 & \frac{\partial c}{\partial A} & \frac{\partial c}{\partial d} & 0 & 0 & 0 & 0 & 0 \\
0 & 0 & -1 & \frac{\partial g}{\partial A} & \frac{\partial g}{\partial d} & 0 & 0 & 0 & 0 & 0 \\
0 & 0 & 0 & -1 & 0 & 0 & 0 & 0 & 0 & 0 \\
0 & 0 & 0 & 0 & -1 & 0 & -1 & 0 & 0 & 0 \\
\frac{\partial N}{\partial p_0} & \frac{\partial N}{\partial c} & 0 & 0 & 0 & -1 & 0 & 0 & 0 & 0 \\
0 & 0 & 0 & 0 & 0 & F_{\zeta\zeta}(f) & -1 & F_{\eta\zeta}(f) & 0 & 0 \\
\frac{\partial T_\eta}{\partial p_0} & \frac{\partial T_\eta}{\partial c} & \frac{\partial T_\eta}{\partial g} & 0 & 0 & 0 & 0 & -1 & 0 & \frac{\partial T_\eta}{\partial v_\eta} \\
0 & 0 & 0 & 0 & 0 & F_{\zeta\eta}(f) & 0 & F_{\eta\eta}(f) & -1 & 0 \\
0 & 0 & 0 & 0 & 0 & 0 & 0 & 0 & \frac{2i\pi f}{v_m} & -1
\end{bmatrix}
\begin{Bmatrix}
\Delta p_0 \\ \Delta c \\ \Delta g \\ \Delta A \\ \Delta d \\ \Delta N \\ \Delta u_\zeta \\ \Delta T_\eta \\ \Delta u_\eta \\ \Delta v_\eta
\end{Bmatrix}
=
\begin{Bmatrix}
0 \\ 0 \\ 0 \\ \frac{\partial A}{\partial z} \\ -1 \\ 0 \\ 0 \\ 0 \\ 0 \\ 0
\end{Bmatrix}
\Delta z. \quad (55)
$$

The upper four equations show the linearized *Hertzian contact mechanics* and the excitation by the fluctuating longitudinal curvature. The three following equations deal with the *vertical dynamics*. The bottom three equations represent the *tangential (lateral) dynamics* of the wheelset-track system.

Some aspects of eq. (55) shall be discussed in a little bit more detail. The fourth equation is the *vertical kinematic relation*

$$
\Delta d + \Delta u_\zeta = \Delta z, \quad (56)
$$

which means that a vertical irregularity Δz can either be equalized by an elastic deformation Δd or by the relative vertical displacement Δu_ζ of wheel and rail. The last equation is the *lateral kinematic relation*,

$$
\Delta v_\eta(t) = \frac{d\Delta u_\eta(t)}{dt}, \quad (57)
$$

formulated in the frequency domain. For a more general model the *kinematic equations* become much more complicated (see Subsection 4.2).

At first it seems surprising that two different excitation terms occur on the right hand side. The vertical excitation by Δz is obvious. There is however also a *curvature excitation* $\frac{\partial A}{\partial z}\Delta z$. Given a vertical irregularity not only the height but also the curvature varies, which is important for short wavelengths as in this case the contact ellipse and the normal pressure also fluctuate. The effect of this curvature fluctuation will be discussed looking at the fluctuation of the normal force ΔN. One gets from eq. (55)

$$\Delta N = c_H \left(1 + \overbrace{\frac{1}{c_H}\frac{\partial N}{\partial A}\left[\frac{1}{2}\left(\frac{2\pi}{L}\right)^2 + \frac{1}{4r_0^2} \right]}^{\text{curvatur effect } \frac{\partial A}{\partial z}} \right) \Delta z \tag{58}$$

$$\underbrace{\phantom{\Delta N = c_H \left(1 + \frac{1}{c_H}\frac{\partial N}{\partial A}\left[\frac{1}{2}\left(\frac{2\pi}{L}\right)^2 + \frac{1}{4r_0^2} \right] \right)}}_{F_{curv}(\frac{1}{L})}$$

F_{curv} is called a *curvature filter* $F_{curv}(\frac{1}{L})$ in Müller (1998). In comparison to a relation where only c_H is taken into account, the effectiveness of the profile amplitude Δz is influenced by F_{curv}, which is plotted in Fig. 37 (full line). The amplitude is reduced by this filter effect as $\partial N/\partial A$ is negative.

Apart from the curvature filter effect, there is an additional phenomenon which reduces the effectiveness of the profile irregularity for short wavelengths. While for long wavelengths the assumption A3 that the size of the contact patch is sufficiently small compared with the mean radii of curvature is true, for short wavelengths L it becomes doubtful. If L/a is too small, Hertzian contact theory can no longer be applied. We will assume that L/a is always high enough (e.g. $L/a \geq 3$) to justify the usage of Hertzian theory.

But even then effectiveness of the profile has to be changed. This will be called the *amplitude filter* effect. Remington (1976) proposed a filter for acoustical investigations by integrating the vertical irregularity over the contact patch to get a mean value. According to Hempelmann (1994) F_{ampl} is approximated by

$$F_{ampl}\left(\frac{1}{L}\right) \simeq \frac{1}{1 + 1.33\left(\frac{2a}{L}\right)^2 + 0.375\left(\frac{2a}{L}\right)^4 - 0.031\left(\frac{2a}{L}\right)^6}. \tag{59}$$

F_{ampl} is also plotted in Fig. 37 (dotted line).

General linearized wheel-rail dynamics

Co-ordinate systems and contact parameters Two different co-ordinate systems are used to describe the situation near the contact point, see Fig. 38. The co-ordinate system $(C_N; x, y, z)$, which is related to the nominal contact point C_N (reference state), is used for relative displacements whereas the co-ordinate system $(C; \xi, \eta, \zeta)$ is used for the description of rail irregularities and contact mechanics. This is slightly different from Subsection 4.2, where only one co-ordinate system $(C; \xi, \eta, \zeta)$ has been used.

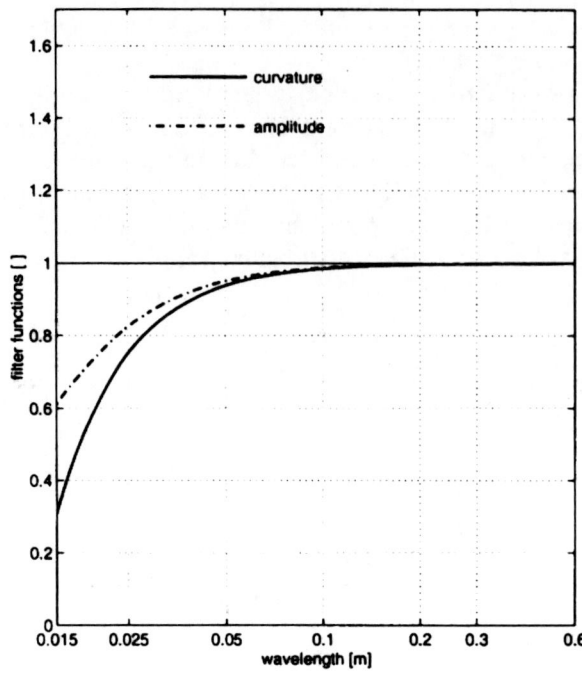

Figure 37. Curvature and amplitude filter for the reference data (Müller, 1998)

The contact geometry generally changes due to relative displacements between wheel and rail. The relative displacements are described using the co-ordinate system shown in Fig. 38(left).

$$
\left\{
\begin{array}{c}
\Delta u_x \\
\Delta u_y \\
\Delta u_z \\
\Delta \varphi_x \\
\Delta \varphi_y \\
\Delta \varphi_z
\end{array}
\right\}
=
\left\{
\begin{array}{c}
\Delta u_x^{wh} \\
\Delta u_y^{wh} \\
\Delta u_z^{wh} \\
\Delta \varphi_x^{wh} \\
\Delta \varphi_y^{wh} \\
\Delta \varphi_z^{wh}
\end{array}
\right\}
-
\left\{
\begin{array}{c}
\Delta u_x^r \\
\Delta u_y^r \\
\Delta u_z^r \\
\Delta \varphi_x^r \\
\Delta \varphi_y^r \\
\Delta \varphi_z^r
\end{array}
\right\},
\tag{60}
$$

or

$$
\Delta \mathbf{u} = \Delta \mathbf{u}^{wh} - \Delta \mathbf{u}^r.
\tag{61}
$$

The amplitude of fluctuations in geometric contact parameters caused by fluctuations in relative displacements and by a harmonic rail irregularity are

Δr	- amplitude of fluctuating longitudinal wheel radius
ΔR^r	- amplitude of fluctuating lateral rail radius
$\Delta \frac{1}{R}$	- amplitude of fluctuating longitudinal rail curvature
$\Delta \alpha$	- amplitude of fluctuating lateral contact location
Δd	- amplitude of fluctuating penetration between wheel and rail
$\Delta \xi^c$	- amplitude of fluctuating longitudinal contact location

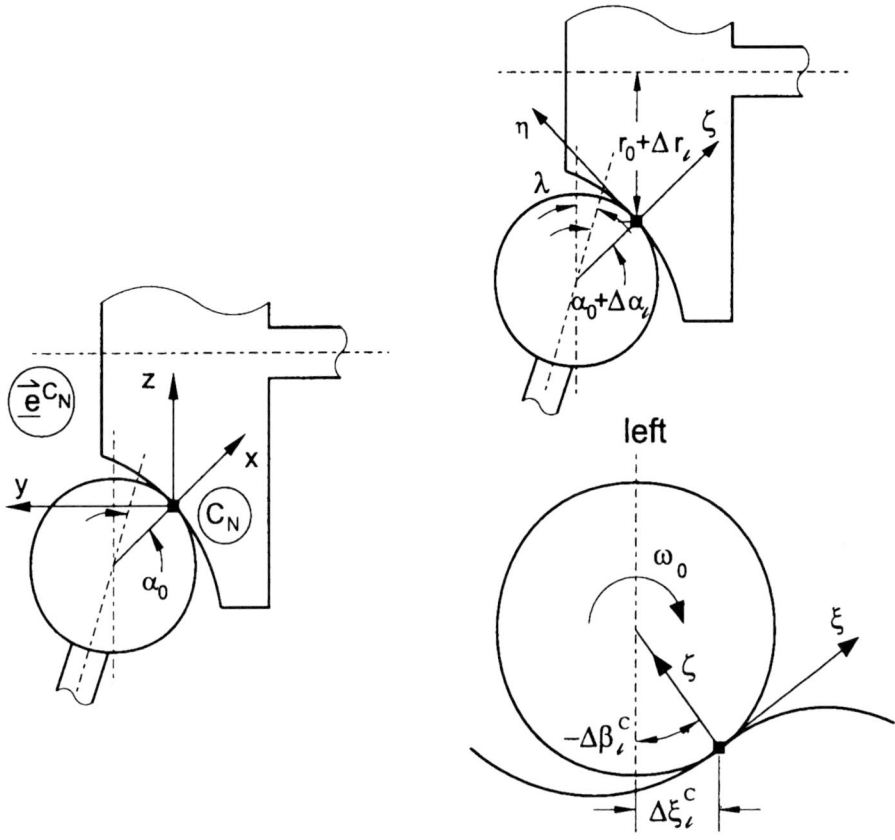

Figure 38. Co-ordinate system for relative displacements (left) and for contact parameters (right)) (from Müller, 1998)

ΔA - amplitude of fluctuating mean curvature in longitudinal direction
ΔB - amplitude of fluctuating mean curvature in lateral direction.

They are explained in Fig. 39.

Generalisation of rail irregularities A smooth and a corrugated running band can differ not only in amplitude but also in slope and curvature, see Fig. 40. This has already been considered by Valdivia (1987) and recently more detailed by Müller (1998). The deviation is described in the co-ordinate system, which is fixed to the nominal contact point (see Fig. 38) as

$$\Delta z(x,y) = (\Delta z_w h_1 + \Delta z_\varphi h_2(y) + \Delta z_\kappa h_3(y)) \; e^{i\frac{2\pi}{L}\,x}. \tag{62}$$

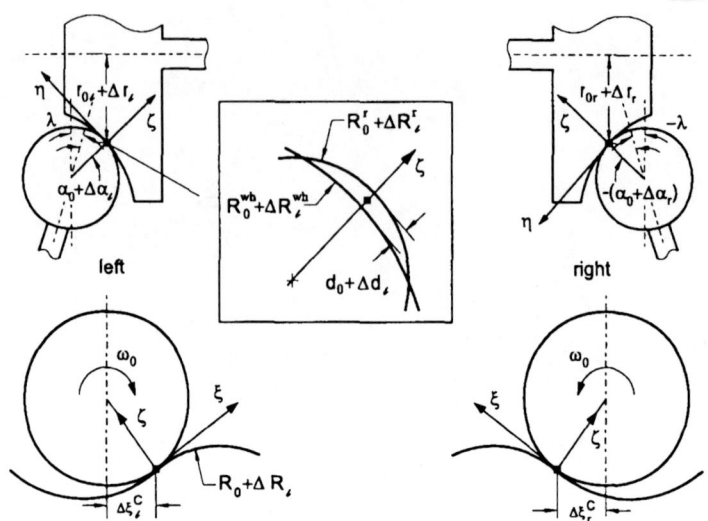

Figure 39. Parameters of contact geometry (Müller, 1998)

The formulation in eq. (62) can describe profile irregularities where minima and maxima are not in a line along the rail but shifted laterally, which is often observed on corrugated rails.

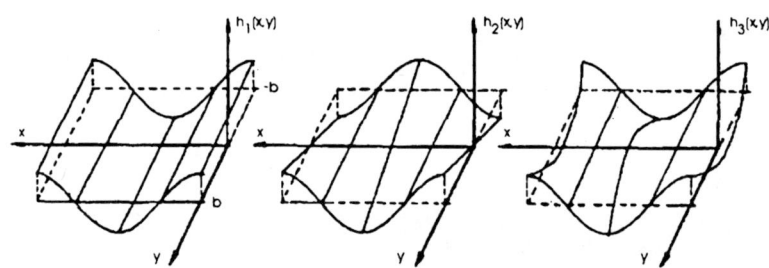

Figure 40. Approximation of possible fluctuations in amplitude, slope and curvature (Groß-Thebing, 1993. $h_1(x,y) = h_1\,e^{i2\pi x/L}$, $h_2(x,y) = h_2(y)\,e^{i2\pi x/L}$, $h_3(x,y) = h_3(y)\,e^{i2\pi x/L}$,)

Kinematics of normal contact For the simplified vertical model there were only two kinematic equations, eqs. (56) and (57). The situation now is more complicated, as the following parameters of contact geometry

$$\mathbf{g}^T = \{\Delta d, \Delta A, \Delta B, \Delta \alpha, \Delta \xi^c\}$$

and their fluctuation have to be considered. They depend on the fluctuating relative displacements and on the fluctuating disturbances:

$$
\left\{ \begin{array}{c} \Delta d \\ \Delta A \\ \Delta B \\ \Delta \alpha \\ \Delta \xi^c \end{array} \right\} =
\left[\begin{array}{cccccc}
0 & \alpha_0 & -1 & 0 & 0 & 0 \\
0 & -\frac{\alpha_0(R_0^{wh}+R_0^r)}{4\,r_0{}^2(R_0^{wh}-R_0^r)} & -\frac{1}{4r_0{}^2} & -\frac{\alpha_0\,R_0^{wh}\,R_0^r}{4\,r_0{}^2(R_0^{wh}-R_0^r)} & 0 & 0 \\
0 & 0 & 0 & 0 & 0 & 0 \\
0 & \frac{1}{R_0^{wh}-R_0^r} & \frac{\alpha_0}{R_0^{wh}-R_0^r} & \frac{R_0^r}{R_0^{wh}-R_0^r} & 0 & 0 \\
1 & 0 & 0 & 0 & r_0\,\alpha_0 r_0
\end{array} \right]
\left\{ \begin{array}{c} \Delta u_x \\ \Delta u_y \\ \Delta u_z \\ \Delta u_{\varphi x} \\ \Delta u_{\varphi y} \\ \Delta u_{\varphi z} \end{array} \right\} +
$$

$$
+ \left[\begin{array}{ccc}
1 & 0 & -\frac{1}{2} \\
\left(\frac{1}{2}\left(\frac{2\pi}{L}\right)^2 + \frac{1}{4\,r_0{}^2}\right) & \frac{\alpha_0\,R_0^{wh}\,R_0^r}{2\,r_0{}^2\,b_0(R_0^{wh}-R_0^r)} & -\frac{1}{2}\left(\frac{1}{2}\left(\frac{2\pi}{L}\right)^2 + \frac{1}{4\,r_0{}^2}\right) \\
0 & 0 & -\frac{\frac{3}{b_0{}^2}}{R_0^r} \\
0 & 0 & -\frac{R_0^r}{b b_0\left(R_0^{wh}-R_0^r\right)} \\
i r_0\frac{2\pi}{L} & 0 & -i r_0\frac{\pi}{L}
\end{array} \right]
\left\{ \begin{array}{c} \Delta z_w \\ \Delta z_\varphi \\ \Delta z_\kappa \end{array} \right\} \quad (63)
$$

or in vector-matrix-notation

$$\Delta \mathbf{g} = \mathbf{G}_{displ}\,\Delta \mathbf{u} + \mathbf{G}_{dist}\,\Delta \mathbf{z}. \tag{64}$$

The determination of the coefficients of both matrices $\mathbf{G}_{displ}$ and $\mathbf{G}_{dist}$ is very troublesome. It can be found in Müller (1998) and will not be discussed here in detail. As an example only the influence of a relative displacement Δu_y on Δd and $\Delta \alpha$ is discussed, see Fig. 41. Points A and E indicate the undisplaced situation whereas B, C, D and F are points of the displaced situation. d is the penetration between wheel and rail. Due to a lateral contact angle α_0 a small value Δd is observed. $\Delta \alpha$ always occurs, independently of α_0.

Kinematics of tangential contact Kinematics of tangential contact means that the linear relations for all creepages depending on fluctuating relative displacements and disturbances must be known. This results in an equation

$$\Delta \nu = \mathbf{C}_{displ}\,\Delta \mathbf{u} + \mathbf{C}_{dist}\,\Delta \mathbf{z}. \tag{65}$$

Again the matrices $\mathbf{C}_{displ}$ and $\mathbf{C}_{dist}$ are given in Müller (1998):

$$\mathbf{C}_{displ} =$$

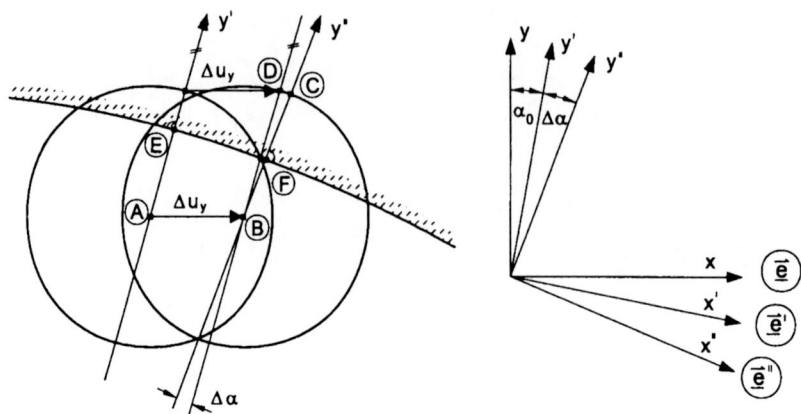

Figure 41. Influence of a relative displacement Δu_y on Δd and $\Delta \alpha$ (from Müller, 1998)

$$
\begin{bmatrix}
\frac{i\Omega}{v_m} & -\frac{\omega_0\,\alpha_0\,(R_0^{wh}+R_0^r)}{2v_0\,(R_0^{wh}-R_0^r)} & -\frac{1}{2}\frac{\omega_0}{v_0} & -\frac{\omega_0\,\alpha_0\,R_0^{wh}\,R_0^r}{v_m\,(R_0^{wh}-R_0^r)}\cdot & 0 & 0 \\[2mm]
0 & \frac{i\Omega}{v_m}\cos(\lambda+\alpha_0) & \frac{i\Omega}{v_m}\sin(\lambda+\alpha_0) & 0 & 0 & -\frac{v_0}{v_m}\cos(\lambda+\alpha_0) \\[2mm]
0 & -\frac{\omega_0\,\cos(\lambda+\alpha_0)}{v_m\,(R_0^{wh}-R_0^r)} & -\frac{\omega_0\,\alpha_o\,\cos(\lambda+\alpha_0)}{v_m\,(R_0^{wh}-R_0^r)} & -\frac{\omega_0\,R_0^r\,\cos(\lambda+\alpha_0)}{v_m\,(R_0^{wh}-R_0^r)} & -\frac{i\Omega}{v_m}\sin(\lambda+\alpha_0) & \frac{i\Omega}{v_m}\cos(\lambda+\alpha_0)
\end{bmatrix}
$$

$$
\text{and}\qquad \mathbf{C}_{dist} =
\begin{bmatrix}
\frac{1}{2}\frac{\omega_0}{v_m} & \frac{\omega_0\,\alpha_0\,R_0^{wh}\,R_0^r}{b_0\,v_m\,(R_0^{wh}-R_0^r)} & -\frac{1}{4}\frac{\omega_0}{v_m} \\[3mm]
-i\frac{r_0}{L}\frac{2\pi\,\omega_0}{v_m}\sin(\lambda+\alpha_0) & 0 & i\frac{r_0}{L}\frac{2\pi\,\omega_0}{v_m}\sin(\lambda+\alpha_0) \\[3mm]
0 & \frac{\omega_0\,R_0^r\,\cos(\lambda+\alpha_0)}{b_0\,v_m\,(R_0^{wh}-R_0^r)} & 0
\end{bmatrix}.
$$

In both matrices it has been distinguished between the vehicle speed v_0 and a mean speed $v_m = \frac{1}{2}(v_0 + r_0\,\omega_0)$, where ω_0 is the angular velocity of the wheel-set. For the examples always $v_m = v_0$ is assumed.

Contact mechanics and structural mechanics Contact mechanics have already been dealt with in Section 2 and structural mechanics in Section 3. The results are assumed to be available.

Transformation of contact forces into the rail system The contact forces refer to the contact co-ordinate system $\mathbf{e}^C$, as described in Fig. 38. These forces and moments act on rail and wheel and cause relative displacement between both bodies. The wheel and rail dynamics are described in co-ordinate systems $\mathbf{e}^{Wh}$ and $\mathbf{e}^R$, which are in general different from the contact system $\mathbf{e}^C$. Contact forces therefore have to be transformed into the wheel and rail coordinate systems. This can be done by setting up the equilibrium equations between the forces acting on wheel or rail respectively and the contact forces. One approach is to apply the principle of virtual work.

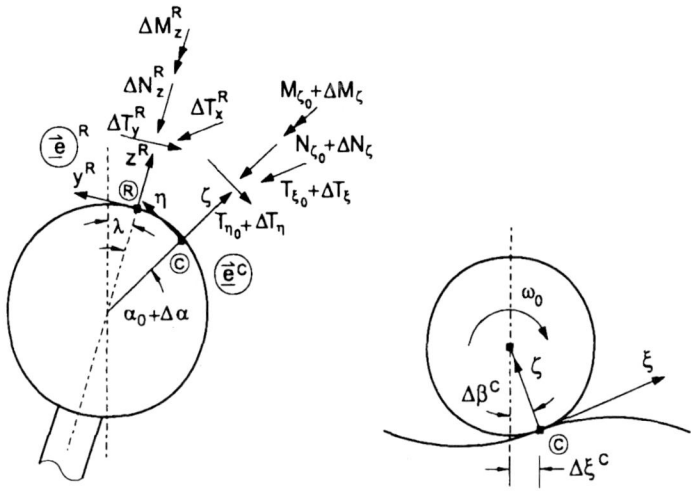

Figure 42. Forces on the left rail. Contact forces $N_{\xi 0}$ etc. are given for a fluctuating contact angle (from Müller, 1998)

In Fig. 42 the contact forces as well as statically equivalent forces acting in the rail coordinate system are shown. The principle of virtual work reads

$$(T_{x0}^R + \Delta T_x^R)\,\delta u_x^R + (T_{y0}^R + \Delta T_y^R)\,\delta u_y^R + (T_{z0}^R + \Delta T_z^R)\,\delta u_z^R +$$
$$(M_{x0}^R + \Delta M_x^R)\,\delta\varphi_x^R + (M_{y0}^R + \Delta M_y^R)\,\delta\varphi_y^R + (M_{z0}^R + \Delta M_z^R)\,\delta\varphi_z^R$$
$$= (T_{\xi 0} + \Delta T_\xi)\,\delta u_\xi + (T_{\eta 0} + \Delta T_\eta)\,\delta u_\eta + (T_{\zeta 0} + \Delta T_\zeta)\,\delta u_\zeta + (M_{\zeta 0} + \Delta M_\zeta)\,\delta\varphi_\zeta. \quad (66)$$

or

$$\delta \mathbf{u}^{R^T}\left(\mathbf{f}_0^R + \Delta\mathbf{f}^R\right) = \delta\mathbf{u}^{C^T}\left(\mathbf{f}_0^C + \Delta\mathbf{f}^C\right). \quad (67)$$

It is not evident that new problems arise with eqs. (66) and (67). The virtual displacements $\delta\mathbf{u}^C$ also may contain terms with fluctuating effects: One therefore has to consider that the transformation matrix between $\delta\mathbf{u}^C$ and $\delta\mathbf{u}^R$ itself depends on $\Delta\mathbf{u}^R$:

$$\delta\mathbf{u}^C = \mathbf{G}_R(\Delta\mathbf{u}^R)\,\delta\mathbf{u}^R. \quad (68)$$

A detailed analysis of this effect can be found in Mauer (1988). In Müller (1998) only the additional expressions due to linearization are given. If now eq. (68) is inserted into eq. (67) one gets on the right hand side additional expressions containing $\mathbf{f}_0^C$. This is called an *initial load effect*. The corresponding matrix is well known from finite element method as *initial stress matrix*. It is an effect of a second order theory which arises because for initial loads (loads of the reference state) equilibrium has to be established with respect to the deformed state. In Fig. 42 e.g. a fluctuating lateral contact angle $\alpha_0 + \Delta\alpha$ has to be considered for $N_{\xi 0}$ and $T_{\eta 0}$.

The basic concept of high frequency dynamics now has to be modified slightly to include initial load effects. Instead of one interaction loop between contact mechanics and structural dynamics as shown in Fig. 35, there are two loops.

4.3   Stability analysis of high frequency dynamics

For the intended dynamic analysis a harmonic excitation is assumed and harmonic fluctuating state variables are obtained. Therefore it is assumed that a complete stationary state of motion exists. It could, however, be possible that already the linear dynamic system is unstable. Instabilities of this kind are known in vehicle dynamics as wheel-set or car-body hunting, e.g. Wickens (1965).

In order to be able to analyse the stability of the wheel-set motion, the kinematic relations of the wheel-set and the constitutive relations of rolling contact mechanics must be available. Inserting both sets of equations into the equations of motion of a single wheel-set, the following equations are obtained for a wheel-set with primary suspension, neglecting gravitational stiffness and spin creep as well as some other secondary effects:

$$\underbrace{\begin{bmatrix} m & 0 \\ 0 & \Theta \end{bmatrix}}_{\text{mass matrix}} \frac{d^2}{dt^2}\left\{\begin{matrix} y \\ \psi \end{matrix}\right\} + \underbrace{\frac{2}{v}\begin{bmatrix} f_y & 0 \\ 0 & e^2 f_x \end{bmatrix}}_{\text{damping matrix}} \frac{d}{dt}\left\{\begin{matrix} y \\ \psi \end{matrix}\right\} + \underbrace{\begin{bmatrix} k_y & -2f_x \\ \frac{2e\lambda}{r_0}f_y & e^2 k_x \end{bmatrix}}_{\text{stiffness matrix}}\left\{\begin{matrix} y \\ \psi \end{matrix}\right\} = \left\{\begin{matrix} 0 \\ 0 \end{matrix}\right\}, \quad (69)$$

where

$$y = \text{lateral displacement of the wheel-set,}$$
$$\psi = \text{yaw angle,}$$
$$m = \text{mass of the wheel-set,}$$
$$\Theta = \text{moment of inertia,}$$
$$r = \text{rolling radius,}$$
$$2e = \text{gauge, lateral distance of longitudinal suspension,}$$
$$f_x(f_y) = \text{longitudinal (lateral) creep coefficient,}$$
$$k_x(k_y) = \text{longitudinal (lateral) primary stiffness,}$$
$$\lambda = \text{conicity and}$$
$$v = \text{vehicle speed.}$$

The main source of instability is the unsymmetrical stiffness matrix.

Several authors have tried to find out whether a kind of unstable wheel hunting is possible in the high frequency range. As long as the track is considered rigid Gasch et al.

(1983), instabilities have been found at running speeds as low as 40 km/s. Further work from Valdivia in Valdivia (1987), however, has shown the same system to be stable as soon as damping provided by the pad is considered.

The model considered here has been investigated by Baskar et al. (1997a and 1997b) and by Müller and Knothe (1997). The problem is that the equations (e.g. eq. (55)) are formulated in the frequency domain. Contact mechanical equations can be transformed into the time domain as has been shown in Section 2. It is, however, much more difficult for the track receptances. Therefore it has been decided to perform the stability analysis in the frequency domain using the Nyquist criterion. For details we refer to the references already mentioned.

The result is unique: For all conditions examined only instabilities related to low frequency hunting motions hunting motions were observed. This can be explained easily: The main reason for unstable wheel-set hunting is unsymmetrical stiffness matrix in eq. (69) which results from the rolling radii difference (factor λ) due to lateral displacement and from the yaw angle. For high frequency vibrations similar unsymmetric expressions also exist. They are, however, so small that pad damping is always sufficient to guarantee stability.

Of course, medium frequency instabilities have been observed as soon as a negative slope of the creep force-creep curve at relatively high creepages is introduced Clark et al. (1988). This is a roll-slip phenomenon which is responsible for corrugations of type 4 listed in Table 1. It becomes relevant only in curves and for driven wheel-sets.

4.4 Long-term wear behaviour

The calculation of wear on the running surface of the rail is based on a proposal given in Krause and Poll (1986). The removed mass per wheel passage n is proportional to the frictional work W_{fric} and k_0 is a proportionality factor

$$\frac{\partial m^{rem}}{\partial n} = k_0 \, W_{fric}(x, \eta). \tag{70}$$

This is a global law. It is easy to get the local law for the change of the profile height z:

$$\frac{\partial z(x, \eta)}{\partial n} = -\frac{k_0}{\rho} w_{fric}(x, \eta), \tag{71}$$

where $w_{fric}(x, \eta)$ is the frictional work density. The height of the profile can be separated into a constant and a fluctuation part

$$z(x, \eta) = z_0 + \Delta z(\eta) \, e^{i \frac{2\pi}{L} x} = z_0 + \mathbf{h}^T(\eta) \Delta \mathbf{z} \, e^{i \frac{2\pi}{L} x}, \tag{72}$$

where $\mathbf{h}(\eta)^T = \left\{ 1, \frac{\eta}{b}, \frac{3}{2} \left(\frac{\eta}{b} \right)^2 - \frac{1}{2} \right\}$ are orthogonal Legendre polynomials and $\Delta \mathbf{z}^T = \{\Delta z_w, \Delta z_\varphi, \Delta z_\kappa\}$ is the corresponding vector. The same approach is used for $\Delta w_{fric}(x, \eta)$. Then it follows from eq. (71)

$$\frac{\partial \Delta \mathbf{z}}{\partial n} = -\frac{k_0}{\rho} \Delta \mathbf{w}_{fric}. \tag{73}$$

The frictional work density vector $\Delta \mathbf{w}_{fric}$ is proportional to $\Delta \mathbf{z}$, where the proportionality matrix is denoted as $\mathbf{W}_{\Delta z}$. Then a system of three ordinary differential equations follows:

$$\frac{k_0}{\rho} \mathbf{W}_{\Delta z}\, \Delta \mathbf{z} + \mathbf{I} \frac{\partial}{\partial n} \Delta \mathbf{z} = 0. \tag{74}$$

The solution is straightforward:

$$\Delta \mathbf{z} = \Delta \mathbf{z}_1\, e^{\lambda_1 n} + \Delta \mathbf{z}_2\, e^{\lambda_2 n} + \Delta \mathbf{z}_3\, e^{\lambda_3 n}. \tag{75}$$

λ_i are called *local wear eigenvalues*. The real parts $\mathrm{Re}(\lambda_i)$ are called *local wear rates* or *local corrugation growth rates (LCGR)*. Irregularities will be amplified if one of the *local corrugation growth rates* $\mathrm{Re}(\lambda_1)$, $\mathrm{Re}(\lambda_2)$ or $\mathrm{Re}(\lambda_3)$ is positive, and suppressed if all are negative. Δz_k are the corresponding wear eigenvectors.

The solution depends among other things on the wavelength of the harmonic profile irregularity, the vehicle speed and on the position within a sleeper bay.

4.5   Results in principle

A general result of the coupled analysis of structural dynamics, contact mechanics and long-term wear behaviour will be discussed for the simpler system which has been used to obtain eq. (55). The only wear degree of freedom which has been considered is a fluctuation in amplitude. The receptances which have been used for the UIC 60 rail are shown in Figs. 28 and 29. The wheel has been considered rigid. A lateral reference creep $\nu_\eta = 0.5\,\%$ has been assumed.

In Fig. 43 the local corrugation growth rates are shown for an exitation above the sleeper and midspan. For reasons of clearness the vertical direct receptance of the track is shown again in Fig. 44. The maximum of the corrugation growth rate appears near 1000 Hz for an excitation at a sleeper whereas at midspan the corrugation growth rate is even negative. The explanation is simple: For an excitation at the sleeper the system is dynamically extremely stiff near 1100 Hz. Unavoidable vertical irregularities therefore result in high fluctuations of the normal force. Fluctuating wear has its maximum in the trough so that corresponding small irregularities are increased. - For an excitation at midspan, however, the vertical direct receptance is high, the track is extremely flexible. Therefore the fluctuating normal forces remain very low and no increase of irregularities is observed.

The **first conclusion**, which can be drawn at this stage is that a stiff situation in vertical direction should be avoided. It is, however, not enough. The situation at the sleepers and the situation at midspan cannot be considered separately. And there is an infinite number of other positions between these two positions: The local corrugation growth rate for the whole sleeper bay is shown in Fig. 45. We need a model which is able to consider the mutual influence of the different positions. This will be discussed in Sect. 5.

A **second conclusion** is that the high corrugation growth rate near 1100 Hz (at a sleeper) is a *frequency constant mechanism*. Therefore, for different vehicle speeds differ-

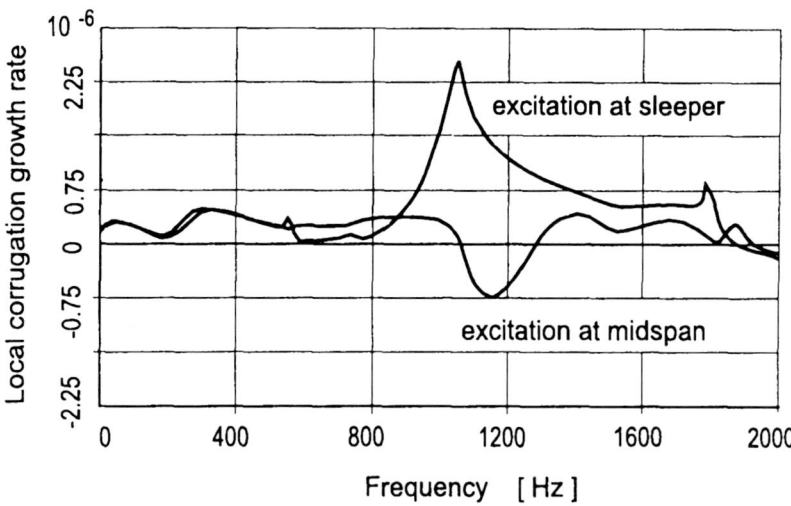

Figure 43. Local corrugation growth rate for excitation above sleeper (thick line) and at midspan (thin line) (from Hempelmann (1994), where also the data can be found)

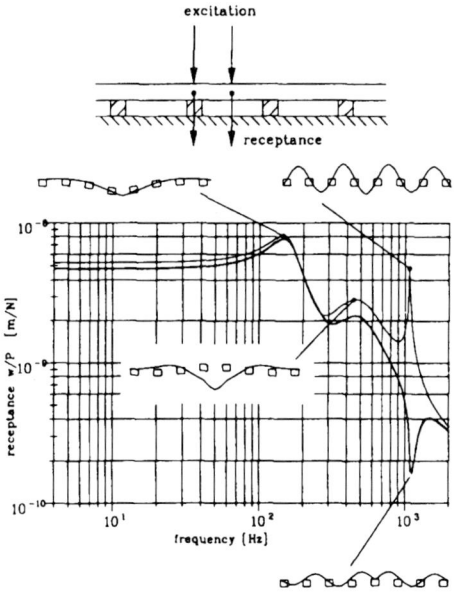

Figure 44. Vertical direct receptance of a ballasted track with UIC 60 rail (Ripke and Knothe, 1991)

ent wavelengths would appear. This, however, has not been observed in praxis. We shall discuss it in the next section.

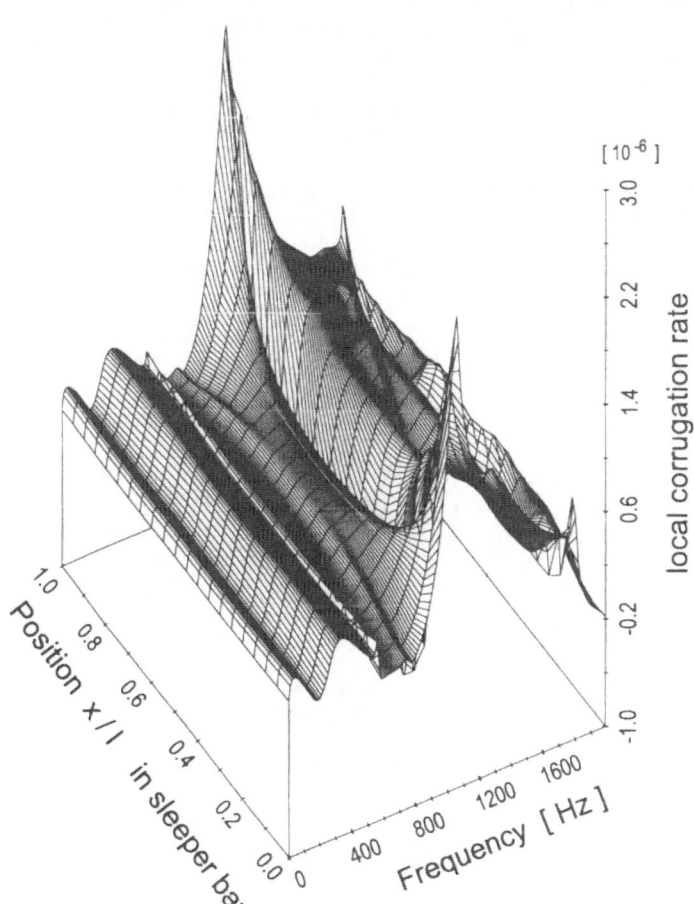

Figure 45. Local corrugation growth rate over a sleeper bay (from Hempelmann, 1994)

5   Numerical results, consequences and open questions [5]

Determination of corrugation pattern. Numerical results of linearized rail corrugation analysis. Corrugation growth rate and corrugation pattern. Comparison with results of non-linear investigations. Discussion of other possible long term mechanisms. Practical consequences and open questions.

5.1   Determination of corrugation pattern

It has been shown at the end of the last section that the *corrugation growth rate* is a local quantity, i. e., it depends on the position of the wheel in the sleeper bay. It is therefore necessary to describe corrugation more *globally*. We shall call the result of such a global description of corrugation the *corrugation pattern*. The first determination of corrugation pattern has been performed by Hiss and has been published in Hempelmann et al. (1991). A simpler possibility has been proposed in Müller (1998).

We assume that the development of the profile is governed by the solution of the differential eq. (74). For reasons of simplicity we only regard vertical height profile irregularities and restrict to the fluctuating part. Then eq. (72) reads

$$\frac{k_0}{\rho}\Delta w_{fric}(x) + \frac{\partial \Delta z_w(x)}{\partial n} = 0. \tag{76}$$

The height fluctuation (Δz_w) is proportional to the frictional work Δw_{fric} per unit length.

After Δn passages the profile has changed slightly. The new profile is now the input of the differential eq. (76). The profile after $N \times \Delta n$ passages is calculated by solving eq. (76) N times. This must be explained in more detail.

We assume that the initial profile irregularity is periodic with respect to the sleeper distance L_S. Then a given vector of initial profile amplitudes $\Delta z_w(x)$ can be Fourier analyzed as

$$\Delta z_w(x) = \sum_{j=-\infty}^{\infty} \Delta z_j e^{ij\frac{2\pi}{L_S}x}, \tag{77}$$

where

$$\Delta z_j = \frac{1}{L_S} \int_0^{L_S} \Delta z_w(x) e^{-ij\frac{2\pi}{L_S}x} dx. \tag{78}$$

Every single Fourier part Δz_j is an input for the high frequency system dynamics. The differential eq. (76) must be solved at **every** position x within the sleeper bay for **all** wavelengths $L_{Sj} = L_S/j$. The equation (76) can be written as

$$\frac{k_0}{\rho}w_{\Delta z}(L_{Sj}, x)\Delta z_j + \frac{\partial \Delta z_j}{\partial n} = 0.$$

---

[5] The section is mainly based on Hempelmann (1994) and on Müller (1998).

$w_{\Delta z}$ is a proportionality factor which equals the first diagonal part of the proportionality matrix $\mathbf{W}_{\Delta z}$ which has been introduced in eq. (74). At a position x the solution for a single Fourier part $\Delta z_j(L_{Sj}, x)$ after Δn wheel passages is

$$\Delta z_j(\Delta n, x) = \Delta z(0)\, e^{\lambda_j(L_{Sj}, x)\Delta n},$$

where $\lambda_j(L_{Sj}, x)$ is the local corrugation growth rate and $\Delta z_j(0) = \Delta z_j$ is the Fourier coefficient of the initial profile.

The profile after Δn passages can be determined as

$$\Delta z_w(x, \Delta n) = \sum_{j=-\infty}^{\infty} \Delta z(\Delta n, x) e^{-i\frac{2\pi}{L_{Sj}}x} = \Delta z(0)\, e^{\lambda_j(L_{Sj}, x)\Delta n}\, e^{-i\frac{2\pi}{L_{Sj}}x}, \qquad (79)$$

which can be Fourier analyzed again and used as a new input.

The procedure is explained in Fig. 46.

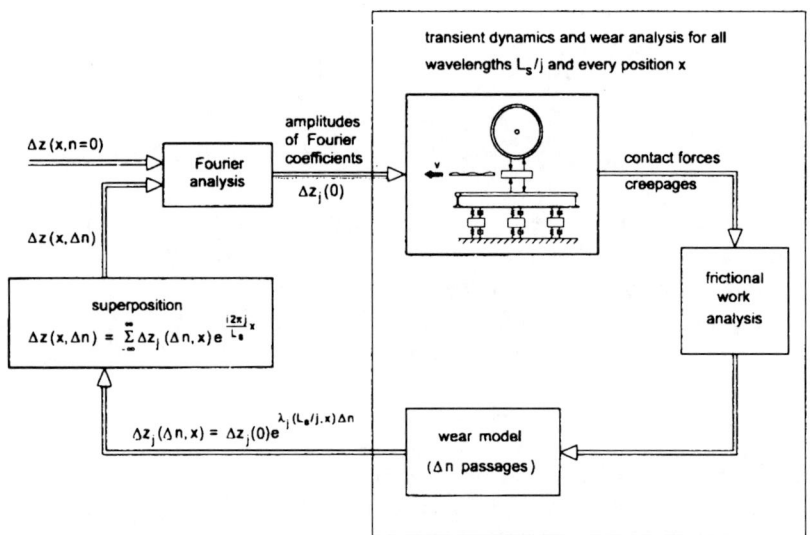

Figure 46. Procedure for the determination of corrugation pattern

5.2   Numerical results for corrugation growth rates

Now some numerical results of the long term corrugation process will be presented and discussed. Section 5.2 investigates how contact mechanical effects affect fluctuating wear. In section 5.2 the additional effects of structural dynamics will be considered.

Analysis of contact mechanical effects on fluctuating wear The *corrugation growth rates* (CGRs) $\text{Re}(\lambda_i)$ in eq. (79) govern the process of fluctuating wear on the rail. As we are first interested in the influence of contact mechanical effects on the wear, the dynamic motion of wheel and rail is omitted. In this case the CGRs are independent of the position within the sleeper bay and the vehicle speed. First, we confine ourselves to the simplified model, where only wear behaviour of the profile height is considered.

Corrugation growth rate (CGR) for the profile height As only wear of the height has been taken into account, we have got

$$\frac{\partial \Delta z_w(x)}{\partial n} = -\frac{k_0}{\rho} \Delta w_{fric}(x). \tag{80}$$

The solution for one Fourier part has been written

$$\Delta z_j(\Delta n, x) = \Delta z_j(0)e^{\lambda_j(Ls_j, x)\,\Delta n}. \tag{81}$$

In this simple case, i.e. only fluctuating height of the profile and no effects of structural mechanics are considered, an analytical expression can be given for the wear eigenvalue. (We omit the Index j).

$$\lambda = \underbrace{-\frac{k_0}{\rho}\left[\frac{\partial w_{w,fric}}{\partial v_\xi}\frac{1}{2r_0} - i\frac{\partial w_{w,fric}}{\partial v_\xi}\frac{2\pi}{L}\sin\lambda + \alpha_0 + \frac{\partial w_{w,fric}}{\partial d}\right]F_{amp}}_{\lambda^{(1)}}$$

$$\underbrace{-\frac{k_0}{\rho}\left[\frac{\partial w_{w,fric}}{\partial A}\left(\frac{1}{2}\left(\frac{2\pi}{L}\right)^2 + \frac{1}{4r_0^2}\right)\right]F_{amp}}_{\lambda^{(2)}} \tag{82}$$

Corrugation is promoted if $\text{Re}(\lambda)$ is positive.

Since at the moment we are interested in contact mechanical effects on wear, λ is divided into two parts. The second part $\lambda^{(2)}$ contains the influence of curvature in longitudinal direction, whereas all the other effects are comprised in the first term $\lambda^{(1)}$.

The influence of non-steady state contact mechanics now can be explained. In Fig. 47 the CGRs are compared with or without non-steady state effects. For reasons of clarity the amplitude filter F_{amp} has been neglected.It is evident that non-steady state contact mechanics dominates the wear process.

Most important is the second part $\lambda^{(2)}$. The frequency response of this second part is given in Fig. 48. Two aspects should be pointed out. First, down to 3 cm the fluctuation ΔA of the mean curvature in longitudinal direction causes an increase of the rate when the wavelength of the profile irregularity decreases. The phase of the fluctuating wear is such as to amplify incipient corrugation. The fluctuating wear is maximum towards the trough and minimum towards the peaks. However, for wavelengths smaller than 3 cm the CGR becomes smaller again because the fluctuating wear part $\frac{\partial w_{w,fric}}{\partial A}$ is increasingly out of phase with the fluctuation in the height of the profile. By this means existing

CGR now three CGRs are obtained, which are plotted in Fig. 49. The CGR Re$\{\lambda_3\}$ is the most important. The corresponding wear eigenvector is a combination of a fluctuation in profile height and in curvature. Qualitatively the curve is similar as in Fig. 47, however, the maximum is higher. The maximum appears at 3 cm. Again the formation of this maximum is an interaction of two effects, as already explained.

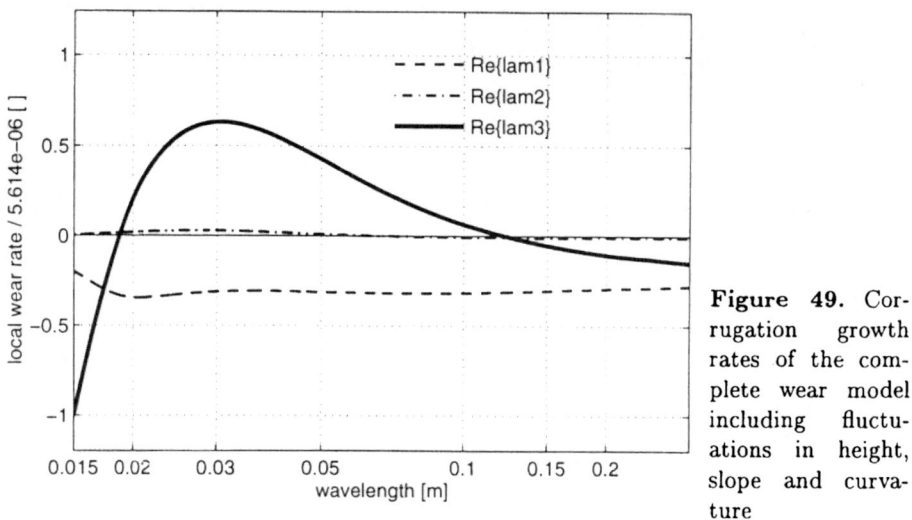

Figure 49. Corrugation growth rates of the complete wear model including fluctuations in height, slope and curvature

Influence of reference creep: For the investigations shown in Fig. 47 a reference state with lateral as well as longitudinal constant reference creep ($\nu_{\xi 0} = \nu_{\eta 0} = 0.15\%$) has been assumed. In Fig. 50 it can be seen, how the CGR changes when the reference creepages change. If either the constant longitudinal or lateral reference creepage is zero, then the maximum is smaller and the wavelength band where profile irregularities are amplified is larger. If, however, there is no constant longitudinal or lateral reference creep (but only spin creep), then the CGR is extremely small.

Influence of normal forces: Fig. 51 the CGR Re$\{\lambda_3\}$ is plotted for various wheel loads $N_{\zeta 0}$. If the wheel load becomes smaller the curve is shifted to the left such that the maximum occurs for smaller wavelengths. It is therefore not possible to state clearly whether the CGR is decreased or increased by an increase in wheel loads.

Influence of conformity: When the conformity of transverse wheel profile with the profile of the rail head shifts, the CGR Re$\{\lambda_3\}$ is also shifted. This is demonstrated in Fig. 52, where Re$\{\lambda_3\}$ is plotted for the reference data if the lateral wheel radius R_0^{wh} is varied. The resulting conformities can be described by a conformity factor K which is defined as

$$K = \frac{R_0^{wh}}{R_0^{wh} - R_0^{r}}. \tag{83}$$

Fig. 52 shows that the curve of the CGR Re$\{\lambda_3\}$ is shifted to shorter wavelengths when conformity increases.

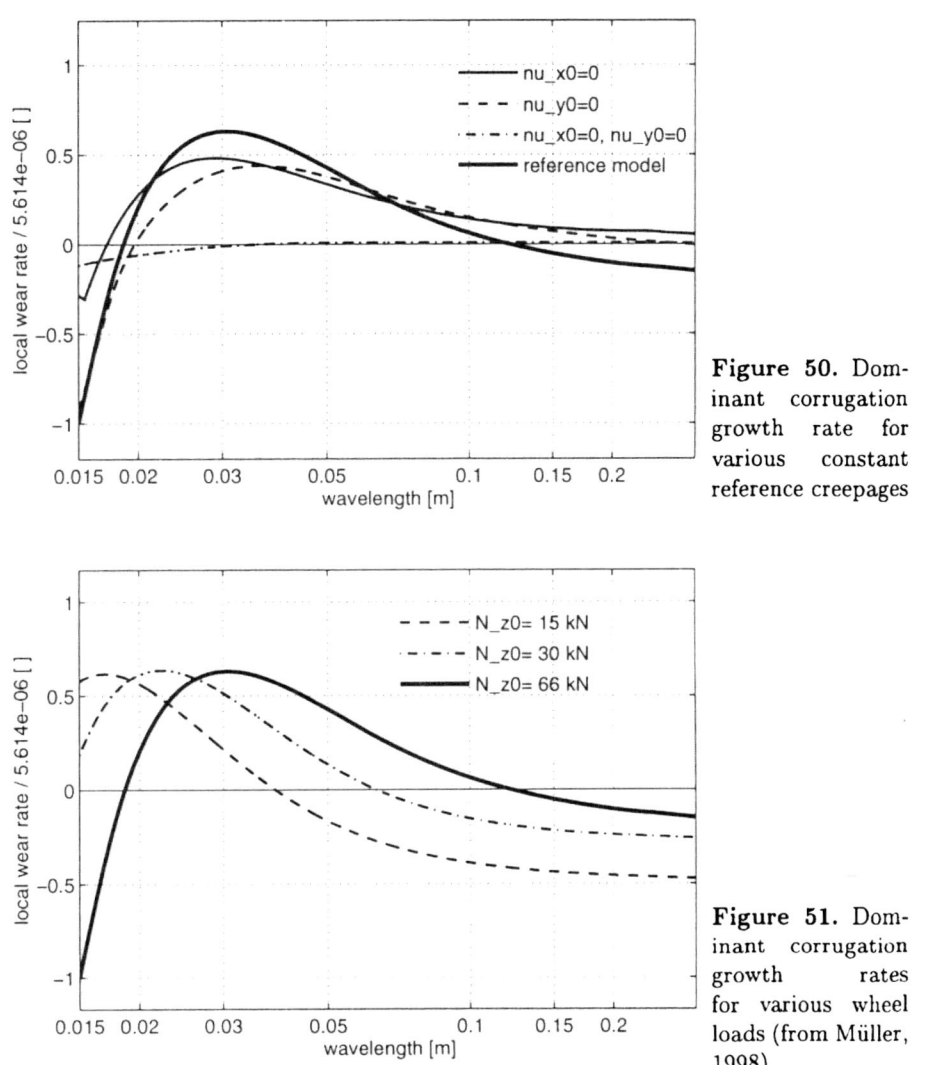

Figure 50. Dominant corrugation growth rate for various constant reference creepages

Figure 51. Dominant corrugation growth rates for various wheel loads (from Müller, 1998)

It should be mentioned in this connection that any prediction about the behaviour of profile irregularities with wavelengths in the order of the longitudinal length of the contact area are questionable, as the elliptical contact patch becomes a poor approximation for this wavelength range. A more sophisticated contact model would be helpful which can take non-Hertzian contact into account.

Influence of structural dynamics on local corrugation growth rates When structural dynamics of the rail are taken into account, then the CGRs depend on the position within the sleeper bay and on the vehicle speed. Therefore they are now called *local corrugation growth rates (LCGR)*.

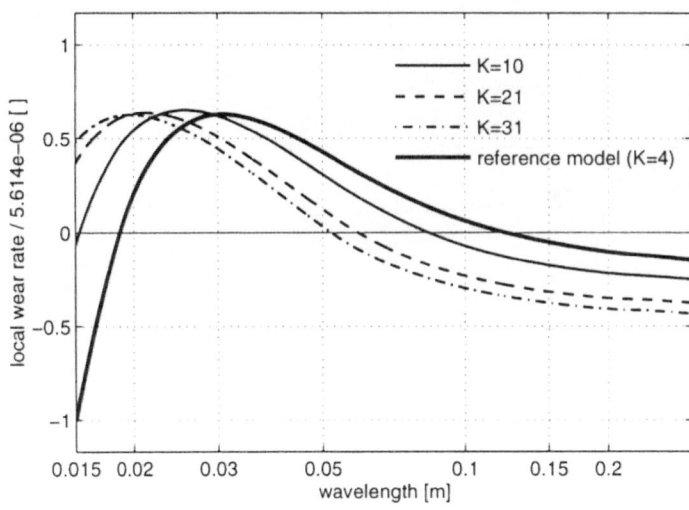

Figure 52. Dominant corrugation growth rate $\text{Re}\{\lambda_3\}$ for various conformities (from Müller, 1998

Solving the eigenvalue problem of eq. (74) gives three local wear rates which are plotted in Fig. 53 for the reference data. Structural dynamics has evidently changed the shape of the curve. The maximum of $\text{Re}\{\lambda_3\}$ at approximately 2.8 cm is caused by the minimum of the vertical direct receptance of the rail at 1100 Hz. The local minimum of $\text{Re}\{\lambda_3\}$ near 5 cm corresponds to the high lateral receptance at approximately 550 Hz.

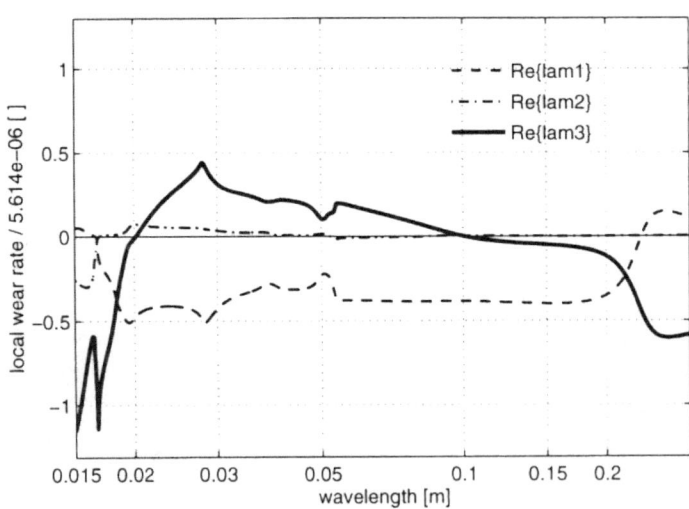

Figure 53. Local corrugation growth rate at a sleeper for $v_m = 30$ m/s (simplified structural dynamics)

Nevertheless the characteristic feature remains the same. In both cases we have one dominant local wear eigenvalue which promotes corrugation with a wavelength between 2 and 10 cm.

Structural dynamics effects depend on the frequency f:

$$f = \frac{v_m}{L} = \frac{\text{vehicle speed}}{\text{wavelength of the profile irregularity}}$$

Since corrugation is only promoted if one of the LCGRs is positive, we restrict ourselves to the dominant LCGR Re$\{\lambda_3\}$.

In Fig. 54 the dominant LCGRs are shown for $v_0 = 30$ m/s and $v_0 = 50$ m/s. For the thick line the wheel is near the sleeper, while the wheel is midspan for the thin line. The dashed curve is the CGR when structural dynamics of the rail are omitted. To explain the differences it is helpful to look at the dashed curve and at the receptances of the rail which are again shown in Fig. 55. The dashed line has a maximum between 2.4 and 5 cm. When a wheel-set is rolling over profile irregularities, a vehicle speed of 30 m/s leads to an excitation between 750 – 1200Hz and for 50 m/s between 1250 – 2000 Hz.

- **30 m/s, at sleeper (thick line)**: The maximum at 2.8 cm results from the anti-resonance of the first vertical pinned-pinned-mode near 1100 Hz. The minimum at 5 cm corresponds to the minimum of the lateral direct receptance below 550 Hz.
- **50 m/s, midspan (thin line)**: At midspan the rail is laterally very soft near 1600 and 1800 Hz. This leads to two maxima of the LCGR at approximately 2.7 and 3.2 cm. Even the maximum of the lateral pinned-pinned-mode at 550 Hz is visible as a sharp peak of LCGR at 9 cm.

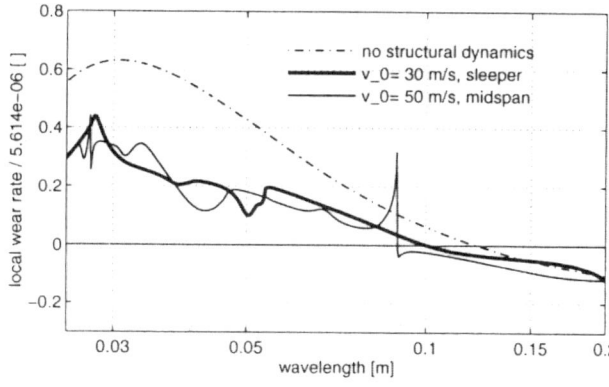

Figure 54. LCGRs Re$\{\lambda_3\}$ over a sleeper for $v_0 = 30$ m/s and at midspan for $v_m = 50$ m/s

In the lower part of Fig. 55 again the corrugation growth rates of contact mechanical effects are given, now depending on the frequency. The open question of Section 4 now can be answered.

Structural dynamics provide frequency constant mechanisms, however contact mechanical effects act as a filter which only allows for corrugation growth in a certain frequency range corresponding to wavelengths between 2 and 10 cm.

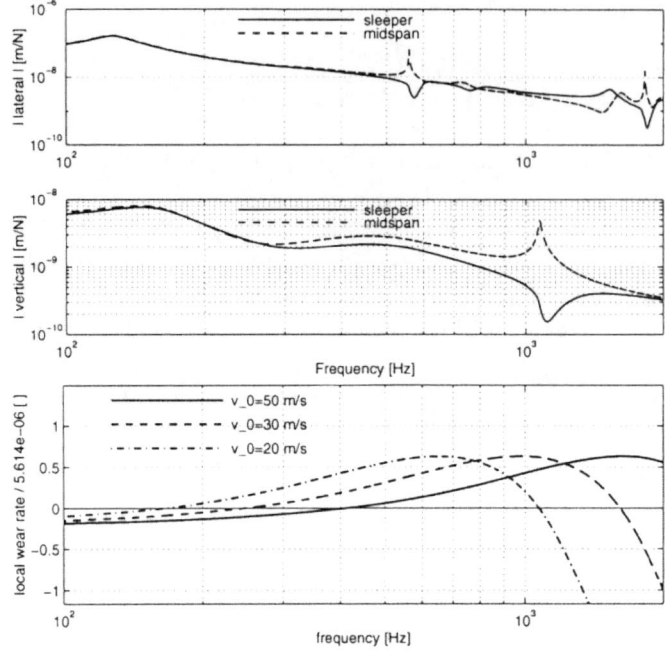

Figure 55. Lateral (a) and vertical (b) direct receptance of the rail and CGR for different vehicle speed (c)

5.3   Development of corrugation pattern

In the following some numerical results for the profile development on the rail are presented if system parameters are varied. Instead of local corrugation growth rates, *corrugation pattern* are now determined. An initial profile irregularity is assumed which is described by an artificial amplitude spectrum. From this spectral density a discrete amplitude spectrum can be developed. The corresponding phase spectrum is assumed to be randomly distributed. When the profile irregularity is assumed to be periodic with respect to the sleeper distance, it can be Fourier analyzed with eqs. (77) and (78). The structural dynamics of the wheel-set is initially omitted.

Effect of vehicle speed In Figs. 56 the wear patterns within a sleeper bay are presented for $v_0 = 50$ m/s and $v_0 = 20$ m/s for 750 000 wheel-set passages. The width of the running band is normalized with respect to the lateral semi-axis of the contact ellipse. A better survey is obtained looking at the *discrete amplitude spectra* of the profile height $\Delta z(x, y = 0)$, see Fig. 57. The irregularity of these amplitude spectra to some extend is traced back to the fact that the highest wavelength is only the sleeper distance $L_s = 60$ cm.

For both vehicle speeds the maxima of the amplitude spectra are different:

- For $v_0 = 50$m/s the maximum values of the amplitude spectra are in the range of 2.5 – 3.5 cm. This corresponds to lateral resonances of the rail at 1600 and 1800 Hz, see Fig. 29. The vertical pinned-pinned-mode anti-resonance at 1100 Hz results in a third maximum near 5 cm.

- For $v_0 = 20$m/s maximum amplitudes are found between 3 and 8 cm which correspond to frequencies between 660 and 300 Hz. Near 300 Hz there is a minimum of vertical direct receptance (see Fig. 55), whereas at 550 Hz there is a resonance of the lateral direct receptance. The effect of the pinned-pinned-mode anti-resonance cannot be seen, as it is completely suppressed by the filter effects of contact mechanics.

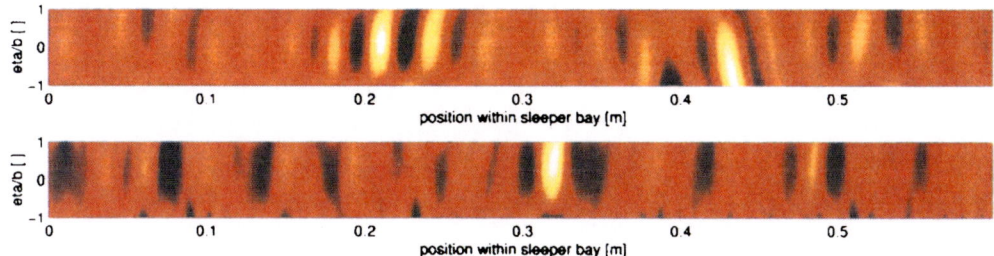

Figure 56. Wear pattern within a sleeper bay after 750 000 wheel-set passages ($v_0 = 50$ m/s and $v_0 = 20$ m/s)

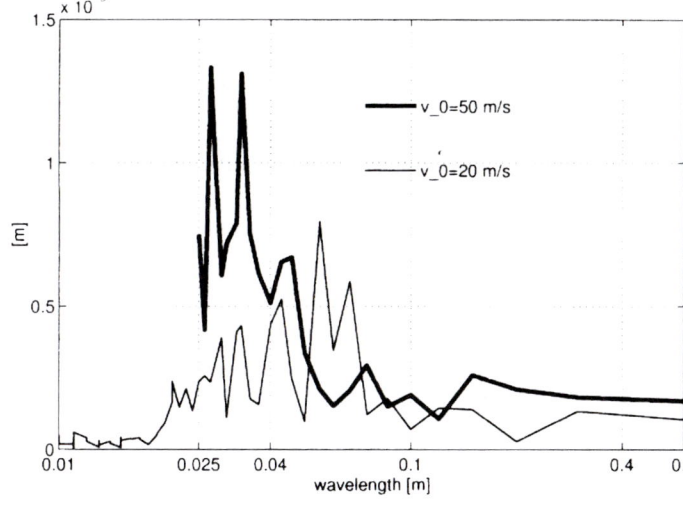

Figure 57. Amplitude curve of $\Delta z(x, y = 0)$ after 750 000 wheel passages for $v_0 = 50$ m/s and $v_0 = 20$ m/s (from Müller, 1998)

Effect of conformity The CGRs $\text{Re}\{\lambda_3\}$ for different conformities have already been shown in Fig. 52. For two different conformities $K = 4$ and $K = 21$ the amplitude spectra are considered in Fig. 58. For K=4 two pronounced maxima are found (2.7 cm corresponding to 1800 Hz, 3 cm corresponding to 1600 Hz) and a smaller one (4.6 cm corresponding to 1100 Hz). Since for $K = 21$ the CGR Re$\{\lambda_3\}$ is shifted to smaller wavelengths, effects at frequencies smaller than 1700 Hz are filtered out. The corrugation process is now completely dominated by the lateral rail receptance at 1800 Hz, which can be seen clearly in the corrugation pattern.

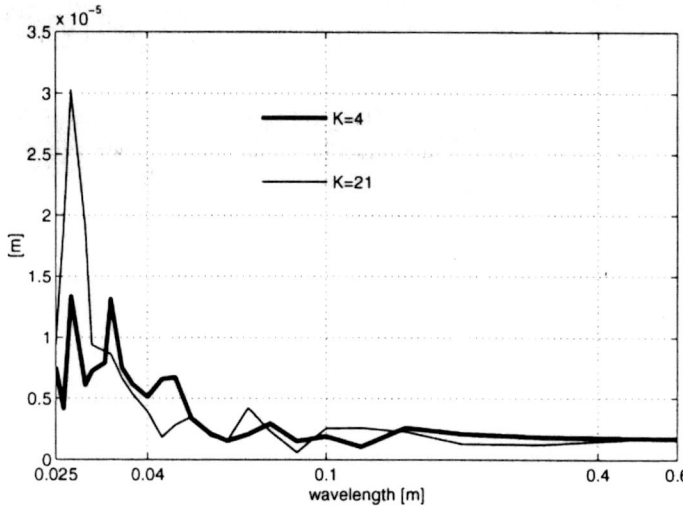

Figure 58. Amplitude spectra of $\Delta z(\eta = 0)$ after 750 000 wheel passages for two different conformities ($v_0 = 50$ m/s) (from Müller, 1998)

Effect of pad stiffness The effect of pad stiffness has been carefully investigated by Hempelmann (1994). We shall only give the receptances for different pad stiffness (the damping has been assumed to be constant). Only the amplitude spectra near the sleeper and at midspan are shown (Fig. 59). The reference value of the pad stiffness has been assumed as 280 MN/m. It has been reduced to 70 MN/m and increased to 700 and 1120 MN/m. It can be seen that for extremely stiff pads the second maximum is shifted to higher frequencies and a second minimum (anti-resonance) is found at 400 or 500 Hz respectively.

It should be expected that this second anti-resonance results in distinct corrugations with wavelengths near $L = v_0/f$. Such a situation has been found on a former DR (Deutsche Reichsbahn) track near Calau south of Berlin, Fig. 60. The vehicle speed was $v_0 = 100$ km/h (27 m/s). The measured receptances, which were used for the determination of track data, have already been shown in Fig. 32. The pronounced anti-resonance in the measured vertical direct receptances near 400 Hz corresponds to the corrugation wavelength of 8 cm. The measured corrugation maximum near 30 cm is questionable, as a straight edge of only 1 m length has been used for measurements. It could also be possible that effects of parametric excitation not included in the model are

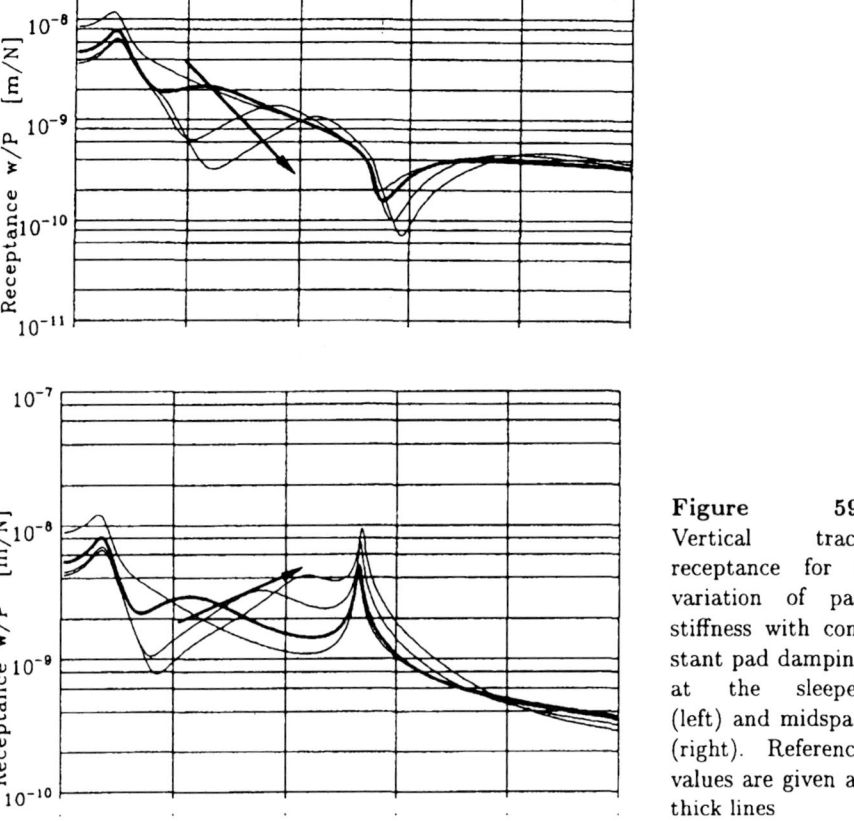

Figure 59. Vertical track receptance for a variation of pad stiffness with constant pad damping at the sleeper (left) and midspan (right). Reference values are given as thick lines

responsible for this peak. Again no effect of the vertical pinned-pinned-mode can be seen, as the corresponding wavelength is suppressed by the effects of contact mechanics.

5.4   Influence of elastic wheel-set

Only one example will be shown where the wheel-set dynamics has been included. An elastic wheel-set (DB type 92) has been included in the analysis. In Fig. 61 lateral and vertical direct receptances of a UIC 60 rail and a type 92 wheel-set are shown. Looking first at the vertical receptances it can be seen that mostly the wheel-set is stiffer. Therefore the rail receptance will dominate. The situation is different for the lateral direct receptances. Here in certain frequency ranges the wheel receptances are evidently higher. It has to be expected that the wheelset dominates the corrugation behaviour.

This can be seen in Fig. 62. As soon as the elastic wheel-set is included in the analysis two pronounced maxima are observed in the amplitude spectrum of corrugations. The

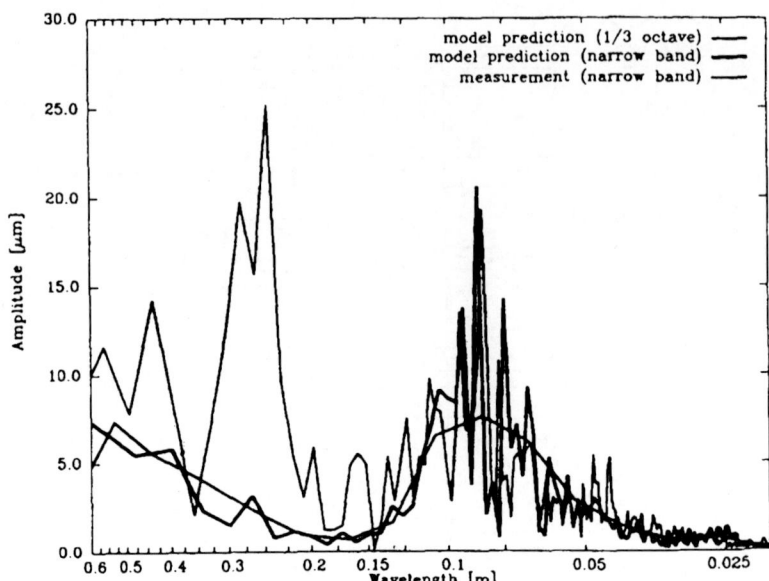

Figure 60. Comparison of measured and predicted corrugation amplitude spectra (from Hempelmann, 1994). (Track data have been taken from the measurements shown in Fig. 32)

reason are the high values of lateral rail receptances near 1000 Hz and 350 Hz which extend over a comparatively broad frequency range. Nevertheless, the wheel-set motion can only be expected to have a strong influence on the corrugation process if the service conditions and the vehicle data only vary slightly. Most of the passing wheel-sets must have similar dynamic characteristics and the same speed. Otherwise, amplified profile irregularities caused by one wheel-set type would be likely to be worn by another.

5.5   Comparison with non-linear analysis

The basic assumption of the corrugation analysis is linearity. Few authors have tried to find out whether or not the corrugation initiation is influenced by non-linearities. An interesting result can be found in Ripke (1995), see Fig. 63. The fluctuation of the normal force is shown for a wheel rolling with 40 m/s over a corrugated rail ($L = 3.3$ cm and $\Delta\hat{z} = 25$ μm. Comparing the results for non-linear (above) and linearized contact mechanics it can be seen that a lift-off of the rail near the sleeper and much higher normal forces are observed when non-linear contact mechanical relations are used. The consequence is that linear analysis would be allowed for a corrugation depth up to only 40 μm!

A detailed nonlinear analysis has also been performed by Igeland and Ilias (1997) and by Igeland and Ilias (1996, 1996 and 1997). Again the importance of nonlinear

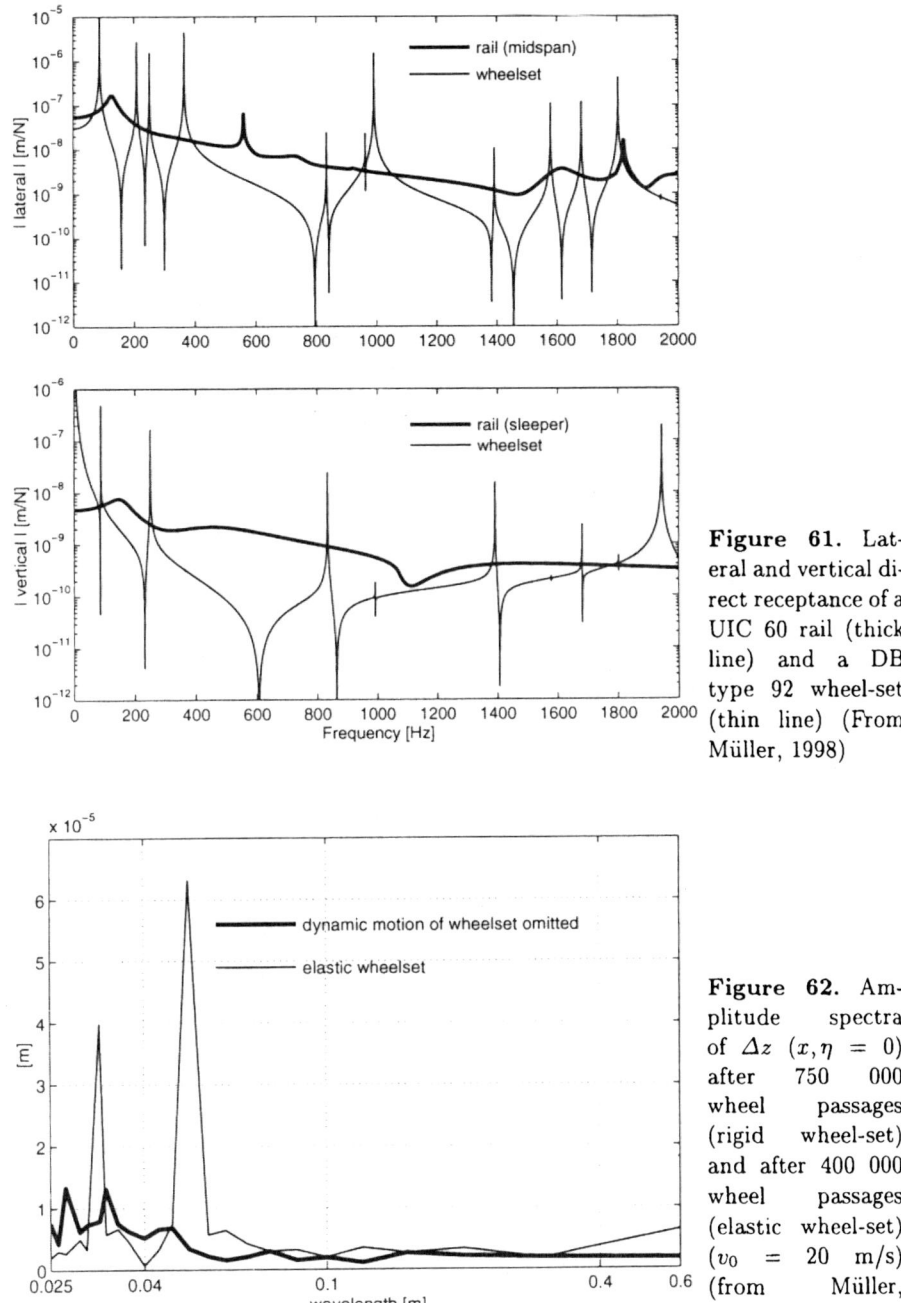

Figure 61. Lateral and vertical direct receptance of a UIC 60 rail (thick line) and a DB type 92 wheel-set (thin line) (From Müller, 1998)

Figure 62. Amplitude spectra of Δz $(x, \eta = 0)$ after 750 000 wheel passages (rigid wheel-set) and after 400 000 wheel passages (elastic wheel-set) $(v_0 = 20 \text{ m/s})$ (from Müller, 1998)

investigations has been confirmed, not only concerning a possible take-off of the wheel for moderate corrugation amplitudes but also a shift of the contact point.

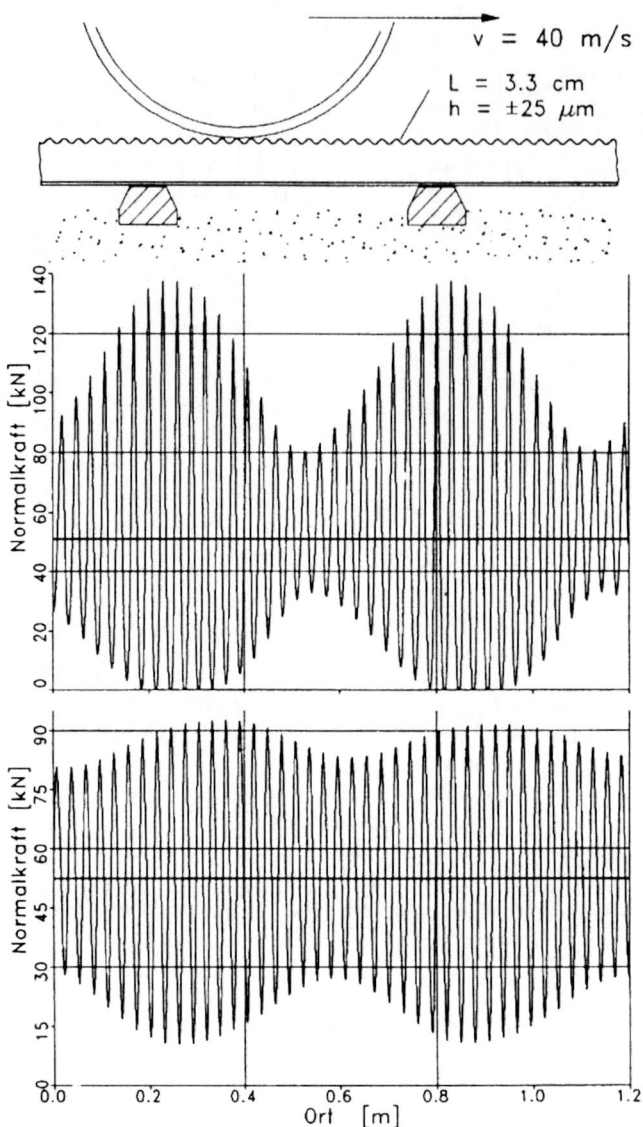

Figure 63. Results of time-step integration for a wheel-set running on corrugated rails ($v = 40$ m/s, $L = 3.3$ cm, $\Delta\hat{z} = 25$ μm; from Ripke, 1995)

At the moment an obvious disadvantage of non-linear analysis is that they are relatively time consuming and that they are restricted to vertical and longitudinal structural dynamics of the discretely supported rails.

5.6 Other long-term mechanisms

The main disadvantage of our model is that wear is the only long-term mechanism. The considerations in Section 1 suggest at least 3 other mechanisms:

- *Plastic deformation* should be considered as a mechanism which is responsible for geometric modifications of the running band. It is likely that plastic deformations are higher at the peaks. The analysis, however, is extremely troublesome. Among others one has to look for plastic shake-down phenomena as discussed by Johnson, now however for corrugated rails. Realistic data of rail steel has to be taken into account. Not very much is known about this (see e.g. McDowell, 1995).
- *Long-term variation of material* is another possible effect. Plastic deformation and the formation of WEL are responsible for modified material characteristics. Increased hardness of WEL by a factor of 4 (cf. Section 1) also means that the wear resistance is higher. This does not matter as long as the surface layer is the same on the whole running band. However, as WEL is worn away in the trough and is intensified on the hills, it has to be assumed that the wear resistance of parts with different surface layers is also different.
- A third effect is the *long term variation of micro-roughness* of corrugated running bands. Not very much is known about this phenomenon. The measurements of Baumann (1998) have shown that the R_a-values in the troughs are twice the values at the peaks! Surprisingly this affects high frequency dynamics: It has been shown (Knothe and Theiler, 1997) that micro-roughness of the rail surface acts like an elastic layer on the rail surface. Such a layer has the same effect as a lower Young's modulus. In consequence, slope of the creep-force creep curve should be lower in the troughs by a factor of 2 compared with the slope on the hills.

One of the problems concerning different long-term effects is that the dissipated energy now has to be split up into two or more parts. Nearly nothing is known about how this has to be done.

5.7 Conclusions

Theoretical conclusions

1. The process of initiation of corrugations is a feed-back loop between high-frequency dynamics of the wheel rolling over the rail and long-term wear behaviour.
2. If effects of structural mechanics are omitted and only contact mechanical effects are considered, then a wavelength constant mechanism is obtained. An increase of small initial irregularities is only possible in a wavelength range between 2 and 10 cm. This range can be shifted slightly by parameters influencing contact mechanics like normal force or conformity.
3. Within this band effects of structural mechanics are responsible for an extreme increase of irregularities at certain wavelengths depending on vehicle speed. This, is a frequency constant mechanism.

4. In Grassie's and Kalousek's classification of rail corrugation (see Section 1) type 5 (booted-sleepers-corrugation) and type 6 (roaring-rail.corrugations) should be summarized as short pitch corrugation. Type 4 (rutting corrugation), however, is completely different as the dynamics are governed by stick-slip phenomena.
5. Most important effects of structural dynamics are vertical anti-resonances of the rail (track) which result in high normal force fluctuations or lateral resonances of the rail resulting in high creepage fluctuations.
6. It seems that in most cases track parameters and service conditions are more responsible for rapid corrugation growths than wheel parameters. Only in the case of uniform traffic does wheel-set data become more important.

Some practical conclusions

1. For existing track, predictions and proposals of remedies are extremely difficult, as all informations on track data and traffic conditions must be available.
2. Discrete support is the main cause of initiation of rail corrugations on high speed lines, as pinned-pinned-modes provide low values of vertical receptances and high values of lateral receptances.
3. Pad stiffness in vertical direction should be as low as possible. It should, however, be mentioned that for noise abatement high pad stiffness is desirable.
4. High reference creepages are another cause of short pitch rail corrugation. Misalignment of wheel-sets, extensive wheel-set hunting and poor track quality should be avoided.
5. High initial irregularities in the frequency band susceptible for corrugations should be avoided. Grinding of newly laid rails is the most common measure.
6. If corrugated rails are ground attention has to be paid to some rules: It is not enough to remove geometrical irregularities. Fluctuations in material behaviour like WEL have to be removed too. If possible, residual stresses should also be removed.

5.8 Open questions

- The *linear models* of structural mechanics and contact mechanics are more or less sufficient for the investigation of corrugation initiation.
- The behaviour of rail pads as one of the most important track component is insufficiently described up to now.
- The influence of a *moving wheel-set model* instead of a moving excitation model has to be checked. Probably an approximate consideration of *parametric excitation* due to moving wheel-set on discretely supported rails is necessary.
- Up to which corrugation depth is it allowed to neglect completely *nonlinear effects*?
- More important is the consideration of *other long-term effects* besides wear as mentioned in 5.6. This is a difficult task as usually no simple physical laws are available and the distribution of dissipated energy to different long-term mechanisms is not clear.
- Finally the continuation of *validation of the theory* presented in this lecture is the most challenging task.

References

Baumann, G., Fecht, H., and Liebelt, S. (1996a). Formation of white etching layers on the rail treads. *Wear* 191:133–140.

Baumann, G., Grohmann, H.-D., and Knothe, K. (1996b). Wirkungsketten bei der Ausbildung kurzwelliger Riffeln auf Schienenlaufflächen. *ETR* 45(12):792–798.

Baumann, G. (1998). *Untersuchungen zu Gefügestrukturen und Eigenschaften der "Weißen Schichten" auf verriffelten Schienenlaufflächen*. Dissertation, TU Berlin.

Bhaskar, A., Johnson, K., Wood, G., and Woodhouse, J. (1997a). Wheel-rail dynamics with closely conformal contact. Part 1: Dynamic modelling and stability analysis. *Proc. Instn. Mech. Engrs.* 211(F):11–26.

Bhaskar, A., Johnson, K., and Woodhouse, J. (1997b). Wheel-rail dynamics with closely conformal contact. Part 2: Forced response, results and conclusions. *Proc. Instn. Mech. Engrs.* 211(F):27–40.

Böhmer, A., Klimpel, T., and Knothe, K. (2000). Dynamik und Festigkeit von gummiringgefederten Radreifen. *ZEV+DET Glasers Annalen* 124: (accepted for publication).

Clark, R., Scott, G., and Poole, W. (1988). Short wave corrugations—An explanation based on stick–slip vibrations. In *Applied Mechanics Rail Transportation Symposium, AMD Vol. 96, RTD Vol. 2*, 141–148. ASME.

Engl, A., Meinke, P., and Stöckel, H. (1983). Corrugations on bearers as effects of short time dynamics investigated in the long term process. In Knothe, K., and Gasch, R., eds., *Rail Corrugations. Papers presented at the Symposium on Rail Corrugation Problems held at Berlin in June 1983*, ILR-Bericht, Nr. 56, 41–70. Berlin: TU Berlin, Institut für Luft- und Raumfahrt.

Frederick, C. O., and Bugden, W. G. (1983). Corrugation research on British Rail. In Knothe, K., and Gasch, R., eds., *Rail Corrugations. Papers presented at the Symposium on Rail Corrugation Problems held at Berlin in June 1983*, ILR-Bericht, Nr. 56, 7–33. Berlin: TU Berlin, Institut für Luft- und Raumfahrt.

Frederick, C. O., and Sinclair, J. C. (1992). A rail corrugation theory which allows for contact patch size. In Knothe, K., ed., *Rail Corrugations. Papers presented at the Symposium on Rail Corrugation Problems held at Berlin on April 17, 1991*, ILR-Bericht Nr.59, 1–27. Berlin: TU Berlin.

Frederick, C. (1987). A rail corrugation theory. In *Proc. of the Second International Symposium on Contact Mechanics and Wear of Rail/Wheel Systems held at Kingston/RI, July 1986*, 181–211. Waterloo/Ontario: University of Waterloo Press.

Gasch, R., and Knothe, K. (1989). *Strukturdynamik, Band 2. Kontinua und ihre Diskretisierung*. Berlin e.a.: Springer.

Gasch, R., Groß-Thebing, A., Knothe, K., and Valdivia, A. (1983). Linear, self-excited vibrations as initiating mechanism of corrugations. In Knothe, K., and Gasch, R., eds., *Rail Corrugations. Papers presented at the Symposium on Rail Corrugation Problems held at Berlin in June 1983*, ILR-Bericht, Nr. 56, 207–230. Berlin: TU Berlin, Institut für Luft- und Raumfahrt.

Grassie, S., and Kalousek, J. (1993). Rail corrugations. Characteristics, causes, and treatments. *Journal of Rail and Rapid Transit, Proc. Instn. Mech. Engrs., part F* 207:57–68.

Grassie, S., Gregory, R., Harrison, D., and Johnson, K. (1982). The dynamic response of railway track to high frequency vertical excitation. *J. Mech. Engng. Sci.* 24(2):77–90.

Groß-Thebing, A. (1993). Lineare Modellierung des instationären Rollkontaktes von Rad und Schiene. In *VDI Fortschritt-Berichte (zugleich Dissertation TU Berlin)*, Reihe 12, Nr. 199. Düsseldorf: VDI-Verlag.

Hempelmann, K., and Knothe, K. (1989). Eigenschwingungsberechnungen von Eisenbahn-radsätzen mit optimalen Ansatzfunktionen. In *VDI Fortschritt–Berichte*, Reihe 11, Nr. 114. Düsseldorf: VDI–Verlag.

Hempelmann, K., Hiss, F., Knothe, K., and Ripke, B. (1991). The formation of wear patterns on rail tread. *Wear* 144:179–195.

Hempelmann, K. (1994). Short pitch corrugation on railway rails – A linear model for prediction. In *VDI Fortschritt–Berichte (zugleich Dissertation TU Berlin)*, Reihe 12, Nr. 231. Düsseldorf: VDI–Verlag.

Igeland, A., and Ilias, H. (1996). Rail head wear calculations based on high frequency wheelset/track interaction - a comparison between different models. In Zobory, I., ed., *Proc. of the 2nd Mini Conference on Contact Mechanics and Wear of Rail/Wheel Systems, Budapest, 29-31 July 1996*, volume , 304-314.

Igeland, A., and Ilias, H. (1997). Rail head corrugation growth predictions based on non-linear high frequency vehicle/track interaction. *WEAR* 213:90–97.

Ilias, H., and Knothe, K. (1992). Ein diskret-kontinuierliches Gleismodell unter dem Einfluß schnell bewegter, harmonisch schwankender Wanderlasten. In *Fortschritt–Berichte VDI, Reihe 12, Nr. 177.* Düsseldorf: VDI–Verlag.

Ilias, H. (1996). Nichtlinear Wechselwirkungen von Radsatz und Gleis beim Überrollen von Profilstörungen. In *VDI Fortschritt–Berichte (zugleich Dissertation TU Berlin)*, Reihe 12, Nr. 297. Düsseldorf: VDI–Verlag.

Ilias, H. (1998/99). The influence of railpad stiffness on wheelset/track interaction and corrugation growth . *Accepted for publication in Journal of Sound and Vibration.*

Jezequel. L. (1981). Response of periodic systems to a moving load. *J. Appl. Mech.* 48:613–618.

Knothe, K., and Grassie, S. (1993). Modelling of railway track and vehicle/track interaction at high frequencies. *Vehicle System Dynamics* 22(3–4):209–262.

Knothe, K., and Liebelt, S. (1995). Determination of temperatures for sliding contact with application for wheel-rail systems. *Wear* 189:91–99.

Knothe, K., Willner,F., and Strzyżakowski, Z. (1994) Rail vibrations in the high frequency range. *J. Sound Vibr.*, 169(1):111–123, 1994.

Knothe, K., and Theiler, A. (1997). Normal and tangential contact problem with rough surfaces. In Zobory, I., ed., *Proc. of the 2nd Mini Conference on Contact Mechanics and Wear of Rail/Wheel Systems held at Budapest in August 1996*, 34 43. Budapest: Technical University of Budapest.

Knothe, K., Yu, M., and Ilias, H. (1997). Studie zu Eigenschaften von Zwischenlagen. Studie im Auftrag der DB AG, Institut für Luft- und Raumfahrt, TU Berlin.

Knothe, K. (1998). Modulare Behandlung des Rollkontakts von Rad und Schiene. In *Böhm, F. und Knothe, K. (Hrsg.), Hochfrequenter Rollkontakt der Fahrzeugräder.* WILEY-VCH. 39–57.

Knothe, K. (1999). Gleisdynamik und Wechselwirkung zwischen Fahrzeug und Fahrweg. *ZAMM* 79(11):723–737.

Koch, H. W. (1932). Messungen von Schwingungen am Eisenbahnoberbau. *Organ für die Fortschritte des Eisenbahnwesens* 87(21):389–399.

Kose, K. (1998). Berechnung der Eigenschwingungen und Rezeptanzen von Eisenbahnradsätzen. In *Fortschritt-Berichte VDI, Reihe 12, Nr. 347 (zugleich Dissertation TU Berlin).* VDI-Verlag Düsseldorf.

Krause, H., and Poll, G. (1986). Wear of wheel–rail surfaces. *Wear* 113(1):103–122.

Lang, W., and Roth, G. (1993). Optimale Kraftschlußausnutzung bei Hochleistungs-Schienenfahrzeugen. *ETR* 42:61–66.

Mauer, L. (1988). *Die modulare Beschreibung des Rad/Schiene-Kontaktes im linearen Mehrkörperformalismus.* Dissertation, TU Berlin.

McDowell, D. (1995). Stress dependence of cyclic ratchetting behaviour of two rail steels. *International Journal of Plasticity* 11:397–421.

Meinke, P., and Morys, B. (1998). Entstehung und Verstärkung von Radunrundheiten bei hohen Fahrgeschwindigkeiten. In *Tagungsband zur Bahn-Bau '98, Verband Deutscher Eisenbahn-Ingenieure, 28.-30.10.98*, 110–126.

Meinke, P., and Szolc, T. (1995). On dynamics of rotating wheel-set/rail systems in medium frequency range. In Bogacz, R., and Popp, K., eds., *Dynamical Problems in Mechanical Systems. Proceedings of the 4th German -Polish Workshop, August 1995, Berlin*. Warszawa: Polska Akademia Nauk.

Mombrei, W. and Rode, W. (1998). Kenntnisse zu aktuellen Problemen am Eisenbahnrad. *Eisenbahningenieur* 49:50–53.

Morys, B., and Kuntze, H. (1997). Simulation analysis and active compensation of the out-of-round phenomena at wheels of high speed trains. In *Proceedings of the World Congress of Railway Research held at Florence/Italy, 16-19 November*, volume D, 95–105.

Morys, G. (1998). *Zur Entstehung und Verstärkung von Unrundheiten an Eisenbahnrädern bei hohen Geschwindigkeiten*. Ph.D. Dissertation, Universität Karlsruhe .

Müller, S., and Knothe, K. (1997). Stability of wheelset-track dynamics for high frequencies. *Archive of Applied Mechanics* 67:353–363.

Müller, S. (1998). Linearized Wheel-Rail Dynamics – Stability and Corrugation. In *Fortschritt-Berichte VDI (also Dissertation TU Berlin)*, Reihe 12, Nr. 368. Düsseldorf: VDI- Verlag.

Remington, P. (1976). Wheel-rail-noise, part I-IV. *J. Sound Vibr.* 46(3):359–451.

Ripke, B., and Knothe, K. (1991). Die unendlich lange Schiene auf diskreten Schwellen bei harmonischer Einzellasterregung. In *VDI Fortschritt-Berichte*, Reihe 11, Nr. 155. Düsseldorf: VDI-Verlag.

Ripke, B. (1992). Anpassung der Modellparameter eines Gleismodells an gemessene Gleisrezeptanzen. ILR–Mitt. 274, Institut für Luft- und Raumfahrt, TU Berlin.

Ripke, B. (1995). Hochfrequente Gleismodellierung und Simulation der Fahrzeug-Gleis-Dynamik unter Verwendung einer nichtlinearen Kontaktmechanik. In *VDI-Fortschritt-Berichte (zugleich Dissertation TU Berlin)*, Reihe 12, Nr. 249. Düsseldorf: VDI-Verlag.

Scholl, W. (1987). Darstellungen des Körperschalls in Platten durch Übertragungsmatrizen und Anwendung auf die Berechnung der Schwingungsformen von Eisenbahnschienen. In *VDI-Fortschritt-Berichte (zugleich Dissertation TU Berlin)*, Reihe 11, Nr. 93. Düsseldorf: VDI-Verlag.

Schwedler, J. (1882). Discussion on iron permanent way. In: Ch. Wood. Iron permanent way. *Minutes of Proc. of the Instn. of Civ. Engnrs., London* 67:95–118.

Tassilly, E., and Vincent, N. (1991). A linear model for the corrugation of rails. *J. Sound Vibr.* 150:25–45.

Umlauf, V. (1994). Simulation hochfrequenter Schienendynamik unter Verwendung eines FE-Übertragungsmatrizenverfahrens. Diplomarbeit, Technische Universität Berlin.

Valdivia, A. (1987). Die Wechselwirkung zwischen hochfrequenter Rad-Schiene-Dynamik und ungleichförmigem Verschleiß - ein lineares Modell (englischer Titel: The interaction between high-frequency wheel-rail dynamics and irregular rail wear - a linear model. In *Fortschritt-berichte VDI, Reihe 12, Nr. 93 (zugleich Dissertation TU Berlin*. Düsseldorf: VDI-Verlag.

Valdivia, A. (1988). Die Wechselwirkung zwischen hochfrequenter Rad-Schiene-Dynamik und ungleichförmigem Schienenverschleiß. Ein lineares Modell. In *VDI Fortschritt-Berichte (zugleich Dissertation TU Berlin*, Reihe 12, Nr. 93. Düsseldorf: VDI-Verlag.

Wickens, A. (1965). The dynamics of railway vehicles on straight track: fundamental consideration of lateral stability. *Proc. I. Mech. Engrs.* 180(3F):1–16.

Widmayer, H. (1983). Measurement on the corrugation trial track of DB. In Knothe, K., and Gasch, R., eds., *Rail Corrugations. Papers presented at the Symposium on Rail Corrugation Problems held at Berlin in June 1983*, ILR-Bericht, Nr. 56, 35–38. Berlin: TU Berlin, Institut für Luft- und Raumfahrt.

Wu, Y. (1997). Einfache Gleismodelle zur Simulation der mittel- und hochfrequenten Fahrzeug/Fahrweg-Dynamik. In *Fortschritt-Berichte VDI (zugleich Dissertation TU Berlin)*, Reihe 12, Nr. 325. Düsseldorf: VDI–Verlag.

MODELLING OF TYRE FORCE AND MOMENT GENERATION

H.B. Pacejka
Delft University of Technology, Delft, The Netherlands

Abstract. The paper gives an overview of different types of mathematical models describing tyre force and moment generating properties that may be used in vehicle dynamics simulation studies. First a physical model based on the brush concept is treated in detail. Then, an empirical model is discussed. This model that is known by the name ˙Magic Formula˙ gives a relatively good approximation of measured data. Finally, the non-steady-state behaviour is briefly discussed. The low frequency transient tyre performance is modelled by a first-order differential equation in which the relaxation length serves as parameter. The higher frequency behaviour is modelled by introducing the dynamics of the tyre belt.

1. Introduction

The performance of a tyre as a force and moment generating structure is a result of a combination of several aspects. Factors which concern the primary tasks of the tyre may be distinguished from factors which involve (often important) secondary effects. In Table. 1.1 these factors have been brought in matrix form. A further distinction is made between (quasi) steady-state and vibratory behaviour and, additionally, between symmetric (or in-plane) and anti-symmetric (or out-of-plane) aspects. The primary task factors appear in bold letters. The remaining factors are considered as secondary factors.

Table 1.1. Tyre factors.

	primary task functions and secondary effects		
	(quasi) steady state		transient / vibratory state
symmetric (in-plane)	**load carrying** **braking/driving** rolling resistance	radial deflection tangential slip and distortion	**cushioning** dynamic coupling natural vibrations
anti-symmetric (out-of-plane)	**cornering** pneumatic trail overturn. couple	lateral and spin/turnslip and distortion	phase lag destabilisation natural vibrations

input vector   *output vector*

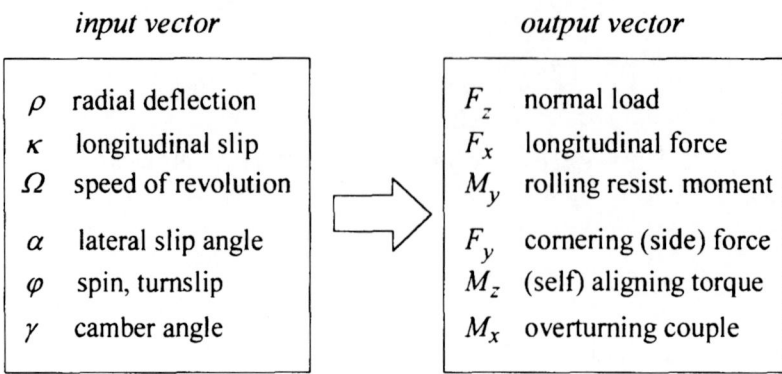

ρ	radial deflection		F_z	normal load
κ	longitudinal slip		F_x	longitudinal force
Ω	speed of revolution		M_y	rolling resist. moment
α	lateral slip angle		F_y	cornering (side) force
φ	spin, turnslip		M_z	(self) aligning torque
γ	camber angle		M_x	overturning couple

Fig. 1.1. Input/output quantities (road surface considered flat).

The primary requirements to transmit forces in the three perpendicular directions (F_x, F_y, F_z) and to cushion the vehicle against road irregularities involve secondary factors like lateral and longitudinal distortions and slip. Although regarded as secondary phenomena, some of the quantities involved are crucial for the generation of the deformations and the associated forces and will be treated as input variables into the system. Figure 1.1 presents the 'vectors' of input and output components. In this diagram the tyre is assumed to be uniform and to move over a flat road surface. The input vector stems from motions of the wheel relative to the road. In more complex situations it may be required to introduce additional input quantities resulting e.g. from variations in road surface geometry and from tyre non-uniformities like out-of-roundness, stiffness variations and 'built-in' forces due to non-symmetric structure of the tyre.

The forces and moments are considered as output quantities. It is sometimes beneficial to assume these forces to act on a rigid disc with inertial properties equal to those of the tyre when considered rigid. These forces may differ from the forces acting between road and tyre because of the dynamic forces acting on the tyre when vibrating relative to the wheel rim.

The discussion on the force generation and dynamic properties of the tyre will be conducted along the two main lines: symmetric and anti-symmetric behaviour. Interaction between these main groups of input motions complicates the situation (combined slip). These interactions become important if at least one of the input motions, or more precisely, one of the associated slip components becomes relatively large. For small deviations from the straight ahead motion a linear description of the behaviour may be given. Then, it is advantageous to recognize the fact that the responses to the symmetric and anti-symmetric motions of the assumedly symmetric wheel-tyre system can be considered as uncoupled.

2. Tyre Models (Introductory Discussion)

Several types of mathematical models of the tyre have been developed during the last half century. Each type for a specific purpose. Different levels of accuracy and complexity may be introduced in the various categories of utilization. This often involves entirely different ways of approach. Figure 2.1 roughly illustrates how the intensity of various consequences associated with different ways of attacking the problem tend to vary. From left to right the model is based less on full scale tyre experiments and more on the theory of the behaviour of the physical structure of the tyre. In the middle, the model will be simpler but possibly less accurate while at the far right the descriptions becomes complex and less suitable for application in the simulation of vehicle motions and more fit for the analysis of detailed tyre performance in relation with its construction.

At the left-hand category, we have mathematical tyre models which describe measured tyre characteristics through tables or mathematical formulae and certain interpolation schemes. These formulae have a given structure and possess parameters which are usually assessed with the aid of regression procedures to yield a best fit to the measured data. A well-known empirical model is the Magic Formula tyre model treated in Chapter 4. This model is based on a sin(arctan) formula which not only provides an excellent fit for the F_y, F_x and M_z curves but in addition features coefficients which have clear relationships with typical shape and magnitude factors of

from experimental data only	using similarity method	through simple physical model	through complex physical model
fitting full scale tyre test data by regression techniques	distorting, rescaling and combining basic characteristics	using simple mechanical representation, possibly closed form solution	describing tyre in greater detail, computer simulation, finite element method

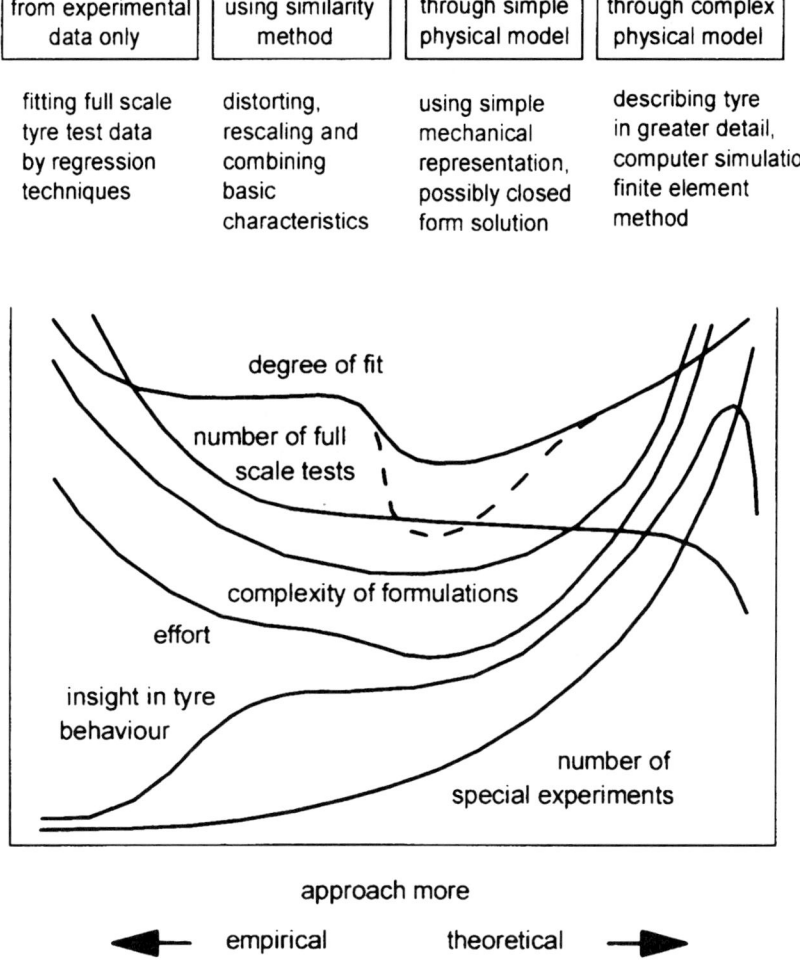

Fig. 2.1. Four categories of possible types of approach to develop a tyre model.

the curves to be fitted.

The similarity approach (second category) is based on the use of a number of basic characteristics typically obtained from measurements. Through distortion, rescaling and multiplications, new relationships are obtained to describe certain off-nominal conditions. This method (not discussed here, cf. Pacejka and Sharp, 1991) is particularly useful for application in vehicle simulation models that requires rapid (e.g. real time) computations. Depending on the type of the physical model chosen, a simple formulation may already provide sufficient accuracy for limited fields of application. The HSRI model depicted in Fig.2.2 developed by Dugoff, Fancher and Segel (1970) and later improved by Bernard et al. (1977) is a good example. The figure illustrates the considerable simplification with respect to a more realistic representation of tyre deformation (Fig.2.3) that is needed to keep the resulting mathematical formulation manageable for vehicle dynamics simulation purposes and still include important matters like the representation of combined slip and a coefficient of friction that may drop with speed of sliding.

The model of Fig.2.3 exhibits carcass flexibility and shows a more realistic parabolic pressure distribution. For such a model, (approximate) analytical solutions are feasible only when pure side slip (possibly including camber) occurs and the friction coefficient is considered constant (e.g. Fiala, 1954).

Relatively simple physical models of this third category like the 'brush model' of Fig.2.2 are especially useful to get a better understanding of tyre behaviour. In Chapter 3 this model will be

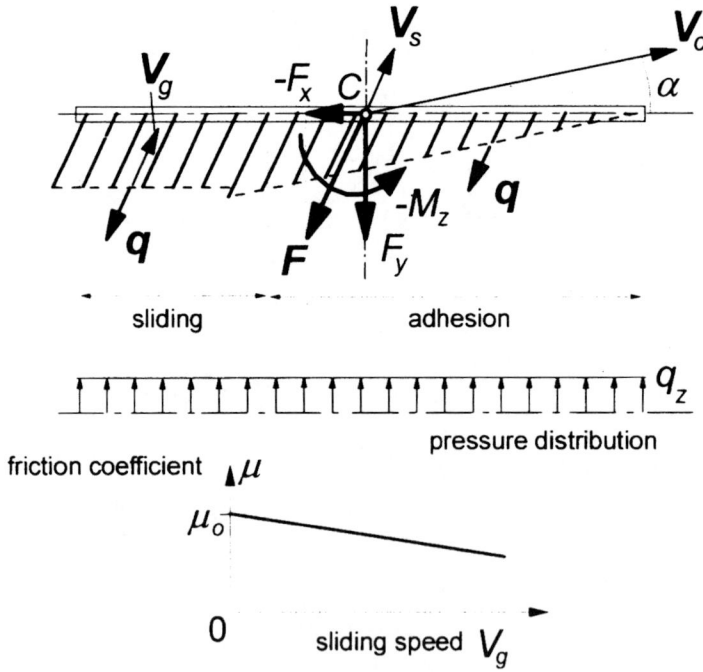

Fig. 2.2.   a. The brush type tyre model at combined longitudinal (brake) slip and lateral slip in case of equal longitudinal and lateral stiffnesses.
            b. The linearly decaying friction coefficient.

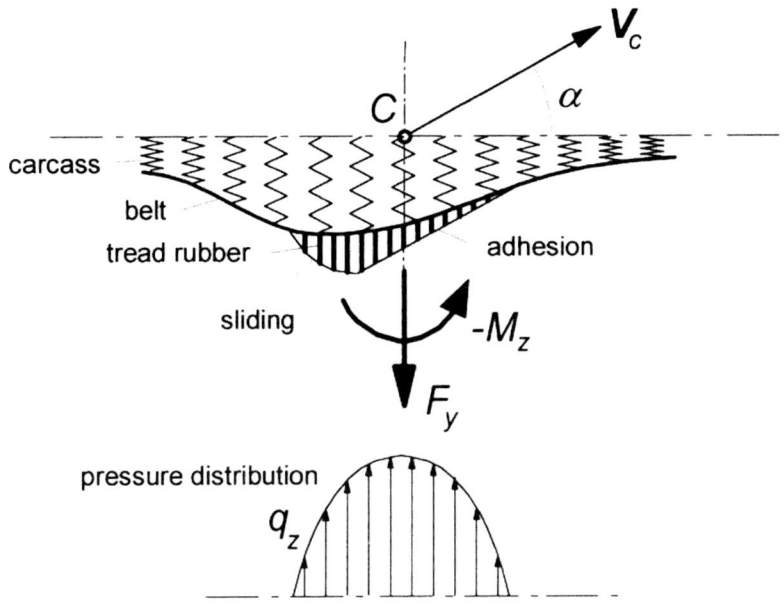

Fig. 2.3.   Tyre model with flexible carcass at steady-state rolling with slip angle α.

discussed at length.

The right-most group of Fig.2.1 is aimed primarily at more detailed analysis of the tyre. The complex finite element based models belongs to this category. A simpler representation of carcass complance that is experienced in the lower part of the tyre near the contact patch considerably speeds up the computation. Also the way in which the tread elements are handled is crucial. The computer simulation tread-element-following method is attractive and allows considerable freedom to choose pressure distribution and friction coefficient functions of sliding velocity and local pressure. The physical model that forms the basis of the latter method has been depicted in Fig.2.4. Influence (Green) functions (cf. Savkoor, 1970) may be used to describe the carcass horizontal compliance in the contact zone and possibly several rows of tread elements may be considered to move through the contact patch. One element per row is followed while it travels through the length of contact (or several elements through respective sub-zones). During such a passage the carcass deflection is kept constant, the motion of the single mass-spring (tread element) system that is dragged over the ground is computed, the frictional forces are integrated, the total forces and moment determined and the carcass deflection is up-dated. Instead of using the dynamic way of solving for the deflection of the tread element while it runs through the contact patch, an iteration process may be employed. The model is capable of handling non-steady state conditions. For further study we refer to the original work of Willumeit (1969), Pacejka et al. (1972), Sharp et al. (1986), Gipser et al. (1997) and Mastinu et al. (1997).

Although it is possible to develop a model for non-steady-state conditions by purely empirical means, most relatively simple and more complex transient and dynamic tyre models are based

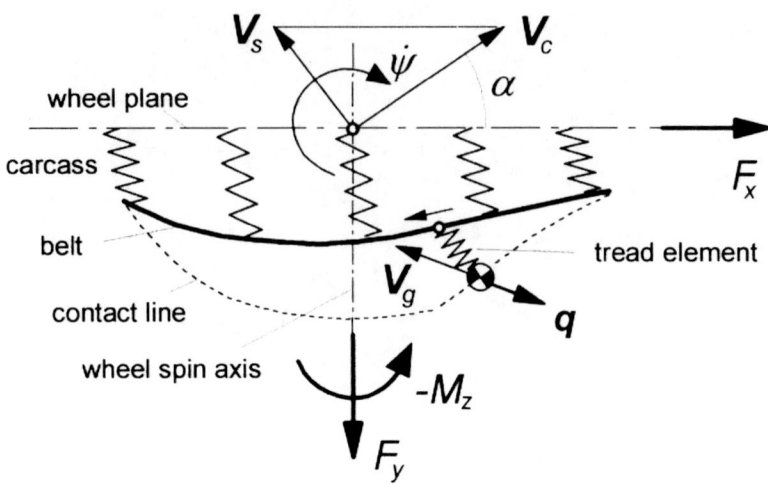

Fig. 2.4.   Computer simulation tyre model with flexible carcass, arbitrary pressure distribution and
            friction coefficient functions. Forces acting on a single tread element mass during one passage
            through the contact length are integrated to obtain the total forces and moment F_x, F_y and M_z.

on the physical nature of the tyre. It is of interest to note that for a proper description of tyre
behaviour at time-varying conditions an essential property must be represented in the physical
models belonging to both right-hand categories of Fig.2.1. That is the lateral and sometimes also
the fore and aft compliance of the carcass. Less complex non-steady-state tyre models feature
only carcass compliance without the inclusion of elastic tread elements. In steady-state models
the introduction of such a flexibility is often not required. Only to properly represent the
self-aligning torque in case of a braked or driven wheel, carcass lateral compliance is needed.
Tyre inertia becomes important at higher speeds and frequencies of the wheel motion.
Development and properties of non-steady-state tyre models are discussed in Section 5 and
references to some sources in the literature have been made.
      Conditions become more demanding when for example: (1) the wheel motion gives rise to
larger values of slip which no longer permits an approximate linear description of the force and
moment generating properties; (2) combined slip occurs, possibly including wheel camber and
turn slip; (3) large camber occurs which may necessitate the consideration of the dimensions of
the tyre cross section; (4) the friction coefficient can not be approximated as a constant quantity
but may vary with sliding velocity and speed of travel as occurs on wet or icy surfaces; (5) the
wavelength of the path of contact points at non-steady-state conditions can no longer be
considered large, which may require the introduction of the lateral and longitudinal compliance
of the carcass; (6) the wavelength becomes relatively short which may necessitate the
consideration of a finite contact length (retardation effect) and possibly the contact width (at turn
slip and camber); (7) the speed of travel is large so that tyre inertia becomes of importance in
particular its gyroscopic effect; (8) the frequency of the wheel motion has reached a level that
requires the inclusion of the first or even higher modes of vibration of the belt; (9) the vertical
profile of the road surface contains very short wavelengths with appreciable amplitudes as would

occur in the case of rolling over a short obstacle or cleat, then amongst other things the tyre enveloping properties should be accounted for; (10) motions become severe (large slip and high speed) which may necessitate modelling of the effect of warming up of the tyre involving possibly the introduction of the tyre temperature as a model parameter.

3. Tyre Brush Model

The brush model consists of a row of elastic bristles that touches the road plane and can deflect in a direction parallel to the road surface. These bristles may be called tread elements. Their compliance represents the elasticity of the combination of carcass, belt and actual tread elements of the real tyre. As the tyre rolls, the first element that enters the contact zone is assumed to stand perpendicularly with respect to the road surface. When the tyre rolls freely (that is without the action of a driving or braking torque) and without side slip, camber or turning, the wheel moves along a straight line parallel to the road and in the direction of the wheel plane. In that situation, the tread elements remain vertical and move from the leading edge to the trailing edge without developing a horizontal deflection and consequently without generating a fore and aft or side force. A possible presence of rolling resistance is disregarded. When the wheel speed vector V shows an angle with respect to the wheel plane, side slip occurs. When the wheel speed of revolution Ω multiplied with the effective rolling radius r_e is not equal to the forward component of the wheel speed $V_x = V\cos\alpha$ we have fore and aft slip. Under these conditions, depicted in Fig.3.1, horizontal deflections are developed and corresponding forces and moment arise. The tread elements move from the leading edge (at the right hand side of the pictures) to the trailing edge. The tip of the element will as long as the available friction allows adhere to the ground (that is, it will not slide over the road surface). At the same time, the base point of the element remains in the wheel plane and moves backwards with the linear speed of rolling V_r (that is equal to Ωr_e) with respect to wheel axis or better: with respect to the contact centre C. With respect to the road, the base point of the element moves with a velocity that is designated as the slip speed V_s of the wheel.

In the lower part of the figure the model is shown at pure side slip. The slip changes from very small to relatively large. We observe that the deflection increases while the element moves further through the contact patch. The deflection rate is equal to the assumedly constant slip speed. The resulting deflection varies linearly with the distance to the leading edge and the tips form a straight contact line that lies in a direction parallel to the wheel speed vector V. The figure also shows the maximum possible deflection that can be reached by the element depending on its position in the contact region. This maximum is governed by the (constant) coefficient of friction μ, the vertical force distribution q_z and the stiffness of the element c_{py}. In the figure, the pressure distribution and consequently also the maximum deflection v_{max} have been assumed to vary according to a parabola. As soon as the straight contact line intersects the parabola, sliding will start. The remaining part of the contact line will coincide with the parabola for the maximum possible deflection. At increasing slip angle, the side force that is generated will increase. The distance of its line of action behind the contact centre is termed the pneumatic trail t. The aligning torque arises through the non-symmetric shape of the deflection distribution and will be found by multiplying the side force with the pneumatic trail. As the slip increases, the deformation shape becomes more symmetric and, as a result, the trail gets smaller. This is

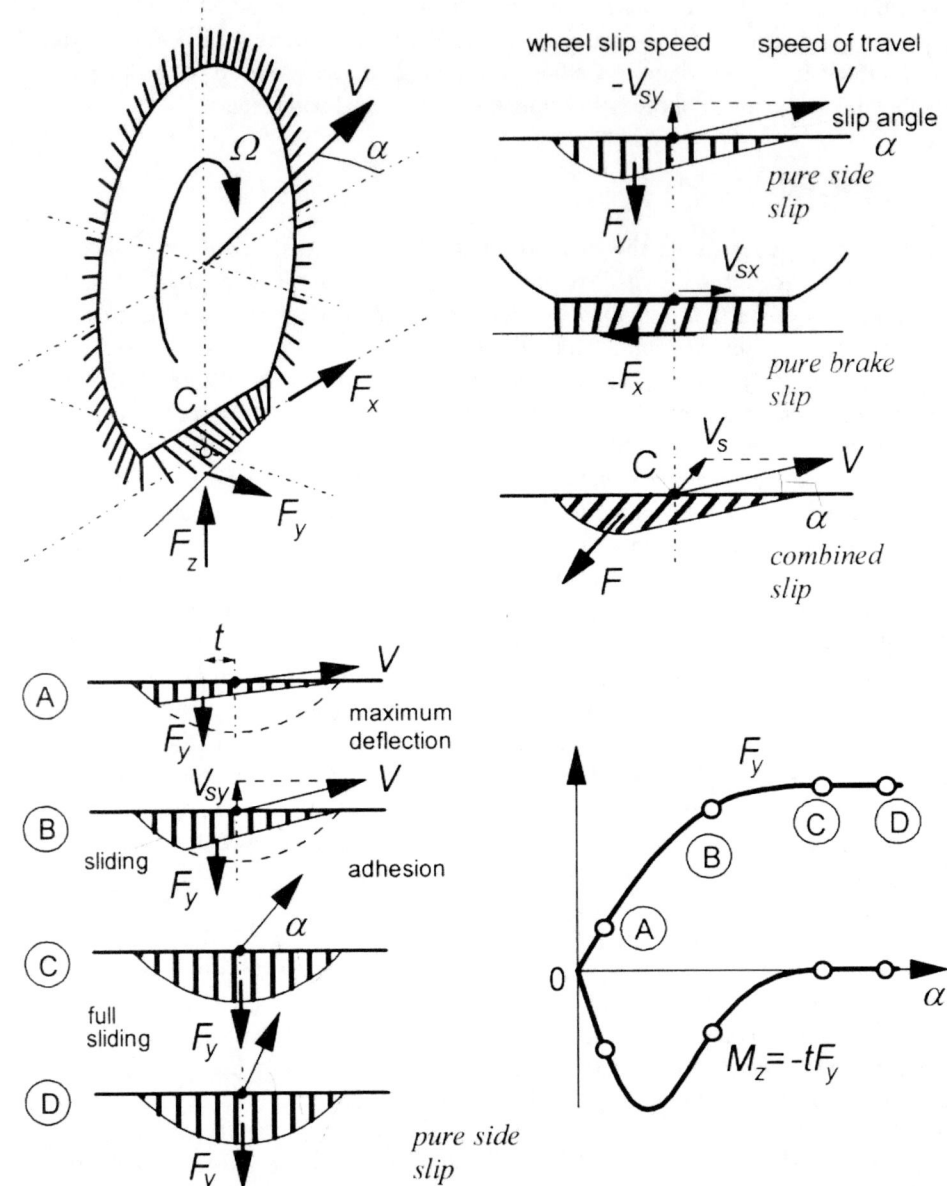

Fig. 3.1. The brush tyre model.
Top left: view of driven and side slipping tyre.
Top right: the tyre at different slip conditions.
Bottom left: the tyre at pure side slip, from small to large slip angle.
Bottom right: the resulting side force and aligning torque characteristics.

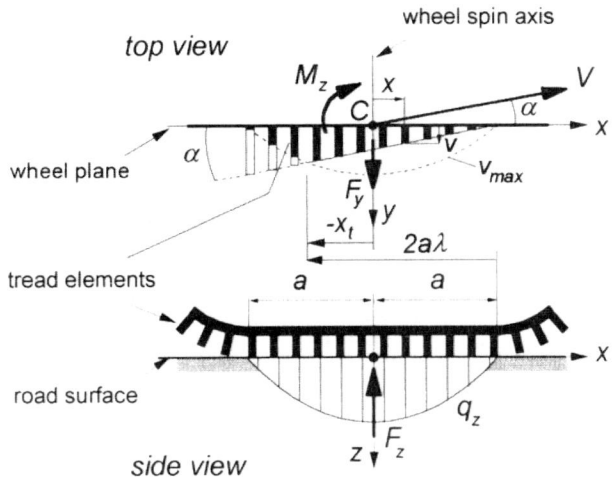

Fig. 3.2. Brush model moving at pure side slip shown in top and side view.

because the point of intersection moves forward, thereby increasing the sliding range and decreasing the range of adhesion. This continues until the wheel speed vector runs parallel to the tangent to the parabola at the foremost point. Then, the point of intersection has reached the leading edge and full sliding starts to occur. The shape has now become fully symmetric. The side force attains its maximum and acts in the middle so that the moment vanishes. That situation remains unchanged when the slip angle increases further. The resulting characteristics for the side force and the aligning torque have been depicted in the same figure. In the part to follow next, the mathematical expressions for these relationships will be derived, first for the case of pure side slip.

3.1. Pure side slip

The brush model moving at a constant slip angle α has been depicted in greater detail in Fig. 3.2. It shows a contact line which is straight and parallel to the velocity vector V in the adhesion region and curved in the sliding region where the available frictional force becomes lower than the force which would be required for the tips of the tread elements to follow the straight line further. For this simple model, the deformation of the tread element at the leading edge vanishes. Consequently, the lateral deformation in the adhesion region reads

$$v = (a-x)\tan\alpha \qquad\qquad (3.1)$$

where a denotes half the contact length.

In case of vanishing sliding, which will occur for $\alpha\to0$ or for $\mu\to\infty$, expression (3.1) is valid for the entire region of contact. With the lateral stiffness c_{py} of the tread elements per unit length of the assumedly rectangular contact area the following integrals and expressions for the cornering force F_y and the aligning torque M_z hold:

$$F_y = c_{py} \int_{-a}^{a} v\,dx = 2c_{py}a^2\alpha$$
(3.2)

$$M_z = c_{py} \int_{-a}^{a} vx\,dx = -\frac{2}{3}c_{py}a^3\alpha$$

Consequently, the cornering stiffness and the aligning stiffness become respectively:

$$C_{F\alpha} = (\partial F_y/\partial\alpha)_{\alpha=0} = 2c_{py}a^2$$
(3.3)

$$C_{M\alpha} = -(\partial M_z/\partial\alpha)_{\alpha=0} = \frac{2}{3}c_{py}a^3$$

Next, we will consider the case of finite μ and a pressure distribution which gradually drops to zero at both edges. For the purpose of simplicity we assume a parabolic distribution of the vertical force per unit length as expressed by

$$q_z = \frac{3F_z}{4a}\cdot\left\{1-\left(\frac{x}{a}\right)^2\right\}$$
(3.4)

where F_z represents the vertical wheel load. Hence, the largest possible side force distribution becomes:

$$|q_{y,max}| = \mu q_z = \frac{3}{4}\mu F_z \cdot \frac{(a^2 - x^2)}{a^3}$$
(3.5)

In Fig.3.2 the maximum possible lateral deformation $v_{max} = q_{y,max}/c_{py}$ has been indicated. For the sake of abbreviation the following composite tyre model parameter is introduced:

$$\theta_y = \frac{2c_{py}a^2}{3\mu F_z}$$
(3.6)

The distance from the leading edge to the point where the transition from the adhesion to the sliding region occurs is written as $2a\lambda$ and is determined by the factor λ. The value of this non-dimensional quantity is found by realising that at this point, where $x = x_t$, the deflection in the adhesion range becomes equal to that of the sliding range. Hence, with Eqs.(3.1, 3.5, 3.6) the following equality holds:

$$|q_y| = c_{py}(a-x_t)|\tan\alpha| = |q_{y,max}| = \frac{c_{py}}{2a\theta_y}(a-x_t)(a+x_t)$$
(3.7)

and thus for $\lambda = (a-x_t)/2a$ we obtain the relationship with the slip angle α:

$$\lambda = 1 - \theta_y|\tan\alpha|$$
(3.8)

From this equation, the angle α_{sl}, where total sliding starts ($\lambda = 0$), can be calculated:

$$\tan\alpha_{sl} = \frac{1}{\theta_y}$$
(3.9)

As the distribution of the deflections of the elements has now been established, the total force F_y and the moment M_z can be assessed by integration over the contact length (like in Eq.(3.2)).

For convenience we introduce the notation for the slip:

$$\sigma_y = \tan \alpha \tag{3.10}$$

The resulting formula for the force reads:
if $|\alpha| \le \alpha_{sl}$

$$
\begin{aligned}
F_y &= \mu F_z (1 - \lambda^3) \operatorname{sgn} \alpha \\
&= 3\mu F_z \theta_y \sigma_y \left\{ 1 - |\theta_y \sigma_y| + \tfrac{1}{3} (\theta_y \sigma_y)^2 \right\}
\end{aligned} \tag{3.11}
$$

and if $|\alpha| \ge \alpha_{sl}$ (but $< \tfrac{1}{2}\pi$)

$$F_y = \mu F_z \operatorname{sgn} \alpha \tag{3.11a}$$

and for the moment:
if $|\alpha| \le \alpha_{sl}$

$$
\begin{aligned}
M_z &= -\mu F_z \lambda^3 a (1 - \lambda) \operatorname{sgn} \alpha \\
&= -\mu F_z a\, \theta_y \sigma_y \left\{ 1 - 3|\theta_y \sigma_y| + 3(\theta_y \sigma_y)^2 - |\theta_y \sigma_y|^3 \right\}
\end{aligned} \tag{3.12}
$$

and if $|\alpha| \ge \alpha_{sl}$ (but $< \tfrac{1}{2}\pi$)

$$M_z = 0 \tag{3.12a}$$

The pneumatic trail t, which indicates the distance behind the contact centre C where the resultant side force F_y is acting, becomes:
if $|\alpha| \le \alpha_{sl}$

$$t = -\frac{M_z}{F_y} = \frac{1}{3} a \frac{1 - 3|\theta_y \sigma_y| + 3(\theta_y \sigma_y)^2 - |\theta_y \sigma_y|^3}{1 - |\theta_y \sigma_y| + \tfrac{1}{3}(\theta_y \sigma_y)^2} \tag{3.13}$$

and if $|\alpha| \ge \alpha_{sl}$ (but $< \tfrac{1}{2}\pi$)

$$t = 0 \tag{3.13a}$$

These relationships have been shown graphically in Fig.3.3. At vanishing slip angle expression (3.13) reduces to

$$t = t_o = -\left(\frac{M_z}{F_y} \right)_{\alpha \to 0} = \frac{1}{3} a \tag{3.14}$$

This value is smaller than normally encountered in practice. The introduction of an elastic carcass will improve this quantitative aspect. Then, the more realistic value of $t \approx 0.5a$ may be achieved (cf. Pacejka 1966 for a stretched string model to represent the elastic carcass).

Another point in which the simple model deviates considerably from experimental results concerns the effect of changing the vertical wheel load F_z. With the assumption that the contact length $2a$ changes quadratically with radial tyre deflection ρ and that F_z depends linearly on ρ, so that $a^2 \sim F_z$, it can be easily shown that for the brush model F_y and M_z vary proportionally with F_z and $F_z^{3/2}$ respectively. Experiments, however, show that F_y varies less than linearly with F_z.

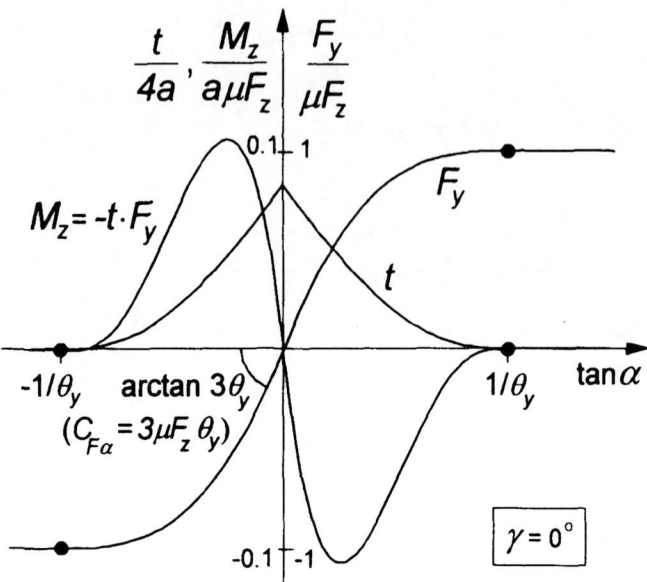

Fig. 3.3. Characteristics of the simple brush model: side force, aligning torque and pneumatic trail vs slip angle.

In most cases, the F_y vs F_z characteristic, obtained at a small value of the slip angle, even shows a maximum after which the cornering force drops with increasing wheel load. Obviously, the same holds for the cornering stiffness C_{F_α}. Also in this respect the introduction of an elastic carcass (in particular when its lateral stiffness decreases with increasing normal load) improves the agreement with experiments. When considering a deflected cross section of a tyre with side walls modelled as membranes under tension encapsulating pressurised air, such a decrease in lateral stiffness can be found to occur in theory (cf. Pacejka, 1981b).

3.2. Pure Longitudinal Slip

For the brush-type tyre model with tread elements flexible in longitudinal direction, the theory for longitudinal (braking or driving) force generation develops along similar lines as those set out in the preceding section, where the side force and aligning torque response to slip angle has been derived. To simplify the discussion, we restrict ourselves here to non-negative values of the forward speed V_r and of the speed of revolution Ω.

In Fig.3.4 a side-view of the brush model has been shown. The so-called slip point S is introduced. This is an imaginary point attached to the wheel rim and is located, at the instant considered, a distance equal to the effective rolling radius r_e (defined at free rolling) below the wheel centre. At free rolling, by our definition, the slip point S does not move. Then, it forms the centre of rotation of the wheel rim. We may think of a slip circle with radius r_e that in case of free rolling rolls perfectly, that is: without sliding, over an imaginary road surface that touches the slip circle in point S. When the wheel is being braked, point S moves forward with the longitudinal slip velocity V_{sx}. When driven, the slip point moves backwards with consequently

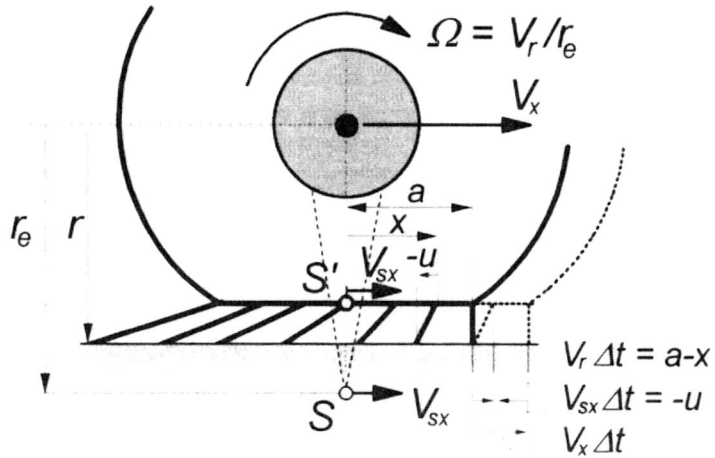

$\Omega = V_r/r_e$

V_x

a

X

S' V_{sx} $-u$

r_e r

S V_{sx}

$V_r \Delta t = a\text{-}x$

$V_{sx} \Delta t = \text{-}u$

$V_x \Delta t$

Fig. 3.4. Side view of the brush tyre model at braking (no sliding considered).

a negative slip speed. In the model, a point S' is defined that is attached to the base line at its centre (that is, at the base point of the tread element below the wheel centre, cf. Fig.3.4). By definion, the velocity of this point is the same as that of point S. That means that S' also moves with the same slip speed V_{sx}. It is assumed that the tread elements attached at their base points to the circumferentially rigid carcass, enter the contact area in vertical position. At free rolling with slip speed V_{sx} (of both points S and S') equal to zero, the orientation of the elements remains vertical while moving from front to rear through the contact zone. Consequently, no longitudinal force is being transmitted and we have a wheel speed of revolution:

$$\Omega = \Omega_o = \frac{V_x}{r_e} \tag{3.15}$$

Here it is assumed that the longitudinal component of the speed of propagation of the contact centre C is equal to the longitudinal component of the speed of the wheel centre ($V_{cx} = V_x$). This will occur on a flat road surface at vanishing camber and/or turning velocity: $\gamma\psi = 0$. When Ω differs from its value at free rolling Ω_o the wheel is being braked or driven and the longitudinal slip speed V_{sx} becomes:

$$V_{sx} = V_x - \Omega r_e \tag{3.16}$$

In the model, the base points of all the tread elements move with the same speed V_{sx}. A base point progresses backwards through the contact zone with a speed V_r called the linear speed of rolling. Apparently, we have:

$$V_r = \Omega r_e = V_x - V_{sx} \tag{3.17}$$

An element the tip of which adheres to the ground and the base point is moved towards the rear over a distance $a\text{-}x$ from the leading edge (for which a time span $\Delta t = (a\text{-}x)/V_r$ is needed) has developed a deflection in longitudinal direction:

$$u = -V_{sx} \cdot \frac{a - x}{V_r} \tag{3.18}$$

We may write the longitudinal deflection u in terms of the 'practical' longitudinal slip defined as

$$\kappa = -V_{sx}/V_x \tag{3.19}$$

yielding:

$$u = -(a - x)\frac{V_{sx}}{V_x - V_{sx}} = (a - x)\frac{\kappa}{1 + \kappa} \tag{3.20}$$

In terms of the alternative definition of longitudinal slip: the 'theoretical' slip, to be used in the subsequent section and defined as

$$\sigma_x = -\frac{V_{sx}}{V_r} = \frac{\kappa}{1 + \kappa} \tag{3.21}$$

(note, we restrict ourselves to non-negative speeds of rolling: $V_r \geq 0$ and $\kappa \geq -1$) we obtain:

$$u = (a - x)\sigma_x \tag{3.22}$$

In Section 3.! we found with Eqs.(3.1, 10) for the lateral deflection at pure side slip ($\sigma_y = \tan \alpha$):

$$v = (a - x)\sigma_y \tag{3.1a}$$

Comparison of the equations shows that the longitudinal deformations u will be equal in magnitude to the lateral deformations v if $\sigma_y = \tan \alpha$ equals $\sigma_x = \kappa/(1+\kappa)$. For equal tread element stiffnesses ($c_{px} = c_{py}$) and friction coefficients ($\mu_x = \mu_y$) in lateral and longitudinal directions, the slip force characteristics in both directions will be identical when $\tan \alpha$ and $\kappa/(1+\kappa)$ are used as abscissa, cf. Fig.3.5. Also, Eq.(3.11) holds for the longitudinal force F_x if the subscripts y are replaced by x and $\tan \alpha$ by κ.

   Obviously, total sliding will start at $\sigma_x = \kappa/(\kappa + 1) = \pm 1/\theta_x$ or in terms of the practical slip at

$$\kappa = \kappa_{sl} = \frac{-1}{1 \pm \theta_x} \tag{3.23}$$

with

$$\theta_x = \frac{2}{3}\frac{c_{px}a^2}{\mu F_z} \tag{3.24}$$

Linearization for small values of slip κ yields a deflection at coordinate x:

$$u = (a - x)\kappa \tag{3.25}$$

and a fore and aft force

$$F_x = 2c_{px}a^2\kappa \tag{3.26}$$

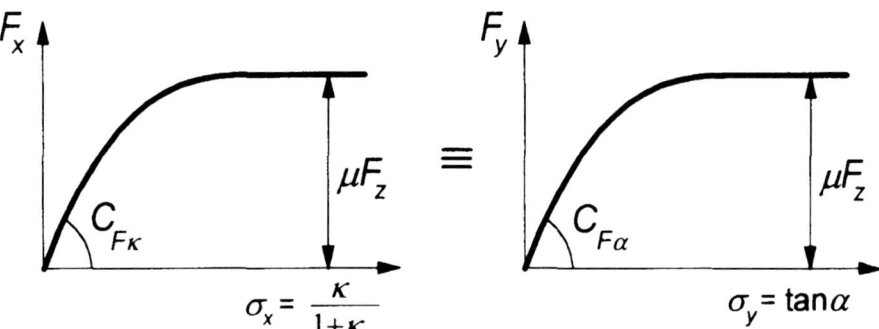

Fig.3.5. Equality of the two pure slip characteristics for an isotropic tyre model if plotted against the theoretical slip.

with c_{px} the longitudinal tread element stiffness per unit length. This relation contains the longitudinal slip stiffness

$$C_{F\kappa} = (\partial F_x / \partial \kappa)_{\kappa=0} = 2c_{px}a^2 \qquad (3.27)$$

For equal longitudinal and lateral stiffnesses $(c_{px} = c_{py})$ we obtain equal slip stiffnesses $C_{F\kappa} = C_{F\alpha}$. In reality, however, appreciable differences between the measured values of $C_{F\kappa}$ and $C_{F\alpha}$ may occur (say $C_{F\kappa}$ about 50% larger than $C_{F\alpha}$) which is due to the lateral (torsional) compliance of the carcass of the actual tyre. Still, it is expected that qualitative similarity of both pure slip characteristics remains.

3.3. Interaction between Lateral and Longitudinal Slip (combined slip)

For the analysis of the influence of longitudinal slip (or longitudinal force) on the lateral force and moment generation properties we shall, for the sake of mathematical simplicity, restrict ourselves to the case of equal longitudinal and lateral stiffnesses of the tread elements (isotropic model), i.e.:

$$c_p = c_{px} = c_{py} \qquad (3.28)$$

and equal and constant friction coefficients

$$\mu = \mu_x = \mu_y \qquad (3.29)$$

Again a parabolic pressure distribution is considered.

Figure 3.6 depicts the deformations which arise when the tyre model which runs at a given slip angle α is driven or braked. Due to the equal stiffness in all horizontal directions and the isotropic friction properties, the deflections are directed opposite to the slip speed vector V_s, also in the sliding region. In this latter region, the tips of the elements slide over the road with sliding speed V_g directed opposite to the local friction force q (per unit contact length). The whole deformation history of a tread element while running through the contact area is a one-dimensional process along the direction of V_s.

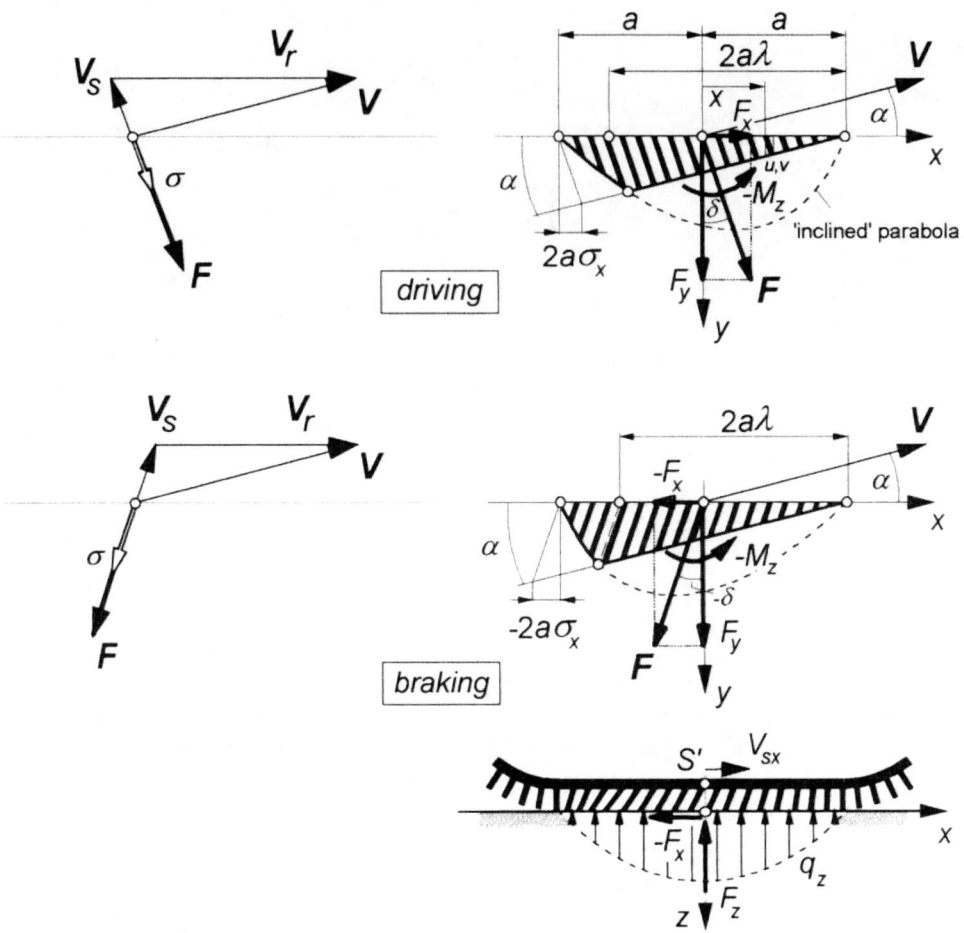

Fig. 3.6. Vector diagram and deformation of the brush model running at a given slip angle for the cases of driving and braking.

The velocity of progression of a base point through the contact length is the rolling speed V_r (again assumed non-negative). The deflection rate of an element in the adhesion region is equal to the slip speed V_s. The time which elapses from the point of entrance to the point a distance x in front of the contact centre equals

$$\Delta t = \frac{a - x}{V_r} \tag{3.30}$$

In this position, the deflection of an element that is still in adhesion becomes in vectorial form

$$e = \begin{pmatrix} u \\ v \end{pmatrix} = V_s \Delta t = -\frac{V_s}{V_r}(a - x) \tag{3.31}$$

It seems natural at this stage to introduce the alternative (theoretical) slip quantity again, but now in vectorial form:

$$\boldsymbol{\sigma} = \begin{pmatrix} \sigma_x \\ \sigma_y \end{pmatrix} = -\frac{V_s}{V_r} = -\frac{1}{V_r}\begin{pmatrix} V_{sx} \\ V_{sy} \end{pmatrix} \tag{3.32}$$

with the linear speed of rolling

$$V_r = V_x - V_{sx} \tag{3.33}$$

The relations of these theoretical slip quantities with the practical slip quantities $\kappa\,(= -V_{sx}/V_x)$ and $\tan\alpha\,(= -V_{sy}/V_x)$ are:

$$\sigma_x = -\frac{\kappa}{1 + \kappa}$$

$$\tag{3.34}$$

$$\sigma_y = \frac{\tan\alpha}{1 + \kappa}$$

The deflection of an element in the adhesion region now reads

$$e = (a - x)\boldsymbol{\sigma} \tag{3.35}$$

from which it is apparent that longitudinal and lateral deflections are governed by σ_x and σ_y respectively and independent of each other. This would not be the case if the deflections were expressed in terms of the practical quantities κ and α!

The local horizontal contact force acting on the tips of the elements (per unit contact length) reads

$$\boldsymbol{q} = c_p(a - x)\,\boldsymbol{\sigma} \qquad\qquad (\textit{adhesion region}) \tag{3.36}$$

As soon as

$$q = |\boldsymbol{q}| = \sqrt{q_x^2 + q_y^2} > \mu q_z \tag{3.37}$$

the sliding region is entered. Then the friction force becomes

$$\boldsymbol{q} = -\frac{V_s}{V_s}\mu q_z = -\frac{\boldsymbol{\sigma}}{\sigma}\mu q_z \qquad\qquad (\textit{sliding region}) \tag{3.38}$$

where

$$V_s = \sqrt{V_{sx}^2 + V_{sy}^2} \tag{3.39}$$

and

$$\sigma = \sqrt{\sigma_x^2 + \sigma_y^2} \tag{3.40}$$

Similarly, the magnitude of the deflection of an element becomes:

$$e = |e| = \sqrt{u^2 + v^2} \tag{3.41}$$

The point of transition from adhesion to sliding region is obtained from the condition:

$$c_p e = \mu q_z \tag{3.42}$$

or

$$c_p \sigma(a - x_t) = \frac{3}{4} \mu F_z \frac{a^2 - x_t^2}{a^3} \tag{3.43}$$

which yields

$$x_t = \frac{4}{3} \frac{c_p a^3 \sigma}{\mu F_z} - a = a(2\theta\sigma - 1) \tag{3.44}$$

or in similar terms as Eq. (3.8):

$$\lambda = 1 - \theta\sigma \tag{3.45}$$

where analogous to expressions (3.6) and (3.24) for the isotropic model parameter θ reads:

$$\theta = \theta_y = \theta_x = \frac{2}{3} \frac{c_p a^2}{\mu F_z} \tag{3.46}$$

From Eq.(3.45) the slip σ_{sl} at which total sliding starts can be calculated. We get analogous to (3.9)

$$\sigma_{sl} = \frac{1}{\theta} \tag{3.47}$$

The total force F (magnitude $|F|$ of F) now easily follows in accordance with (3.11):

$$F = \mu F_z (1 - \lambda^3) = \mu F_z \{3\theta\sigma - 3(\theta\sigma)^2 + (\theta\sigma)^3\} \qquad \text{for } \sigma \le \sigma_{sl}$$

$$F = \mu F_z \qquad \text{for } \sigma \ge \sigma_{sl} \tag{3.48}$$

and obviously follows the same course as those shown in Fig.3.5. The force vector F acts in a

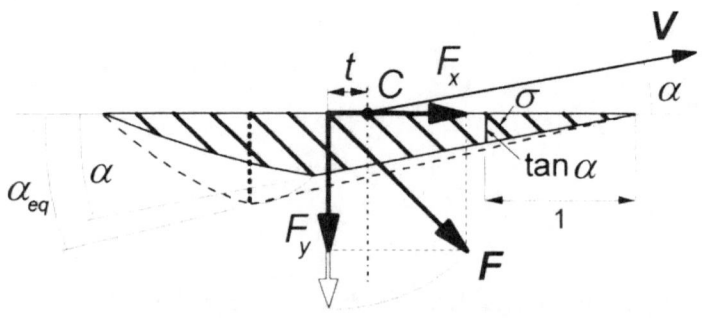

Fig. 3.7. Equivalent side slip angle producing the same pneumatic trail t.

direction opposite to V_s or $-\sigma$. Hence

$$F = F \frac{\sigma}{\sigma}$$ (3.49)

from which the components F_x and F_y may be obtained.
The moment $-M_z$ is obtained by multiplication of F_y with the pneumatic trail t. This trail is

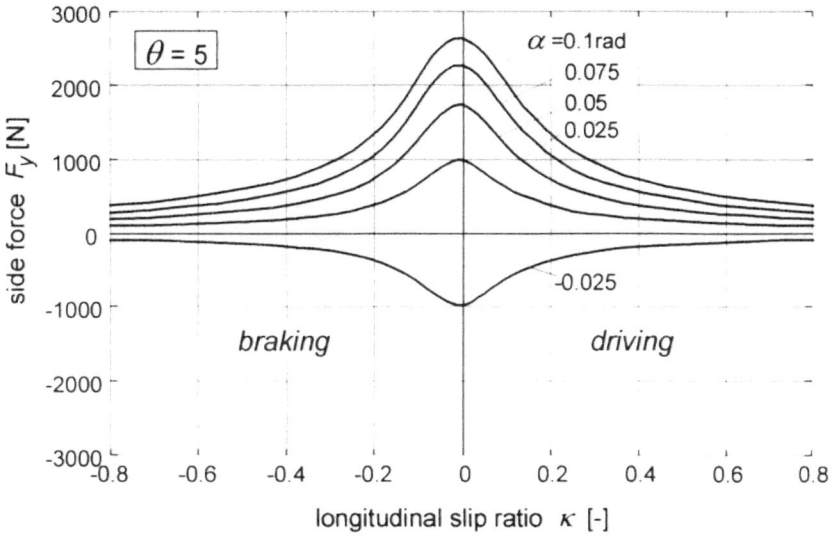

Fig.3.8. Reduction of side force due to the presence of longitudinal slip.

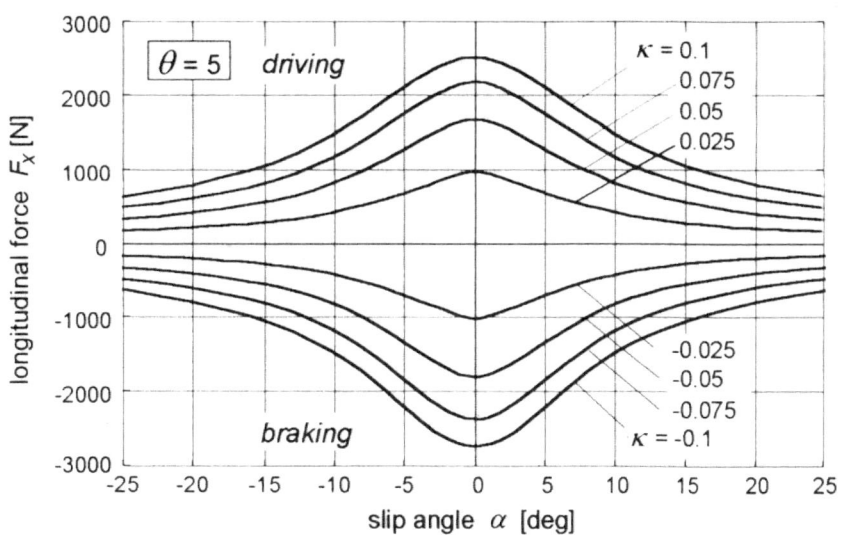

Fig.3.9. Reduction of longitudinal force due to the presence of side slip.

easily found when we realize that the deflection distribution over the contact length is identical with the case of pure side slip if $\tan\alpha_{eq} = \sigma$ (cf. Fig. 3.7). Consequently, the formula (3.13) represents the pneumatic trail at combined slip as well if $\theta_y \sigma_y$ is replaced by $\theta\sigma$. We have with (3.13):

$$M_z = -t(\sigma) \cdot F_y \qquad (3.50)$$

In Figs.3.8 and 3.9 the dramatic reduction of the pure slip forces (the side force and the longitudinal force respectively) that occurs as a result of the simultaneous introduction of the other slip component (the longitudinal slip and the side slip respectively) have been indicated. We observe an (almost) symmetric shape of these interaction curves. The peak of the side force versus longitudinal slip curves at constant values of the slip angle appears to be slightly shifted towards the braking side. This phenomenon will be further discussed in connection with the alternative representation of the same results according to Fig.3.10. At very large longitudinal slip that is when $|V_{sx}|/V_x \to \infty$ the side force approaches zero and the same occurs for the longitudinal force when the lateral slip $\tan\alpha$ goes to infinity ($\alpha \to 90°$) at a given value of the

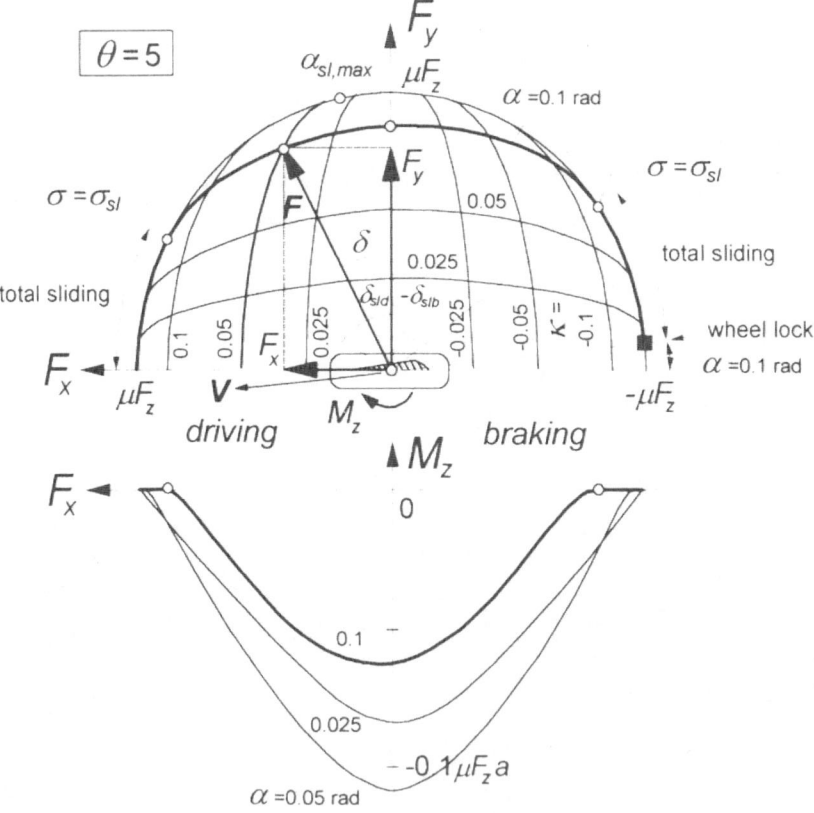

Fig. 3.10. Cornering force and aligning torque as functions of longitudinal force at constant slip angle α or longitudinal slip κ.

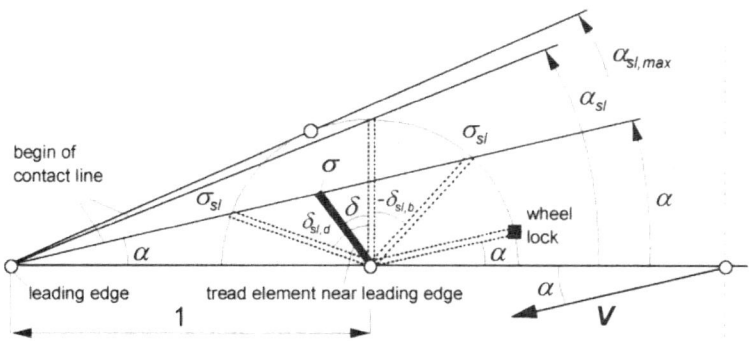

Fig.3.11. The situation near the leading edge. Various deflections of an element at a distance equal to 1 from the leading edge are shown corresponding with points in the upper diagram of Fig.3.13.

longitudinal slip because obviously in that case with V_x–0 also the longitudinal slip speed must vanish. For a locked wheel $V_{sx}=V_x$ and $\kappa=-1$ we have $F_y=\mu F_z \sin\alpha$ and $F_x=-\mu F_z \cos\alpha$.

In the diagram of Fig.3.10 the calculated variations of F_y and M_z with F_x have been plotted for several fixed values of α. Also the curves for constant κ have been depicted. For clarification of the nature of the F_y - F_x diagram the deflection of an element near the leading edge has been shown in Fig.3.11. Since the distance from the leading edge has been defined for this occasion to be equal to unity the deflection e of the element equals the slip σ. The radius of the circle denoting maximum possible deflections is equal to σ_{sl}. The two points on the circle where at the α considered the sliding boundary σ_{sl} is attained correspond to the points on the α curve of the force diagram of Fig.3.10. This also explains the slightly inclined nature of the α curves. At braking, F_y appears to be a little larger than at driving. This is at least true for the two cases of Fig.3.6, one at driving and the other at braking, showing the same slip angle and the same magnitude of the deviation angle δ of the slip velocity vector and thus of the force vector with respect to the y axis. Then, the slip speeds V_s are equal in magnitude, but at braking the speed of rolling V_r is obviously smaller. When considering the definition of the theoretical slip σ (3.32) and the functional relationship (3.48) with the force F it becomes clear that F must be larger in the case of braking because of the then larger σ. Finally, the case of wheel lock is pointed at. Then, the force vector F which in magnitude is equal to μF_z, is directed opposite to the speed vector of the wheel V which then coincides with the slip speed vector V_s.

Experimental evidence (e.g. Fig.3.14) supports the nature of the theoretical curves for the forces as shown in Fig.3.10. Often, the shape appears to be more asymmetric than predicted by the simple brush model. This may be due to a slight increase in the contact length while braking (making the tyre stiffer) and by the brake force induced slip angle of the contact patch at side slip. This is accomplished through the torsion of the carcass induced by the moment about the vertical axis that is exerted by the braking force which has shifted its line of action due to the lateral deflection connected with the side force.

The moment curves presented in the lower diagram of Fig.3.10 show a more or less symmetrical bell shape. As expected, the aligning torque becomes equal to zero when total sliding occurs ($\sigma \geq \sigma_{sl}$). Later on, we will see that the computed moment characteristics may appreciably deviate from experimentally obtained curves.

At this stage, we will first apply the knowledge gained so far to the analysis of a practical situation that occurs with a wheel that is braked or driven (at constant brake pressure and throttle respectively) while its slip angle is varied. In Fig.3.12 the force diagram is shown in combination with the corresponding velocity diagram. At a given braking force $-F_x$ and wheel speed of travel V, the slip angle α is changed from zero to 90°. The variation of slip speed V_s and rolling speed V_r may be followed from case 1 where $\alpha = 0$ to case 3 where total sliding starts and further to case 4 where V_r and thus Ω vanishes and the wheel becomes locked. A further increase of α (at constant brake pedal force) as in case 5 will necessarily lead to a reduction in braking force $-F_x$ unless the wheel is rotated in opposite direction ($\Omega<0$) as represented by case 5'. In the cases of driving indicated by Roman numerals the driving force F_x can be maintained irrespective of the value of α (with $|\alpha| < 90°$).

The nature of the resulting F_y - α characteristics at given driving or braking effort is shown in Fig.3.13. Plotting of F_y versus $\sin\alpha$ is advantageous because the portion where wheel lock occurs is then represented by a straight line.

Another important advantage of putting $\sin\alpha$ instead of $\tan\alpha$ along the abscissa is that (after having completed the diagram for negative values of $\sin\alpha$ resulting in an oddly symmetric graph) the complete range of α is covered: the speed vector V may swing around over the whole range of 360°.

For illustration we have shown in Figs.3.14 and 15 experimentally assessed characteristics. The force diagrams correspond reasonably well with the theoretical observations. The moment curves, however, deviate considerably from the theoretical predictions (compare Fig.3.15 with Fig.3.10). It appears that according to this figure M_z changes its sign in the braking half of the diagram. This phenomenon can not be explained with the simple tyre brush model that has been employed thus far.

The introduction of a laterally flexible carcass seems essential for a proper representation of the self-aligning torque acting on a driven or braked wheel. In Fig.3.16 a possible extension of the brush model is depicted. The carcass line is assumed to remain straight and parallel to the wheel plane in the contact region. A lateral and longitudinal compliance with respect to the wheel plane is introduced. In addition, a possible initial off set of the line of action of the longitudinal force with respect to the wheel centre plane is regarded. Such an off set is caused by asymmetry of the construction of the tyre or by the presence of a camber angle.

With this model the moment M_z is composed of the original contribution M_z' established by the brush model and those due to the forces F_y and F_x which show lines of action shifted with respect to the contact centre C over the distances u_c and $v_o + v_c$ respectively. The self-aligning torque now reads:

$$M_z = M_z' - F_x\left(v_o + v_c\right) + F_y u_c$$
$$= M_z' - c F_x F_y - F_x v_o$$

(3.51)

where the compliance coefficient c has been introduced that is defined by:

$$c = \frac{\epsilon_y}{C_{cy}} - \frac{\epsilon_x}{C_{cx}}$$

(3.52)

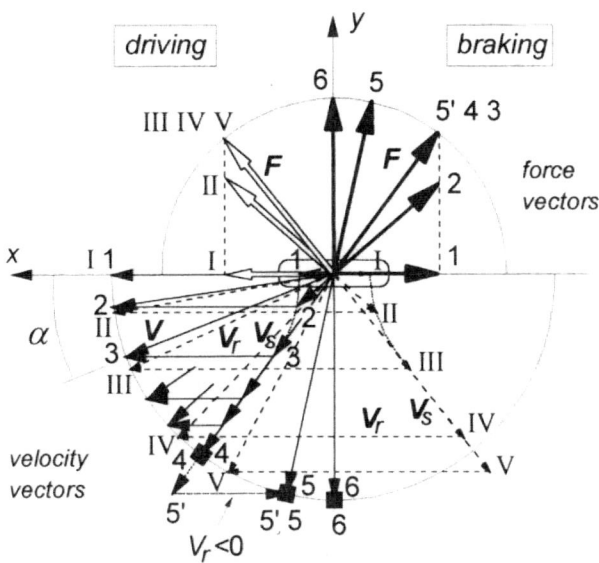

Fig. 3.12.   The F_x - F_y diagram extended with the corresponding velocity diagram. The brake (pedal) force is kept constant while the slip angle is increased. The same is done at a constant driving effort. In the case of braking the speed of rolling V_r decreases until the wheel gets locked. In the special case 5' the wheel is rotated backwards to keep the slip speed V_s and thus the force vector F in the original direction.

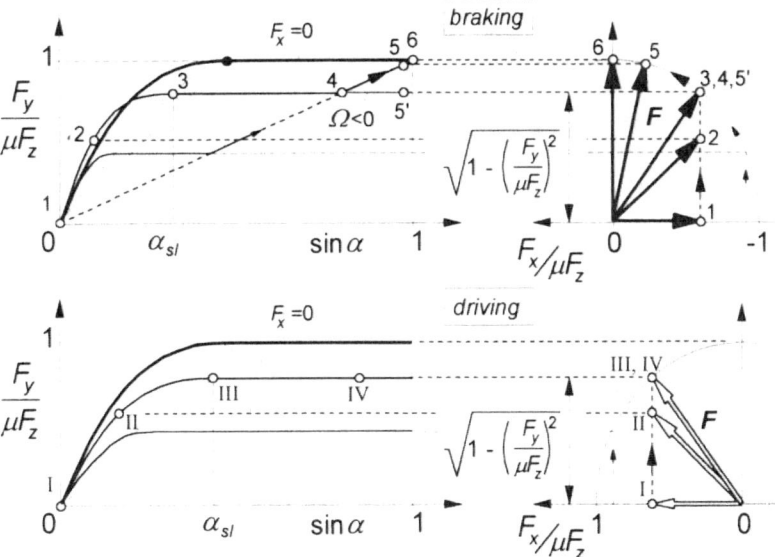

Fig. 3.13.   Tyre characteristics at constant brake (pedal) force and driving force. Numerals correspond to those of Fig. 3.12.

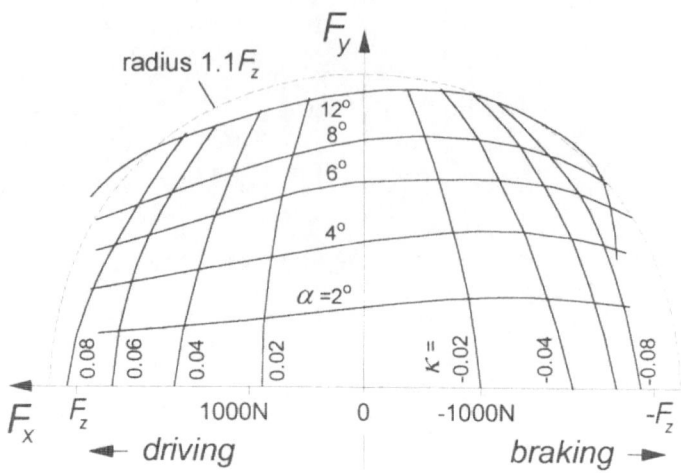

Fig. 3.14. F_y - F_x characteristics for 6.00-13 tyre measured on dry internal drum with diameter of 3.8m (from Henker, 1968, cf. Pacejka, 1981b).

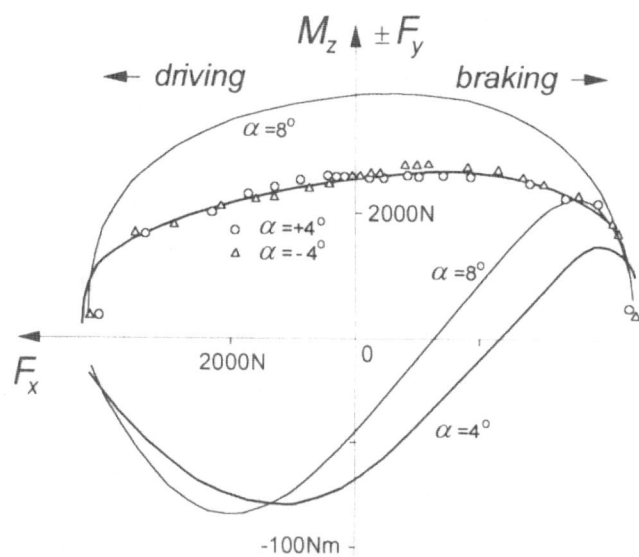

Fig.3.15. Combined slip side force and moment versus longitudinal force characteristics measured at two values of the slip angle for a 7.60-15 tyre on a dry flat surface (from Nordeen and Cortese, 1963, cf. Pacejka, 1981b).

Here, C_α and C_{cy} denote the longitudinal and lateral carcass stiffnesses respectively and ϵ_x and ϵ_y the effective fractions of the actual displacements. These fractions must be considered because of concurrent lateral and longitudinal rolling of the tyre lower section against the road surface

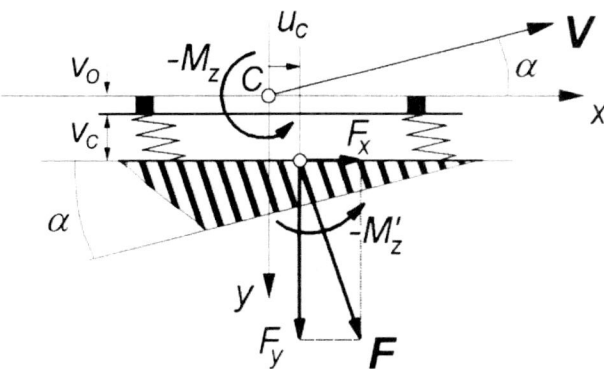

Fig.3.16. Extended tyre brush model showing offset and deflection of carcass line (straight and parallel to wheel plane).

under the action of a lateral and fore and aft force respectively which will change the normal force distribution in the contact patch and thereby reduce the actual displacements of the lines of action of both horizontal forces. The resulting calculations can be performed in a direct straightforward manner because the slip angle of the extended model is the same as the one for the internal brush model. The effect of the introduction of a torsional and bending stiffness of the carcass and belt is certainly significant. The model that would result, however, is a lot more complex and closed form solutions are no longer possible. The tread element following simulation method is quite suitable to assess the response of such more complex models but will not be treated here.

The combined slip response of the simple extended model of Fig.3.16 is given in the Figs.3.17 and 3.18. It is observed that in Fig.3.17 the aligning torque changes its sign in the braking range. This is due to the term in Eq.(3.51) with the compliance coefficient c. The resulting qualitative shape is quite similar to the experimentally found curves of Fig.3.15. In Fig.3.18 the effect of an initial offset of $v_o = 5$mm has been depicted. Here carcass compliance has been disregarded and only the last term of (3.51) has been added. We see that a moment is generated already at zero slip angle. The curves found in Fig.3.10 for non-zero side slip are then simply added to the inclined straight line belonging to $\alpha = 0$. The type of curves that result are often found experimentally. The effect of lateral compliance may then be very small or cancelled by the effect of the fore and aft compliance of the carcass (second term of right-hand member of (3.52)).

3.4. Camber and Turning (Spin)

For the study of horizontal cornering, one should not only consider side slip but also the influence of two other effects, which in most cases (except for the motorcycle) are of much less importance than side slip. The introduction of these two input variables which completes the description of the out-of-plane tyre force and moment generation, are: the wheel camber or tilt angle γ between the wheel plane and the normal to the road, and the turn slip $\dot\psi/V_r$. Both are contributions to the total spin φ. In the steady-state case the turn slip equals the curvature

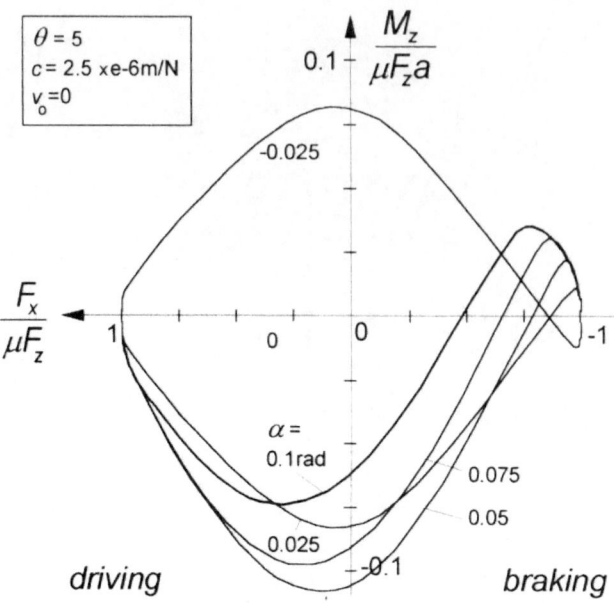

Fig.3.17.The influence of lateral carcass compliance on the aligning torque for a braked or driven wheel according to the extended brush model.

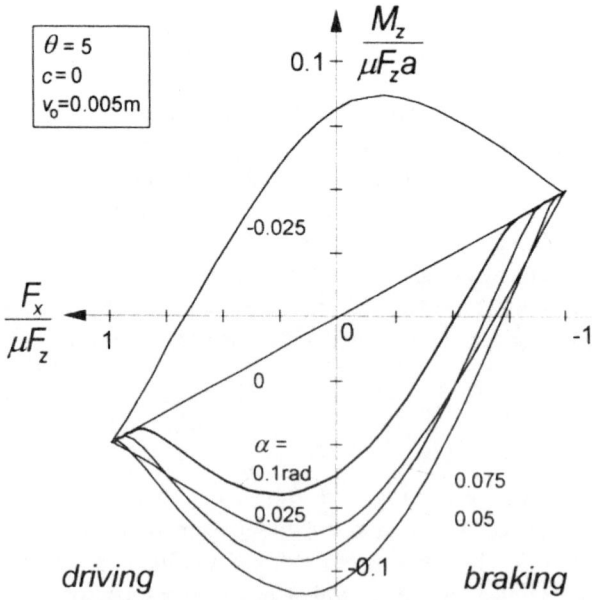

Fig.3.18. The influence of an initial lateral off set v_o of F_x on the aligning torque for a braked or driven wheel according to the extended brush model.

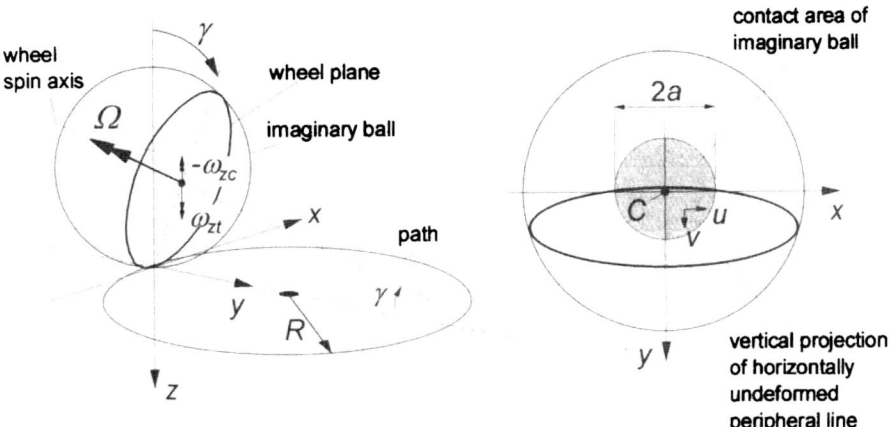

Fig.3.19. Wheel rolling at a camber angle while turning along a circular path without side slip. At the right: a top-view of the peripheral line of the non-rolling wheel considered as being a part of an imaginary ball pressed against a flat surface.

$1/R$ of a circular path with radius R. For homogeneous rolling bodies (solid rubber ball, steel railway wheel) the mechanisms to produce side force and moment as a result of camber and turning are equal as they both originate from the same spin motion. For a tyre with its rather complex structure the situation may be different for the two components of spin.

As depicted in Fig. 3.19, the wheel is considered to move tangentially to a circular path with radius R while the wheel plane shows a constant camber angle γ and apparently the slip angle is kept equal to zero. The wheel is picturised here as a part of an imaginary ball. When lifted from the ground, the intersection of wheel plane and ball outer surface forms the peripheral line of the tyre. When loaded vertically, the ball and consequently the peripheral line are assumed to show no horizontal deformations, which in reality will approximately be the case for a homogeneous ball showing a relatively small contact area. For a tyre when compressed on a frictionless surface the peripheral line may show a lateral distortion which is attributed to the torsion of the lower part of the belt when loaded and at the same time to the high bending stiffness of the belt in horizontal direction. This causes the peripheral line to become almost straight when viewed from above. Only a relatively small lateral (camber) force is then needed to make the contact line perfectly straight when rolling over a surface with friction. This explains the relatively low camber stiffness of car and truck tyres with respect to motorcycle tyres with an almost circular cross section and as expected with respect to the turn slip stiffness.

In the analysis, we will first restrict ourselves to the case of steady-state pure spin, i.e.: with $\alpha = \kappa = 0$. Then

$$V_{sx} = 0, \ V_{sy} = 0, \ V_r = V_c, \ \omega_z = \dot{\psi} - \Omega \sin\gamma = V_c \left(\frac{1}{R} - \frac{1}{r_e} \sin\gamma \right) \tag{3.53}$$

The spin is defined as

$$\varphi = -\frac{\omega_z}{V_c} \tag{3.54}$$

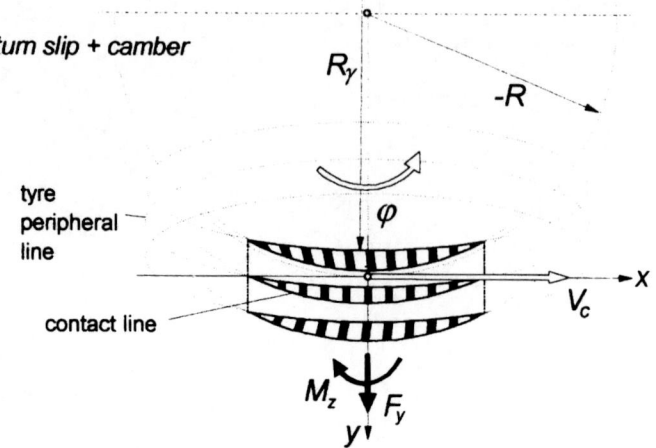

Fig.3.20. Top-view of cambered tyre model rolling in a curve with radius $-R$.

To account for the distortion of the peripheral line due to loading at camber a reduction factor ϵ_y for the radius of curvature R_y of the peripheral line is introduced. This factor will be equal or close to zero for a railway wheel or a motorcycle tyre and is expected to be closer to unity for a steel belted car or truck tyre. With ϵ_y we define a total corrected spin

$$\varphi = -\frac{1}{V_c}\{\dot{\psi} - (1 - \epsilon_y)\Omega\sin\gamma\} = -\frac{1}{R} + \frac{1 - \epsilon_y}{r_e}\sin\gamma \qquad (3.55)$$

The last term represents the curvature $-1/R_y$ of the tyre peripheral line touching a frictionless surface at a cambered position. When disregarding a possible uniform offset of this line we obtain for its initial defection:

$$y_o = y_{yo} = -\frac{1 - \epsilon_y}{2r_e}(a^2 - x^2)\sin\gamma \qquad (3.56)$$

With an assumedly same reduction factor for longitudinal initial contact point positions, the sliding velocities of the contact points in x and y directions in a range of sliding appear to read in the case of pure spin:

$$V_{gx} = -\left(\frac{\partial u}{\partial x} - y\varphi\right)V_c$$

$$V_{gy} = -\left(\frac{\partial v}{\partial x} + x\varphi\right)V_c \qquad (3.57)$$

In the range of adhesion where the sliding velocities vanish, the deflection gradients become:

$$\frac{du}{dx} = y\varphi, \qquad \frac{dv}{dx} = -x\varphi \qquad (3.58)$$

Integration yields the following expressions for the horizontal deformations in the contact area:

$$u = yx\varphi + C_1$$
$$v = -\tfrac{1}{2}x^2\varphi + C_2 \tag{3.59}$$

The second expression is, of course, an approximation of the actual variation which is according to a circle. The approximation is due to the assumption made that the deflections v are much smaller than the path radius R. The constants of integration follow from boundary conditions which depend on the tyre model employed and on the slip level. As an example, consider the simple brush model with horizontal deformations through elastic tread elements only. The contact area is assumed to be rectangular with length $2a$ and width $2b$ and filled with an infinite number of tread elements. In Fig.3.20 three rows of tread elements have been shown in the deformed situation. For this model the following boundary conditions apply:

$$x = a: \quad v = u = 0 \tag{3.60}$$

With the use of (3.59) the formulae for the deformations in the adhesion zone starting at the leading edge read:

$$u = -y(a - x)\varphi$$
$$v = \tfrac{1}{2}(a^2 - x^2)\varphi \tag{3.61}$$

After introducing c'_{px} and c'_{py} denoting the stiffness of the tread rubber per unit area in x and y direction respectively and assuming small spin and hence vanishing sliding, we can calculate the lateral force and the moment about the vertical axis by integration over the contact area. We obtain:

$$F_y = \tfrac{4}{3}c'_{py}a^3 b \varphi = C_{F\varphi}\varphi$$
$$M_z = \tfrac{4}{3}c'_{px}a^2 b^3 \varphi = C_{M\varphi}\varphi \tag{3.62}$$

or in terms of path curvature (if $\alpha \equiv 0$ or constant) and camber angle (γ small):

$$F_y = -C_{F\varphi}\left(\frac{1}{R} - \frac{1-\epsilon_\gamma}{r_e}\gamma\right) = C_{F\varphi}\frac{-1}{R} + C_{F\gamma}\gamma$$
$$M_z = -C_{M\varphi}\left(\frac{1}{R} - \frac{1-\epsilon_\gamma}{r_e}\gamma\right) = C_{M\varphi}\frac{-1}{R} + C_{M\gamma}\gamma \tag{3.63}$$

In case of pure turning, the force acting on the tyre is directed away from the path centre and the moment acts opposite to the sense of turning. Consequently both the force and the moment try to reduce the path curvature $1/|R|$. In case of pure camber, the force on the wheel is directed towards the point of intersection of the wheel axis and the road plane, while the moment tries to turn the rolling wheel towards this point of intersection. No resulting force or torque is expected to occur when $(1-\epsilon_\gamma)\sin\gamma = r_e/R$. For the special case that $\epsilon_\gamma = 0$, this will occur when the point of intersection and the path centre coincide. As the lateral deflection shows a symmetric distribution, the moment must be caused solely by the longitudinal forces. The generation of the moment may be explained by considering three wheels rigidly connected to each other, mounted on one axle. The wheels rotate at the same rate but in a curve the wheel centres travel different distances in a given time interval and when cambered, these distances are equal but the effective rolling radii are different. In both situations opposite longitudinal slip occurs, which results in a braked and a driven wheel (in Fig.3.20 the right and the left-hand wheel respectively) and consequently in a couple M_z.

Up to now we have dealt with the relatively simple case of complete adhesion. When sliding is allowed by introducing a limited value of the coefficient of friction μ, the calculations become quite complicated. When a finite width $2b$ is considered, complete adhesion will only occur for vanishing values of spin. We expect that sliding will start at the left and right rear corners of the contact area, since in these points the available horizontal contact forces reduce to zero and the longitudinal deformations u would become maximal in the hypothetical case that $\mu \to \infty$. The zones of sliding grow with increasing spin and will thereby cause a less than proportional variation of F_y and M_z with φ. The case of finite contact width is too difficult to handle by a simple analysis. For this, and for the case of additional braking, the tread element following simulation method (similar to the model of Fig.2.4) may be employed.

For now we assume a thin tyre model with $b = 0$. If, as before, a parabolic pressure distribution is assumed with a similar variation of the maximum possible lateral deflection v_{max}, it is obvious from Eq.(3.61) showing that the lateral deflection is also (approximately) parabolic, that no sliding will occur up to a certain critical value of spin φ_{sb} where the adhesion limit is reached throughout the contact length. Up to this point F_y varies linearly with φ and M_z remains equal to zero.

According to Eq. (3.63) with ϵ_γ assumed to take a value that is minimally equal to zero, spin due to camber theoretically cannot exceed the value $1/r_e$. Consequently, at larger values of spin turn slip must be involved. Beyond the critical value φ_{sl} the situation becomes quite complex. The discussion may be simplified by considering a turn table on top of which the wheel rolls with its spin axis fixed. The condition of adhesion is satisfied when the deflections remain within the boundaries given by the parabola's $\pm v_{max}$ as indicated in Fig.3.21a. In the same figure the corresponding situation at camber has been indicated at the same value of spin with curvature $1/R_\gamma$ of the deflected peripheral line $(y_{yo}, Eq.(3.56))$ on a $\mu=0$ surface equal to the path curvature $1/R$.

Sliding occurs as soon as the points of the table surface can cross the adhesion boundary $\pm v_{max}$. This occurs simultaneously for all the points in the front half of the contact line when the path curvature exceeds the curvature of the adhesion boundary. Once the contact point arrives in the rear half of the contact zone (at $x = 0$), the point can maintain adhesion because now the point on the table moves towards the inside of the adhesion boundaries. The point follows a circle until the opposite boundary is reached at $x=x_{s2}$ (cf. Fig.3.21b) where the deformation v is opposite in sign and reaches its maximum value v_{max}, after which sliding occurs again. With increasing turn slip $\varphi (= -1/R)$ this latter sliding zone grows. At the same time the side force F_y decreases and the torque M_z, that arises for $\varphi > \varphi_{sb}$, increases until the situation is reached where R and F_y approach zero and M_z attains its maximum value (tyre standing still and rotating about its vertical axis). In Fig.3.21b the radius R has been considered large with respect to half the contact length a. Then the circle segments may be approximated by parabola's which makes the analysis a lot simpler. In the lower part of the figure a camber equivalent graph has been depicted. The curved contact line in the adhesion range is then converted to a straight (horizontal) line. The graph is obtained from the turn slip graph by subtracting from all the curves the deflection v_φ that would occur if full adhesion can be maintained (e.g. at $\mu \to \infty$). We use this graph as the basis for the calculation of the force and moment response to spin. First for

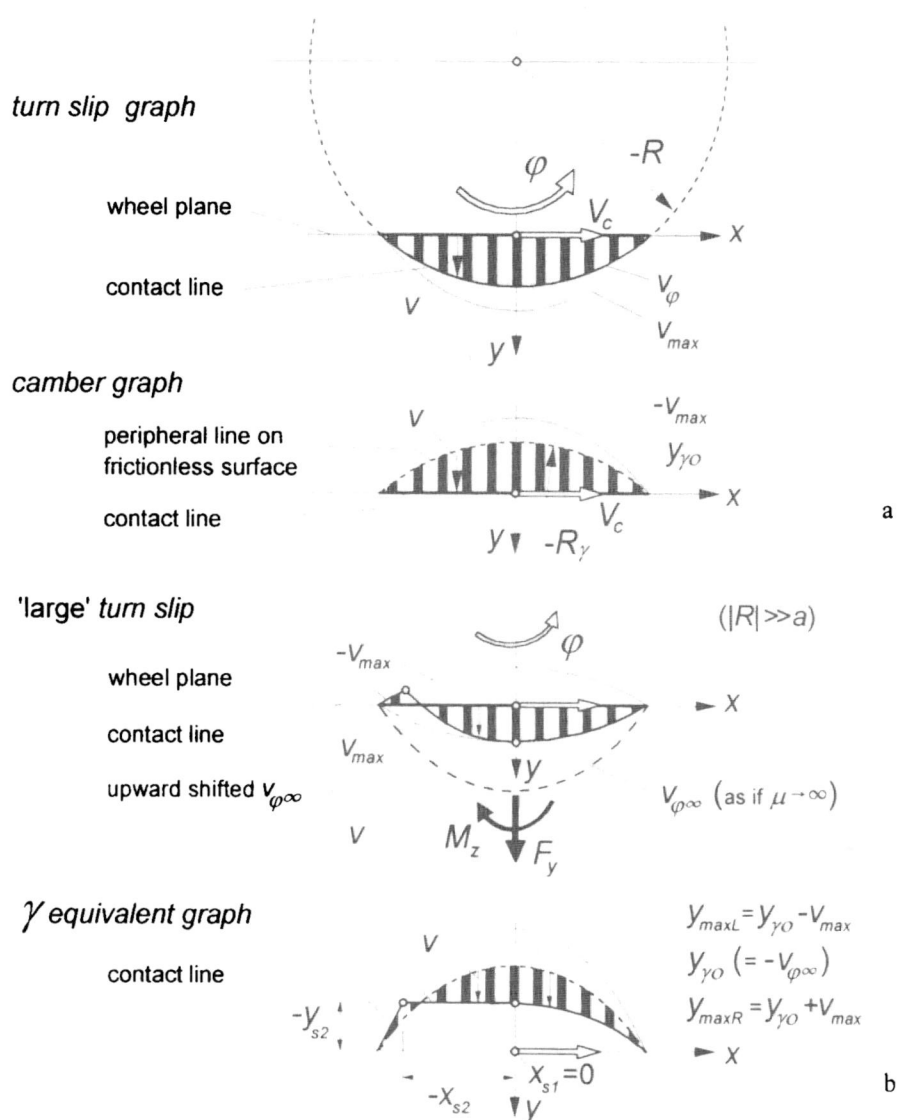

Fig.3.21. The tyre brush model with zero width rolling while turning or at a camber angle (a) at full
adhesion. Turning at large spin showing sliding at the front half and at the rear end (b,c), with
parabolic approximation of the circular path in (c).

the case of pure spin and then for the case of combined spin and side slip.

In Fig.3.22 the resulting pure spin characteristics have been presented according to analytical
expressions. It is of interest to note that although the parabola does not resemble the circular path
so well at smaller radii, the resulting response seems to be acceptable at least if the deflections
remain sufficiently small.

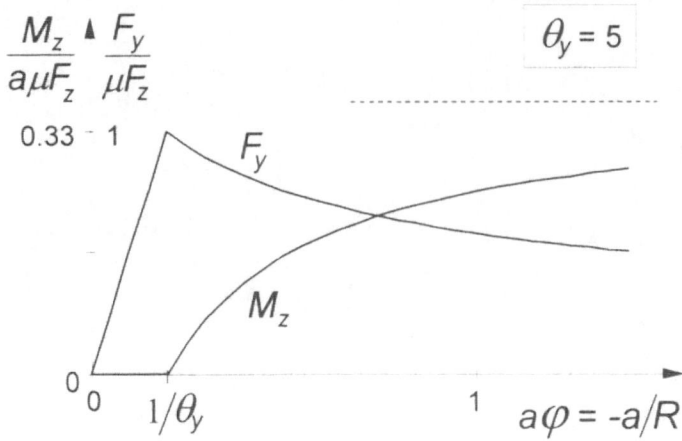

Fig.3.22. Force and moment versus non-dimensional spin for single row brush model.

Especially at larger levels of spin, the width of the contact patch will have a considerable effect on the torque and indirectly on the side force because of the consumption of some of the friction by the longitudinal forces. Also, the spin torque may be responsible for an additional distorsion of the carcass which results in a further change of the effective slip angle. Amongst other things, these matters can be taken into account in the tread element simulation model. An example of the result of calculations using the tread element following simulation method is depicted in Figs.3.23 and 3.24 for the single row model and in Figs.3.25 and 26 for the two-row model. The latter model shows equal longitudinal and lateral element stiffnesses and a distance between the two rows $2b = a$. It can be observed that for the single row model, in accordance

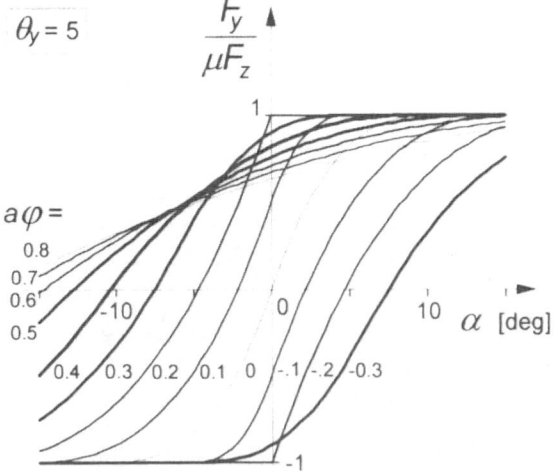

Fig.3.23. Side force characteristics of the single row brush model up to large levels of spin.

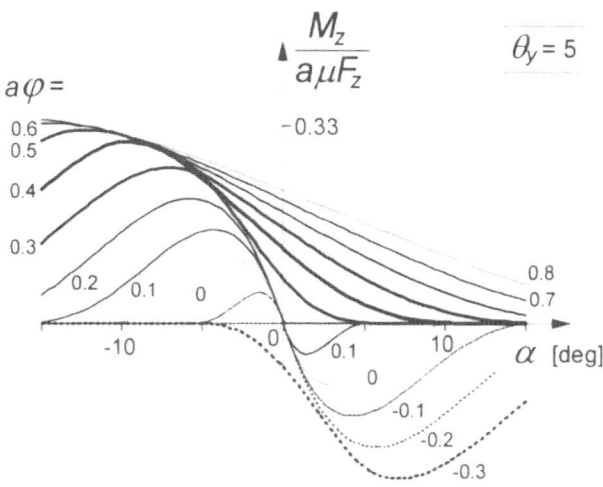

Fig.3.24. Aligning torque characteristics of the single row brush model up to large levels of spin

with Fig.3.22, the force at zero side slip (α=0) first increases with increasing spin and then decays. The aligning torque first remains zero and then starts to increase. The diagrams for the case of finite contact width indicate that without side slip, that is through spin alone, the side force can not reach its maximum level anymore. Figure 3.26 shows that the model with contact width develops a torque directly starting from zero spin. Furthermore, it appears that this model can generate much larger spin torques than the model with only one row of elements.

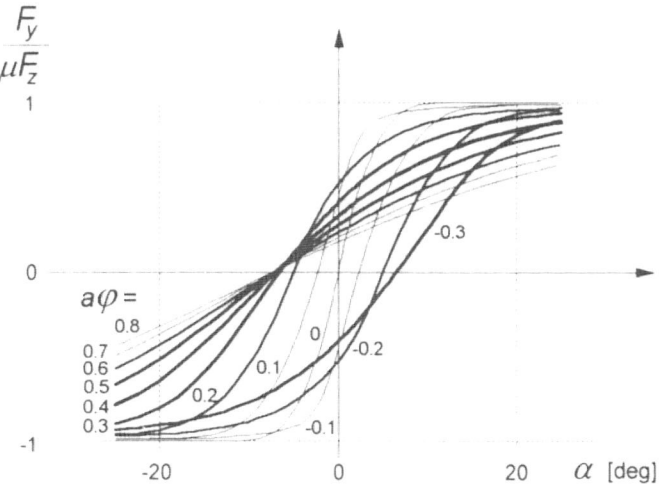

Fig.3.25. Side force characteristic for a double row brush model computed with tread element following simulation model. Same parameters as for model of Fig.3.23. same stiffnesses of elements in x and y direction, half tread width $b = 0.5a$.

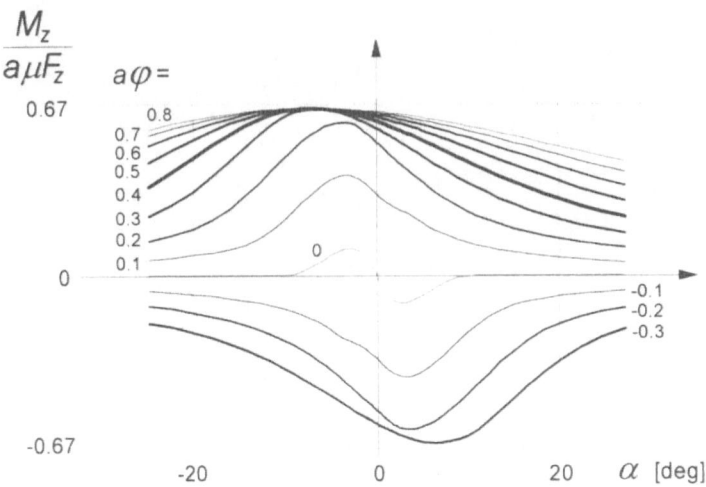

Fig.3.26. Aligning torque characteristic for double row brush model computed with tread element
following model. Same parameters as for model of Fig.3.24, same stiffnesses of elements in
x and y direction, half tread width $b = 0.5a$.

4. An Empirical Steady-State Tyre Model (the 'Magic Formula')

A widely used empirical tyre model is based on the so-called Magic Formula. The development
of the model was started in the mid-eighties. In a cooperative effort TU-Delft and Volvo
developed several versions (Bakker et al. 1987, Bakker et al. 1989, Pacejka et al. 1993). In these
models the combined slip situation was modelled from a physical view point. In 1993 Michelin
(cf. Bayle et al. 1993) introduced a purely empirical method using Magic Formula based
functions to describe the tyre horizontal force generation at combined slip. This approach was
adopted by DVR (the Delft Vehicle Dynamics Research Center, a joint venture of TU-Delft and
TNO- Delft). In the newest version of 'Delft Tyre' the original description of the aligning torque
is altered to accommodate a relatively simple combined slip extension. The pneumatic trail is
introduced as a basis to calculate this moment about the vertical axis, cf. Pacejka (1996). A
complete description of the model is given in the Appendix.

    We refer to Pacejka and Bakker (1993) for a detailed treatment of the pure slip part of this
model (that is: either lateral or longitudinal slip). For the side force and the fore and aft force
that part of the model remained unchanged. The formula reads:

$$y = D \sin[C \arctan \{Bx - E(Bx - \arctan Bx)\}] \tag{4.1}$$

with

$$Y(X) = y(x) + S_V$$
$$x = X + S_H \tag{4.1a}$$

where

Y: output variable F_x or F_y
X: input variable α or κ

and

B stiffness factor
C shape factor
D peak value
E curvature factor
S_H horizontal shift
S_V vertical shift

The Magic Formula $y(x)$ typically produces a curve that passes through the origin $x = y = 0$, reaches a maximum and subsequently tends to a horizontal asymptote. For given values of the coefficients B, C, D and E the curve shows an anti-symmetric shape with respect to the origin. To allow the curve to have an off-set with respect to the origin two shifts S_h and S_v have been introduced. A new set of coordinates $Y(X)$ arises as shown in Fig.4.1. The formula is capable of producing characteristics that closely match measured curves for the side force F_y and the fore and aft force F_x as functions of their respective slip quantities: the slip angle α and the longitudinal slip κ.

Figure 4.1 (upper part) illustrates the meaning of some of the factors through a typical side force characteristic. Obviously, coefficient D represents the peak value (with respect to the x-

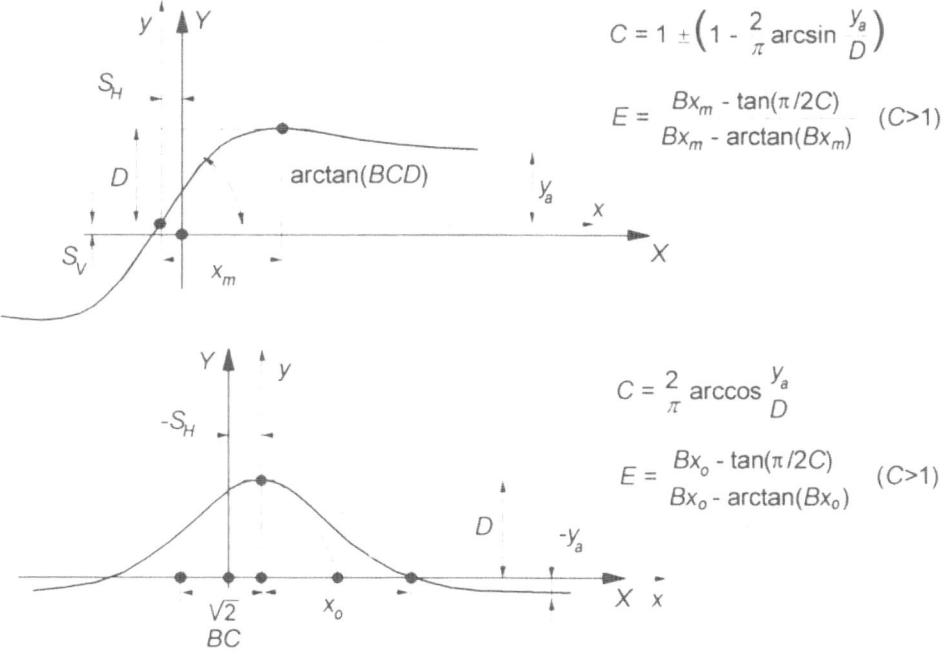

$$C = 1 \pm \left(1 - \frac{2}{\pi}\arcsin\frac{y_a}{D}\right)$$

$$E = \frac{Bx_m - \tan(\pi/2C)}{Bx_m - \arctan(Bx_m)} \quad (C>1)$$

$$C = \frac{2}{\pi}\arccos\frac{y_a}{D}$$

$$E = \frac{Bx_o - \tan(\pi/2C)}{Bx_o - \arctan(Bx_o)} \quad (C>1)$$

Fig.4.1.   Curves produced by the sine and cosine versions of the Magic Formula, Eqs.(4.1) and (4.5). The meaning of curve parameters have been indicated.

axis) and the product BCD corresponds to the slope at the origin ($x = y = 0$). The shape factor C controls the limits of the range of the sine function appearing in formula (4.1) and thereby determines the shape of the resulting curve. The factor B is left to determine the slope at the origin and is called the stiffness factor. The offsets S_H and S_V appear to occur when ply-steer and conicity effects and possibly the rolling resistance cause the F_y and F_x curves not to pass through the origin. Also, wheel camber will give rise to a considerable offset of the F_y vs α curves. Such a shift may be accompanied by a significant deviation from the pure anti-symmetric shape of the original curve (cf. Fig.3.25). To accommodate such an asymmetry, the curvature factor E is made dependent of the sign of the abscissa (x).

$$E = E_o + \Delta E \cdot \mathrm{sgn}(x) \tag{4.2}$$

Also the difference in shape that is expected to occur in the F_x vs κ characteristic between the driving and braking ranges can be taken care of.

The various factors are functions of normal load and wheel camber angle. Several parameters appear in these functions. A suitable regression technique is used to determine their values from measured data according to a quadratic algorithm for the best fit (cf. van Oosten and Bakker, 1993). One of the important functional relationships used is the variation of the cornering stiffness (or approximately: $BCD_y = K_y = \partial F_y/\partial\alpha$ at $\alpha = -S_h$) with F_z.

$$BCD_y = p_1 \sin[2 \arctan(F_z/p_2)] \cdot (1 - p_3|\gamma|) \tag{4.3}$$

For zero camber, the cornering stiffness attains its maximum p_1 at $F_z = p_2$. In Fig.4.2 the basic relationship has been depicted. Apparently, for a cambered wheel the cornering stiffness decreases with increasing $|\gamma|$. We refer to the Appendix for a complete listing of the formulae.

The aligning torque M_z can now be obtained by multiplying the side force F_y with the pneumatic trail t and adding the usually small (except with camber) residual torque M_{zr} (cf. Fig.4.3). We have:

$$M_z = -t \cdot F_y + M_{zr} \tag{4.4}$$

with the pneumatic trail

$$t(\alpha_t) = D_t \cos[C_t \arctan\{B_t\alpha_t - E_t(B_t\alpha_t - \arctan(B_t\alpha_t))\}] \tag{4.5}$$

where

$$\alpha_t = \alpha + S_{Ht} \tag{4.5a}$$

and the residual torque:

$$M_{zr}(\alpha_r) = D_r \cos[\arctan(B_r\alpha_r)] \tag{4.6}$$

with

$$\alpha_r = \alpha + S_{Hf} \tag{4.6a}$$

It is seen that both parts of the moment are modelled using the Magic Formula, but instead of the sine function, the cosine function is employed. In that way a hill-shaped curve is produced.

In the lower part of Fig.4.1 the basic properties of the cosine based curve have been indicated

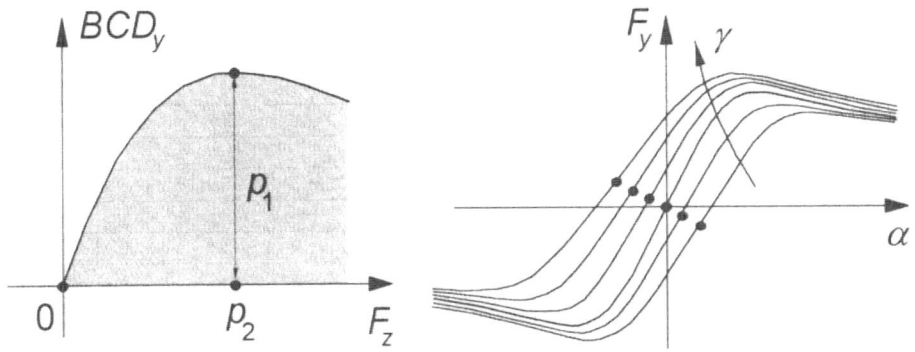

Fig.4.2. Cornering stiffness versus vertical load and the influence of wheel camber, Eq. (4.3).

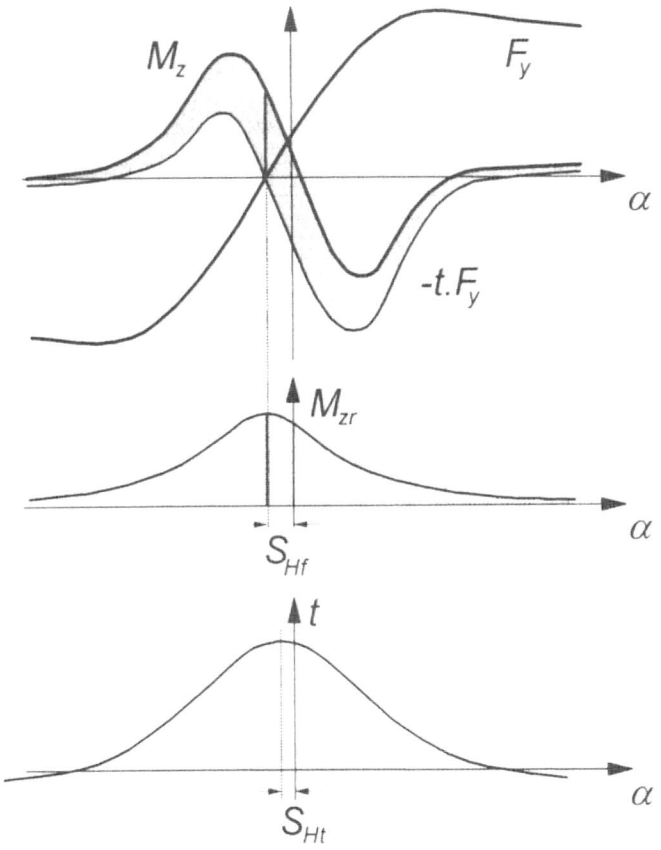

Fig.4.3. The aligning torque characteristic composed of a part directly attributed to the side force and a part due the so-called residual torque (due to tyre conicity and camber).

(subscripts of factors have been deleted again). Again, D is the peak value, C is a shape factor determining the level of the horizontal asymptote and now B influences the curvature at the peak (illustrated with the inserted parabola). Factor E modifies the shape at larger values of slip. The residual torque attains its maximum D, at the slip angle where the side force becomes equal to zero. This is accomplished through the horizontal shift S_{Hf}. The peak of the pneumatic trail occurs at $\alpha = -S_{Ht}$. This formulation has proven to give very good agreement with measured curves. The advantage with respect to the earlier versions, where formula (4.1) is used for the aligning torque as well, is that we have now directly assessed the function for the pneumatic trail which is needed to handle the combined slip situation. In case of the possible presence of large camber angles (motorcycles) it is better to use in (4.4) the side force F_y that would arise at $\gamma = 0$. The shift S_{Hf} in (4.6a) would then correspond to the value of $-\alpha$ where the side force vanishes at zero camber. Also the side force function (4.1) and the cornering stiffness function (4.3) have been modified to better approximate large camber response for motorcycle tyres, cf. de Vries (1998).

In the paper of Pacejka and Bakker (1993) the tyre's response to combined slip was modelled by using physically based formulae. A newer more efficient way is purely empiric. This method was developed by Michelin and published by Bayle, Forissier and Lafon (1993). It describes the effect of combined slip on the lateral force and on the longitudinal force characteristics. Weighting functions G are introduced which when multiplied with the original pure slip functions (4.1) produce the interaction effects of κ on F_y and of α on F_x. The weighting functions have a hill shape. They take the value one in the special case of pure slip (κ or α equal to zero). When, for example, at a given slip angle a from zero increasing brake slip is introduced, the relevant weighting function for F_y may first show a slight increase in magnitude (becoming larger than one) but will soon reach its peak after which a continuous decrease follows. The cosine version of the Magic Formula is used to represent the hill shaped function:

$$G = D \cos[C \arctan\{Bx - E(Bx - \arctan Bx)\}] \qquad (4.7)$$

Here, G is the resulting weighting factor and x is either κ or α (possibly shifted). The coefficient D represents the peak value (slightly deviating from one if a horizontal shift of the hill occurs), C determines the height of the hill's base and B influences the sharpness of the hill. As an extension to the original function published by Bayle et al., the part with shape factor E has been added. This appears to improve the approximation especially in view of the strict condition that the weighting function G must remain positive for all slip conditions.

For the side force we get the following formulae:

$$F_y = G_{y\kappa} \cdot F_{yo} + S_{Vy\kappa} \qquad (4.8)$$

$$G_{y\kappa} = \frac{\cos[C_{y\kappa}\arctan\{B_{y\kappa}\kappa_S - E_{y\kappa}(B_{y\kappa}\kappa_S - \arctan B_{y\kappa}\kappa_S)\}]}{\cos[C_{y\kappa}\arctan\{B_{y\kappa}S_{Hy\kappa} - E_{y\kappa}(B_{y\kappa}S_{Hy\kappa} - \arctan B_{y\kappa}S_{Hy\kappa})\}]} \qquad (4.9)$$

with

$$\kappa_S = \kappa + S_{Hy\kappa} \qquad (4.9a)$$

and further

$$B_{y\kappa} = r_{By1} \cos[\arctan\{r_{By2}(\alpha - r_{By3})\}]$$ (4.10)

$$C_{y\kappa} = r_{Cy1}$$ (4.11)

$$E_{y\kappa} = r_{Ey1} + r_{Ey2} df_z$$ (4.12)

$$S_{Vy\kappa} = D_{Vy\kappa} \sin[r_{Vy5} \arctan(r_{Vy6}\kappa)]$$ (4.13)

$$D_{Vy\kappa} = \mu_y F_z \cdot (r_{Vy1} + r_{Vy2} df_z + r_{Vy3}\gamma) \cdot \cos[\arctan(r_{Vy4}\alpha)]$$ (4.14)

In Eq.(4.8) F_{yo} denotes the side force at pure side slip obtained from Eq.(4.1); $S_{Vy\kappa}$ is the vertical 'shift' sometimes referred to as the κ-induced ply-steer. This function varies with longitudinal slip κ as indicated in (4.13). Its maximum $D_{Vy\kappa}$ decreases with increasing magnitude of the slip angle and changes with vertical load. The quantity df_z denotes the non-dimensional increment of the wheel load with respect to the nominal load. The factor $B_{y\kappa}$ influences the sharpness of the hill shaped weighting function (4.9). As indicated, the hill becomes more shallow (wider) at larger slip angles (then $B_{y\kappa}$ decreases according to (4.10)). The denominator of $G_{y\kappa}$ makes $G_{y\kappa}$ $= 1$ at $\kappa = 0$. The horizontal shift $S_{Hy\kappa}$ of the weighting function accomplishes the slight increase that the side force experiences at moderate braking before the peak of $G_{y\kappa}$ is reached and the decay of F_y commences. This 'shift' may be made dependent of the vertical load.. The combined slip relations for F_x are similar. However, a vertical shift was not needed to be included. In Fig.4.4 a three-dimensional graph is shown indicating the variation of F_x and F_y with both α and κ.

Regarding the aligning torque, physical insight is used to model the situation at combined slip. The arguments α_t and α_r (including a shift) appearing in the functions for pneumatic trail and residual torque are replaced by equivalent slip angles incorporating the effect of κ on the composite slip. Besides, an extra term is introduced to account for the fact that a moment arm s arises for F_x as a result of camber γ and lateral tyre deflections through F_y. This may give rise to a sign change of the aligning torque in the range of braking (cf. Fig.3.15). We write:

$$M_z = -t(\alpha_{t,eq}) \cdot F_y + M_{zr}(\alpha_{r,eq}) + s(F_y, \gamma) \cdot F_x$$ (4.15)

$$\alpha_{t,eq} = \arctan \sqrt{\tan^2\alpha_t + \left(\frac{K_x}{K_y}\right)^2 \kappa^2} \cdot \mathrm{sgn}(\alpha_t)$$ (4.16)

and similar for $\alpha_{r,eq}$. The complete set of steady-state formulae has been listed in the Appendix. Parameters p, q, r and s of the model are non-dimensional quantities. In addition, user scaling factors λ have been introduced. With that tool the effect of changing friction coefficient, cornering stiffness, camber stiffness etc. can be quickly investigated in a qualitative way without having the need to implement a completely new tyre data set. Scaling is done in such a way that realistic relationships are maintained. For instance, when changing the cornering stiffness and the friction coefficient in lateral direction (through λ_{Ky} and $\lambda_{\mu y}$), the abscissa of the pneumatic trail characteristic is changed in a way similar to that of the side force characteristic.

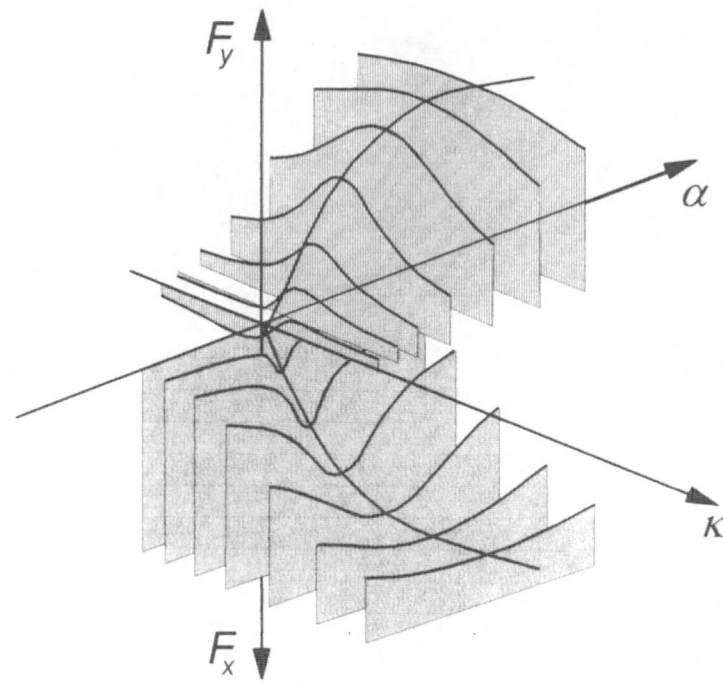

Fig.4.4.   Combined slip lateral and longitudinal force 3-D diagram showing the interaction of both slip
components.

5. Transient and Oscillatory Motion Response

Steady-state tyre models will loose their accuracy when the motion of the wheel shows variations
in time. The fact that a deflection has to be built up to create a force calls for a model that
contains carcass compliance. For the assessment of the side force and aligning torque response
to varying slip angle and turn slip, a model based on the physical description of the lateral
deflection of a stretched string restrained with respect to the wheel centre plane by an elastic
foundation is often used. This model was originally developed by Von Schlippe (1942). A
portion of the string touches the ground. There, frictional forces are transmitted. Sliding is
assumed not to occur, at least in the range close to the leading edge. We refer to Pacejka (1966,
1981a) for details on the string model.

Figure 5.1 depicts the deflected string model in top view. The input wheel motions are
represented by the slip angle α and the turn slip φ (camber disregarded here). The latter quantity
is the rate of change of the yaw angle with respect to travelled distance. In case the slip angle
remains equal to zero, the turn slip equals the path curvature. The deflection of the string can
be calculated in a rather simple manner starting from the leading edge (where the in fact existing
very short sliding zone is neglected and adhesion is assumed to occur) until the transition point

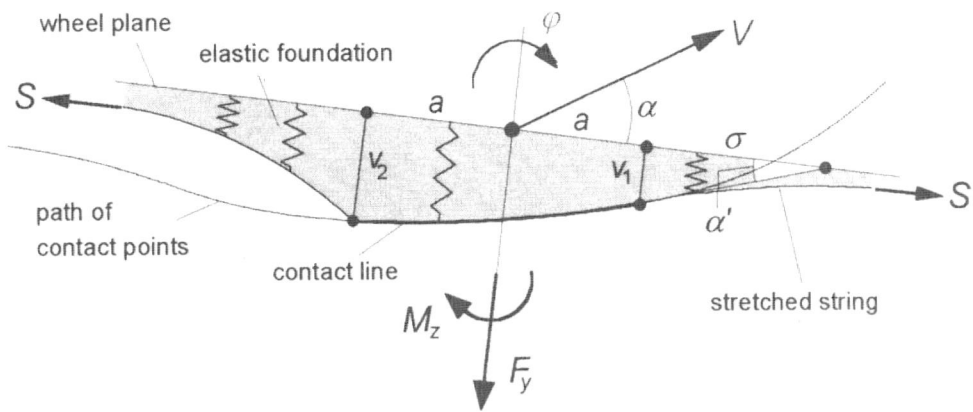

Fig. 5.1. Stretched string tyre model for non-steady-state side force and moment response. The relaxation length σ has been indicated.

where sliding starts. For the deflection v_1 at the leading edge we have the following first-order ordinary differential equation:

$$\frac{1}{V_x}\frac{dv_1}{dt} + \frac{v_1}{\sigma} = \tan \alpha - a\,\phi \qquad (5.1)$$

The symbol σ denotes the relaxation length and a stands for half the contact length.

The slip input quantities are assumed relatively small to allow the equation to remain linear. Since it can be proven that the slope of the string deflection changes continuously around the leading edge (no kink), the string deflection angle α' in the foremost portion of the contact line directly relates to the deflection itself.

$$\tan \alpha' = \frac{v_1}{\sigma} \qquad (5.2)$$

At steady-state side slip conditions, this angle becomes equal to the slip angle α of the wheel. For not too small wavelengths of the motion, the curvature of the contact line may be neglected. Then, the string deflection will be approximately equal to the deflection that would arise at steady-state side slip with slip angle equal to the deflection angle calculated with Eq.(5.1). As a consequence, we may use the steady-state functions $F_y(\alpha')$ and $M_z(\alpha')$ according to e.g. the Magic Formula dealt with in the preceding section, to find the force and moment response to varying slip angle.

The prime effect of the differential equation is that a first-order lag of the output (tyre deflections and forces) with respect to the input (the wheel slip) arises. This lag is the main cause of typical wheel transient motion and oscillatory behaviour, e.g. the unstable wheel shimmy phenomenon. In Fig.5.2 areas of instability have been shown calculated for the simple (linearised) trailing wheel system. In the closed area self sustained oscillations occur which may be extremely violent and dangerous, cf. Pacejka (1966). The area shrinks when more damping

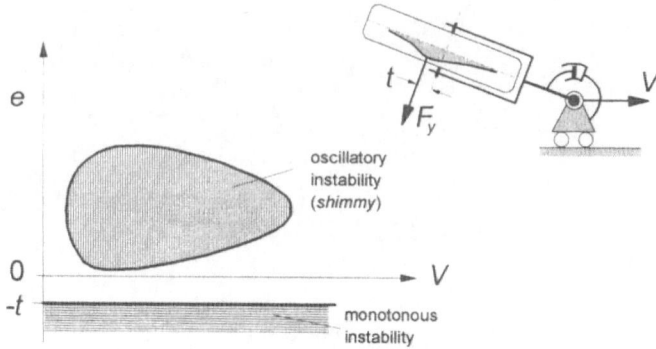

Fig.5.2. Possible unstable areas for the simple trailing wheel system.

around the swivel axis is introduced. Light trucks, aircraft and motorcycles sometimes suffer from this phenomenon.   When operating in the larger range of slip not only the non-linear force and moment characteristics are to be used but also the gradual decay of the relaxation length with increasing deflection angle is to be taken into account (cf. Pacejka and Takahashi, 1992).

The differential equation (5.1) may also be used to study the effect of varying the load while cornering. Essential is then that the load dependency of the relaxation length $\sigma(F_z)$ is taken into account. Due to the gradient $\partial\sigma/\partial F_z$ a dynamic loss in average side force will occur when the side slipping wheel rolls over an undulated road surface. Figure 5.3 illustrates how F_y changes due to a sinusoidal variation of F_z while the wheel moves at a constant slip angle of 6 degrees. Here, with the relatively large slip angle, it was necessary to take the relaxation length's decay with deflection angle into account. Two different frequencies have been considered. One very low (almost steady state) and the other relatively high. This simulated result shows a clear drop of the average side force at 8 Hz. In the diagram of Fig.5.4 the average loss in side force at a small slip angle is presented as a function of frequency at three different amplitudes of the load variation. This loss should be minimized by (actively) controlling and limiting the variation of vertical load. It should then be realized that the effect is non-linear. Small to moderate amplitudes of the load will do no harm. Moreover, there is a wavelength effect (that is: more serious at shorter wavelengths, meaning: higher frequencies). At very low frequencies only the so-called static loss remains which is a result of the non-linear (degressive) change of cornering stiffness with vertical load (cf. Fig.4.2).

After multiplying left and right members of (5.1) with the longitudinal wheel speed V_x and σ, while disregarding turnslip φ we find:

$$\sigma\frac{dv_1}{dt} + |V_x|v_1 = -\sigma V_{sy} \tag{5.3}$$

where V_{sy} represents the lateral slip velocity of the wheel which is related to the slip angle through:

$$\tan\alpha = -\frac{V_{sy}}{|V_x|} \tag{5.4}$$

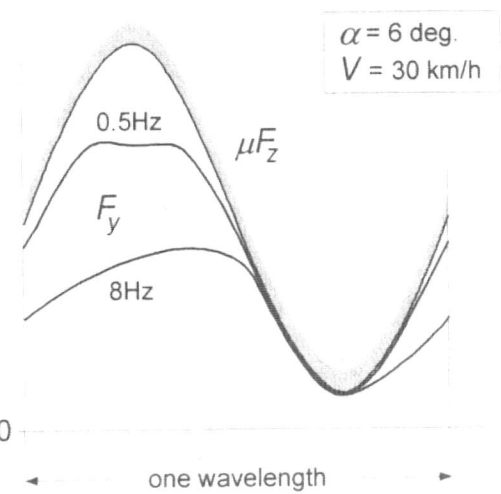

$\alpha = 6$ deg.
$V = 30$ km/h

0.5Hz

μF_z

F_y

8Hz

0

one wavelength

Fig.5.3. Side force variation due to sinusoidal vertical wheel load fluctuations.

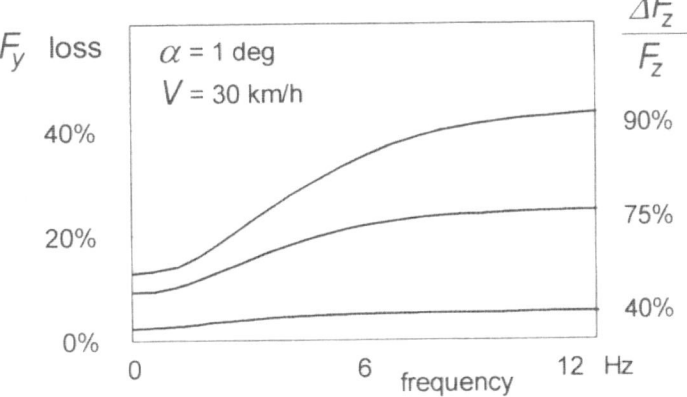

F_y loss

$\alpha = 1$ deg
$V = 30$ km/h

40%

20%

0%

0   6   12   Hz
frequency

$\dfrac{\Delta F_z}{F_z}$

90%

75%

40%

Fig.5.4. Static (at 0Hz) and dynamic loss in average side force at three amplitudes of vertical load
fluctuations.

Note that we have introduced the absolute value of the forward speed which allows us to use the
differential equation also when the speed changes sign. This version (5.3) of differential equation
(5.1) makes it possible to start and stop from and to stand-still where $V_x=0$. This would not have
been the case when an instantaneous steady-state response is assumed and the slip angle itself
from (5.4) is employed as input into the steady-state tyre force and moment formulation.
Obviously, we then have to deal with the singularity at $V_x=0$. From Eq.(5.3) it can be seen that
at vanishing forward speed V_x an integrator arises which changes the tyre into a spring. In

reality, this is in fact the case.

An equation similar to Eq.(5.3) can be derived for the response of the longitudinal deflection u_1 of the carcass to variations of the longitudinal wheel slip speed V_{sx}. A different relaxation length will then apply. A deformation gradient κ' may then be defined analogous to $\tan\alpha'$. From that gradient the longitudinal force $F_x(\kappa')$ is obtained. In case of time-varying combined slip, we may use the combined slip steady-state force and moment functions with α' and κ' used as the input slip quantities. Then, the two relaxation lengths showing decaying functions of both deflection gradients may be employed for greater accuracy.

Equation (5.3) holds for cases up to a limited level of slip where still adhesion occurs near the leading edge. However, the model may still be used for larger values of slip. But when the slip is reduced after having reached such high levels (for instance releasing the brake after wheel lock), an erroneous result will occur: it takes too long for the deflection to recover. This error may be avoided by using the relaxation length's dependency on the deflection gradient. Then, the relaxation length (in (5.2) and (5.3) is continuously adapted such that the lateral deflection v_1 remains equal to the side force divided by the tyre lateral stiffness. Reference is made to Pacejka and Takahashi (1992) and Higuchi (1997) for further information on this method. However, numerical difficulties may be encountered at violent vertical wheel oscillations. Cases of oscillating wheel motions at high levels of slip (e.g. situations where at violent braking, ABS control and at the same time vertical load variations on rough roads give rise to fluctuations of wheel slip) may call for a different model approach. In Van der Jagt et al. (1989) a model is used in which the carcass fore and aft compliance has been introduced explicitly. This implies that the relaxation length concept has been abandoned. A further development of this approach also for lateral slip variations has been published by Pacejka and Besselink (1997). In Fig.5.5 the model is depicted. For reasons of computational causality a mass representing the lower portion of the tread band (belt) had to be included. The slip speed of that mass with respect to the road surface is then considered as input (via Eq.(3.19) and (5.4)) into the steady-state Magic Formula force model. Of course, the wheel/tyre polar moment of inertia has to be accounted for in the analysis.

When higher frequencies, shorter wavelengths and/or high speeds are to be considered, the inclusion of tyre inertia and contact line curvature may become necessary. As the aligning torque is especially sensitive to inertia effects, Delft Tyre takes the gyroscopic couple due to lateral tyre deformation rates into account, cf. Pacejka (1981a). Figure 5.7 shows an example of theoretically assessed frequency response functions of the aligning torque to yaw oscillations (small angles, linear model with first belt vibrational modes considered; from same reference). The earlier mentioned phase lag can be seen to occur at relatively low values of speed and frequency. At higher speed the lag changes into a lead because of the effect of the gyroscopic couple that arises due to tilt angle variations of the belt with respect to the rim.

A versatile tyre model for vehicle simulation studies developed in a project called SWIFT (Short Wavelength Intermediate Frequency Tyre model) is now available. It is based on the use of the Magic Formula and is able to handle situations up to frequencies not much higher than 50 Hz. Running over short unevennesses forms a difficult problem to model. Effective road plane changes both in level and slope, and varying effective rolling radius (determined experimentally at very low speed) are employed as inputs to the relatively simple dynamic tyre model, cf. Zegelaar (1996) and Maurice (1998). Figure 5.6 depicts this more advanced dynamic model.

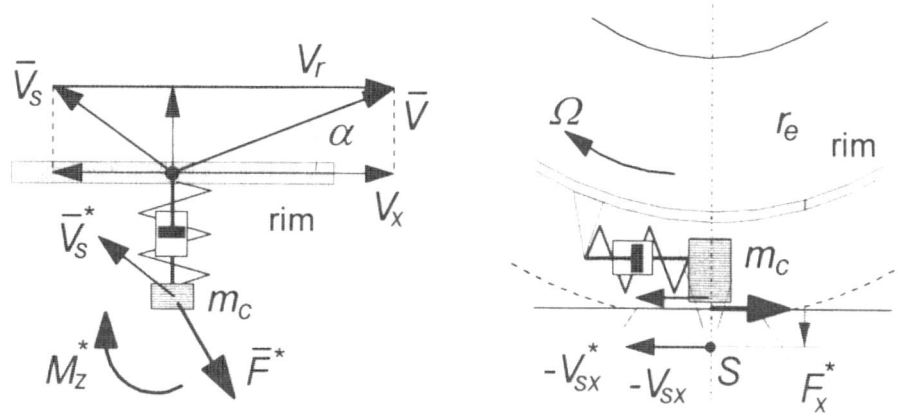

Fig.5.5. In-plane dynamic tyre/wheel model showing carcass tangential compliance.

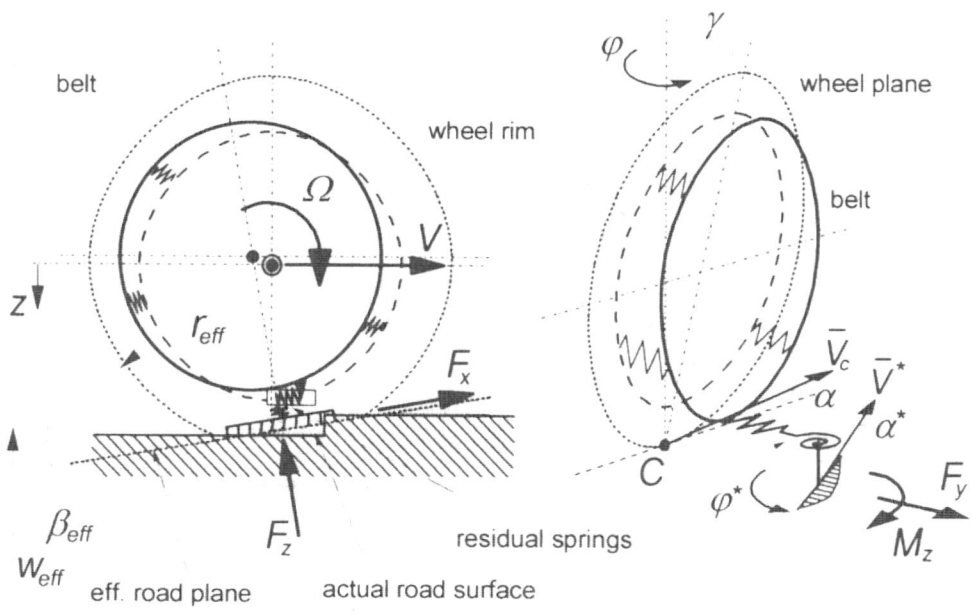

Fig.5.6. 'SWIFT' dynamic tyre model, running on uneven roads while moving at possibly large slip and frequencies < 50Hz.

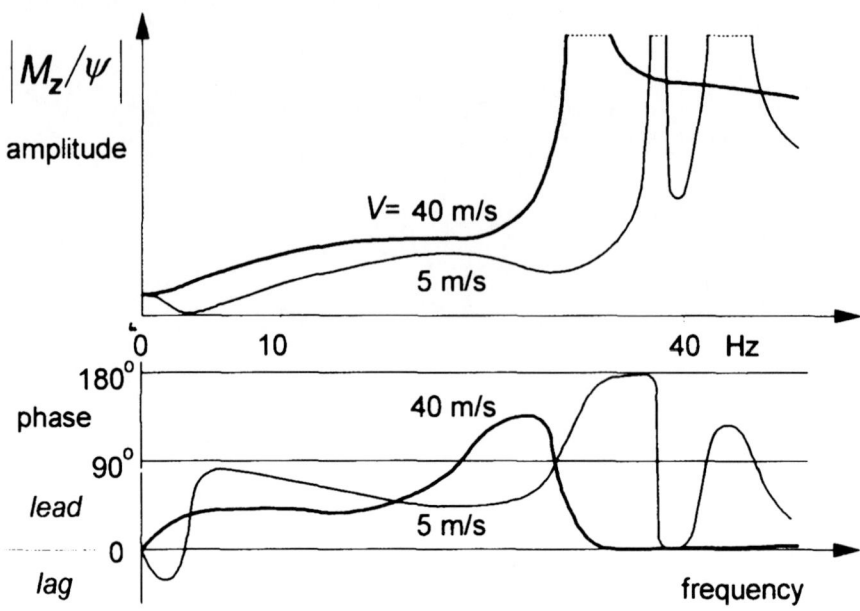

Fig.5.7. Aligning torque amplitude and phase frequency response to (small) steer angle variations
according to a linear dynamic out-of-plane tyre model.

References

Bakker, E., Nyborg, L., and Pacejka, H.B. (1987). 'Tyre modelling for use in vehicle dynamics studies'.
    SAE 870421.
Bakker, E., Pacejka, H.B., and Lidner, L. (1989). 'A new tyre model with application in vehicle dynamics
    studies'. *Proceedings of the 4th Autotechnologies Conference*, Monte Carlo, June 1989. *SAE* 890087.
Bayle, P., Forissier. J.F., and Lafon, S. (1993). 'A new tyre model for vehicle dynamics simulations'.
    Automotive Technology International '93, pp. 193-198.
Bernard, J.E., Segel, L., and Wild, R.E. (1977). 'Tire shear force generation during combined steering and
    braking maneuvres'. *SAE* 770852.
Dugoff, H., Fancher, P.S., and Segel, L. (1970). 'An analysis of tire traction properties and their influence
    on vehicle dynamic performance'. *Proceedings FISITA/SAE*, Brussels, *SAE* 700377.
Fiala, E. (1954). 'Seitenkräfte am rollenden Luftreifen'. *VDI-Zeitschrift* 96, p.973.
Gipser, M., Hofer, R., and Lugner, P. (1997). 'Dynamical Tyre Response to Road Unevennesses'.
    Proceedings of the 2nd Tyre Colloquium, Berlin, Sept.1996, Suppl. to *Vehicle System Dynamics* 27.
Higuchi, A., and Pacejka, H.B. (1997). 'The relaxation length concept at large wheel slip and camber'.
    Proceedings of the 2nd Tyre Colloquium, Berlin, Sept.1996, Suppl. to *Vehicle System Dynamics* 27.
Jagt, P. van der, Pacejka, H.B., and Savkoor, A.R. (1989). 'Influence of tyre and suspension dynamics on
    the braking performance of an anti-lock system on uneven roads'. *Proceedings of EAEC Conf. '89*,
    C382/047, IMechE 1989.

Mastinu. G., Gaiazzi, S., Montanaro, F., and Pirola, D. (1997). 'A Semi-Analytical Tyre Model for Steady-and Transient State Simulations'. *Proceedings of the 2nd Tyre Colloquium*, Berlin. Sept.1996, Suppl. to *Vehicle System Dynamics* 27.

Maurice, J.P., Zegelaar, P.W.A., and Pacejka, P.W.A. (1998). 'The Influence of Belt Dynamics on Cornering and Braking Performance of Tyres'. *Proceedings of the 15th IAVSD Symposium*. Budapest. Aug. 1997, Suppl. to *Vehicle System Dynamics* 29.

Oosten, J. van, and Bakker. E. (1993). 'Determination of Magic Formula tyre model parameters'. *Proceedings of the 1st Tyre Colloquium*, Delft. Oct. 1991. Suppl. to *Vehicle System Dynamics* 21.

Pacejka, H.B. (1966). '*The wheel shimmy phenomenon*'. Dissertation. University of Technology Delft.

Pacejka, H.B., and Fancher, P.S. (1972). 'Hybrid simulation of shear force development of a tyre experien cing longitudinal and lateral slip'. *Proceedings XII'th FISITA Congress*. London. p. 3/78.

Pacejka, H.B. (1981a). 'In-plane and out-of-plane dynamics of pneumatic tyres'. *Vehicle System Dynamics* 10. pp. 221-251.

Pacejka, H.B. (1981b). 'Analysis of tyre properties, Yaw and Camber Analysis'. Chapter 9 of *Mechanics of Pneumatic Tires*, ed. S.K. Clark. US Gov. Printing Office 1981-0-308-639. Washington D.C.

Pacejka, H.B., and Sharp, R.S. (1991). 'Shear force development by pneumatic tyres in steady state conditions: A review of modelling aspects'. *Vehicle System Dynamics* 20. pp. 121-176.

Pacejka, H.B., and Takahashi. T. (1992). 'Pure slip characteristics of tyres on flat and on undulated road surfaces'. *Proceedings of AVEC '92*, Yokohama.

Pacejka, H.B., and Bakker. E. (1993). 'The Magic Formula tyre model'. *Proceedings of the 1st Tyre Colloquium*, Delft, Oct.1991, Suppl. to *Vehicle System Dynamics* 21.

Pacejka, H.B. (1996). 'The Tyre as a Vehicle Component'. *Proceedings XXVI FISITA Congress*. Prague.

Pacejka, H.B., and Besselink, I. (1997). 'A consistent Magic Formula tyre model with dynamic extension'. *Proceedings of the 2nd Tyre Colloquium*, Berlin. Sept.1996, Suppl. to *Vehicle System Dynamics* 27

Savkoor, A.R. (1970). 'The lateral flexibility of pneumatic tires and its application to the lateral rolling contact problem'. *Proceedings FISITA/SAE*. Brussels. SAE 700378.

Schlippe, B. von, and Dietrich, R. (1942). 'Zur Mechanik des Luftreifens'. Zentral für Wiss. Ber.. Berlin.

Sharp, R.S., and El-Nashar. M.A. (1986). 'A generally applicable digital computer based mathematical model for the generation of shear forces by preamatic tyres'. *Vehicle System Dynamics* 15. p.187-210.

Vries. E.J.H. de, and Pacejka, H.B. (1998). 'Motorcycle Tyre Measurements and Models'. *Proceedings 15th IAVSD Symposium*. Budapest, Aug. 1997. Suppl. to *Vehicle System Dynamics* 29.

Willumeit, H.P. (1969), '*Theoretische Untersuchungen an einem Modell des Luftreifens unter Seiten- und Umfangskraft*'. Dissertation. Technische Universität Berlin.

Zegelaar, P.W.A., and Pacejka, H.B.(1996). 'In-plane dynamics of tyres on uneven roads'. *Proceedings 14th IAVSD Symposium*, Ann Arbor. Aug. 1995, Suppl. to *Vehicle System Dynamics* 25.

Appendix

Delft Tyre steady-state part: the *Magic Formula* tyre model '99.

Non-dimensional model parameters p, q, r and s have been introduced. For this we define: unloaded tyre radius R_o, nominal load F_{zo}, adapted nominal load $F'_{zo} = \lambda_{Fzo} F_{zo}$, normalized vertical load increment $df_z = (F_z - F'_{zo})/F'_{zo}$.

Instead of using as input the slip angle α (in radians, from Eq.(5.4)) and the longitudinal slip ratio $\kappa = -V_{sx}/|V_x|$ one may take the transient slip quantities $\tan\alpha'$ (5.2) and κ' resulting from Eq.(5.1) and a similar differential equation for the longitudinal deflection u_1. Note, that in spite of (5.4) where $V_{sy} = V_y$ one should take for $\cos\alpha$ appearing in the equations for the aligning torque to properly handle the case of backward driving ($V_x < 0$): $\cos\alpha = V_x/V$. The velocity of the wheel centre or better of the contact centre V should be downward limited, e.g.: $V = \max(V, 0.01)$.

User Scaling Factors

For the user's convenience a set of scaling factors λ is available to examine the influence of changing a number of important overall parameters. The default value of these factors is set equal to one.

Longitudinal Force (pure longitudinal slip)

$$F_{xo} = D_x \sin[C_x \arctan\{B_x\kappa_x - E_x(B_x\kappa_x - \arctan(B_x\kappa_x))\}] + S_{Vx} \tag{A1}$$

$$\kappa_x = \kappa + S_{Hx} \tag{A2}$$

$$C_x = p_{Cx1} \cdot \lambda_{Cx} \qquad (>0) \tag{A3}$$

$$D_x = \mu_x \cdot F_z \tag{A4}$$

$$\mu_x = (p_{Dx1} + p_{Dx2} df_z) \cdot \lambda_{\mu x} \qquad (>0) \tag{A5}$$

$$E_x = (p_{Ex1} + p_{Ex2} df_z + p_{Ex3} df_z^2) \cdot \{1 - p_{Ex4} \operatorname{sgn}(\kappa_x)\} \cdot \lambda_{Ex} \qquad (\leq 1) \tag{A6}$$

$$K_x = F_z \cdot (p_{Kx1} + p_{Kx2} df_z) \cdot \exp(-p_{Kx3} df_z) \cdot \lambda_{Kx} \qquad (= B_x C_x D_x = \frac{\partial F_{xo}}{\partial \kappa_x} \text{ at } \kappa_x = 0) \tag{A7}$$

$$B_x = K_x/(C_x D_x) \tag{A8}$$

$$S_{Hx} = (p_{Hx1} + p_{Hx2} df_z) \cdot \lambda_{Hx} \tag{A9}$$

$$S_{Vx} = F_z \cdot (p_{Vx1} + p_{Vx2} df_z) \cdot \lambda_{Vx} \cdot \lambda_{\mu x} \tag{A10}$$

Lateral Force (pure side slip)

$$F_{yo} = D_y \sin[C_y \arctan\{B_y\alpha_y - E_y(B_y\alpha_y - \arctan(B_y\alpha_y))\}] + S_{Vy} \tag{A11}$$

$$\alpha_y = \alpha + S_{Hy} \tag{A12}$$

$$\gamma_y = \gamma \cdot \lambda_{\gamma y} \tag{A13}$$

$$C_y = p_{Cy1} \cdot \lambda_{Cy} \qquad (>0) \tag{A14}$$

$$D_y = \mu_y \cdot F_z \tag{A15}$$

$$\mu_y = (p_{Dy1} + p_{Dy2} df_z) \cdot (1 - p_{Dy3} \gamma_y^2) \cdot \lambda_{\mu y} \qquad (>0) \tag{A16}$$

$$E_y = (p_{Ey1} + p_{Ey2} df_z) \cdot \{1 - (p_{Ey3} + p_{Ey4} \gamma_y) \operatorname{sgn}(\alpha_y)\} \cdot \lambda_{Ey} \qquad (\leq 1) \tag{A17}$$

$$K_y = p_{Ky1} F_{zo} \sin[2 \arctan\{F_z/(p_{Ky2} F_{zo} \lambda_{Fzo})\}] \cdot (1 - p_{Ky3}|\gamma_y|) \cdot \lambda_{Fzo} \cdot \lambda_{Ky}$$
$$\qquad (= B_y C_y D_y = \partial F_{yo}/\partial \alpha_y \text{ at } \alpha_y = 0) \tag{A18}$$

$$B_y = K_y/(C_y D_y) \tag{A19}$$

$$S_{Hy} = (p_{Hy1} + p_{Hy2} df_z) \cdot \lambda_{Hy} + p_{Hy3} \gamma_y \tag{A20}$$

$$S_{Vy} = F_z \cdot \{(p_{Vy1} + p_{Vy2} df_z) \cdot \lambda_{Vy} + (p_{Vy3} + p_{Vy4} df_z) \gamma_y\} \cdot \lambda_{\mu y} \tag{A21}$$

Aligning Torque (pure side slip)

$$M_{zo} = -t \cdot F_{yo} + M_{zr} \tag{A22}$$

$$t(\alpha_t) = D_t \cos[C_t \arctan\{B_t\alpha_t - E_t(B_t\alpha_t - \arctan(B_t\alpha_t))\}] \cdot \cos\alpha \tag{A23}$$

$$\alpha_t = \alpha + S_{Ht} \tag{A24}$$

$$M_{zr}(\alpha_r) = D_r \cos[\arctan(B_r\alpha_r)] \cdot \cos\alpha \tag{A25}$$

$$\alpha_r = \alpha + S_{Hf} \quad (= \alpha_f) \tag{A26}$$

$$S_{Hf} = S_{Hy} + S_{Vy}/K_y \tag{A27}$$

$$\gamma_z = \gamma \cdot \lambda_{\gamma z} \tag{A28}$$

$$B_t = (q_{Bz1} + q_{Bz2} df_z + q_{Bz3} df_z^2) \cdot (1 + q_{Bz4} \gamma_z + q_{Bz5}|\gamma_z|) \cdot \lambda_{Ky}/\lambda_{\mu y} \quad (>0) \tag{A29}$$

$$C_t = q_{Cz1} \qquad (>0) \tag{A30}$$

$$D_t = F_z \cdot (q_{Dz1} + q_{Dz2} df_z) \cdot (1 + q_{Dz3} \gamma_z + q_{Dz4} \gamma_z^2) \cdot (R_0/F_{zo}) \cdot \lambda_t \tag{A31}$$

$$E_t = (q_{Ez1} + q_{Ez2} df_z + q_{Ez3} df_z^2) \cdot \{1 + (q_{Ez4} + q_{Ez5} \gamma_z) \arctan(B_t C_t \alpha_t)\} \quad (\leq 1) \tag{A32}$$

$$S_{Ht} = q_{Hz1} + q_{Hz2} df_z + (q_{Hz3} + q_{Hz4} df_z) \gamma_z \tag{A33}$$

$$B_r = q_{Bz10} B_y C_y \tag{A34}$$

$$D_r = F_z \cdot \{(q_{Dz6} + q_{Dz7} df_z) \cdot \lambda_{Mr} + (q_{Dz8} + q_{Dz9} df_z) \gamma_z\} \cdot R_0 \cdot \lambda_{\mu y} \tag{A35}$$

Longitudinal Force   (combined)

$$F_x = D_{x\alpha} \cos[C_{x\alpha} \arctan\{B_{x\alpha} \alpha_S - E_{x\alpha}(B_{x\alpha} \alpha_S - \arctan(B_{x\alpha} \alpha_S))\}] \tag{A36}$$

$$\alpha_S = \alpha + S_{Hx\alpha} \tag{A37}$$

$$B_{x\alpha} = r_{Bx1} \cos[\arctan(r_{Bx2} \kappa)] \cdot \lambda_{x\alpha} \qquad (>0) \tag{A38}$$

$$C_{x\alpha} = r_{Cx1} \tag{A39}$$

$$D_{x\alpha} = F_{xo}/\cos[C_{x\alpha} \arctan\{B_{x\alpha} S_{Hx\alpha} - E_{x\alpha}(B_{x\alpha} S_{Hx\alpha} - \arctan(B_{x\alpha} S_{Hx\alpha}))\}] \tag{A40}$$

$$E_{x\alpha} = r_{Ex1} + r_{Ex2} df_z \tag{A41}$$

$$S_{Hx\alpha} = r_{Hx1} \tag{A42}$$

Lateral Force   (combined)

$$F_y = D_{y\kappa} \cos[C_{y\kappa} \arctan\{B_{y\kappa} \kappa_S - E_{y\kappa}(B_{y\kappa} \kappa_S - \arctan(B_{y\kappa} \kappa_S))\}] + S_{Vy\kappa} \tag{A43}$$

$$\kappa_S = \kappa + S_{Hy\kappa} \tag{A44}$$

$$B_{y\kappa} = r_{By1} \cos[\arctan\{r_{By2}(\alpha - r_{By3})\}] \cdot \lambda_{y\kappa} \qquad (>0) \tag{A45}$$

$$C_{y\kappa} = r_{Cy1} \tag{A46}$$

$$D_{y\kappa} = F_{yo}/\cos[C_{y\kappa} \arctan\{B_{y\kappa} S_{Hy\kappa} - E_{y\kappa}(B_{y\kappa} S_{Hy\kappa} - \arctan(B_{y\kappa} S_{Hy\kappa}))\}] \tag{A47}$$

$$E_{y\kappa} = r_{Ey1} + r_{Ey2} df_z \tag{A48}$$

$$S_{Hy\kappa} = r_{Hy1} + r_{Hy2} df_z \tag{A49}$$

$$S_{Vy\kappa} = D_{Vy\kappa} \sin[r_{Vy5} \arctan(r_{Vy6} \kappa)] \cdot \lambda_{Vy\kappa} \tag{A50}$$

$$D_{Vy\kappa} = \mu_y F_z \cdot (r_{Vy1} + r_{Vy2} df_z + r_{Vy3} \gamma) \cdot \cos[\arctan(r_{Vy4} \alpha)] \tag{A51}$$

Aligning Torque (combined)

$$M_z = -t \cdot F_y' + M_{zr} + s \cdot F_x \tag{A52}$$

$$t = t(\alpha_{t,eq}) = D_t \cos[C_t \arctan\{B_t \alpha_{t,eq} - E_t(B_t \alpha_{t,eq} - \arctan(B_t \alpha_{t,eq}))\}] \cdot \cos \varepsilon \tag{A53}$$

$$F_y' = F_y - S_{Vy\kappa} \tag{A54}$$

$$M_{zr} = M_{zr}(\alpha_{r,eq}) = D_r \cos[\arctan(B_r \alpha_{r,eq})] \cdot \cos \alpha \tag{A55}$$

$$s = \{s_{sz1} + s_{sz2}(F_y/F_{zo}) + (s_{sz3} + s_{sz4} df_z) \gamma\} \cdot R_o \cdot \lambda_s \tag{A56}$$

$$\alpha_{t,eq} = \arctan\sqrt{\tan^2 \alpha_t + \left(\frac{K_x}{K_y}\right)^2 \kappa^2} \cdot \mathrm{sgn}(\alpha_t) \tag{A57}$$

$$\alpha_{r,eq} = \arctan\sqrt{\tan^2 \alpha_r + \left(\frac{K_x}{K_y}\right)^2 \kappa^2} \cdot \mathrm{sgn}(\alpha_r) \tag{A58}$$

ROLLING NOISE

C. Stanworth

Acoustic Consultant, Muir of Ord, Ross-shire, UK

1   Introduction

Whenever rolling of one surface over another takes place, there will always be noise generation. In the cases where the rolling-contact surfaces are sufficiently smooth, or where the rolling is sufficiently slow the level may be so low as to be inaudible.

The statement covers all cases, for example   ball over plane
wheel over plane
wheel-on-wheel
steel wheel on rail
tyre on road

Note, however, that airborne noise is not the only effect of rolling. Vibration other than that which is responsible for the airborne noise will also be caused within the rolling member(s) and in the surface or track on which rolling takes place. This may be reflected in two ways. On the one hand as vibration and noise in the vehicle supported by the rolling members (wheels). This noise can be due in large part to vibration passing up through the vehicle suspension and structure, appearing within the body shell as re-radiated noise. This will add to the rolling noise arriving more directly by airborne path.

On the other hand, the vibration generated in the surface or track over which the rolling takes place can be transmitted to and through the structure or ground supporting it. This may result, for example, in ground-borne vibration, which is a whole separate and complex subject area. Since, however, one of the resultant effects of ground-borne vibration may be re-radiated noise within adjacent buildings, this can be thought of as an aspect of traffic (and hence rolling-) noise. Rumble noise is common near some underground railways.

However, neither structurally-transmitted internal vehicle noise nor vibration and/or wayside noise within buildings due to a ground transmission path is covered in this presentation.

The presentation is concerned with that noise arising from rolling which is transmitted to the nearby hearer by a purely airborne path. The bulk of the presentation is devoted to the noise generated by wheel on rail, the familiar railway noise, but some mention will be made of road vehicle tyre noise.

2   The General Features of Rolling Noise

The noise due to rolling, and to any other process is characterised by two features: its level and its spectrum.

2.1   The Level of Rolling Noise. There are two very obvious effects contributing to the level of rolling noise.
-        the faster the rolling process, the higher the noise level

-        the rougher the surfaces involved with the rolling, the greater the noise.

Effect of speed: railways Figure 1 (simplified from Hemsworth, 1987) illustrates the way in which the wayside noise due to wheel-on-rail of a passing train varies with speed. The levels are taken at a distance of 25m from the nearest rail adjacent to level, continuously welded track in good condition in open surroundings.

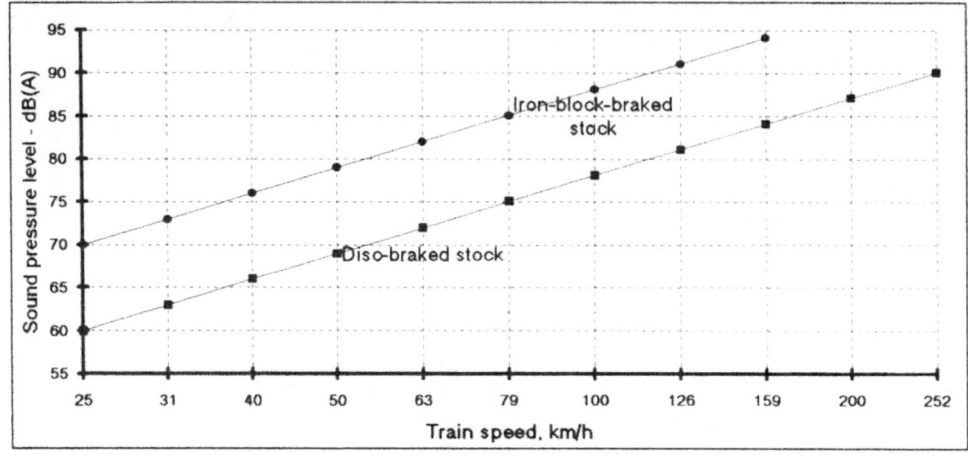

Figure 1 Wayside noise level from trains at 25m running over continuously welded rail.

The two curves represent two different types of train. The upper curve is typical of trains using iron block brakes, acting in a clasp fashion on the running surfaces of the wheels. The second curve, some 10 dB(A) lower, is typical of trains equipped with disc brakes, having no friction braking effort applied to the wheel running surfaces.

The upper curve level L dB(A) at speed v km/h is given by :

$$L = L_0 + 30\log_{10}(v/v_0) \qquad \qquad \qquad \qquad \qquad 1$$

where L_0 = 88dB(A) and v_0 = 100 km/h

The lower curve level is uniformly some 10 dB(A) less at a given speed.

The crucial difference between these two types of rolling stock rests in the wheel roughness, which is significantly higher for the tread-braked stock (Stanworth, 1987).

In summary, tread-braked stock is some 10 dB(A) noisier than disc-braked stock. In both cases the noise level increases by about 9 dB(A) for each doubling of train speed. Whilst figure 1 illustrates the general way in which wayside wheel-rail noise varies with speed: within the two distinct types of stock portrayed, there will be variations resulting from individual design characteristics which will be discussed later.

Wheel-rail noise is almost always the dominant contributor to railway noise. Traction noise will only be more significant at the lower end of the speed range. (Also mentioned later.)

Effect of speed: tyre noise   Rolling noise of road vehicles also increases rapidly with pass-by speed. The effect for the range of vehicles is illustrated in Figure 2, adapted from Tyler (1987).

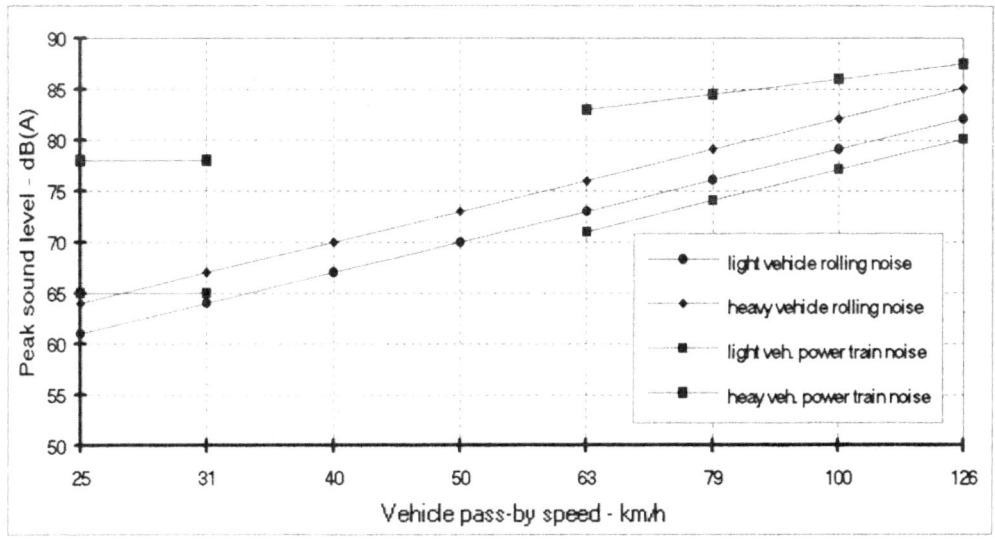

Figure 2. Peak pass-by noise of road vehicles at 7.5m. (adapted from Tyler, 1987)

Thus under given conditions, the rolling noise of heavy vehicles is some 3dB(A) higher than that of light vehicles. These values will, however, be subject to large variation, depending on the tyre tread, the amount of wear, tyre carcase construction and the character of the roadway surface. It is clear that for light vehicles the rolling noise is usually comparable to the power-train noise, whereas for heavy vehicles, the power noise is dominant, except sometimes at the highest speeds.
The relationship between speed and noise level in this heavy vehicle case is:

$$L = L_0 + 30\log_{10}(v/v_0) \qquad \text{(ie. +9dB(A) per doubling of speed). .} \qquad \qquad 2$$

where $L_0 = 82$ dB(A) and $v_0 = 100$ km/h. The coefficient 30 is close to the lowest value available. It may rise above 40 for the heavy vehicles running on figured concrete carriageway surfaces, resulting in a rate of noise increase of more than 12 dB(A) per doubling of speed.
Other effects arise from the structure of the tyres. Radial-ply tyres are quieter than cross-ply: steel belted tyres are quieter than those with a completely textile structure by a few dB.

Roughness effects: railways   It is noted above that tyre noise increases more rapidly with speed for rougher surfaces.
In the railway case, rougher rolling surfaces also lead to higher noise levels, but the rate of increase with speed remains the same. The principal effect responsible for higher roughness is rail corrugation, where longitudinal surface rippling of the rail surface develops over time, with the passage of traffic (Stanworth, 1987). The pitch of the effect is often irregular, and can vary from a few tens of mm. to a

few hundreds of mm., with a peak-to-trough depth of some tens of microns. Figure 3 illustrates the way in which wayside noise is affected.

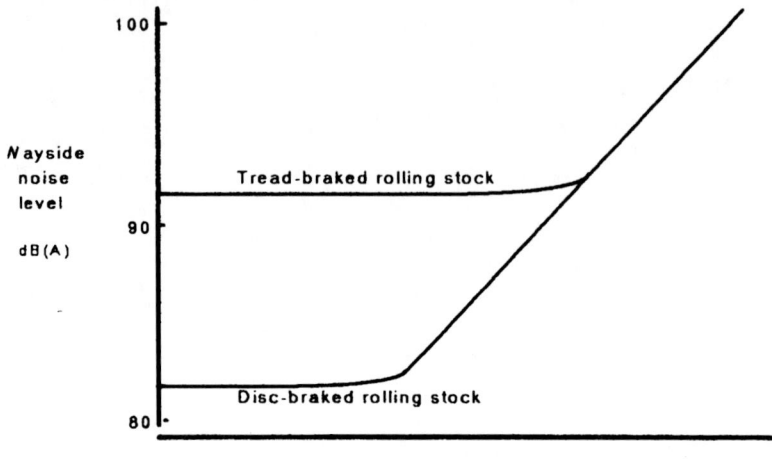

Fi gure 3. Effect of corrugation severity on the wayside noise level of trains.
(160 km/h, 25m to wayside)

For smooth rails having low levels of rail corrugation (peak-to-trough amplitude less than 10 microns) there is insignificant effect on wayside noise level, which is determined by the type of rolling stock. As the corrugation severity increases the effect on wayside noise level begins to appear for the smoother disc-braked wheels. The wayside noise then rises towards that for tread-braked stock as the corrugation develops.

When the corrugation severity is sufficiently high, the type of rolling stock becomes irrelevant, the effective total roughness being dictated by that of the rails.

Note that no attempt has been made to quantify the concept of "corrugation severity". Corrugation takes several forms, and the feature(s) of it responsible for increased noise is/are not certain.

2.2   The Spectrum of Rolling Noise

Noise spectra: railways   Figure 4 (from Stanworth, 1987) illustrates the noise spectra to be expected at the wayside of passing trains. It is clear that the spectra of iron-block braked rolling stock are significantly different from the spectra of disc braked rolling stock.

Although below about 500Hz there is little difference, at the higher frequencies from around 500hz to 4kHz, there is a quite substantial difference, and this is the region in which the additional noise energy of the block-braked rolling stock occurs. It is also in this region that the ear is most sensitive, and the difference between the two types of stock is subjectively readily apparent.

It is notable that by 3 or 4kHz and above the noise energy is falling off rapidly with increasing frequency.

Tyre noise spectra (Tyler, 1987) are much more variable, although it is notable that tread pattern does not always influence noise level greatly (on dry roads), even though bold tread patterns may

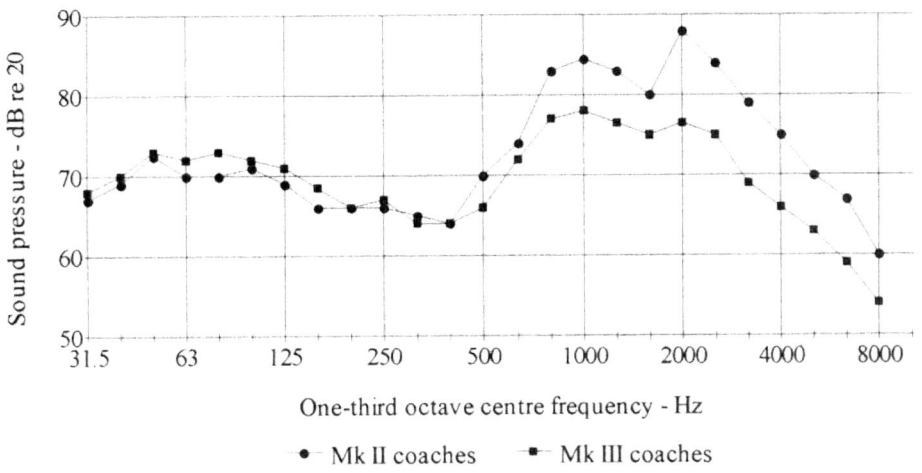

One-third octave centre frequency - Hz

• Mk II coaches   ■ Mk III coaches

Figure 4   Spectra of passing trains (160 km/h on c.w.r.) (Mk II stock is iron-block braked;
Mk III stock is disc-braked)

influence the spectra subjectively. Figure 5 demonstrates the narrow-band spectra of "smooth" and
"traction" tyres. These are remarkably similar, save for the peak at around 400Hz for the traction
blocks hitting the rolled asphalt surface. The dominant effects on tyre noise arise from running over
wet roads, or concrete surfaced roads which are acoustically "hard".

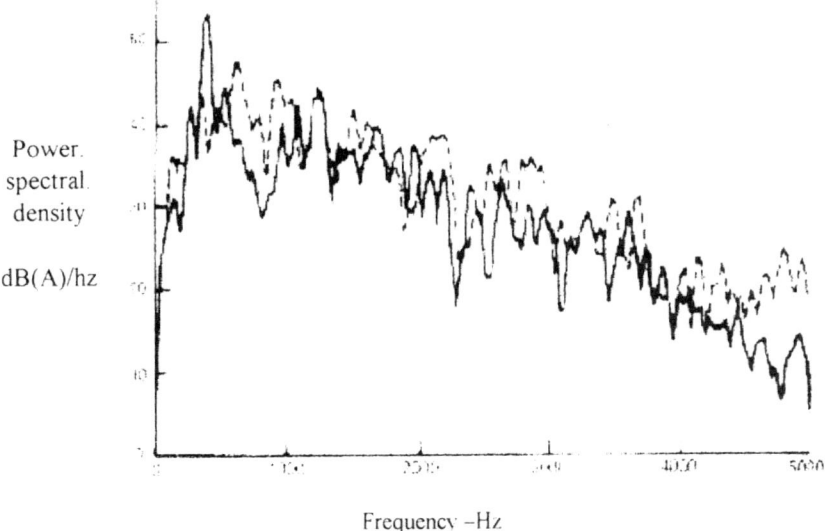

Frequency –Hz

Figure 5 Narrow-band frequency spectra of "smooth" (broken line) and "traction" tyres

2.3   The Polar Diagram of Rolling Noise

Polar diagram   It is necessary to point out a further characteristic of noise emission which is distinctly different between railways and road vehicles. Railway wheels emit noise having a strongly marked polar diagram. Almost no noise is emitted in the plane of the wheels, whereas there is very strongly-directed radiation laterally in the axial direction of the wheelset. (Stanworth, 1987)

By contrast, the radiation from road vehicle tyres is very much more strongly spherical in character. The consequence of this is quite important. Since trains radiate little noise forwards, it is difficult to hear trains coming for people working on the track, whereas, fortunately, vehicles coming along the road are more easily heard.

3   Over-riding requirements of transport systems.

3.1   Railways

Railways were developed to exploit the low rolling resistance of the steel wheel on the steel rail to enable large loads to be hauled economically at a sufficiently high operating speed with a limited amount of power. In their design, construction and operation, however, they had to remain a safe form of transport.

The low rolling resistance and low adhesion mean trains are unable to accelerate and brake at high rates, even though trains are continuously-braked, (ie. all vehicles have their brakes applied simultaneously) in order to make best use of the limited adhesion available. A control (signalling) system has to be imposed to ensure safe spacing between the trains. The signalling system is also needed because railways are a serial system: overtaking is not generally possible.

This accounts for the first, obvious characteristic of railway noise. It is composed of isolated (but quite loud) events, separated by long periods of silence. This characteristic affects the subjective perception of railway noise.

It is commercial necessity to give a comfortable ride to passengers, and a ride that does not cause damage to freight. It is therefore necessary for railway track to be laid to a high standard of line and level, and for the running surface of the rail to be as smooth as possible. The vehicles need to run smoothly and steadily down the track, so their wheels have to be perfectly round and concentric on the axles.

For the vehicles follow the track effectively, the design of the wheel transverse profiles is of

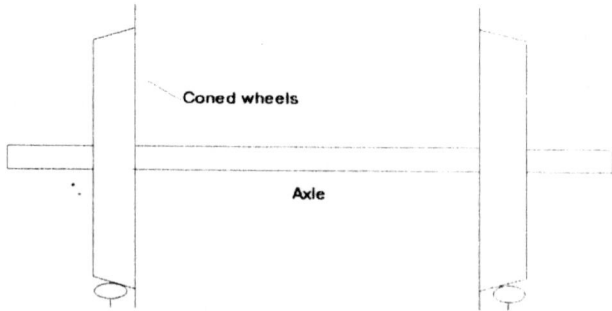

Figure 6. Illustration of a coned railway wheelset running along straight track.

crucial importance. Conventional railway wheels are mounted rigidly on their axle to form what is known as a *wheelset*. See the diagram in figure 6. This also shews that the running surfaces of the wheels are <u>coned</u>. The effect of coning is to ensure that the wheels run directly down the centre of straight track. If the wheelset is displaced to one side, that wheel which has its flange closer to the rail has a larger running circle diameter than the opposite wheel. Since both wheels are forced to rotate at the same speed, the wheelset is steered back towards the track centreline.

3.2   Road Traffic

Road tyres equally have to support and steer their vehicles safely and comfortably, and carry freight satisfactorily. There is an inevitable conflict between the needs of low rolling resistance and good roadholding, particularly in adverse weather. Vehicles intended for rough terrain need different tyres to those confined mostly to paved roads.

The different types of tyre are then responsible for different noise characteristics, depending on the degree of wear, the form of tyre construction, the type of road surface, and the weather conditions encountered.

Road traffic noise is very much more weather-sensitive than is true for railways.

4   Railway Noise Models

4.1   The Remington Noise Model

The study of railway noise generation springs almost entirely from the early work of Paul Remington at Bolt, Beranek and Newman in the USA, supported by the US Department of Transportation . The model due to Remington (1975, 1987) is illustrated in figure 7: it still forms a good basis for considering the mechanisms of wheel-rail noise generation.

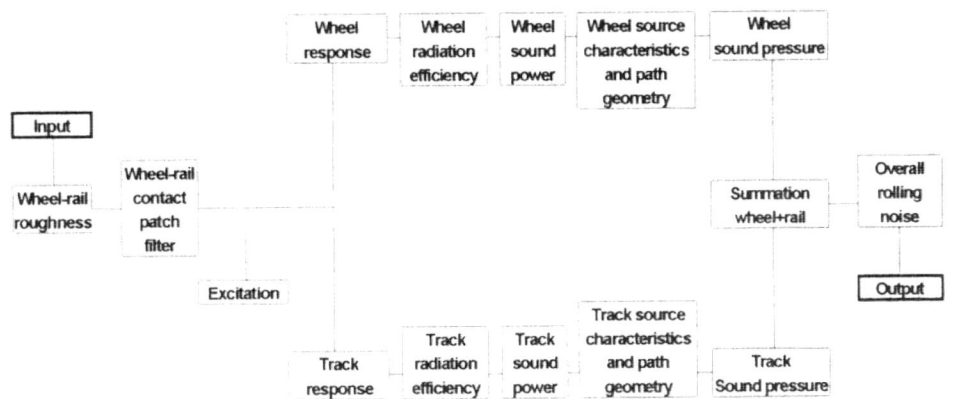

Figure 7 Model of wheel-rail noise generation due to Remington.

The simple model assumes it is solely combined wheel and rail roughness that is the cause of the noise generated during rolling. No distinction is made between wheel and rail roughness – only their sum is significant. However, the character (spectrum) of that roughness <u>is</u> important.

Wheel and rail roughness.   There is plenty of evidence to support the assumption that roughness is important. It has been said already that iron block-braked wheels are noisier than disc braked wheels. This was recognised in the late 1960s/early 1970s when disc braked rolling stock started to be introduced routinely. It was very obviously quieter than tread braked stock, with a more acceptable quality of noise.

It was clear that the running surfaces of the wheels were visibly different. This led to pioneering wheel roughness measurement by the Deutsche Bundesbahn, who shewed (Stanworth, 1987) that the iron block-braked stock had peripheral rippling of the wheel running surface. These ripples had an approximately constant "wavelength", varying between some 5 and 20cm, with a peak-to-trough depth quite consistently about 30 microns. By contrast, the disc-braked wheels, which had no brake effort applied to the running surfaces, were quite smooth. This roughness on tread-braked wheels develops fully on newly-reprofiled wheels quite quickly, needing only a few service brake applications. Thereafter the noise increases no further: it has to be assumed that the phenomenon is self-limiting.

Equally compelling evidence of the rôle of roughness came from the effect of rail corrugation, already noted by way of introduction. Rail corrugation, the audible effect which is known in english as "roaring rail", has been familiar since railways began. Corrugation consists of rippling of the rail surface. The effect is approximately periodic with a "wavelength" of some 40mm. upwards. Whilst it is a wear effect associated with the passage of traffic, it is not self-limiting as wheel corrugation seems to be. Rail corrugation does not occur everywhere, but where it does happen, it appears to increase without limit. Severe corrugation may attain a peak-to-trough amplitude of 0.1mm.. Such a severe effect leads to track damage, as well as an enhancement of noise level by 20 dB(A) or more.

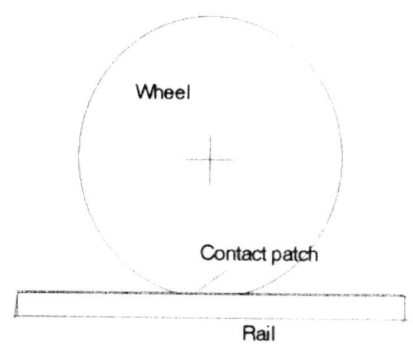

Figure 8. Illustration of contact patch

The contact-patch filter.   Accepting that rail roughness plays a powerful rôle in noise generation, it is interesting to note that all forms of roughness are not significant. As an example, minor pitting of the wheels and/or rails causes no audible effect; by contrast, asperities are always significant. This is because of the influence of the contact patch filter. The effect of the contact patch (figure 8.) is to "average out" many of the roughness features with a wavelength smaller than the length of the contact. Minor pits are bridged by the contact patch, but asperities are almost always significant. The contact patch has a length of some 1.5 cm. for the usual size of railway wheel. It is instructive to examine the spectra of noise with this in mind. Figure 4 illustrates the spectra of trains passing at 160 km/h (= 44 m/s), and shews the spectrum tailing off rapidly above some 3 kHz. At a train speed of 44 m/s, a frequency of 3 kHz (= a period of 330 microsec.) corresponds to a wavelength on the rail of 44x0.00033 m., or 14.5 mm.. This is surprisingly close to 1.5cm for a very simple calculation.

Wheel behaviour. This filtered roughness now forms the excitation to both the wheels and the rails. Consider first the upper (wheel) branch of figure 7. Railway wheels have a very rich population of available modes of response, each with its distinct natural frequency. Figure 9 (Stanworth, 1987) illustrates one of the many. At their natural frequencies the wheels will respond.readily to any corresponding frequency input. The roughness of wheels and rails is

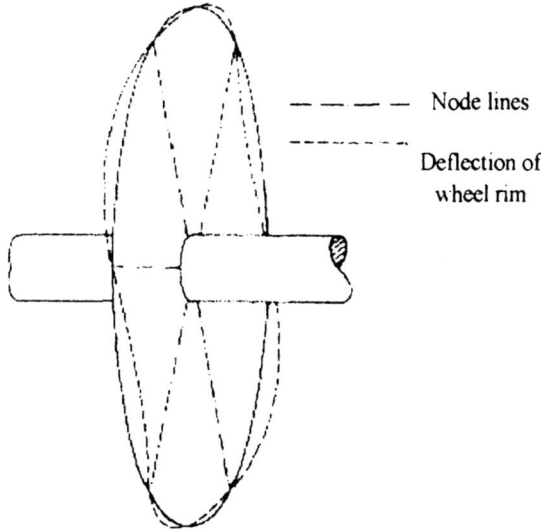

Figure 9 Diagram of wheel vibration in 3 nodal diameter node

"broad band" and contains all frequencies within the (filtered) band, although some may be stronger than other. A wide range of modes is thus excited.

Not all of these modes radiate strongly: the radiation will be different in different directions. For many modes the wheel disc deflects out of plane. any modes which have balanced amounts of phase and anti-phase energy from one side of the wheel will radiate little axially of the wheelset. Radiation from opposite sides of the wheels will be equal and opposite, resulting in complete cancellation in the plane of the wheels. (This is why trains are not heard approaching.)

Thus, while the filtered roughness excitation may make the wheels vibrate, the noise heard in the "far field" (a few diameters from the wheels), representing the sound power radiated, is determined by the wheels' acoustic characteristics. Additional effects may be introduced by any screening structures interposed in the propagation path.

Track behaviour. The lower branch of figure 7 referred originally not to the "track", but only to the rails. More recent work (Thompson, 1991) has recognised that like the rails, the sleepers are an important component of the noise generation process, so the "track" side of the diagram now has to be split again into two sub-branches. (The same split has not been necessary on the "wheels" side: the vibration energy flowing upwards through the suspension into the vehicle is (almost always) insufficient to influence the externally radiated noise, although it is very important for internal noise.)

Vibration energy that flows into the rails thus divides between the rails and the structure (usually rail seating pads, baseplates, sleepers and ballast) which supports them. Whilst the sleepers-and-baseplates structure will behave modally in a way similar to the rails, the rails have a distinctly different behaviour.

Modern railway is laid with continuously welded rail (cwr), in which the 20 yard (18.3 m.) lengths of rail from the steel works are butt-welded into long lengths for transport to site on special trains. Once laid on the sleepers, they are then site welded by the Thermit process into lengths that may extend up to several kilometres.

From an acoustic point of view, cwr is an infinite structure. It is axiomatic that infinite structures do not resonate. Vibration energy entering the rails is therefore dispersed along the rails, which act as waveguides. However, at each baseplate some of the vibration energy will bleed into the baseplate and sleeper, causing them to respond in a manner depending on their damping and natural frequencies. Behaviour of the sleepers and baseplates is analogous to that of the wheels, but with only the top surface exposed above acoutically-absorbent ballast, there will be somewhat less scope for field cancellation in determining the resultant acoustic power.

Damping of the rail turns out to be quite light. Those modes of vibration transmission that do not couple well to the sleepers are capable in some instances of propagation for a few kilometres. The phenomenon is known as singing rail.

Wheel- and track-noise summation. Whilst the radiation from railway vehicles is confined to the wheels, and the wheels are large enough to be capable of effective radiation of appropriate modes, radiation from the rails and track is more diffused: long lengths may be involved.

Finally, however, the vehicle and track noise sources combine to produce the resultant wheel-rail noise of the passing train, which will then be subject to all the influences in the propagation path beyond the trackside.

It is interesting that the sound energy radiated comes <u>about</u> equally from the wheels and from the track. This rough balance means that if either track or wheels were made completely silent, the wayside noise would be reduced by only some 3 dB(A), a barely noticeable amount. This is an important message when considering noise reduction options.

TWINS. The study of wheel-rail noise generation is very much an active subject. The principal worker in the field is David Thompson (1993, 1997). His studies over the years have led to the development of the TWINS suite of programmes. These programmes embody David Thompson's development of Paul Remington's original conceptual model. The work has followed David Thompson through his career, starting in British Rail Research (now AEA Technology [rail]), then to TNO/TPD, Delft, and now to Southampton University, UK. A selection of David Thompson's work has been quoted in the references cited. It is only part of the developing whole.

5   Techniques of Rolling Noise Reduction

5.1   Roughness Control – Rails.

The dominant rôle played by wheel and rail roughness in the theoretical analysis of wheel-rail noise generation has been made clear. In considering the ways in which wheel-rail noise might be reduced it is absolutely necessary to start by making sure that the rails and wheels are as smooth as possible.

On this topic, the rail roughness is by far the most significant place to begin.

Figure 3 depicts the way in which rail corrugation has a completely over-riding effect on the wayside noise if it is sufficiently severe. At is most serious, corrugation of the rail surface can raise the wayside noise by 20 dB(A) or more, equal to an increase of wayside noise energy by a factor of more than 100. The subjective impact is even greater, since the character of the railway noise becomes more obtrusive as the population of higher frequencies in the spectrum increases.

Although research on corrugation formation has been carried out for many years, and is still taking place, no sure way of preventing formation of corrugation wear seems to have been identified. Corrugation does not happen at all places along a route, but where it does occur the only option currently available is to remove the rippled surface by grinding. Examination of the corrugated surface in detail reveals changes in the chemical structure of the steel during the process of ripple formation.

Grinding of the rail surface (by the use of a specially constructed rail-grinding machine) to restore simply the unrippled profile will not be sufficient. It is necessary to grind sufficiently deeply to restore a uniform metallurgical state along the surface. Otherwise, the presence of an incipient corrugation pattern in the surface metallurgy will lead to rapid regrowth of an audible disbenefit. The surface finish immediately following grinding always shews the marks left behind by the grinding stones: being much smaller than the contact patch length, however, they are "filtered out" and have no acoustic influence. After the passage of a few trains, the familiar bright, smooth wear-band of the running line is restored. On good rails the rippling will be less than 10 microns deep.

Note that severe corrugation is not only an acoustic disbenefit. The resultant high level of rail vibration may be sufficient to damage the track components, causing baseplates to shatter, or baseplate fixing to come adrift from sleepers.

5.2   Roughness Control - Wheels

Given smooth rails, it is possible to judge reliably the effect of wheel roughness. Figure 4 demonstrates the way in which the brake system is capable of altering the wayside noise. The historic form of railway braking involves iron block brakes clasped onto the wheel treads. It is still the most common form of braking, in particular for freight vehicles. Unfortunately their usage leads to the formation of wheel tread corrugation, although in this case the corrugation is self-limiting: it does not increase without limit as seems to happen on rails. The corrugation, once formed, does not, however, "roll out" during subsequent unbraked running.

Trains fitted alternatively with disc brakes are readily observed to run under given conditions on uncorrugated track around 10 dB(A) more quietly. Not only are disc-braked vehicles quieter, the quality of their noise is more acceptable, since the reduction in noise takes place in that part of the spectrum to which the ear is most sensitive. (Drum brakes are equally effective in keeping treads smooth, although their use in railway practice is uncommon. Electric braking, using the traction motors of electric trains to generate, also does not cause the problem.) There are good reasons why disc brakes are not uniformly adopted.

- Disc brakes are more expensive than tread brakes.
- It is not usually a useful retrofit option.
- Disc brakes work best at high speed. (Tread brakes are not a viable option for fast trains.)
- Tread brakes work best at low speed.
- Therefore the two types of brake cannot be mixed on the same train.   The vehicle pooling arrangements among the mainland european railways mean that it would be extremely difficult to introduce disc-braked stock into wide use within a large existing pool of tread braked wagons.

In addition, there is evidence that disc-braked vehicles are more vulnerable to fouling of the wheel-rail system (leading to wheelspin and skidding). Tread brakes have the advantage that brake applications refresh the wheel tread. Attempting to adopt the advantages of discs whilst enjoying this benefit of tread brakes by using both types together does not reduce noise. Even 20% of brake effort invested in tread brakes results in wheel corrugation as severe as full use of iron block brakes. Some administrations have applied so-called scrubber blocks to disc-braked wheelsets in order to maintain a clean but uncorrugated surface whilst reducing tread damage due to spinning/sliding.

The desirable option is the use of effective traction control to reduce spinning (which leads to rail burns) and wheelslide protection to eliminate wheel flats. Of the two, wheel flats are the more serious, since they aie audible features both inside and outside the train during its whole journey.

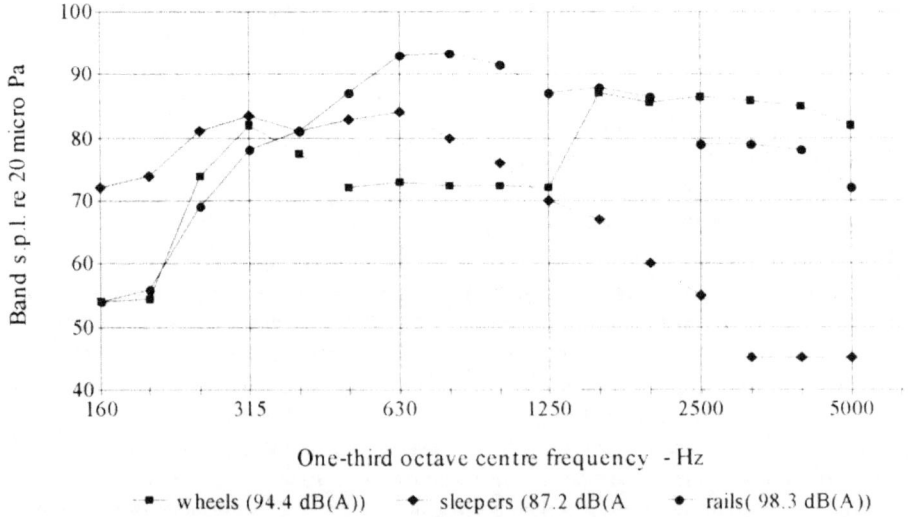

Figure 10 Example of noise components from wheel, rail and sleeper for freight wagon on concrete-sleepered track

5.3   Acoustic Optimisation of Wheel and Rail

If all has been done to make the wheels and rails as smooth as possible, it is necessary to consider what options there might be to inhibit or restrain the response of smooth wheels and smooth track.

Since there is rough equality between wheels and track in their total contribution to wayside noise, (although there will be cases where one or the other carries a slightly higher share) it is necessary to reduce the noise from each in parallel. Note, however, that the track is dominant at frequencies up to 1 kHz and the wheel is dominant at higher frequencies. Within the track, the sleepers are the principal source at the lower frequencies of the track régime, with the rail contributing the higher frequencies. Figure 10 illustrates one of very many calculations made by David Thompson using his TWINS programme. In this particular case the wheels are the dominant source, though not overwhelmingly so. Changes in the system parameters can alter the balance quite noticeably.

This is illustrated well in figure 11, which shews the result of changing the values of one parameter, the stiffness of the rail seating pad, which lies between the rail foot and the baseplate.

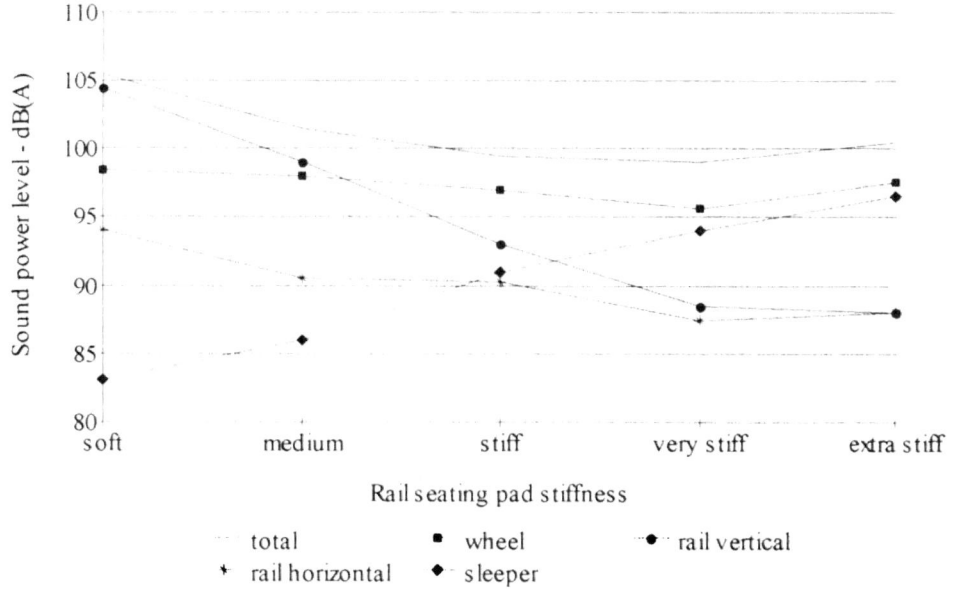

Figure 11 Examples of the effect of rail pad stiffness on the contributions to total noise.

When the rail pad is very soft, there is little vibration coupling between the rail and (most of) the sleepers. The rail is thus allowed freedom to vibrate vertically: the resultant noise output dominates the total noise produced. As the rail pad stiffness increases, the vibration energy that bleeds through into the sleepers also increases. Thus less energy flows outwards from the contact along the rails. Consequently, the noise contribution from the rail falls, whilst that from the sleepers rises. During these changes, the wheel output remains relatively unaffected.

The lateral vibration of the rails consists to a large extent of motion involving lateral flexure of the rail web. It is thus less affected by the stiffness of the rail foot fixing.

These studies suggest that very soft rail pads are not favourable to quiet running, but it is also clear that extremely stiff pads would also be undesirable, since the sleeper and wheel contributions are both dominant and rising.

The message must be that careful selection of system parameters may well be necessary for a quiet railway. It remains to be seen whether a satisfactory compromise for all types of traffic is possible.

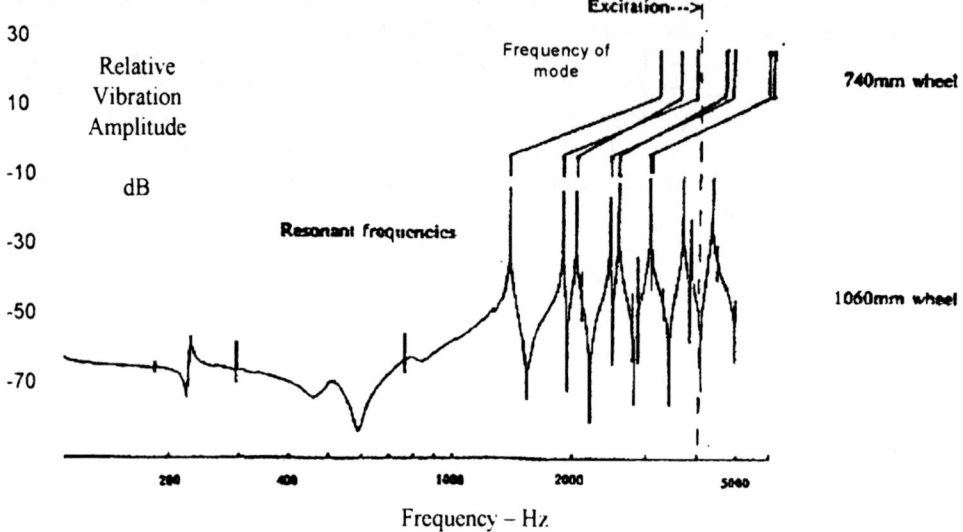

Figure 12. Predicted effect of smaller wheels

Effect of wheel diameter. Throughout the changes examined above, only at the extreme range was the wheel affected. It is worth considering whether there are feasible options for reducing wheel noise. One of the possibilities worth careful consideration is the reduction of wheel diameter.

As the diameter of a wheel decreases, its modes of vibration, whilst retaining much the same form, move to higher frequencies. The effect is illustrated in figure 12 for two practical diameters of wheel. It was pointed out before that the effect of the finite contact patch size was to impose a low-pass filter into the excitation mechanism. Although the contact patch reduces as the wheel diameter reduces, it does so rather slowly. The overall effect of wheel diameter reduction, therefore, is to raise some of the available modes, particularly those influential in noise production, above the frequency range in which excitation is strongly possible. The possibilities of this technique have yet to be exploited fully – wheel diameter reduction has some consequences for the passage of wheelsets through points and crossings. In addition, a full analysis would be required to ensure that any reduction in wheel noise was not offset by an increase of rail/track noise.

It is worth consideration that some wheel profiles excite into vibration much more readily than others. Straight webbed wheels are much less likely to vibrate seriously than S-web wheels. S-web wheels were originally necessary to cope with differential heating when tread brakes were applied. For some reason they are still in use with disc-braked wheels, leading unnecessarily to noisier running.

Damping.   One of the most powerful techniques generally in the field of noise reduction is the application of damping. Damping can be applied quite readily to wheels, either by the use of tuned dampers (tuned to the pre-determined natural frequencies of the wheelsets), or more practically by the use of constrained-layer damping.

Wheel damping is known to be very effective in restricted circumstances, but has yet to demonstrate real application in the reduction of normal rolling noise. This may be because rolling contact already causes significant damping of the wheel and it is difficult to increase this damping by a sufficient margin to cause perceptible advantage.

Damping of the rail has also been suggested as a viable possibility. The difficulty lies in applying sufficient damping. Damping is conventionally applied to a resonant system, so that energy absorption by the damping mechanism can take place over several, sometimes many, cycles of oscillation. Unfortunately railway rails are not a resonant system, the vibration energy is propagating along the rails away from the contact point, so that any given energy packet passes a given absorber only once. Thus, a degree of damping which would damp an oscillation of 1 kHz in a resonant system within (say) 10 cycles (=10ms) would require 10 wavelengths to suffer similar attenuation for that frequency vibration travelling in the rails. This could be several metres, so the effective radiating length would be twice that.

Attempts have also been made to reduce noise by encapsulating rails. In this system, the rails have been laid in troughs and then surrounded to the greatest extent possible with elastomer. Predictably, there was no beneficial effect: the rail vibration was transmitted faithfully to the open surface of the elastomer, which then became the source of the airborne noise.

5.4   Barrier systems.

The one technique that is certain to reduce the wayside noise of trains is to interpose a barrier in the "line-of-sight" between the wheels/track, and the observer, see figure 13. The attenuation

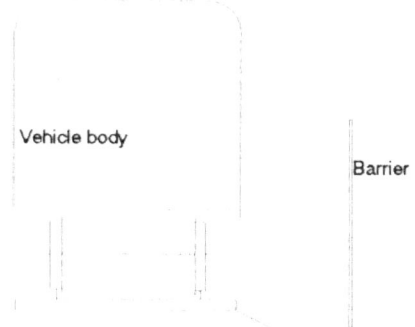

Figure 13 Diagram of lineside noise barrier.

achieved depends on the degree of intrusion into the line-of-sight and there is a wide variety of prediction methods available in various countries that (surprisingly) lead to widely varying predictions of barrier effectiveness (van Leeuwen, 1996). It can be assumed safely, however, that a barrier which intrudes 2m into line-of-sight can be relied on for an attenuation of 10 dB.

Combined barrier__Since lineside barriers are undoubtedly unsightly, a combined system has been developed which is more compact (Jones, R.R.K., 1994).

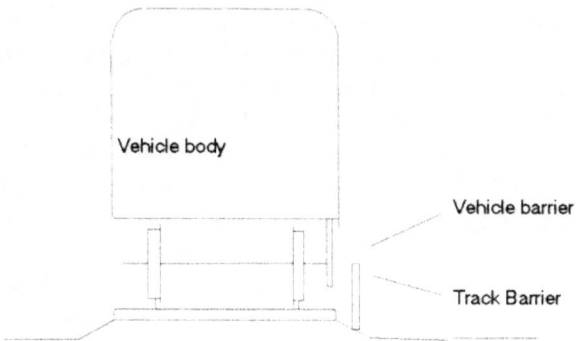

Figure 14. Diagram of combined vehicle//lineside barrier

By mounting a barrier on the vehicle which is just clear of the bogie, and complementing that with a close-mounted barrier at the trackside, it is possible to achieve much the same result much more neatly. The clearances needed are much closer than the diagram suggests.

5.5   Lubrication.

The only practical lubricant that could have widespread use in the wheel-rail system is water, which often occurs naturally. There is an insignificant reduction in the wayside rolling noise.

5.6   Tyre noise reduction.

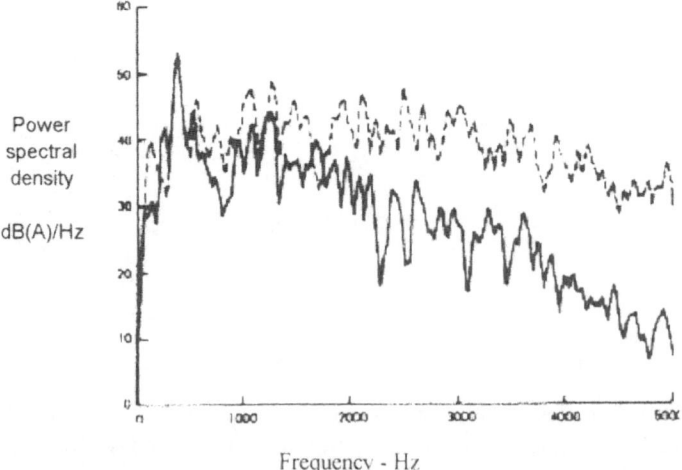

Figure 15 Narrow band spectra from traction tyres rolling on a dry (lower) and wet (upper) motorway surface at 96km/h

Unlike the steel wheel on steel rail case, water on roads causes a very large increase in the wayside noise. This depends on the type of tyre and the character of the road surface, but a typical case is illustrated in figure 15 for traction tyres in on a motorway service (from Nelson and Underwood, 1984). It is clear that wet running increases the noise level very substantially, in this case by some 10 dB.

There are a number of features of tyres which can influence tyre noise, and which need to be observed. As an example, radial ply tyres are a few dB quieter than cross-ply. Within radial ply tyres, steel belted tyres are quieter than textile tread banding.

There are several possible contributory mechanisms for tyre noise generation, including aerodynamic noise, air-pumping, ie. the expulsion of pockets of air trapped in the tread, or simply noise due to vibration of tyre carcase as it rolls.

The fact that smooth tyres on a smooth road make very similar amounts of noise to treaded tyres of identical carcase construction suggests that aerodynamic noise and air pumping can be disregarded. It turns out that the wayside noise spectrum correlates well with the measured vibration spectrum of rolling tyres, adding further credibility to that conclusion. As mentioned above, the form of carcase construction affects the noise generated. This is also consistent with tyre carcase vibration being determinant.

Nelson and Underwood (1984) state that the following factors contribute to quiet rolling:-

• "Tyres should have a random circumferential rib style tread pattern as this allows drainage of water from the contact patch, but does not allow the tyre to vibrate at passing frequencies related to the tread pattern.

• "The road surface should have a random pattern to avoid the effects of tyre (passing) frequencies.

• "The tyre should be of radial ply construction with steel belts in order to reduce the amount of slip in the contact area and to maximise damping in the tyre structure.

• "The tyre size and inflation pressure should be maximised consistently with safety as this will minimise the deflection of the tyre in the contact region and reduce tyre vibration. If the road surface can be porous, as for example with porous macadam, there will be an additional advantage of the removal of surface water with the reduction of tyre splash noise in wet conditions. This also reduces noise in dry conditions by acoustic absorption in the surface voids."

6   Anomalous Mechanisms of Rolling Noise Generation

Discussion so far has assumed "pure" rolling. It has allowed the development of a theory of railway noise generation that assumes purely a roughness input to the vibration generation process. In fact railway wheels rarely have the luxury of pure rolling.

6.1   Rolling of Wheelsets (Straight Track)

Figure 6 sketched the typical railway wheelset with its coned running surfaces, necessary to ensure that the vehicles run along the track in a comfortable, stable manner. However, a consequence of the coning is to ensure that "pure" rolling does not take place. The significance of the contact patch in limiting the spectrum of wayside noise due to its finite length has been pointed out. Of course, the contact patch also has a finite width.

Thus there is a range of rolling circle diameters within the contact patch. In the absence of the wheelset axle, the effect would be to make the wheels roll outwards, since the larger rolling circle diameter is on the flange side of the contact. It is clear that there is circular creep within the contact area exerting equal and opposite outward torques on the bottoms of each wheel.

A glance at figure 9, a diagram of one of the many modes of wheel vibration response reveals that the wheel rim deflects out-of-plane. It would not be surprising if a mode of vibration such as this could excited by stick-slip relief of the circular creep torque within the contact patch. Even the smoothest wheels on the smoothest rails generate significant noise. I believe that mechanisms other than roughness excitation must also be available.

Figure 6 is very much a diagrammatic representation of a wheelset. In reality, the wheel flange is very substantial and is mates to the coned running surface with a curved fillet. Figure 16 gives a better idea of the proportions of wheel tyre and rail head.

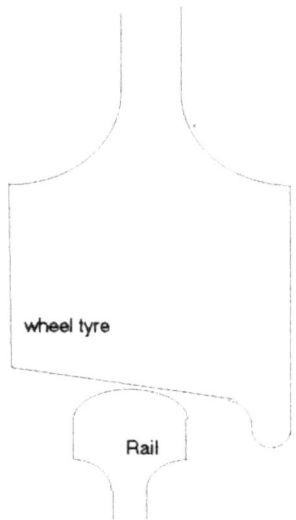

Figure 16 Diagram of railway wheel tyre on rail head

6.2   Rolling of wheelsets (curved track)

Suppose that the wheelset is travelling over the rails, moving into the diagram on a right-hand curve. The wheelset would then move to the left. If the curve were sharp enough, the rail would start to encounter the curved profile into the flange, and the first result is for the effective conicity to increase. Under even sharper curvature the wheel will run its flange root firmly against the rail. The range of conicity within the contact patch is now very high: the circular creep becomes more severe. The acoustic effect is for curve squeal to appear. The types of vibration mode excited are similar to the families of which figure 9 is a member. These modes have the characteristic that the wheel tyre bends out-of-plane, and (clearly) is also subject to torsion.

The resultant wheel-squeal can be very penetrating, and may be a considerable source of annoyance to the wayside public. The traditional railway solution to the problem has been to fit rail flange lubricators ahead of the curve. The lubricators, which are track-based and dispense grease to the wheel flanges as they pass, are often effective, but are not an easy option. Lubricators appear to be difficult to fit correctly: they need careful maintenance and periodic refilling. There are, however, situations where flange lubricators are not seen as suitable. This is particularly so where the track is laid on concrete, and/or is in tunnel. In such places a non-polluting acoustic solution is needed. It is, however, not the most serious wheel squeal mechanism. (Note that whereas lubrication is capable of dealing both with the squeal problem and wear of the wheels and rails, acoustic techniques leave the wear problem unresolved.)

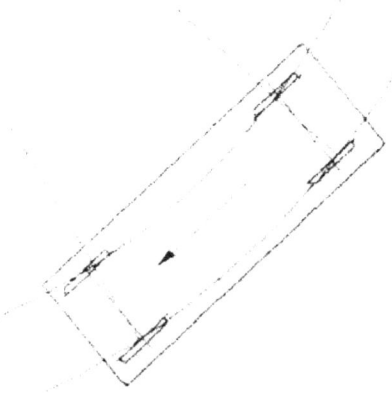

Figure 17 Illustration of the attitude of the wheelsets of a bogie or two-axle vehicle during curving.

6.3   Rolling of Vehicles (Curved Track)

The view of curving described above is a simple one, which assumes that the wheelset is running along the track by itself. Figure 17 illustrates the effect when a two-axle vehicle, or the bogie of a 4-axle vehicle traverses a curve.   Because the vehicle is built with the wheelsets mounted parallel, they will not present radially to a curve, but will be is forced to go round the curve at an angle to the direction of travel. On a sufficiently sharp curve, the leading wheelset then runs with its outer wheel flange root hard against the outside rail (the "high" rail), and the inner wheel of the trailing wheelset having its flange root hard onto the inner rail (the "low" rail). Thus the strongest squeals come from the outer leading wheel and the inner trailing wheel.

By allowing some yaw compliance of the wheelsets in their frame it is possible to ease the passage round curves, but careful balancing of suspension parameters is needed if unstable running on straight track is not to result. A more expensive option is to fit so-called cross-bracing to the bogie suspension. This is a linkage that ensures the orientation of wheelset axes in curves is radial to the curve. In most cases squealing is avoided.

6.4   Rolling Round Tight Curves

In the sharpest curves, (say) with a curve radius less than 200m, the flange root of the outer leading wheel would be in extremely firm contact with the high rail. This would be a consequence both of inadequate conicity and severe mal-presentation of the wheelset. There would in such a case be a finite risk of flange-climbing, leading to derailment.

The solution adopted is to fit a check-rail inside the low rail so that the inside of the inner wheel flange bears on the check rail as the vehicle passes round the curve. There is very firm contact indeed between the inner wheel flange and the check-rail, but no contact between the outer wheel flange root and the high rail. It is interesting to consider the motion between inner wheel and the check rail

There is roughly rolling contact between the wheelset tyres and the running rails. However, because of the inadequate conicity on such a sharp curve, there is certain to be longitudinal

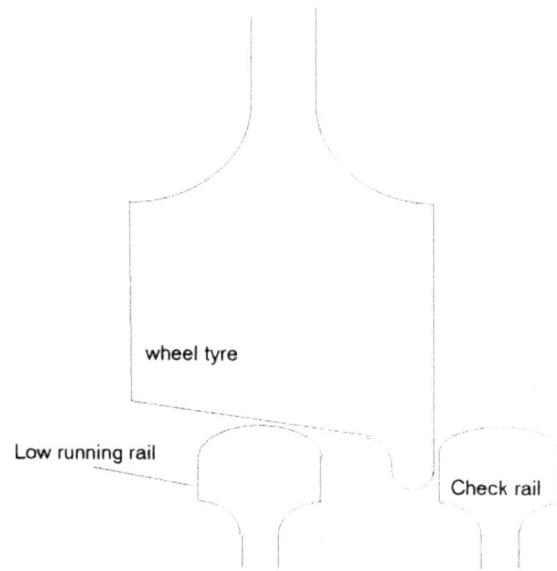

wheel tyre

Low running rail

Check rail

Figure 18 The use of a check rail to steer wheelsets round sharp curves.

slipping on one or both rails, as well as lateral slipping because of mal-presentation. In consequence, the motion between the inner wheel flange and the check rail is one of continuous shearing. In cases where the wheel design is such that there are modes of vibration available that allow radial compliance, stick-slip relief of the shearing motion will excite intense squeal.

One such case occurred on a sinuous line in the United Kingdom. This lead to vociferous complaint, since in still conditions the trains were audible at about 10 km. An acoustic cure was necessary, since it was estimated that a cure by lubrication would need several tonnes of grease per year.

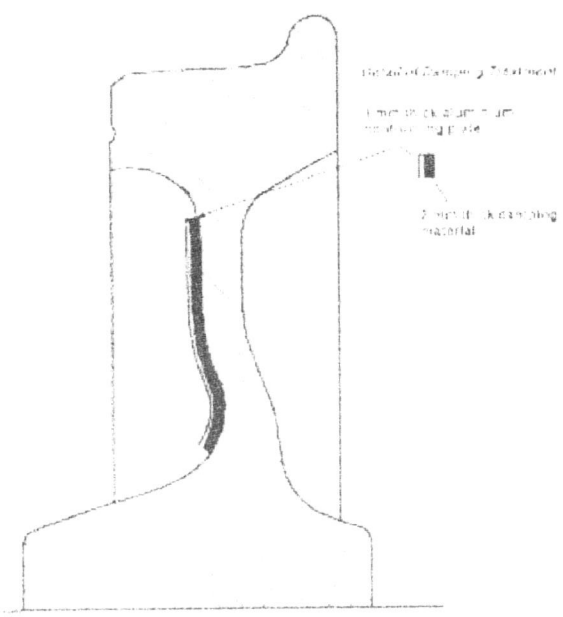

Figure 19 Example of constrained layer damping on the S-shaped web of a wheel

6.5 Solutions for Flange Squeal

Wheel web damping   The technique adopted, which turned out to be a complete cure for the problem addressed, is illustrated in figure 19. Constrained layer damping was fitted to the web of the wheels on their outside, to cover as much of the web area as possible. (the S-web was necessary on this rolling stock because it was tread-braked.) Constrained layer damping consists of a visco-elastic layer bonded to the member to be damped, backed by a so-called constraining layer bonded to the visco-elastic sheet. When the member to be damped distorts in the vibration mode concerned, bending of the whole of the damping structure (member/visco-elastic/constraint) occurs. The visco-elastic layer is forced into shear strain, resulting in energy being absorbed.

The treatment was entirely successful in removing completely the objectionable intense squeal due to check-rail contact on the back of the inside wheel flanges. *It was not effective in preventing the (much less intense) squeal due to outer wheel flange root contact on the high rail of curves with no check-rail.* The reason is that the modes excited where a different family.

Wheel tyre damping.   Another application of damping involved the use of damping on the tyre of the wheel. In this instance the wheels were straight-web wheels running on an underground track system which had no check-rails, and where the rails were mounted onto a concrete tunnel invert. Squeal was therefore due to flange-root contact with the high rail on bends. It would not have been possible to use flange lubricators without attendant fire hazard.

Noise measurements carried out near the vehicle bogies were analysed to reveal the natural frequencies being excited during the squealing, which happened on curves with radii tighter than 300 m. The results were compared with modal analysis of the wheels, which shewed the modes of vibration available, and their respective frequencies. In this way the modes being excited during squealing were revealed. Calculation, backed up with experience, suggested that damping of the wheel tyres should be effective. Constrained-layer damping was fitted as illustrated in figure 20. The result was almost total removal of the squeal noise.

The wheels used in this application had a straight web. Experience shews they would not have squealed on check-rail-fitted curves, since the mode of vibration excited by check-rail contact is not available on straight-web wheels.

A significant general message emerges. *Do not use S-web wheels unless there is a good reason – they are more likely to squeal.* S-web wheels tend to be noisier also in normal straight-track running.

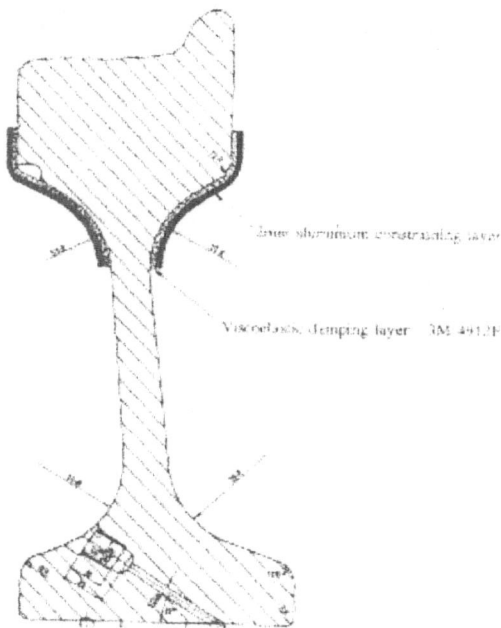

Figure 20 Constrained-layer damping applied to the tyre of a straight web wheel.

Flange-riding   One of the early techniques used to control squeal noise in urban tramway operation was the introduction of flange-riding.

Because tramways in urban streets run on grooved tramway rail, rather than the more familiar Vignole section used on main line track, it is possible to raise the level of the groove bottom surface on sharply curved sections. The wheel is then supported on its flange tip, rather than of the coned tyre surface. The effect is a large increase in the effective conicity, and the wheel steers readily round sharp curves. The technique is most commonly used in the relatively sharp turn-outs used in tramway practice.

An early statement of characteristics of flange squeal was given by Rudd (1976). For an extensive account of the squealing process in tramway operation see the work by Périard (1998). Much of the discussion is also relevant to mainline railway.

Lubrication   Mention of the application of flange lubricators was made above, but a more common, although usually erratic, lubricant is water.

Water sprays have been used to control squealing if several instances, particularly for low-speed freight traffic in freight yards, with generally satisfactory result. Naturally there needs to be an adequate water supply and also adequate drainage. Areas nearby may become hazardously slippery, and track-circuit signalling may suffer. Very cold weather will cause difficulty.

There is some evidence that routine wetting of rails will lead to enhanced corrugation development.

6.6   Impact Noise

Surface defects in the running rails will be audible as trains run over them. These may occur as part of normal practice at various forms of rail joint, or crossings. There may also arise from wheel burns, where spinning wheels on stationary or slowly-moving trains have caused local melting of the rail steel.   The only cure for wheel burns is by weld repair, or by replacement of the section of rail.

Wheel flats are most likely to occur as a result of skidding of the train, usually due to over-braking in poor adhesion conditions. They are more likely on disc-braked rather than tread-braked wheels (tread brakes will eventually remove them).

Unlike track defects, whose effect is only local, wheel-flats are audible as the train passes anywhere along its route. They also form an objectionable background within passenger trains. The only cure is for the wheelset to be re-profiled.

7   How rolling noise is perceived and its overall significance

It is worth considering the extent to which rolling noise, in the context in which we have been considering it, is really a problem.

7.1   The perception of railway noise.

There is a large body of literature, of which Fields and Walker (1982) is probably the most comprehensive and influential, that the response of people in dwellings to routine railway operating noise correlates best with the external façade noise level. The index of noise used is the *equivalent continuous sound level* or L_{Aeq}. The "A" indicates that the index is A-weighted. L_{Aeq} is a mean value of noise level over time, averaged on an energy basis, the effect of which is to bias the index towards the higher levels that occur during the averaging time. L_{Aeq} has been adopted as the index to determine eligibility to noise insulation in most european countries, but the levels and rules of time vary somewhat among countries.

L_{Aeq} is determined by the following relationship:-

$$L_{Aeq} = L_0 + 20\log_{10}v/v_0 + 10\log_{10}n + 10\log_{10}N + 10\log_{10}M - 10\log_{10}d/d_0 - 10\log_{10}t \qquad 3$$

where L_0 is the notional noise level[1] due to the passage of one wheelset at speed v_0 and distance d_0 from the track. n is the number of wheelsets per vehicle, N is the number of vehicles per train, M is the number of trains passing during the averaging time t (seconds) at a speed of v, a distance d away. (There is a great simplification in this, since it is assumed that all trains are the same and all travel at the same speed. In addition there will be additional terms to account for ground and air absorption, and the effect of a façade. Note that whereas the measured wayside level increases at 30 $\log_{10}v$, the value of L_{Aeq} increases at only 20 $\log_{10}v$. As the train goes faster, although the wayside level increases, the exposure time falls.).

A principal conclusion of Fields and Walker (1982) was that the averaging period should be over the full 24hr day (24 x 60 x 60 sec.). However, legislation in virtually all countries has chosen to adopt a lower level at night than during the daytime, although conventions about the length of night and day vary.

7.2   Consequences of the Way Train Noise is Perceived.

Wheel-rail noise dominant.   It is clear that the over-riding factor in equation 3 that determines the effect of trains is noise output of the single wheelset (and its associated track segment) – all of the other factors depend on the train consist, the operating conditions, or the position of the observer. Since virtually all european countries have introduced noise insulation regulations for railways, it has to be concluded that train noise is significant.

Whilst the wheel-rail noise is the most important factor in overall train rolling noise, there are some train design features, not too commonly employed, which are certain to reduce wayside noise. So-called train articulation is one of these, where the ends of adjacent coaches sit on a common bogie (a system used on the TGV of the SNCF).. Clearly this can nearly halve the

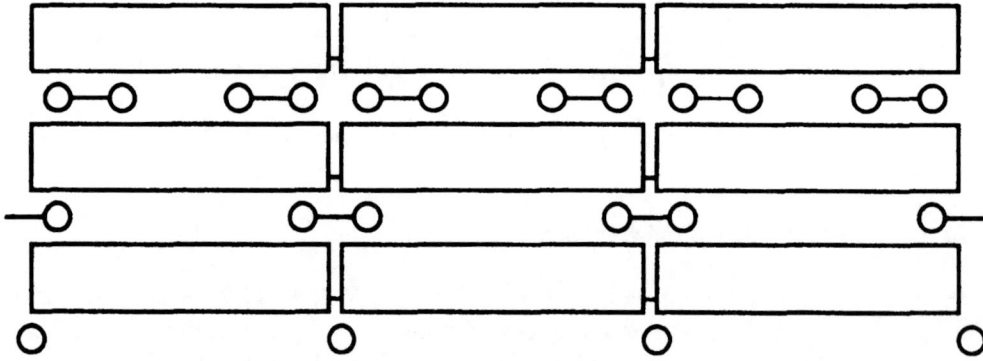

Figure 21 Diagram of different forms of articulation, compared with standard bogie layout.

number of wheels on the train. As a more extreme example, the Talgo train of the Spanish railways RENFE employs articulation of single axes, reducing the total axle count even further. The forms of articulation, compared with a standard bogie layout are shewn diagrammatically in

---

[1] Note that this is the single event level, SEL, and in this instance is the L_{Ae} of a single wheel passage, averaged over 1 second. SEL is also known as L_{AX}.

figure 21. Articulation, however, is not without its penalties, since the axle load is increased, with its own set of consequences for the rolling process, and there is an important loss of operational flexibility with what becomes inevitably a fixed-formation train.

Traction noise. Whilst rolling noise is almost always significant, its significance may not be over-riding. Other noise sources may dominate, so that rolling noise reduction, whilst worthy, may not yield the expected benefit. Among these cases is low speed operation of freight trains,

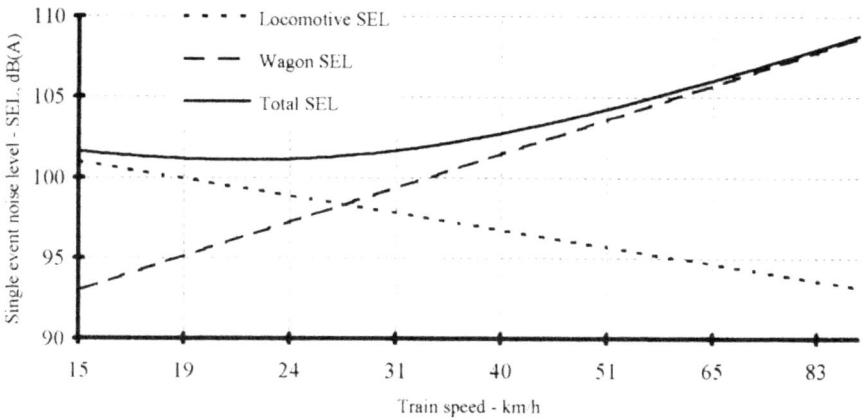

Figure 22 The overall noise due to a diesel-hauled freight train as a function of speed.

where the traction noise can dominate, particularly with diesel haulage. Diesel locomotives on full power have a sound power level virtually independent of their speed along the track. As they go faster, however, the exposure time to the noise that a wayside observer hears is progressively reduced, so that the noise energy experienced reduces. By contrast, the noise from the following wagons increases steadily and eventually dominates the noise climate when the speed is high enough. The effect is illustrated in figure 22, re-plotted and corrected from Hemsworth, (1987), which shews that in this instance, the traction and rolling sources balance at about 21 km/h, which is therefore where the minimum overall effect is heard. Clearly, because the noise effect is dominated by rolling at high speed, reducing the rolling noise is well justified, but do not expect low-speed benefit.

Aerodynamic noise If trains go sufficiently fast, the aerodynamic noise of even the smoothest trains will rise to exceed the steadily-increasing rolling noise. This is because aerodynamic noise increases by the sixth or seventh power of speed, and will eventually exceed traction and rolling sources as the speed rises sufficiently. It is expected that train speeds above around 350 km/h will herald the dominance of aerodynamic noise (King and Bechert, 1979). Routine operation of trains at such a high speed is unlikely to take place for many years, so the stimulus for research into rolling noise reduction remains undiminished.

Tyre noise There is no doubt that the perception of road vehicle noise by wayside householders is one of strong criticism, especially for vehicles with high noise levels.

Interestingly, road vehicle noise becomes objectionable in L_{Aeq} terms at around 5 dB lower than the case for rail traffic. This difference is reflected in noise insulation regulations.

It has been pointed out already that on dry roads the rolling noise of tyred vehicles is a principal source at higher speeds, but only just dominant, when compared with total power-train noise (see figure 2). It was also mentioned that when the roads are wet (see figure 15), the situation changes radically, and rolling noise becomes the principal source at higher speeds. (Tyler, 1987)

...As power-train noise is reduced, however, and road traffic volumes continue to grow, it is likely that tyre noise will assume greater significance in the absence of further work to reduce it.

References.

Fields, J.M. and Walker, J.G. (1982) The response to railway noise in residential areas in Great Britain. In *Journal of Sound and Vibration*, vol 85, pp177-255.

Hemsworth, B. (1987) *Transportation Noise Reference Book*, Butterworth, ed. P.M. Nelson. Chap. 15 – Prediction of Railway Noise.

Jones, R.R.K., (1994) Bogie shrouds and low barriers could significantly reduce wheel/rail noise. In *Railway Gazette international*, pp. 459-462, July 1994

King, F.W. and Bechert, D.(1979) An experimental study of aerodynamic noise generated by high speed trains. In *Noise Control Engineering Journal*.

van Leeuwen, J.J.A.,(1996) Noise prediction models to determine the effect of barriers placed alongside railway lines. In *Journal of Sound and Vibration Research*, vol 193 pt 1, pp269-276,

Nelson, P.M. and Underwood, M.P.C., (1984) "Lorry tyre noise". In *Paper C139 84, Vehicle Noise and Vibration Conference*, Institution of Mechanical Engineers, London.

Périard, F.(1998) Wheel-Rail Noise generation: Curve Squealing by Trams. Thesis, TU Delft.

Remington, P.J. (1975) Wheel-rail noise part IV; rolling noise. In *Journal of Sound and Vibration*, vol 46, pp.419-436

Remington, P.J. (1987) Wheel-rail noise I; theoretical analysis. In *Journal of the Acoustical Society of America*, vol 81, pp1805-1823

Rudd, M.J.(1976), Wheel/rail noise, Part III: Wheel squeal..In *Journal of sound and Vibration*, vol 46, pp381-394.

Stanworth, C.G. (1987) "Transportation Noise Reference Book", Butterworth, ed. P.M. Nelson. Chap. 14 – Sources of Railway Noise.

Thompson, D. (1991) Theoretical modelling of wheel-rail noise generation. In *Proceeding of the institution of Mechanical engineers*, vol 205, pp137-149

Thompsom, D. (1993) , Wheel-Rail Noise Generation, parts I to V. In *Journal of Sound and Vibration*, vol 161 pt 3, pp387-482

Thompson, D. (1997) Experimental Analysis of Wave Propagation in Railway Tracks. In *Journal of Sound and vibration*, vol 203 pt 5, pp867-888

Tyler, J.W. (1987) "Transportation Noise Reference Book", Butterworth, ed. P.M. Nelson. Chap. 7– Sources of Vehicle Noise.

B. Jacobson

Lund University, Lund, Sweden

1 Rheometers for the Non-Newtonian Range

1.1 Introduction

Rheometers able to measure non-Newtonian properties must induce the same stress and the same shear rate in the whole volume of the test liquid. Otherwise some kind of mean value of the shear stress-shear strain rate relationship will be measured by the instrument. The stresses induced in the liquid must also be maintained without introducing a disturbing heat generation in the fluid. This means that the shear stress multiplied by the shear strain rate must not be so large that any considerable temperature increase takes place. Normal mineral oil based lubricants lose typically half their viscosity when the temperature is increased by 15 °C, so the temperature variation allowed in the viscometer during a measurement is only a fraction of a degree if any reasonable accuracy should be expected.

1.2 Concentric Cylinder Rheometer for Non-Newtonian Fluids

The equation of motion and the boundary conditions are

$$\frac{1}{r^2}\frac{\partial}{\partial r}(r^2\tau_{r\phi}) + \frac{1}{r}\frac{\partial \tau_{\phi\phi}}{\partial \phi} + \frac{\partial \tau_{\phi z}}{\partial z} - \frac{1}{r}\frac{\partial p}{\partial \phi} = \rho(\frac{\partial v}{\partial t} + u\frac{\partial v}{\partial r} + \frac{v}{r}\frac{\partial v}{\partial \phi} + \frac{uv}{r} + w\frac{\partial v}{\partial z}) \quad (1)$$

where (u, v, w) is the velocity vector.

The fluid is assumed to move in concentric circles around a common axis at a constant velocity for each fluid element, see figure 1. Furthermore the rheological model

$$\tau = K\dot{\gamma}^n \quad (2)$$

is used.

Then we obtain

$$\Delta\Omega = \Omega_2 - \Omega_1 = \frac{n}{2}[\frac{T}{2\pi LK}]^{1/n}[(1/R_1)^{2/n} - (1/R_2)^{2/n}] \quad (3)$$

with boundary conditions

$$\tau(R_1) = \frac{T}{2\pi R_1^2 L}$$

$$\tag{4}$$

$$\tau(R_2) = \frac{T}{2\pi R_2^2 L}$$

where T is the torque, L the length of the cylinder and R_1 and R_2 the radii of the inner and outer cylinders respectively.

Equation (3) gives

$$T = \frac{2\pi L K (\frac{2\Delta\Omega}{n})^n}{[(1/R_1)^{2/n} - (1/R_2)^{2/n}]^n} = A(\Delta\Omega, n) K \tag{5}$$

where

$$A(\Delta\Omega, n) = \frac{2\pi L (\frac{2\Delta\Omega}{n})^n}{[(1/R_1)^{2/n} - (1/R_2)^{2/n}]^n} \tag{6}$$

The curve fitting of the results is done by minimizing the function

$$f = \sum_{i=1}^{N} (A_i(\Delta\Omega_i, n) K - T_i)^2 \tag{7}$$

to get the best estimation for n and K. N is the number of test measurements, T_i is the measured torque and $\Delta\Omega_i$ is the measured angular velocity corresponding to measurement number i.

If n is fixed, $n = n_1$, it is possible to minimize

$$f = \sum_{i=1}^{N} (A_i(\Delta\Omega_i, n_1) K - T_i)^2 \tag{8}$$

to obtain

$$\frac{\partial f}{\partial K} = \sum_{i=1}^{N} 2(A_i(\Delta\Omega_i, n_1) K - T_i) A_i(\Delta\Omega_i, n_1) = 0 \tag{9}$$

and

$$K = \frac{\sum_{i=1}^{N} T_i A_i(\Delta\Omega_i, n_1)}{\sum_{i=1}^{N} A_i^2(\Delta\Omega_i, n_1)} \tag{10}$$

Starting with $n = n_1$, K is calculated using equation (10) and f with equation (8). A new n-value is chosen so that $f_{new} < f_{old}$. The iteration continues until n is correct to within three significant figures.

The definition of viscosity gives

$$\eta = \frac{\text{shear stress}}{\text{shear strain rate}} \tag{11}$$

and the viscosity at the inner cylinder is η_1.

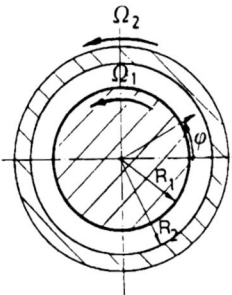

Figure 1. Definition of the geometry for the apparatus.

$$\eta_1 = \frac{nTR_2^{2(1/n-1)}}{4\pi\Delta\Omega L}[(1/R_1)^{2/n} - (1/R_2)^{2/n}] \tag{12}$$

1.3   Luleå High Pressure Chamber

In order to evaluate the relationship between limiting shear strength and pressure for lubricants, a high pressure chamber was built at Luleå Technical University [Höglund 1984]. Pressures up to 2.2 GPa and temperatures up to 200 °C were attainable simultaneously. Thus the limiting shear strength - pressure relationship could be surveyed over a wide range of conditions. Several synthetic and natural lubricants have been tested. The results show that all the mineral oils tested behave in quite a similar way. The synthetic lubricants tested do not show the same type of behaviour, but the relationship between pressure and shear strength is strongly dependent on the chemical structure of the lubricant molecules.

1.4   Introduction

Knowledge of lubricant rheology plays a very important role in the understanding of elastohydrodynamic lubrication. Relationships between pressure, temperature, and viscosity must be known, as well as the limiting shear strength of the lubricant as a function of pressure to make it possible to calculate pressure, oil film thickness, and power loss in an EHL contact. When the pressure between two contacting surfaces is low, simple expressions give good approximations of the lubricant behaviour. An example of such an expression is the well-known Barus equation: $\eta = \eta_0 \exp(\alpha(p - p_0) - \beta(t - t_0))$ given by Barus in 1893. Here η, η_0 are viscosities at pressures p, p_0 and temperatures t, t_0 respectively. The equation for a Newtonian liquid $\tau = \eta(du/dy)$ (given by Newton in 1687) is then used together with the viscosity expression to give the shear stresses as a function of the shear strain rate, and thereby form the basis of the calculation of power loss in the lubricant between the bearing surfaces. However, in an EHL contact the pressures are very high (1-5 GPa), and the equations mentioned earlier predict extremely high shear stresses between the contacting surfaces. Actually, the calculated values of shear stress can be much higher than the contacting parts are able to sustain, even if the surfaces are made of steel. These high shear stresses cannot be accommodated by the molecular structure of the oil.

The fact that the lubricant in an EHL contact is subjected to large pressure variations and high shear rates during very short periods of time (10 microseconds to a few milliseconds) led Smith [Smith 1958/59] to assume that the lubricant shear properties change from Newtonian behaviour to that of an amorphous elastic solid, having a limited shear strength.

This limiting shear strength (τ_L) of the lubricant is found to be linearily dependent on the pressure, i.e.

$$\tau_L = \tau_0 + \gamma p \tag{13}$$

where τ_0 is the shear strength at zero pressure and τ_L is the limiting shear strength. γ is the slope of the limiting shear strength - pressure relationship, $\partial \tau_L / \partial p$. This subject was discussed by Hamrock and Dowson [Hamrock 1981]. Values of τ_0 were found by Jacobson [Jacobson 1970] and are of the order 15 MPa for standard mineral oils. Bair and Winer [Bair 1979] found the value of γ to be in the range 0.05 to 0.1 for different types of mineral oils and synthetic oils. Their measurements covered the range -27 °C to $+40$ °C and pressures up to 1.2 GPa. Using the aforementioned formula, Jacobson and Hamrock [Jacobson 1983] solved the complete elastohydrodynamic problem of a rectangular contact. The value of γ is an essential parameter in the resulting formulae for minimum oil film thickness and coefficient of friction. Obviously, the property γ, which is a function of the oil type, the temperature and eventually also of the pressure, must be known to make it possible to calculate the limiting shear strength of the lubricant.

1.5   Experimental Apparatus

The experimental apparatus is shown in figure 2. The main parts are

- plungers
- a hollow circular cylinder
- seals
- hydrostatic bearings
- a frame (composed of guide bars and steel plates)
- a hydraulic jack
- driving machinery
- measuring equipment

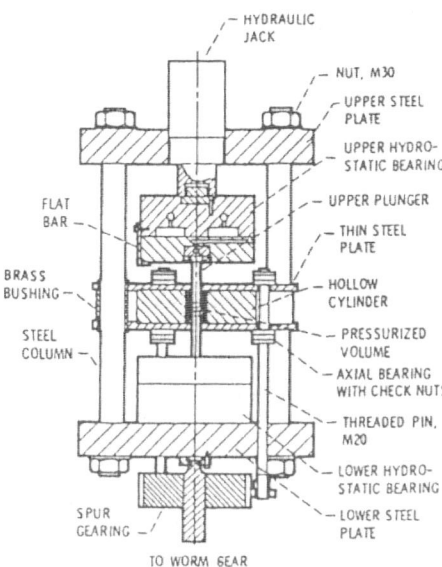

Figure 2. Main parts of the experimental apparatus. Some details are omitted for simplicity.

2   Testing Procedure

First the upper plunger was mounted in the upper hydrostatic bearing and the thin plates
holding the hollow cylinder were moved upwards. The lower plunger was then fitted into
the conical groove in the lower bearing and the threaded plunger-holder was screwed into
the bearing. The seals between the cylinder hole and the plungers were then inserted
about 2 mm into the hole from each side. After fitting the plunger into the cylinder
hole it was inserted to a depth of 9 mm of the cylindrical part, not including the 8 mm
high cross-shaped upper part. Of the lubricant to be tested 2.5 cm^3 was injected into the
cylinder and left for some hours in order for the air bubbles in the lubricant to disappear.
The upper plunger was then lowered and fitted into the hole of the cylinder by the jack.
If the test was to be performed at elevated temperatures the lubricant and cylinder
were heated by means of a thermocord and a stove plate. When the thermocouple in
the lubricant indicated that the required temperature had been reached the hydrostatic
bearings were pressurized and the upper plunger was pressed into the cylinder, thus
increasing the pressure in the lubricant. When the required pressure was obtained using
the jack and the temperature in the lubricant was correct, the a-c-motor was started and
both the thin steel plates and the hollow cylinder started to move up and down together.
The cylinder was only allowed to move up and down for about 20 seconds, because at
longer times the friction between the cylinder wall and the solidified lubricant caused
a temperature rise in the lubricant and the test would no longer be isothermal. After
stopping the cylinder the distance between the hydrostatic bearings was measured.

2.1   Evaluation of the Test Results

By knowing the distance between the bearings and the length of the plungers it is possible to calculate the contact area between the cylinder wall and the pressurized lubricant. The shear stress between the cylinder wall and the lubricant can easily be calculated by means of the relation:

$$\pi \, \tau_L \, d_p \, l = \pi \, \delta p_j \left(\frac{d_j}{d_p}\right)^2 \frac{d_p^2}{4} - F_s \tag{14}$$

where:

τ_L = limiting shear strength
d_p = diameter of plunger
l = length of lubricant column
δp_j = pressure variation in the jack system due to moving the cylinder up and down
d_j = effective diameter of jack
F_s = friction force between seals and cylinder wall

F_s can be calculated from the measured shear stress values, see, for example figure 3 . At low pressures the shear stress values are very low and increase very little with increasing pressure. This is caused by the friction between the seals and the cylinder wall. If a straight line is fitted to these low shear stress values it will give a shear stress of 1.3 MPa at a pressure of 2.5 GPa. Slight variations of this value are caused by the fact that different seals were used in different experiments. Solving for the limiting shear strength:

$$\tau_L = \frac{d_p}{4}\left(\frac{d_j}{d_p}\right)^2 \frac{\delta p_j}{l} - \frac{1.3}{2500}p \tag{15}$$

With: d_p= 10 mm, d_j= 65 mm, this gives:

$$\tau_L = 0.1056\delta p_j/l - 0.00052p$$

The last term is very small and is neglected in the evaluation of the test results. For each testing temperature and pressure the values of δp_j [MPa] and l [m] were determined and τ_L [MPa] was calculated. Different types of lubricant were tested, see table 1. Lubricants Nos. 1, 2, 7, and 8 were supplied by the same manufacturer, Nos. 1, and 2 were made from a paraffinic mineral oil, and Nos. 7 and 8 from a blend of polyalphaolefin and a polyolic-type ester. The rest of the lubricants were supplied by different manufacturers. Lubricant No. 3 was a paraffinic mineral oil. Lubricants Nos. 4 and 5 were based on polyglycol, and lubricant No. 6 was made from synthetic hydrocarbons. A well-known traction fluid was also tested, lubricant No. 9. As a comparison with the above-mentioned lubricants, which all contained a lot of additives, a pure castor oil, No. 10, was also tested. Finally a grease containing lithium soap, lubricant No. 11, was tested. This was done in order to find out if there was a difference between liquid and soap thickened lubricants.

Table 1. Tested Lubricants.

Lubricant	Figure	Type of oil/base	Viscosity (mm²/s)		Viscosity index	Density (kg/m³)
No.	No.		40°C	100°C	VI	15°C
1	5	Mineral oil/paraffinic	68	8.5	95	890
2	6	Mineral oil/paraffinic	460	32.0	100	910
3	7	Mineral oil/paraffinic	119	13.7	112	892
4	8	Synthetic oil/Polyglycol	468	79.5	253	1048
5	9	Synthetic oil/Polyglycol	130	24.0	200	1020
6	10	Synthetic oil/ Synthetic hydrocarbon	143	18.7	148	864
7	11	Synthetic oil/ Polyalphaolefine + Poliolic ester	68	11.0	153	851
8	12	Synthetic oil/ Polyalphaolefine + Poliolic ester	220	26.0	150	863
9	13	Synthetic oil/ Santotrac 50	31.3	5.19	-	898
10	14	Castor oil/vegetable	254	12.0	-	959
11	15	Grease/Lithium soap	200 (base oil)			

Thus different types of lubricants, both mineral oils, synthetic oils and a grease were tested.

The results are shown in figures 3 - 6. It is obvious from the figures that at a certain pressure, depending on the type of lubricant and test temperature, the limiting shear strength suddenly increases from a very low value. As the pressure is increased the limiting shear strength starts to increase rapidly. A straight line has been fitted to the measurement points and it clearly shows the linear relationship between the limiting shear strength of the lubricant and the pressure to which it is subjected.

The pressure at which the shear strength starts to increase and the slope of the straight line are very different for the different lubricants tested. A certain "solidification" pressure can also be obtained from the figures. The values of the slope of the lines, $\partial \tau_L/\partial p$ in the figures, and the "solidification" pressures are given in table 2. From this it can be seen that the value of $\partial \tau_L/\partial p$ decreases with increasing temperature, and that the value of the "solidification" pressure increases.

2.2 Conclusions

The different types of lubricants tested in the high pressure apparatus all have an increasing "solidification" pressure with temperature. The values of $\partial \tau_L/\partial p$ normally decrease with increasing temperature. Lubricants based on mineral oil, tend to behave in a rather similar way, independent of viscosity and manufacturer. The values of the "solidification" pressure are relatively low compared with those of many synthetic lubricants. The lubricants with polyalphaolefin + polyolic type ester, have the highest "solidification" pressures and the lowest values of $\partial \tau_L/\partial p$ of the lubricants tested. Lubricants containing

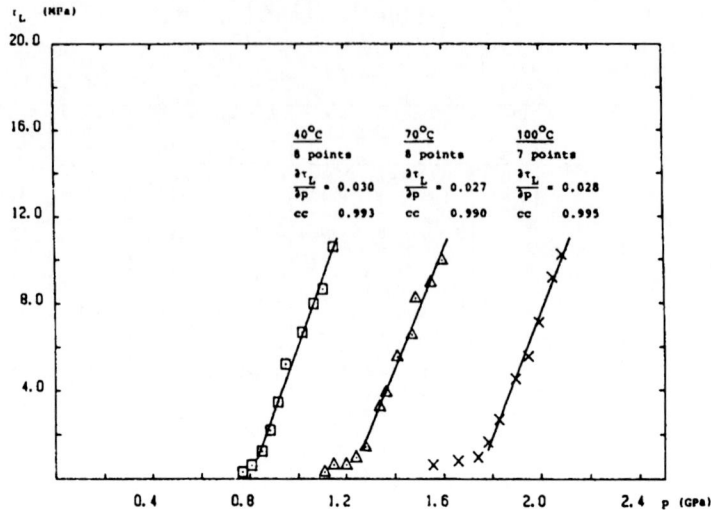

Figure 3. Limiting shear strength for lubricant No. 3, mineral oil, as a function of pressure and temperature.

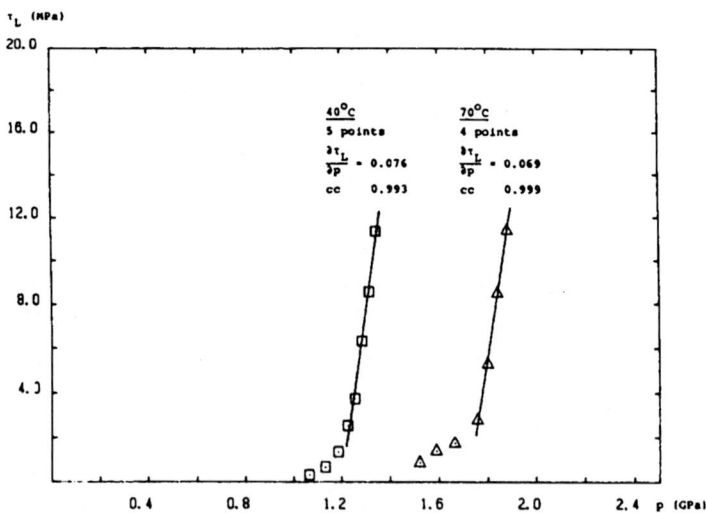

Figure 4. Limiting shear strength for lubricant No. 5, polyglycol-type oil, as a function of pressure and temperature.

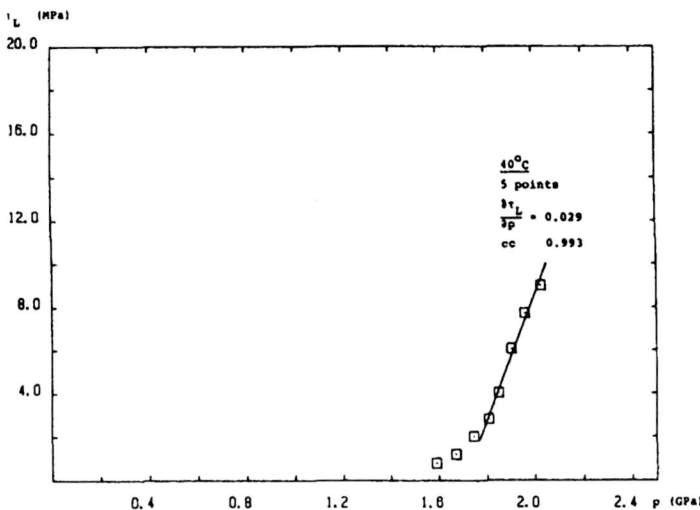

Figure 5. Limiting shear strength for lubricant No. 7, polyalphaolefin + polyolic-type ester oil, as a function of pressure and temperature.

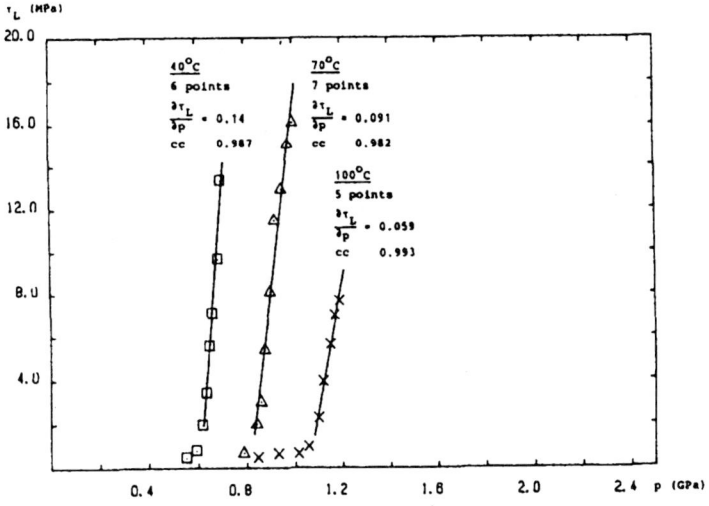

Figure 6. Limiting shear strength for lubricant No. 9, santotrac-type oil, as a function of pressure and temperature.

Table 2. Solidification pressure and shear strength increase with pressure for the tested lubricants.

Lubricant No.	Figure No.	slope $\partial\tau_L/\partial p$ / "solidification" pressure (GPa)			
		40°C	55°C	70°C	100°C
1	5	0.034/0.82	-	0.028/1.19	0.026/1.66
2	6	0.036/0.65	-	0.032/1.03	0.028/1.48
3	7	0.030/0.82	-	0.027/1.24	0.028/1.77
4	8	0.055/1.41	0.041/1.66	-	-
5	9	0.076/1.21	-	0.069/1.73	-
6	10	0.040/1.46	0.033/1.82	-	-
7	11	0.029/1.74	-	-	-
8	12	0.030/1.44	0.026/1.74	-	-
9	13	0.140/0.61	-	0.091/0.82	0.059/1.07
10	14	0.044/1.54	0.036/1.91	-	-
11	15	0.044/0.63	-	0.037/0.96	0.033/1.37

polyglycol, give high values of $\partial\tau_L/\partial p$ but do not otherwise show any similarities. The traction fluid gives the highest value of $\partial\tau_L/\partial p$ and the lowest "solidification" pressure of all the lubricants tested.

3   The Newtonian Elastohydrodynamic Problem

3.1   Introduction

A procedure for the numerical solution of the complete isothermal elastohydrodynamic problem for rectangular contacts is outlined in this chapter. This procedure calls for the simultaneous solution of the elasticity and Reynolds' equations. In the elasticity analysis the conjunction is divided into equal rectangular areas. By using the procedures outlined in the analysis the influence of the dimensionless speed U, load W, and materials parameter G on minimum film thickness is investigated. Ten cases are used to generate the minimum film thickness relationship.

$$\bar{H}_{min} = 3.07\, U^{0.71}\, G^{0.57}\, W^{-0.11} \tag{16}$$

The most dominant exponent occurs in association with the speed parameter; the exponent on the load parameter is very small and negative. The materials parameter also carries a significant exponent, although the range of the parameter in engineering situations is limited.

The recognition and understanding of elastohydrodynamic lubrication represents one of the major developments in the field of tribology in the twentieth century. The revelation of a previously unsuspected regime of lubrication not only explained the remarkable physical action responsible for the effective lubrication of many non-conforming machine elements like gears and rolling element bearings, but also brought order to the complete spectrum of lubrication regimes, ranging from boundary to hydrodynamic lubrication.

3.2   Theory

Reynolds' Equation.   The general approach to the numerical solution of the one-dimensional rectangular or line-contact problem covered in this chapter is similar to the method used by Hamrock and Dowson [Hamrock 1976] in solving the two-dimensional elliptical contact problem in elastohydrodynamic lubrication. The Reynolds equation for one-dimensional flow where side leakage is neglected can be written as

$$\frac{d}{dx}(\frac{\rho h^3}{\eta}\frac{dp}{dx}) = 12u\frac{d}{dx}(\rho h) \tag{17}$$

where $u = (u_a + u_b)/2$ is the mean surface velocity or the entraining velocity in the x-direction.
Letting

$$X = x/b, \ \bar{\rho} = \rho/\rho_0, \ \bar{\eta} = \eta/\eta_0, \ H = h/R, \text{ and } P = p/E' \tag{18}$$

where

$$\frac{1}{R} = \frac{1}{r_a} + \frac{1}{r_b} \tag{19}$$

$$\frac{2}{E'} = \frac{1 - \nu_a^2}{E_a} + \frac{1 - \nu_b^2}{E_b} \tag{20}$$

equation (17) can be rewritten in dimensionless form as

$$\frac{d}{dX}(\frac{\bar{\rho}H^3}{\bar{\eta}}\frac{dP}{dX}) = 24U\sqrt{\frac{2W}{\pi}}\frac{d}{dX}(\bar{\rho}H) \tag{21}$$

where

$$U = \frac{\eta_0 u}{E'R} \tag{22}$$

is the dimensionless speed parameter and

$$W = \frac{w_z}{E'R} \tag{23}$$

is the dimensionless load parameter, and w_z is the load per unit width.
Figure 7 shows the radius of the rollers used in defining equation (19). Convex surfaces, as shown in figure 7, are defined to have positive curvature and concave surfaces, negative curvature. Therefore if the centre of curvature lies within the solid, the radius of curvature is positive; if the centre of curvature lies outside the solid, the radius of curvature is negative.

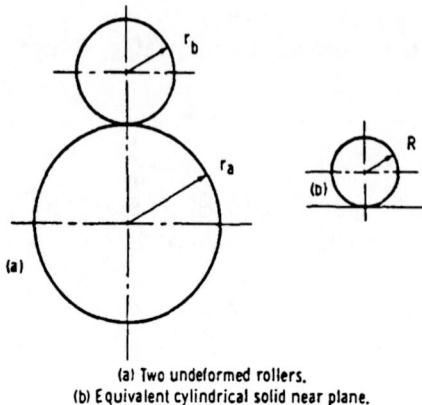

(a) Two undeformed rollers.
(b) Equivalent cylindrical solid near plane.

Figure 7. Rollers and equivalent roller.

Because of the large pressure variation in the lubricant films the viscosity of the lubricant is not constant in elastohydrodynamically lubricated hard contacts. [Barus 1893] proposed the following formula for the isothermal viscosity-pressure dependence of liquids:

$$\eta = \eta_0 e^{\alpha p} \tag{24}$$

In dimensionless form this equation can be written as

$$\bar{\eta} = \frac{\eta}{\eta_0} = e^{GP} \tag{25}$$

where

$$G = \alpha E' \tag{26}$$

is the dimensionless materials parameter.
Substituting equation (25) into equation (21) gives

$$\frac{d}{dX}(\bar{\rho}H^3 e^{-GP}\frac{dP}{dX}) = 24U\sqrt{\frac{2W}{\pi}}\frac{d}{dX}(\bar{\rho}H) \tag{27}$$

Ertel and Grubin [Grubin 1949] were the first to write the pressure and viscosity in terms of a reduced pressure as

$$Q = \frac{q}{E'} = \frac{1}{G}(1 - e^{-GP}) \tag{28}$$

and

$$\frac{dQ}{dX} = e^{-GP}\frac{dP}{dX} \tag{29}$$

Note that, as P approaches high values (infinity), Q approaches the value $1/G$. Substituting equation (29) into equation (27) gives

$$\frac{d}{dX}(\bar{\rho}H^3\frac{dQ}{dX}) = 24U\sqrt{\frac{2W}{\pi}}\frac{d}{dX}(\bar{\rho}H) \tag{30}$$

3.3   Density

For a comparable change in pressure the density change is small as compared with the viscosity change. However, very high pressures exist in elastohydrodynamic films, and the liquid can no longer be considered as an incompressible medium. In the calculations in this chapter the dimensionless density variation for mineral oils given by Dowson and Higginson in [Dowson 1966] is used. The dimensionless density formula can be written as

$$\bar{\rho} = \frac{\rho}{\rho_0} = 1 + \frac{0.6E'P}{1+1.7E'P} \tag{31}$$

where E' is given in gigapascals.

3.4   Film Shape

The film shape can be written simply as

$$h(x) = h_0 + S(x) + \delta(x) \tag{32}$$

where
  h_0 constant
  $S(x)$ separation due to the geometry of the undeformed solids
  $\delta(x)$ elastic deformation

The separation due to the geometry of the two undeformed rollers shown in figure 7(a) can be described by an equivalent cylindrical solid near a plane as shown in figure 7(b). The geometrical requirement is that the separation of the two rollers in the initial and the equivalent situation should be the same at equal values of x. Therefore, using the parabolic approximation we can write the separation due to the undeformed geometry of the two rollers as

$$S(x) = \frac{x^2}{2R} \tag{33}$$

Figure 8 shows a rectangular area of uniform pressure having the width $2\bar{b}$. From Timoshenko and Goodier [Timoshenko 1951] the elastic deformation at point $\bar{x}$ on the surface of a semi-infinite solid subjected to a pressure p at the point x_1 can be written as

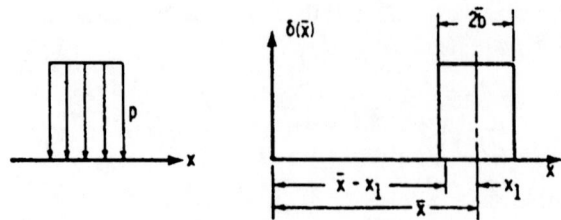

Figure 8. Surface deformation of semi-infinite body subjected to uniform pressure over a rectangular area.

$$\delta(\bar{x}) = -\frac{2}{\pi E'} \int_{-\bar{b}}^{\bar{b}} p \ln(\bar{x} - x_1)^2 \, dx_1 \tag{34}$$

Since the pressure is assumed to be constant over the rectangular area, the pressure can be put in front of the integral. The integration of equation (34) then results in the following:

$$\delta(\bar{x}) = \frac{2}{\pi} PD \tag{35}$$

where

$$D = b\left[(\bar{X} - B) \ln(\bar{X} - B)^2 - (\bar{X} + B) \ln(\bar{X} + B)^2 + 4B(1 - \ln B)\right] \tag{36}$$

and

b semiwidth of Hertzian contact, [m]
B $\bar{b}/b$
$\bar{b}$ b/n
n number of nodes within semiwidth of Hertzian contact

Now the term $\delta(\bar{x})$ in equation (35) represents the elastic deformation at a point $\bar{x}$ due to a rectangular area of uniform pressure p and width $2\bar{b}$. If the contact is divided into a number of equal rectangular areas, the total deformation at a point $\bar{x}$ due to the contributions of the various rectangular areas of uniform pressure in the contact can be evaluated numerically. The total elastic deformation caused by the rectangular areas of uniform pressure within a contact can be written as

$$\delta_k(x) = \frac{2}{\pi} \sum_{i=1,2,3..}^{N} P_i D_j \tag{37}$$

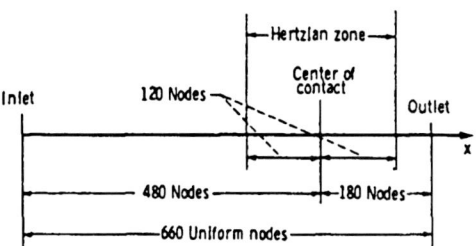

Figure 9. Nodal structure for numerical calculations.

where

$$j = |k - i| + 1 \tag{38}$$

Therefore, substituting equations (33) and (37) into equation (32) while writing the film thickness in dimensionless form gives

$$H_k = \frac{h}{R} = H_0 + \frac{1}{R}\left[\frac{X^2}{2}(\frac{b^2}{R}) + \frac{2}{\pi}\sum_{i=1,2,3..}^{N} P_i D_j\right] \tag{39}$$

Figure 9 shows the uniform distribution of the nodes within the contact. This nodal structure was used in all the calculations.
The following boundary conditions were adopted:

- At the inlet and the outlet the pressure is put equal to zero. This implies that Q is also zero at these positions.

- At the cavitation boundary eq. (40) holds.

$$P = \frac{dP}{dX} = 0 \tag{40}$$

3.5   Force Components

Figure 10 shows the force components acting on the two solids along with the oil film geometry in a portion of the concentrated contact. Conventionally only the z-components of the normal force acting on the solids (w_{az} and w_{bz}) are considered. However, it was felt that the tangential x–components (w_{ax} and w_{bx}), shear forces (f_a and f_b), coefficient of friction μ, and centre of pressure x_{cp} should also be expressed, and quantitative values obtained for each of these expressions. The normal z-component of the force per unit length acting on the solids can be written as

$$w_z = w_{az} = w_{bz} = \int p\, dx \tag{41}$$

Making it non-dimensional employing equation (18), this equation can be written as

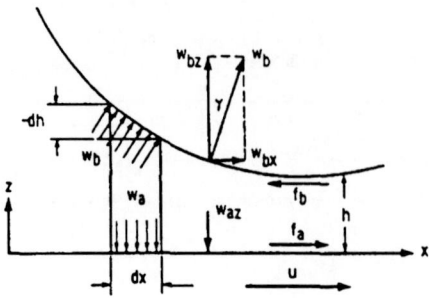

Figure 10. Force components.

$$W = \frac{w_z}{E'R} = \frac{b}{R} \int P\,dX = \frac{8}{\pi}\left(\int P\,dX\right)^2 \tag{42}$$

In the equation (42) W is usually referred to as the dimensionless load parameter. The tangential force component w_{ax} for solid a is zero. The tangential force component per unit length of solid b is not zero and can be written as

$$w_{bx} = -\int p\,dh = -\int p\frac{dh}{dx}\,dx \tag{43}$$

Using integration by parts gives

$$w_{bx} = -[ph]_i^o + \int h\frac{dp}{dx}\,dx \tag{44}$$

where i and o refer to inlet and outlet edge of the computing zone, respectively. However, the pressure at the inlet and outlet edge of the computing zone is zero. Making it non-dimensional employing equation (18), gives

$$W_{bx} = \frac{w_{bx}}{E'R} = \int H\frac{dP}{dX}\,dX \tag{45}$$

The resulting force components per unit length can be written as

$$W_a = \frac{w_a}{E'R} = \sqrt{W_{ax}^2 + W_{az}^2} = W_{az} \tag{46}$$

$$W_b = \frac{w_b}{E'R} = \sqrt{W_{bx}^2 + W_{bz}^2} = \sqrt{W_{bx}^2 + W^2} \tag{47}$$

$$\gamma = \tan^{-1}\left(\frac{w_{bx}}{w_{bz}}\right) = \tan^{-1}\left(\frac{W_{bx}}{W}\right) \tag{48}$$

3.6   Results

Dimensionless Grouping. From the variables of the numerical analysis the following dimensionless groups can be defined.
Dimensionless film thickness:

$$H = \frac{h}{R} \tag{49}$$

Dimensionless load parameter:

$$W = \frac{w_z}{E'R} \tag{50}$$

where w_z is the load per unit length.
Dimensionless speed parameter:

$$U = \frac{\eta_0 u}{E'R} \tag{51}$$

Dimensionless materials parameter

$$G = \alpha E' \tag{52}$$

The dimensionless film thickness for a rectangular contact can thus be written as a function of the other three parameters:

$$H = f(W, U. G) \tag{53}$$

The most important practical aspect of elastohydrodynamic lubrication theory is the determination of the minimum film thickness within the contact. Therefore, in the fully flooded results to be presented, the dimensionless parameters (W, U, and G) will be varied and the effect on minimum film thickness will be studied.

3.7   Influence of Load, Speed and Materials Parameter

Changes in the dimensionless load parameter W can be achieved while keeping the other parameters constant by changing only the applied normal load per unit length w_z in equation (50). Four pairs of data were used to determine an empirical relationship between the dimensionless load and the minimum film thickness.

$$\bar{H}_{min} \propto W^{-0.11} \tag{54}$$

If the surface velocity in the X-direction is changed, the dimensionless speed parameter U is modified as shown in equation (51), but the other dimensionless parameters (W and G) remain constant. Therefore the influence of speed on minimum film thickness can be written as

$$\bar{H}_{min} \propto U^{0.71} \tag{55}$$

There is a considerable change in film thickness as the dimensionless speed is changed, as indicated by equation (55). This illustrates most clearly the dominant effect of the dimensionless speed parameter U on the minimum film thickness in elastohydrodynamically lubricated contacts. A study of the influence of the dimensionless materials parameter G on minimum film thickness has to be approached with caution since in practice it is not possible to change the physical properties of the materials, and hence the value of G, without influencing the other dimensionless parameters considered earlier. Equations (50) to (52) show that as either the materials of the solids (as expressed in E') or the lubricant (as expressed in η_0 and α) are varied, not only does the materials parameter G change, but so do the dimensionless speed U and load W parameters. The effect of the dimensionless materials parameter on minimum film thickness can be written with adequate accuracy as

$$\bar{H}_{min} \propto G^{0.57} \tag{56}$$

3.8   Minimum Film Thickness Formula

The proportionality equations (54), (55), and (56) have established how the minimum film thickness varies with the load, speed, and materials parameters, respectively. This enables a composite dimensionless minimum film thickness formula for a fully flooded, isothermal elastohydrodynamic, rectangular contact to be written as

$$\bar{H}_{min} = 3.07\, U^{0.71}\, G^{0.57}\, W^{-0.11} \tag{57}$$

In dimensional terms this equation is written as

$$\bar{h}_{min} = 3.07\, \frac{\alpha^{0.57}\, R^{0.4} (u\eta_0)^{0.71}}{(E')^{0.03}\, w_z^{0.11}} \tag{58}$$

From this equation we can find that the minimum film thickness depends inversely on the effective elastic modulus E' and load per unit length w_z. Both have small exponents, indicating that the minimum film thickness h_{min} is only slightly affected by the effective elastic modulus and the load per unit length. In contrast to these effects from equation (58) we find that the film thickness depends directly on the pressure-viscosity coefficient of the lubricant α, the geometry R, the surface velocity in the direction of motion u, and the viscosity at atmospheric pressure η_0. From the values of the exponents on these parameters (α, R, u, and η_0) it is clear that they have a dominating effect on the minimum film thickness.

4   Non-Newtonian Fluid Model

A procedure is outlined for the numerical solution of the complete elastohydrodynamic lubrication of rectangular contacts incorporating a non-Newtonian fluid model. The approach uses a Newtonian model as long as the shear stress is less than a limiting shear stress. If the shear stress exceeds the limiting value, the shear stress is set equal to the

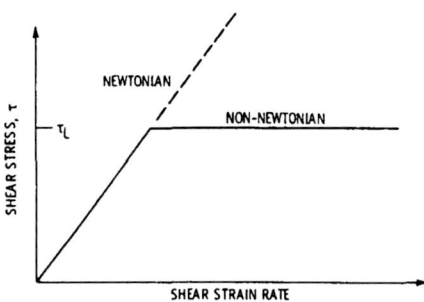

Figure 11. Lubricant model.

limiting value. The numerical solution requires the coupled solution of the pressure, film shape, and fluid rheology equations from the inlet to the outlet of the contact. Isothermal and no-side-leakage assumptions were imposed in the analysis.

The influence of dimensionless speed U, load W, materials parameter G, and sliding velocity U^* and limiting-shear-strength proportionality constant γ on the dimensionless minimum film thickness H_{min} was investigated. Fourteen cases were used in obtaining the minimum film thickness equation for an elastohydrodynamically lubricated rectangular contact incorporating a non-Newtonian fluid model.

$$\bar{H}_{min} = \bar{H}_{min,N}\left\{0.15\exp[-20.6\times10^{-9}(U^*)^{0.60}U^{0.23}(WG^2)^{3.85} + 10.4(\gamma-0.07)] + 0.85\right\}$$
$$\times(1-U^{*2})^{0.71} \tag{59}$$

where

$$\bar{H}_{min,N} = 3.07\,U^{0.71}\,G^{0.57}\,W^{-0.11} \tag{57}$$

Computer plots are also presented that indicate in detail the pressure distribution, film shape, shear stress at the surfaces, and flow throughout the conjunction.

4.1 Introduction

Figure 11 shows the effect of shear stress on shear strain rate for the present model and that of a Newtonian fluid. From this figure it is observed that in the present model, if the Newtonian shear stress exceeds the limiting shear stress, the shear stress is set equal to the limiting shear stress.

The fluid model is Newtonian except when the shear stress reaches the shear strength value. At that point, slippage occurs and the shear stress is equal to the shear strength. Besides the dimensionless load, speed, and materials parameters that were found to have an influence on film thickness in section 3 , when non-Newtonian effects are considered two additional parameters were found to influence the minimum film thickness, namely:

- Sliding velocity

- Limiting-shear-strength proportionality constant.

Fourteen cases were used in obtaining a fully flooded film thickness equation when considering non-Newtonian effects of the liquid. Besides the film thickness calculations that were made, calculations of the force components, shear forces, coefficient of friction, and centre of pressure were also performed. Computer plots are presented that indicate pressure distribution, lubricant film shape, flow, and shear stresses within the conjunction.

4.2   Theory

The non-Newtonian approach will be to consider the flow conditions at the surfaces. The flow has two components, the flow due to velocity (Couette) and the flow due to the pressure gradient (Poiseuille).

In figure 12 we attempt to explain the velocity of the fluid for the five distinct zones that might exist in an elastohydrodynamic conjunction when non-Newtonian effects of the lubricant are considered. In each zone the velocity of the top surface is greater than that of the bottom surface. To better indicate the difference between the zones, values of u_a and u_b will be kept constant for each zone. In figure 12(a), zone 0, the normal Newtonian zone, the shear stresses at the surfaces are less than the limiting shear stress, and no slippage of the fluid at the surfaces occurs. In figure 12(a), zone 1, the Newtonian shear stress at the top surface is larger than the limiting shear stress and slippage occurs at the top (faster) surface. In figure 12(a), zone 3, the shear stresses at both the top and bottom surfaces are outside the limiting range ($\bar{\tau}_a < -\bar{\tau}_L$ and $\bar{\tau}_b > \bar{\tau}_L$) and slippage occurs at both surfaces, but the slippage velocity is less than the velocity at the surfaces. In figure 12(b), zone 2, the Newtonian shear stress at the bottom surface is larger than the limiting shear stress, and slippage occurs at the bottom (slower) surface. In figure 12(b), zone 4, the same sort of situation is present as in zone 3 with the exception that the fluid slippage is greater than the surface velocity.

The relevant equations for the five distinct zones that can occur in an elastohydro-dynamic lubrication contact when non-Newtonian effects are considered are developed next. Before we define these relevant equations for the five zones, we need to define the dimensionless limiting shear strength. For most materials the shear strength varies linearly with pressure over wide pressure ranges and there is normally a certain shear strength at zero pressure. This behaviour can be expressed by

$$\bar{\tau}_L = \bar{\tau}_0 + \gamma P \tag{60}$$

where:
   $\bar{\tau}_0$ dimensionless initial shear strength
   γ limiting-shear-strength proportionality constant

The author [Jacobson 1970] found the values of $\bar{\tau}_0$ for a range of fluids to be between 1×10^{-5} and 1×10^{-4}. For the results given in that paper $\bar{\tau}_0$ was assumed to be 9×10^{-5}. Bair and Winer [Bair 1979] found γ to be between 0.05 and 0.10 for a complete range of

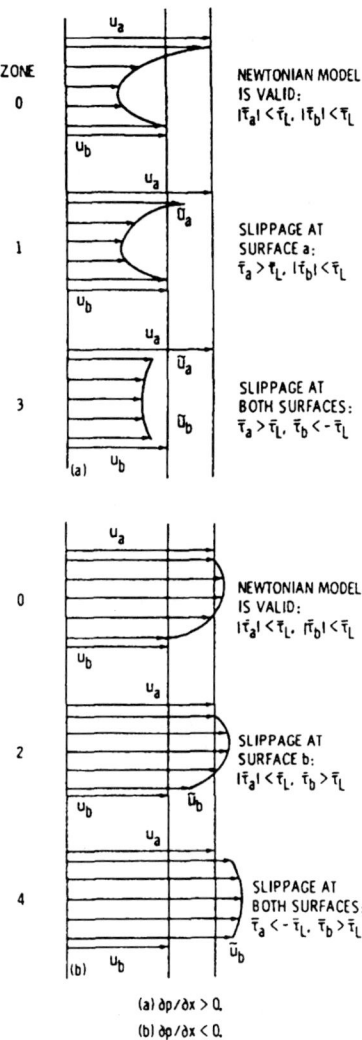

Figur 12. Velocity distributions in the different lubricating zones.

natural and synthetic lubricating oils. Tevaarwerk [Tevaarwerk 1976] used similar values of γ in his traction studies. In the present chapter we consider a range of γ between 0.04 and 0.10.

4.3   Zone 0

The Newtonian model gives $\bar{\tau}_a < \bar{\tau}_L$ and $\bar{\tau}_b < \bar{\tau}_L$. In this zone the rheological model is a Newtonian fluid and no slippage occurs at either surface. The velocity distribution across the lubricant film for this situation is depicted in figures 12(a) and 12(b), zone 0. Reynolds' equation is

$$\frac{d}{dX}\left(\frac{\bar{\rho}H^3}{\bar{\eta}}\frac{dP}{dX}\right) = 24U\sqrt{\frac{2W}{\pi}}\frac{d(\bar{\rho}H)}{dX} \tag{61}$$

This equation will be referred to as the pressure equation since it will change for the respective zones.

4.4   Zone 1

The Newtonian model for zone 1 gives $|\bar{\tau}_a| > \bar{\tau}_L$ and $|\bar{\tau}_b| < \bar{\tau}_L$. The velocity distribution across the film that exists in this zone is shown in figure 12(a). Slippage occurs at the top surface.

The dimensionless mass flow rate per unit length and the pressure equation for this zone can be written as

$$Q = \frac{q}{\rho_0 u_s R} = \bar{\rho}H\left(1 - U^* + \frac{\bar{\tau}_L H}{2\bar{\eta}U} - \frac{H^2}{6\bar{\eta}U}\sqrt{\frac{\pi}{2W}}\frac{dP}{dX}\right) \tag{62}$$

$$\frac{d}{dX}\left(\frac{\bar{\rho}H^3}{\bar{\eta}}\frac{dP}{dX}\right) = 6U\sqrt{\frac{2W}{\pi}}\left[(1 - U^*)\frac{d(\bar{\rho}H)}{dX} + \frac{1}{2U}\frac{d}{dX}(\frac{\bar{\rho}H^2\bar{\tau}_L}{\bar{\eta}})\right] \tag{63}$$

The non-unity terms in parentheses in equation (62) are the Poiseuille terms. Equation (63) is referred to as the pressure equation but can also be viewed as the Reynolds equation for a non-Newtonian fluid when conditions in zone 1 prevail.

4.5   Zone 2

The Newtonian model for zone 2 gives $\bar{\tau}_a < \bar{\tau}_L$ and $\bar{\tau}_b > \bar{\tau}_L$. The velocity distribution across the film that exists in this zone is shown in figure 12(b). Slippage occurs at the bottom surface (slower moving surface).

The dimensionless mass flow rate per unit length and the appropriate pressure equation for this zone can be written as

$$Q = \bar{\rho}H\left(1 + U^* - \frac{\bar{\tau}_L H}{2\bar{\eta}U} - \frac{H^2}{6\bar{\eta}U}\sqrt{\frac{\pi}{2W}}\frac{dP}{dX}\right) \tag{64}$$

$$\frac{d}{dX}\left(\frac{\bar{\rho}H^3}{\bar{\eta}}\frac{dP}{dX}\right) = 6U\sqrt{\frac{2W}{\pi}}\left[(1+U^*)\frac{d(\bar{\rho}H)}{dX} - \frac{1}{2U}\frac{d}{dX}\left(\frac{\bar{\rho}H^2\bar{\tau}_L}{\bar{\eta}}\right)\right] \qquad (65)$$

The non-unity terms in the parentheses in equation (64) are the Poiseuille terms.

4.6   Zone 3

The Newtonian model for zone 3 gives $\bar{\tau}_a > \bar{\tau}_L$ and gives $\bar{\tau}_b \leq -\bar{\tau}_L$. Figure 12(a) shows the velocity distribution across the lubricant film. Slippage occurs at both the top and bottom surfaces, and the slip velocity is less than the velocity of the slow surface (surface b).

The dimensionless mass flow rate per unit length and the appropriate pressure equation can be written as

$$Q = \bar{\rho}H\left(1 - \frac{u_s - \tilde{u}_a}{u_s} - \frac{H\bar{\tau}_L}{6\bar{\eta}U}\right) \qquad (66)$$

$$\frac{dP}{dX} = 4\sqrt{\frac{2W}{\pi}}\frac{\bar{\tau}_L}{H} \qquad (67)$$

The non-unit terms in parentheses in equation (66) are the Poiseuille terms.

4.7   Zone 4

The Newtonian model for zone 4 gives $\bar{\tau}_b > \bar{\tau}_L$ and gives $\bar{\tau}_a < -\bar{\tau}_L$. Figure 12(b) shows the velocity distribution across the lubricant film. Slippage occurs at both the top and bottom surfaces. The difference between zone 3 and zone 4 is that in zone 3 the slip velocity is less than the slower surface velocity and in zone 4 the slip velocity is greater than the faster surface velocity. Note that in order to be in zone 3 from equation (67), the pressure gradient must be positive, and in order to be in zone 4, the pressure gradient must be negative.

4.8   Density, Viscosity, and Film Shape Equations

The equations that define velocity, flow, and pressure for the five zones have been defined. Before proceeding, however, the dimensionless density, viscosity, and film shape need to be expressed. They are the same as defined in section 3.2 to 3.4.

4.9   Results

Pressure and Film Profiles. Representative variations of dimensionless pressure and film thickness, are shown in figure 13. In these figures the inlet region is to the left and the outlet region to the right. Figure 13 shows the variation of dimensionless pressure and film thickness as a function of x-coordinate for four dimensionless sliding velocities. The Hertzian pressure profile is also shown in this figure. The characteristic pressure spike is clearly evident for each dimensionless sliding velocity, but the spike diminishes

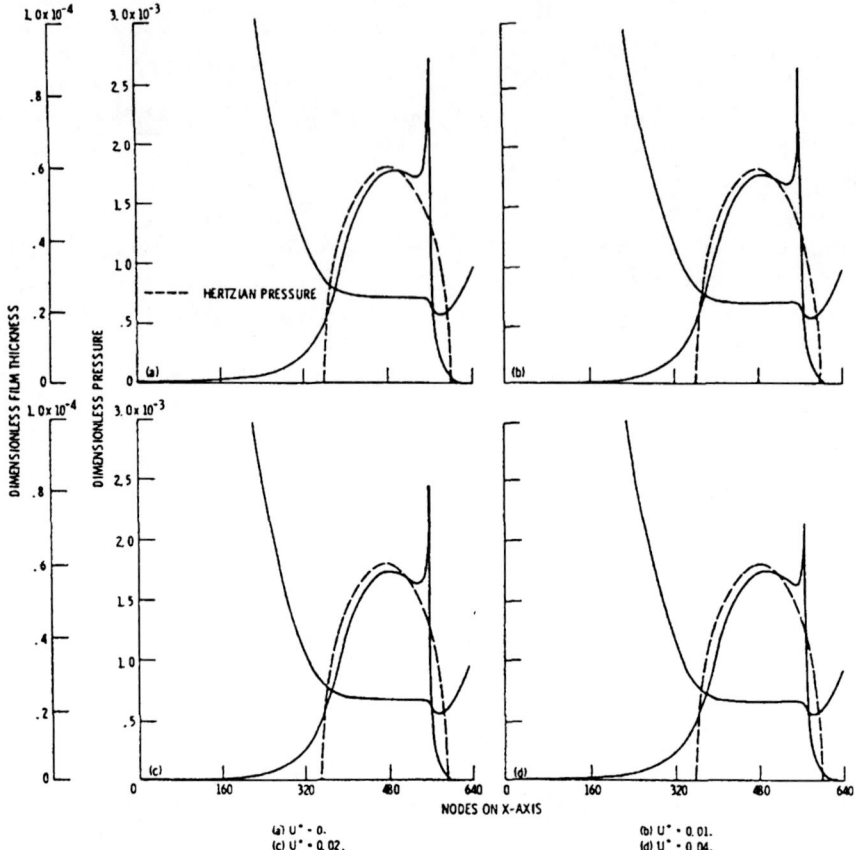

Figure 13. Pressure distributions and film shapes for different sliding velocities U^*. Dimensionless load parameter $W = 2.0478 \times 10^{-5}$; dimensionless speed parameter $U = 1.0 \times 10^{-11}$; dimensionless materials parameter $G = 5000$; limiting-shear-strength proportionality constant $\gamma = 0.07$.

as the sliding velocity increases. Also clearly evident for each sliding velocity is the parallel film shape throughout the central part of the contact, with a minimum film thickness occurring near the outlet of the contact. As the dimensionless sliding velocity increases, the central parallel film thickness decreases more than the minimum film thickness decreases. As a result, the nip decreases with increasing sliding velocity where the nip is that portion of the film shape from the tip of the pressure spike to the outlet of the conjunction.

4.10 Coefficient of Friction

The values of the coefficient of friction for the 14 cases studied are used to write an approximate formula for the coefficient of friction as

$$\bar{\mu}_1 = 1.13 \, (U^*)^{0.81} \, U^{0.26} \, e^{0.28\sqrt{WG^2}} \quad \text{when} \quad \bar{\mu}_1 \leq 0.85\gamma \tag{68}$$

If the coefficient of friction $\bar{\mu}_1$ is greater than 0.85γ, the following approximate expression should be used:

$$\bar{\mu} = 0.85\gamma + 0.021 \tanh[\frac{\bar{\mu}_1}{\gamma} - 0.85] \quad \text{when} \quad \bar{\mu}_1 > 0.85\gamma \tag{69}$$

This coefficient of friction is mainly determined by the shear strength of the lubricant.

4.11 Conclusions

A procedure for the numerical solution of the complete elastohydrodynamic lubrication of rectangular contacts incorporating a non-Newtonian rheological model is outlined. The approach uses a Newtonian model as long as the shear stress is less than the limiting shear stress. If the shear stress exceeds the limiting shear stress, the shear stress is set equal to the limiting shear stress. The limiting shear stress is expressed as a semi-empirical linear function of pressure. The numerical solution therefore requires the coupled solution of the pressure, film shape, and fluid rheology equations from the inlet to the outlet without making any assumptions other than neglecting side leakage.

Fourteen cases were used to generate the minimum film thickness equation for an elastohydrodynamically lubricated rectangular contact incorporating a non-Newtonian rheological model:

$$\bar{H}_{min} = \bar{H}_{min,N} \left\{ 0.15 \exp[-20.6 \times 10^{-9} (U^*)^{0.60} U^{0.23} (WG^2)^{3.85} + 10.4(\gamma - 0.07)] + 0.85 \right\}$$
$$\times (1 - U^{*2})^{0.71} \tag{59}$$

The minimum film thickness equation obtained for an elastohydrodynamically lubricated rectangular contact incorporating a Newtonian fluid rheology model was developed in the previous chapter, where

$$\bar{H}_{min,N} = 3.07\, U^{0.71}\, G^{0.57}\, W^{-0.11} \tag{57}$$

Besides the dimensionless film thickness formula, formulae for the coefficient of friction were developed.

$$\bar{\mu}_1 = 1.13\, (U^*)^{0.81}\, U^{0.26}\, e^{0.28\sqrt{WG^2}} \quad \text{when} \quad \bar{\mu}_1 \leq 0.85\gamma \tag{68}$$

$$\bar{\mu} = 0.85\gamma + 0.021 \tanh[\frac{\bar{\mu}_1}{\gamma} - 0.85] \quad \text{when} \quad \bar{\mu}_1 > 0.85\gamma \tag{69}$$

The effect of sliding velocity on coefficient of friction from various loads indicates that as the load increases, the limiting value of the coefficient of friction is reached for a much lower value of sliding velocity.

5   Thin Film Lubrication of Real Surfaces

5.1   Lubrication of Rough Surfaces

Elastohydrodynamically lubricated gears and bearings have always shown the effect of
running-in. The surface structure is changed by the running, and if that change is slow
and well controlled, the surfaces will run-in until they have become smooth enough not
to wear any more. This light wear thus determines to a large extent the functionality
of the surfaces as long as the running conditions do not change. If the load, speed or
temperature are changed so that asperity tops again touch each other through the oil
film, running-in will start again. The running-in continues until there is no more direct
asperity contact for a given running condition, indicating that the real minimum oil film
thickness between the highest roughness peaks will be very close to zero. If the forces,
stresses and/or generated temperatures during the running-in are too large the surface
contacts on the asperity level will damage the surface and increase the roughness. This
leads to higher local contact pressures between the asperities and thus to a progressively
more severe contact condition, making the surfaces increasingly rough while the wear rate
remains high. There is thus a bifurcation point at which the surfaces can either run-in
and become smooth or become rougher and rougher depending on some small variation
in one or a few of the contact parameters. When the peak to valley surface roughness
for a new surface is very much larger than the existing oil film thickness, a major part
of the load on the contact will be carried by direct solid contact between the asperity
tops and the opposite surface. As the reduced radius of curvature of the asperity contact
is usually very small, the local asperity contact pressures will be very high, and plastic
deformation takes place both as a result of the high normal pressure and as a result of
the large tangential stresses caused by tangential motion of the two surfaces relative to
each other. The sharper the contact points, the lower the force needed to reach the limit
for plastic deformation and thereby the limit for break-up of the protective oxide layers
on the contacting surfaces. The steeper the asperity slopes, the more difficult it is to
generate a separating oil film, so large asperity slopes are bad for cooperating surfaces,
both from a stress point of view and from a lubrication point of view.

When rolling bearings and gears are lubricated with a clean lubricant the initial metal
to metal contact through the oil film will normally stop in less than a million loading
cycles if the surfaces are at all able to run-in. If, after the running-in, the temperature is
increased or the speed is decreased, metallic contact through the oil film will start again
and continue until the surfaces have run in again at the new running conditions. The
running-in continues as long as some asperities reach each other through the oil film,
and the run-in state is characterized by the minimum separation between the highest
roughness tops being very close to zero.

For high quality ball bearings the wear and surface roughness change under clean lubrication conditions stops already when the calculated oil film thickness is half of the composit
surface roughness, indicating that the oil film can separate surfaces with roughness peaks
more than ten times higher than the film thickness. The roughness peaks thus have to
be elastically flattened in the lubricated contacts.

The size and height of each single asperity gives information about the local pressure variation and pressure gradients, and thus about the local shear stresses in the lubricant when the motion of surfaces and the lubricant film thickness is known. When bearings and gears are run, occasional surface asperities are penetrating the oil film and make direct metal-to-metal contact. This results in high local stresses, and controlled wear or plastic deformation on the asperity level takes place until the surfaces are run-in and separated by a continuous lubricant film. For surfaces lubricated with contaminated lubricants the surface roughness will increase and the surface roughness slope will be a function of the particle size, the particle form, the particle hardness and toughness. The contamination will thus determine if the surfaces will be permanently damaged or if it is possible to run-in the surfaces again when the lubricant has been cleaned. For most applications only very soft particles much larger than the film thickness can be accepted and for tough ceramic particles the maximum allowable size is about the same as the oil film thickness if the surfaces should be able to run-in again when the system has been cleaned

For heavily loaded lubricated contacts the total shear deformation of the oil in the high pressure region gives stresses well above the limit for Newtonian behaviour of the lubricant. This means that stresses and lubricant flows in two perpendicular directions in the oil film will be coupled to each other via the limited shear strength of the oil. A pressure gradient in one direction will also influence the flow in the perpendicular direction, much the same as for dry friction between solid bodies.

As soon as the lubricant is compressed into the solid state, even very low sliding speeds superimposed on the rolling speed will introduce high shear stresses in the direction of sliding. These stresses, superimposed on the stresses emanating from the pressure distribution in the contact, drive the lubricant into a strongly non-Newtonian state. When the limiting shear strength of the lubricant is approached, a motion of the surfaces in one direction will directly influence the lubricant flow perpendicular to it. The asperity behaviour in a heavily loaded lubricated contact will therefore be quite dependent on the presence or absence of a sliding motion component superimposed on the main rolling velocity component. If no sliding is present the oil can behave like a Newtonian liquid, building up high local pressures on the asperity tops by squeeze motion which pushes them down and makes the lubricated surfaces conform much more than outside the Hertzian contact. The composite surface roughness of the two contacting surfaces becomes almost zero and the lambda value becomes large. This ideal condition is quickly destroyed if some sliding is introduced between the load-carrying surfaces. The sliding motion in one direction uses all of the available strength of the oil in that direction which makes it very easy for local pressure variations within the contact to push the oil down into the valleys of the surface structure, allowing the asperity tops to reach through the oil film and collide. The interaction between the surface asperity behaviour and the lubricant rheology determines the state of lubrication.

5.2 Asperity Lubricaton

It has been known for more than 40 years that the optimum slope of a slider bearing requires 2 to 3 times thicker oil film at the inlet edge compared to the outlet edge to give maximum load carrying capacity and minimum relative power loss. At the same time the length of the slider bearing in the direction of motion is typically 500 to 1000 times the minimum oil film thickness. Similar slopes are also found for elastohydrodynamic conjunctions when the surfaces are mathematically smooth. Calculations and measurements give slopes of the order 2×10^{-4} for the whole contact and about 10 times more locally at the outlet edge and the side lobes. Grinding and honing type operations, where material is removed by sliding stones with sharp crystals over the surface, give surface roughness mean slopes of the order 5° and maximum slopes of about 25° . That is about 500 times steeper slope than what is required for an optimum bearing. The asperities can thus not build up any pressure distribution of their own until they have been smoothed further either by polishing or running-in. When the tops of the asperities collide through the oil film they plastically deform and wear until metallic contact ceases and the wear stops. Depending on the mean oil film thickness, the resulting surface structure can still be quite rough if the film is thick or very smooth with low surface slope if the mean film is thin. For very thin mean oil film thickness the asperity slopes decrease by the wear until the slope reaches about 0.8° and the surface contact becomes totally elastic. That means that the cooperating surfaces can be elastically deformed to totally conform to each other. Such surfaces are very easy to lubricate and need an oil film of molecular dimensions only to be totally separated and show no metallic contact.

For pure squeeze or rolling motion, the shear strength of the oil is high enough to deform the surfaces elastically so that they become totally conforming and thus need only a very thin oil film to avoid metallic contact. For pure sliding motion, on the other hand, the shear stresses in the high pressure zone of the contact reach the lubricant shear strength and thus no strength is left to keep the oil from flowing out sideways from the asperity contacts, and a mean oil film thickness large enough to prevent the high tops of the aspserities from colliding is necessary. Thus, depending on the details of the surface structures of the cooperating surfaces and the kinematics of the contact, a very large variation can be found in the lambda value needed to lubricate the contact without breaking through the oil film. Lambda values as high as 20 can give occasional asperity contacts, while for well run-in surfaces, lambda as low as 0.3 can stop all metallic contact. It all depends on the lubrication of the asperities and on whether the local pressure hills can elastically make the contacting surfaces more conformable inside the Hertzian contact area compared with the roughness they have outside the contact.

6   Mixed Lubrication

6.1   Introduction

If the simplest rheological model of a lubricant, the Newtonian model, is used in the calculation of the oil film thickness between elastohydrodynamically lubricated smooth or rough surfaces, oil film breakdown can never occur. As soon as the oil film over an asperity becomes thin enough, a local pressure will be built up there, squeezing the asperity into the surface and making sure that no metal-to-metal contact can take place. It is only possible to explain oil film breakdown and surface contact through the oil film if the behaviour of the oil is non-Newtonian.

The surface finish of elstohydrodynamically lubricated gears and bearings has so far
been rough enough to show the effect of running-in. The surface structure is changed by
running and if that change is slow and well controlled, the surfaces will run in until they
become smooth enough to no longer wear. This running-in thus determines the running
conditions and the functionality of surfaces. To ensure that the surfaces can work in a
stable mode without wear or spalling, the running conditions have to be ideal for each
point on the lubricated surfaces all the time. This means that the lubrication analysis
has to be made at an asperity size level. It is not possible to predict the behaviour of a
lubricated, elastically deformed asperity just by knowledge of the mean oil film thickness
and the mean contact pressure. Besides the lubricant film thickness, it is also necessary
to know the local pressure, temperature, shear stress and lubricant reology, as well as
the form and elasticity of the asperity.

The idea behind the experimental part of this chapter is that when hard surfaces are elastohydrodynamically lubricated in pure rolling or pure squeeze motion, the shear stresses
in the lubricant will be so low that it behaves as a Newtonian liquid. The slopes and
heights of the asperities will then induce local pressure peaks which in turn press down
the asperities. This gives a surface with a lower RMS and peak to valley surface roughness (R_{max}) than the free surface outside the Hertzian contact.

If sliding occurs in the contact and the pressure, and thereby the viscosity, is high, the
limiting shear strength of the lubricant is easily reached. The shear strength of the lubricant will almost totally be used in the direction of sliding for the surfaces and thus the
pressure derivative perpendicular to the sliding direction will easily displace the lubricant
sideways. This means that the surface asperities will push the oil perpendicular to the
sliding direction and they will grow to about the same height as those outside the high
pressure region.

6.2   Theoretical Model

A non-Newtonian isothermal lubricant model is used to describe the oil. The lubricant
is assumed to behave in a Newtonian manner up to the point when the shear stress is
equal to the shear strength of the fluid. The maximum shear stress in the oil, the shear
strength of it, is proportional to the pressure according to the equation

$$\tau_L = \tau_0 + \gamma p \qquad (70)$$

When the shear strength is reached, the shear stress can no longer increase. It can only
change direction. This is similar to the behaviour of dry friction where the friction force
is always equal to the normal force multiplied with the coefficient of friction, and the
direction of the friction force is given by the direction of sliding and always counteracts
the motion.

For rough lubricated surfaces sliding along each other, $\partial p/\partial x$ will vary due to both
micro-contacts and the whole of the EHL contact, thus the necessary pressure gradient
$\partial p/\partial y$ to move the lubricant in the y-direction (perpendicular to the surface motion) will
be very small. This means that the pressure distribution will be close to the pressure

distribution for smooth surfaces and the surface roughness in the EHL contact will be almost the same as that outside the high pressure zone.

When the shear stress in the lubricant has reached the shear strength, the shear stress component perpendicular to the surface motion will be of the order

$$\tau_L \frac{V}{(U_1 - U_2)/2} \tag{71}$$

where V is the slip velocity if the lubricant moves with the mean velocity of the surfaces. This shear stress is induced by the pressure gradient in the oil film and can be calculated as

$$\frac{\tau_L V}{(U_1 - U_2)/2} = \frac{h}{2} \frac{\partial p}{\partial y} \tag{72}$$

and the sliding speed perpendicular to the direction of motion

$$V = \frac{U_1 - U_2}{2} \frac{h \partial p / \partial y}{2 \tau_L} \tag{73}$$

If the slip plane in the x-direction is located at the fast surface at the inlet of the contact, the shear stress component in the y-direction there will be negligible and the total pressure gradient in the y-direction will be taken up by the slow surface

$$V_{max} = \frac{U_1 - U_2}{4} \frac{h \partial p / \partial y}{\tau_L} < \frac{U_1 - U_2}{4} \tag{74}$$

This is the maximum sliding speed at the slow surface in the y-direction.

If the oil is sliding in the y-direction at the slow surface with the velocity V_{max}, the oil flow in the y-direction will be given by

$$-q_y < \frac{h^3}{3\eta} \frac{\partial p}{\partial y} + \frac{U_1 - U_2}{4} h \tag{75}$$

If the flattened Hertzian zone has the width b and the slow surface has the velocity U_2, the time b/U_2 will be needed for an asperity on the slow surface to cross the Hertzian contact. The local decrease in film thickness due to the non-Newtonian side flow will be

$$-\frac{b}{U_2} \frac{\partial h}{\partial t} = -\frac{b}{U_2} \frac{2q_y}{\Delta y} < \frac{2b}{U_2 \Delta y} \left(\frac{h^3}{3\eta} \frac{\partial p}{\partial y} + \frac{U_1 - U_2}{4} h \right) \tag{76}$$

$$-\frac{b}{U_2} \frac{\partial h}{\partial t} < \frac{b}{\Delta y} \left(\frac{2h^3}{3\eta U_2} \frac{\partial p}{\partial y} + \frac{(U_1 - U_2) h}{2 U_2} \right) \tag{77}$$

In the high pressure zone $\eta \to \infty$, and $b/\Delta y = n$ is a measure of the number of asperities in the Hertzian contact.

$$-\frac{b}{U_2} \frac{\partial h}{\partial t} < n \frac{(U_1 - U_2)}{2 U_2} h_{min} < \Delta h \tag{78}$$

to avoid collapse if the viscous flow term should be neglected.

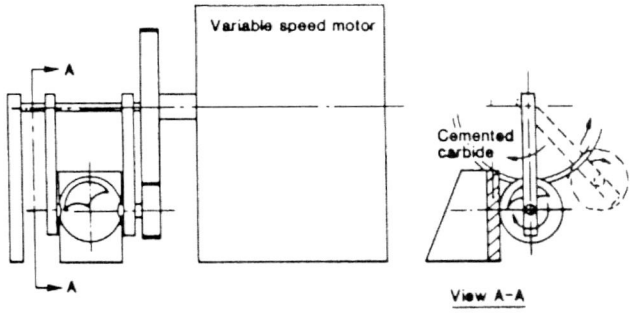

Figure 14. Drawing of the test apparatus.

$$n < \frac{2U_2}{U_1 - U_2} \frac{\Delta h}{h_{min}} \tag{79}$$

If the asperities are roughly circular, the leakage flow will be about twice as much as it would be if the asperities were longitudinal ridges. This means that

$$n \frac{U_1 - U_2}{U_2} h_{min} < \Delta h \tag{80}$$

If the collapse takes place by expansion of earlier compressed asperities through the oil film, $\Delta h \sim R_{max}/2$, where

$$n < \frac{U_2}{U_1 - U_2} \frac{R_{max}}{2h_{min}} \tag{81}$$

to avoid collapse of the oil film.

6.3 Experimental Investigation

An experimental investigation of film collapse in heavily-loaded lubricated contacts was performed. The lubricants were solidified under pressure and behaved like solid films. The geometry of the experiment was such that a lubricant film was built up by squeeze motion when the sliding speed was low. The main result of the investigation is that for different surface roughnesses the solid-like behaviour of the lubricants made it necessary to increase the viscosity of the lubricant at least one order of magnitude to separate the surface asperities. The solid-like behaviour of the lubricant was induced by sliding the surfaces as little as 2.5% of one Hertzian width during the impact time. For rougher surfaces, the required difference in viscosity could be more than a hundredfold.
The purpose of the simple tests reported in the next sections was to determine the minimum oil viscosity and pressure viscosity coefficient needed in squeeze motion and squeeze

motion combined with sliding to separate the impacting surfaces from each other. The separation of the surfaces was measured by registration of the electric resistance through the oil film by a digital oscilloscope.

The surface finish of the impacting ball was very good. It was polished to mirror finish and the surface roughness was probably less than $R_a = 0.008$ μm, which was the surface finish of the finest flat surface used in the experiments.

The impact velocity of the ball was kept reasonably constant by a spring-loaded mechanism that let the ball swing along an arc from a predetermined height.

6.4   Experimental Results

The experiments were performed at room temperature with 5 surfaces and 17 lubricants. The difference in the necessary viscosities for surface separation in pure squeeze motion and combined squeeze and sliding motion is large as the asperity micro-squeeze elastohydrodynamic lubrication only works when the sliding velocity is very low. In the experiments, the viscosity needed to separate the surfaces in pure squeeze motion was 14 to 240 times lower than that needed to separate the surfaces when sliding took place during impact.

This means that the pressure gradients caused by the macro- and micro-pressure distribution moved the oil from the compressed asperities to the valleys in the surface texture as soon as sliding took place.

This indicates that the R_{max} values describe the important surface roughness properties for a pure squeeze situation better than the R_a values as it is the extreme features of the surface which determine the oil film breakthrough. When low sliding speeds are introduced, a much higher viscosity is needed to separate the surfaces.

6.5   Comparison with Theory

The theory of section 6.2, says that the number of asperities within the Hertzian area should be low to avoid contact in transient elastohydrodynamic lubrication. Along the Hertzian width in the direction of motion, the number of asperities determining the boundary between no contact and contact is given by:

$$n = \frac{U_2}{U_1 - U_2} \frac{\Delta h}{h_{min}} \qquad (82)$$

where, U_2 is the slow surface velocity;
        $U_1 - U_2$ is the sliding velocity;
        Δh is collapse distance for the asperity;
        h_{min} is the film thickness when contact starts.

In this theory, the transport time for the slow surface through the Hertzian area is $t = b/U_2$, where b is the Hertzian width. If the impact time in the experiments is used instead of the transport time, the above equation becomes:

$$n = \frac{U_2}{U_1 - U_2} \frac{\Delta h}{h_{min}} = \frac{b}{(U_1 - U_2)t} \frac{\Delta h}{h_{min}} \qquad (83)$$

but $(U_1 - U_2)t = s$ is the sliding distance during the impact time, therefore,

$$n = \frac{b}{s} \frac{\Delta h}{h_{min}} \qquad (84)$$

is the number of asperities which just give collapse. For surfaces 1 and 4 in this investigation, the number of asperities within the Hertzian area was $n_1 = 36$ and $n_4 = 5$.
If the elastohydrodynamic oil film thickness in pure squeeze is assumed to be governed by the same equations as in pure rolling, the film thickness decrease is proportional to:

$$\Delta h \sim \eta^{0.68} \alpha^{0.49}, \qquad (85)$$

and can be compared for the rough and smooth surfaces,

$$\frac{n_1}{n_4} = \frac{(b_1/s_1)(\Delta h_1/h_{min,1})}{(b_4/s_4)(\Delta h_4/h_{min,4})} = \frac{b_1}{b_4} \frac{s_4}{s_1} \left(\frac{\eta_1}{\eta_4}\right)^{0.68} \left(\frac{\alpha_1}{\alpha_4}\right)^{0.49} \frac{R_{max,4}}{R_{max,1}} \qquad (86)$$

Here, $h_{min,4}/h_{min,1}$ is assumed to be equal to $R_{max,4}/R_{max,1}$.
The oils giving no collapse for the two surfaces at low sliding speeds are compared.
$b_1 = b_4$ and $s_4/s_1 = $ sliding distance ratio $= 1$.

$$\frac{n_1}{n_4} = \left(\frac{16300}{26}\right)^{0.68} \left(\frac{2.3}{1.2}\right)^{0.49} \left(\frac{0.22}{1.4}\right) = 7.16 \qquad (87)$$

From direct measurement of the surfaces, $n_1/n_4 = 36/5 = 7.2$.
The absolute values of $h_{min,4}$ and $h_{min,1}$, however, have not been determined.

6.6   Conclusions

Both the theoretical investigations and the experiments show that the slopes and radii of curvature for the asperities are important parameters. For a given roughness, the radius of curvature is inversely proportional to the number of asperities per unit length, and for a given number of asperities the radius of curvature decreases with increasing asperity height. The theory predicts a higher risk of oil film collapse in large machine elements compared to small ones as the machining procedures are often similar and thereby the number of asperities within the Hertzian area increases with the size of the machine element.
In pure squeeze lubrication, and probably in pure rolling lubrication, the necessary viscosity to separate the surfaces is much lower than the viscosity needed if only a slight sliding component is superimposed on the squeeze motion. For the two surfaces 1 and 4 in this investigation, the viscosity had to be increased between 8.5 and 14 times for the fine surface and between 32 and 240 times for the coarse surface to prevent metal-to-metal contact when sliding was introduced. For the coarse surface a total slip distance during the impact time of less than 2.5% of the Hertzian width was enough to collapse the oil film, even when the viscosity was 100 times higher than the viscosity necessary to separate the surfaces in pure squeeze motion.

7   Lubricant Contamination

7.1   Introduction

Through the years, the load carrying capacity, found in laboratory experiments with rolling element bearings, has increased. This has mainly been assumed to depend on the cleaner steels and the finer surfaces of the bearings, but at the same time the filtration of the lubricants used in the tests has become better and better, so some of the contribution to longer lives has come from cleaner lubricants.

In all normal endurance tests, lubricants are virtually free from water. In outdoor applications and when the machines contain water or steam, the lubricant is very easily contaminated with water. It is also well known that free water in a lubricating oil decreases the life of rolling element bearings ten to more than a hundred times, depending on the water content. The reduction of bearing life with water concentration is steepest when the water can be dissolved in the oil, but even at high concentrations more water gives shorter life. This means that it always pays to keep the dirt and water concentration in the lubricant as low as possible.

7.2   Literature Review

The earliest papers on contaminated lubricants for rolling element bearings treated water contamination. In 1968 and 1969 Schatzberg and Felsen published two papers on the effect of dissolved water on rolling element fatigue life [Schatzberg 1968, Schatzberg 1969]. The nominal stress level on the four-ball machine was varied between 6.4 GPa and 9.0 GPa, but of course there was a great deal of plastic deformation in the balls at these high nominal stress levels. Under these high stress conditions and in a laboratory environment, the presence of 100 ppm (0.01 per cent) water dissolved in the squalane lubricant decreased the fatigue life 32 to 48 per cent. Schatzberg and Felsen proposed that water collected in micro-cracks in the ball surface and that this caused corrosion and hydrogen embrittlement, giving a reduction in fatigue life.

Later the same year, Yardley and Crump [Yardley 1969] showed some results from grease lubricated rolling element bearings where they stated that dirt introduced into the bearing by the grease packing procedure or through the grease channels on re-greasing could rapidly destroy the bearing.

In two papers Fitzsimmons together with Cave [Fitzsimmons 1975] and together with Clevenger [Fitzsimmons 1977] investigated the wear of taper roller bearings as a function of the amount of contaminant in the lubricant. They found that :

1.   The wear was proportional to the amount of contaminant.

2.   Tapered roller bearings will continue to wear as long as the particle size of the contaminant is larger than the lubricant film thickness between the bearing surfaces.

3. For bearings to wear significantly, the contaminant particle hardness has to be larger than or equal to the hardness of the bearing material.

In a paper from 1976, Tallian describes an experimental investigation of rolling contact fatigue life for contaminated deep groove ball bearings [Tallian 1976]. Depending on the lubrication system and the type of lubricant and contamination, he obtained different lives for the bearings.

The importance of cleanliness for bearing life is demonstrated by three groups of experiments, where the first group is ultrasonically washed, greased and sealed. The second group is kerosene washed, greased and open. The third group is sump lubricated with oil artificially contaminated with powdered hardening scale. The relative lives of the three groups are 100 per cent, 23 per cent and 2 per cent respectively. The sump lubrication in a differential gear oil with 5 mg of powdered heat treating scale up to 600 μm platelet diameter thus gave a reduction of life of 1:50 or more.

On the other hand, Tallian shows that extreme cleanliness pays off, giving 15-30 times longer life than expected from normal calculations.

In an interesting paper from 1979 Loewenthal and Moyer describe laboratory experiments with deep groove ball bearings lubricated with an uncontaminated and an artificially contaminated lubricant [Loewenthal 1979]. The lubricant type was neopentylpolyol(tetra)ester which is normally used as a gas turbine engine lubricant. The artificial lubricant contaminant contained 88 per cent carbon dust, 11 per cent Arizona test dust and 1 per cent stainless steel particles.

In the tests with uncontaminated lubricant, the lubricant was circulated through a 49 μm absolute filter. This means that a lot of particles produced within the system were circulated through the bearings. Despite this, the 10 per cent life for the test bearings was 672 hours whereas the AFBMA catalogue 10 per cent life was 47 hours. Even if the adjustment for good lubrication is taken into account, the observed life was 2.7 times longer than expected.

When test dirt was added to the oil at a rate of 125 mg per bearing per hour, the L_{10} lives were not changed very much for oils filtered through 3 μm and 30 μm filters, but the L_{50} lives were decreased. As soon as dirt particles were introduced into the lubrication system, the Weibull slope increased considerably and the surfaces of the balls and races started to show progressive surface damage.

Using the coarsest filters in the investigation, 49 μm and 105 μm absolute, the bearing surfaces showed extensive micro-cracking and wear.

In a continuation paper to their 1979 work, Loewenthal, Moyer and Needelman studied the effect of ultraclean and centrifugal filtration on rolling element bearing life [Loewenthal 1982]. This time they used the 3 μm absolute filter in the circulation system, also when no artificial contamination was added. By doing so they increased the L_{10} life of the bearings by a factor of two as compared to the contaminated case with 3 μm filtration, and the L_{10} life increased by 64 per cent relative to the baseline tests with a laboratory clean lubrication system. They also investigated the wear rate as a function of the filtration. The non-failed bearings in the baseline tests with clean oil and 49 μm absolute filtration had 3.9 times the weight loss of the bearings from the ultraclean tests while the bearings from the centrifugal filter, 3, 30, 49 and 105 μm filter tests with contaminated oil had 6.6, 7.4, 12.5, 16.2 and 344.6 times respectively the weight loss of the ultraclean test bearings.

In a paper from 1982 Sayles and Macpherson found that the early failures determining the L_{10} lives for contaminated bearings were mostly surface-initiated. They experimented with standard 25 mm bore, extra light series, single row cylindrical roller bearings with a brass cage. The contamination of the lubricant was produced by a helicopter gear box lubricated with the same oil as the test bearings. The oil containing the wear debris from the gear box was pumped through different filters to the bearing test rig. The cartridge-type filters were of absolute ratings 40, 25, 8, 3 and 1 μm. The amount of particles in the oil was strongly influenced by the filter rating giving 100 particles larger than 2.5 μm per millilitre of oil in 4 hours for the 40 μm filter and in 24 hours for the 3 μm filter. Sayles and Macpherson found that oil film thickness had a minor influence on bearing life, but that the filter rating had a much bigger influence. The finer filters always gave longer bearing lives with typically 6 times longer L_{10} life with a 3 μm filter than with a 40 μm filter. The most striking of their results is that surface damage by dirt particles in the oil does not recover if the oil is cleaned after a short while. They made pre-runs with 40 μm filter for 30 minutes and then the rest of the test was run with 3 μm filtration. Despite the fine filtration during the main part of the test, the bearings failed just as early as if they had been lubricated with the dirty oil all the time. The damage done to the bearing surface by the dirt particles during the first half-hour was enough to cause early failure.

Another influence of wear is the redistribution of forces and stresses in the bearing. Lorösch [Lorösch 1983] showed that experiments with worn bearings always exhibited pitting failure in the non-worn parts as the contact pressure there was higher. He also showed that calculated bearing life was always shorter than the real life of the bearings, the only exception being when dirt particles entering with the oil made sharp indentations in the bearing surfaces. In that case, the real life and the catalogue life of the bearings were similar. Under ideal lubricating and cleanliness conditions, the bearing life increased with decreasing load to a much greater extent than obtained by standard calculations. He also found that the influence of contaminants is less for large bearings than for small ones. Consequently, small bearings have a larger capacity than calculated if they are reasonably clean.

To investigate the influence of dirt, Lorösch made small indentations in the rolling track with hard balls of different sizes. The steeper the edges of the indentations, the shorter the life of the bearing. When a 0.4 mm ball was pressed into the surface to a depth of 13 μm, the life of the bearing decreased by 98.5 per cent to only 1.5 per cent of the life without indentation. His experiments showed that the sharp indentations never disappeared but stayed and gave high stress concentrations for the rest of the bearing life.

Ishibashi, Hoyashita and Sonoda [Ishibashi 1985] studied the influence of different hardnesses for the two surfaces working together. The surface roughness of the two roller surfaces was 3 μm R_{max} and the calculated oil film thickness was about 1 μm for the load range 1.5 GPa to 2.8 GPa. When a 520 HB roller was combined with a case hardened 800 HV roller at 1.5 GPa pressure, they never wore in but metallic contact continued for 800.000 revolutions, after which the surfaces exhibited pitting failures. In contrast to this, when the 525 HB rollers with the same roughness of 3 μm R_{max} were used in equal hardness combinations, the surfaces wore in so that 90 per cent of the time there was no metallic contact and they could be rotated 6.5×10^6 revolutions at a contact pressure of 2 GPa without failure. For rollers with a mirror-like finish, 0.2 μm R_{max}, and equal hardness, the metallic contact stopped after 10^6 revolutions and no pitting occurred within 10^7 revolutions, even at 2.4 GPa pressure. Their observation of metallic contact through the oil film for the first million revolutions is very interesting as the λ-value for the mirror-like rollers was about 15. This means that it does not matter how fine the surfaces are ground, run-in can improve them. This also means that wear particles are always produced in new machinery and should be transported away from the elastohydrodynamic contacts.

When Ishibashi et al. used one case hardened roller together with a 500 HB roller, both with mirror finish, at 2.3 GPa pressure, the life was less than 10^6 revolutions before severe pitting occurred. This means that the occasional asperity contacts, even at $\lambda = 15$, between a soft and a hard steel surface will decrease the life considerably compared to when the surfaces are of similar hardness. If they are of similar hardness they will wear in and the surfaces will be so fine that metallic contact stops. In the experiments of Ishibashi et al., a pitting limit of p_{max}=0.45 HB was found, which is considerably higher than the conventional pitting limit of p_{max}=(0.2-0.3) HB.

From the above review it is obvious that the water and dirt content in the rolling element bearing lubricant should always be as low as possible. Even minute amounts of water of the order per mille and dirt particles of the order parts per million are enough to considerably decrease the life of bearings. If, on the other hand, it is not possible to reach those low contamination levels, life is always increased if the dirt level is lowered.

7.3   Theoretical Results

In addition to the experimental work described above, theoretical calculations have been published in recent years in which an attempt has been made to predict the life reduction caused by contamination.

Previous experimental work indicates that, under certain conditions, dents, formed by overrolling of debris, act as sites of stress concentrations and as such are responsible for the initiation of surface fatigue associated with a life reduction of the bearings.

In 1988 Ko and Ioannides [Ko 1988] made a finite element calculation of the debris denting and obtained as a result the residual stresses left in the steel after the indentation. They could therefore calculate the decrease in fatigue life caused by the indentation. The FEM calculations also gave as a result the actual form of the indentation and the height of the ridge around the dent. Under this ridge is a region of tensile residual stress which not only gives a lower endurance strength but the ridge itself causes high stress variations when it is overrolled. This induces a large reduction in life.

Ioannides, Jacobson and Tripp [Ioannides 1988] introduced the decrease of endurance strength with hydrostatic tensile stresses in the steel caused by residual stresses or hoop stresses from the mounting of the bearing. The calculations show a slight increase in bearing life for clean and undented conditions without hoop stresses, but a sharp decrease in life when tensile stresses and hoop stresses are introduced into the calculations. For a hoop stress of 325 MPa, the life decreases to only 1 per cent of the life for a bearing with the same load but without hoop stresses. The indentations in bearings caused by overrolling of wear debris also induce tensile residual stresses close to the raceway surface and therefore give a large reduction in life, not only by causing high stress variations but also by decreasing the local endurance strength of the bearing steel.

7.4   Conclusions

When dirt particles in lubricating oils are considered, it is obvious that the cleaner the oil the better it is. Even dirt particles much smaller than the mean film thickness cause wear if they are hard. If they are large enough to penetrate the oil film thickness, they cause local stresses at the surface and thereby shorten the life of the bearing considerably.

Finally, the reason why water can decrease bearing life as much as it does is not well understood, but as small a concentration as 0.01 per cent is enough to decrease the bearing life to half its original value. The reduction of bearing life with water concentration is steepest when the water can be dissolved in the oil, but even at high concentrations, more water gives shorter life. This means that it always pays to keep the dirt and water content in the lubricant as low as possible.

References

[Bair 1979]   Bair, S. and Winer, W.O., "Shear Strength Measurements of Lubricants at High Pressure," Trans. ASME Journal of Lubrication Technology, Vol. 101, 1979, pp. 251-257.

[Barus 1893]   Barus, C., "Isotherms, Isopiesics, and Isometrics relative to Viscosity", Am. J. Sci., 45, pp. 87-96, 1893.

[Dowson 1966]   Dowson, D., and Higginson, G.R., "Elastohydrodynamic Lubrication - The Fundamentals of Roller and Gear Lubrication", Pergamon Press, 1966.

[Fitzsimmons 1975]   Fitzsimmons, B., and Cave, B.J., "Lubricant contaminants and their effect on bearing performance", Society of Automotive Engineers, 750583.

[Fitzsimmons 1977]   Fitzsimmons, B., and Clevenger, H.D., "Contaminated lubricants and tapered roller bearing wear", ASLE Trans., Vol.20, 2, 1977, pp. 97-107.

[Grubin 1949]   Grubin, A.N., and Vinogradova, I.E., "Fundamentals of the Hydrodynamic Theory of Lubrication of Heavily Loaded Cylindrical Surfaces", Investigation of the Contact Machine Components, Kh. F. Ketova, ed., Translation of Russian Book No. 30, Central Scientific Institute for Technology and Mechanical Engineering, 1949, Chapter 2.

[Hamrock 1976]   Hamrock, B.J., and Dowson, D., "Isothermal Elastohydrodynamic
                     Lubrication of Point Contacts. Part I-Theoretical Formulation", J.
                     of Lubrication Technology, vol. 98, no. 2, April 1976, pp. 223-229.

[Hamrock 1981]   Hamrock, B.J., and Dowson, D., "Ball Bearing Lubrication". Wi-
                     ley, 1981, pp. 269-275.

[Höglund 1984]   Höglund E., "Elastohydrodynamic Lubrication, Interferometric
                     Measurements, Lubricant Rheology and Subsurface Stresses,"
                     Doctoral Thesis 1984:32D, Luleå University of Technology.

[Ioannides 1988]   Ioannides, E., Jacobson, B., and Tripp, J., "Prediction of rolling
                     bearing life under practical operating conditions", Proc. of the
                     15th Leeds-Lyon Symposium on Tribology, Leeds, September 1988.

[Ishibashi 1985]   Ishibashi, A., Hoyashita, S. and Sonoda, K., "Remarkable increase
                     in pitting limit of through hardened steel rollers under rolling
                     with sliding conditions", Proc. of the JSLE International Tribology
                     Conference, July 1985, Tokyo, Japan, pp. 929-934.

[Jacobson 1970]   Jacobson, B.O., "On the Lubrication of Heavily Loaded Spherical
                     Surfaces Considering Surface Deformation and Solidification of the
                     Lubricant," Acta Polytechnica Scandinavica, Mech. Eng. Series
                     No.54, 1970.

[Jacobson 1983]   Jacobson, B.O., and Hamrock, B.J., "Non-Newtonian Fluid Model
                     Incorporated into Elastohydrodynamic Lubrication of Rectangular
                     Contacts," NASA Technical Memorandum No. 83318, 1983.

[Ko 1988]   Ko, C.N., and Ioannides, E., "Debris denting - the associated resid-
                     ual stresses and their effect on the fatigue life of rolling bearings:
                     An analysis.", Proc. of the 15th Leeds-Lyon Symposium on Tri-
                     bology, Leeds, September 1988.

[Loewenthal 1979]   Loewenthal, S.H., and Moyer, D.W., "Filtration effects on ball
                     bearing life and condition in a contaminated lubricant", ASME
                     Trans., Journal of Lubrication Technology, Vol. 101, April 1979,
                     pp. 171-176.

[Loewenthal 1982]   Loewenthal, S.H., Moyer, D.W., and Needelman, W.M., "Effect
                     of ultra-clean and centrifugal filtration on rolling element bearing
                     life", ASME Trans., Journal of Lubrication Technology, Vol. 104,
                     July 1982, pp. 283-292.

[Lorösch 1983]   Lorösch, H-K, "Research on longer life for rolling element bear-
                     ings", ASLE Lubrication Engineering, Vol. 41, 1, 1983, pp. 37-43.

[Sayles 1982]   Sayles, R.S., and Macpherson, P.B., "Influence of wear debris on rolling contact fatigue", Rolling Contact Fatigue Testing of Bearing Steels, ASTM STP 771, J.J.C. Hoo, Ed., American Society for Testing of Materials, 1982, pp. 255-274.

[Schatzberg 1968]   Schatzberg, P., and Felsen, I.M., "Effects of water and oxygen during rolling contact lubrication", WEAR, 12, 1968, pp. 331-342.

[Schatzberg 1969]   Schatzberg, P., and Felsen, I.M., "Influence of water on fatigue failure location and surface alteration during rolling contact lubrication", ASME Trans., Journal of Lubrication Technology, Vol. 91, 2, 1969, pp. 301-307.

[Smith 1958/59]   Smith, F.W., "Lubricant Behaviour in Concentrated Contact Systems-The Castor Oil-Steel System," Wear, Vol. 2, 1958/59, pp. 250-263.

[Tallian 1976]   Tallian, T., "Prediction of rolling contact fatigue life in contaminated lubricant: Part II - Experimental", ASME Trans., Journal of Lubrication Technology, 1976, pp. 384-392.

[Tevaarwerk 1976]   Tevaarwerk, J.L., "The shear of elastohydrodynamic oil films", Ph.D. thesis, Cambridge, England 1976.

[Timoshenko 1951]   Timoshenko, S., and Goodier, J.N., "Theory of Elasticity", 2nd ed., McGraw-Hill Book Co., 1951.

[Yardley 1969]   Yardley, E.D., and Crump, W.J.J., "Some failures of grease lubricated rolling element bearings", Proc. Inst. Mech. Engs., Vol. 184, 1969, pp. 63-73.

Printed by Books on Demand, Germany